Debris Flow

Debris Flow

Mechanics, Prediction and Countermeasures

2nd Edition

Tamotsu Takahashi

CRC Press
Taylor & Francis Group
Boca Raton London New York

CRC Press is an imprint of the
Taylor & Francis Group, an **informa** business

A BALKEMA BOOK

CRC Press
Taylor & Francis Group
6000 Broken Sound Parkway NW, Suite 300
Boca Raton, FL 33487-2742

First issued in paperback 2018

CRC Press/Balkema is an imprint of the Taylor & Francis Group, an informa business

© 2014 Taylor & Francis Group, London, UK

Typeset by MPS Limited, Chennai, India

No claim to original U.S. Government works

ISBN-13: 978-1-138-00007-0 (hbk)
ISBN-13: 978-1-138-07367-8 (pbk)

Library of Congress Cataloging-in-Publication Data

Takahashi, Tamotsu, 1939–
   [Dosekiryu no kiko to taisaku. English]
   Debris flow : mechanics, prediction and countermeasures / Tamotsu Takahashi. — 2nd edition.
     pages cm
   Includes bibliographical references and index.
   ISBN 978-1-138-00007-0 (hardback : alk. paper) — ISBN 978-0-203-57674-8 (eBook PDF)
1. Debris avalanches.   2. Rock mechanics.   I. Title.
   QE599.A1T3513   2014
   551.3'07—dc23

                                             2013050615

Published by: CRC Press/Balkema
               P.O. Box 11320, 2301 EH Leiden, The Netherlands
               e-mail: Pub.NL@taylorandfrancis.com
               www.crcpress.com – www.taylorandfrancis.com

**Visit the Taylor & Francis Web site at
http://www.taylorandfrancis.com**

**and the CRC Press Web site at
http://www.crcpress.com**

Table of contents

Preface

A typical debris flow is a torrential flow of a mixture of water, mud and debris that suddenly pushes ahead with a vanguard of huge, jostling and roaring boulders. It is certainly a very fearful phenomenon that causes disasters, but it is also truly a wonder of nature exciting the curiosity of researchers as to how such a phenomenon can arise. The phenomena themselves had been recognized since ancient times in Japan and given various mnemonic names to make people aware of the dangers. Although there were several detailed witness records around in 1965 when I began working for the Disaster Prevention Research Institute of Kyoto University (hereafter called DPRI), the characteristics and mechanisms of debris flows were still vague, and it was called a 'phantasmal disaster'.

My first field investigation of such disasters was the 'Okuetsu torrential rain disaster' in 1965, which was brought about by a locally concentrated rainfall at the boundary between Fukui and Gifu Prefectures in central Japan. I found many wooden houses were buried up to the second floor by debris that consisted of particles a few tens of centimeters in diameter that had run down the mountain at the back of the town. Yet the frames of those houses were not completely destroyed. This fact told us that the thick deposit of big stones was laid gently. I wondered how such a phenomenon could occur.

In 1968, a debris flow occurred in the Shiramizudani experimental watershed of Hodaka Sedimentation Observatory, DPRI. I was greatly impressed by the traces of debris flow; it passed through a shallow and narrow stream channel incised in the wide riverbed leaving almost no large boulders in the central part, whereas along both sides of the channel big boulders were left in an orderly fashion as if they were artificially arranged levees. I wondered why such a phenomenon occurred.

The same rain storm that generated the Shiramizudani debris flow caused the 'Bus tumbling down accident at the Hida River'. Two buses among the queue of vehicles obliged to stop by a landslide ahead were attacked by a debris flow from the side, and the buses with a total of 104 victims on board tumbled into the Hida River that was parallel to the road. This accident clearly showed the dangerous situation of the mountain road, and it triggered the subsequent reinforcement of those roads by the governments. I felt keenly that the prediction of debris flow occurrence to help early refuge was an urgent theme to be investigated.

A similar accident happened in 1971 at Aioi, Hyogo Prefecture, Japan. In this case, a bus was pushed down to the valley bottom from the hillside road by a small-scale landslide. The facts I paid attention to in this case were that grasses on the route of earth block's motion survived although they were leveled down to the ground, and the surface of the road pavement on which the bus traveled was not much damaged. These facts revealed that the landslide in this case had little ability to erode the ground, yet it had the power to push away a bus. A month later, a debris flow occurred in a small ravine in Shodo Island. Our main object of investigation was to make clear whether the collapse of the road embankment at the headwater triggered the debris flow or whether the debris flow that was generated by the erosion of the riverbed just downstream of the road caused the collapse of the embankment. Setting our conclusion to this question aside, I would like to mention that the upstream part of the ravine was completely eroded to the bedrock. These two examples at Aioi and Shodo Island show that some debris flows have enormous ability to erode the ground but others do not. What is the cause of the difference between the two?

As described above, through the field investigations, I was much interested by the phenomena of debris flow, which had been little studied, so little was known about the mechanism of occurrence, flowage and deposition of debris flow. Thus, I made a resolution to investigate debris flow. I was, however, at that time involved mainly in research on the characteristics of flood flow in river. Therefore, my fully-fledged debris flow research commenced after 1975.

Because my main interest was the curious behaviors of debris flow, I commenced my work with the aim of making clear the physical mechanisms of debris flow phenomena. Then, being a staff member of DPRI, I developed my work on the engineering subjects such as the prediction of hazards and the functions of various structural countermeasures. The decision of specific themes was also influenced by the needs of the times; it was in line with the change of governmental policy against riverine hazards. In Japan, the imbalance between the property increment due to urban development and the necessary investment for river conservancy to cope with the enhanced vulnerability became larger and larger after the mid-1960s and the urban flood hazard problems became serious. A governmental program named 'comprehensive mitigation measures against flood damage' was launched in 1977. This program placed emphasis on the non-structural (in other words, soft) countermeasures in addition to the structural (in other words, hard) countermeasures as an effective strategy. As a part of the program, the policy to cope with debris flow disasters by evacuation was put forth. The designation of debris flow prone ravines was essential for the implementation of the plan, and the method to identify the potential debris flow ravines proposed by us was adopted. Since then, I have made a strong effort to research both the fundamental mechanics and rather practical problems such as the delineation of hazardous area, the determination of the critical rainfall for evacuation, etc.

The background of the importance of research on the mitigation of sediment disasters is as follows. In Japan, the number of casualties due to water related disasters has experienced a big change from more than 2,000 per annum in the decade after 1945 to less than 50 per annum in recent years. The improvement of major rivers has lessened large-scale flooding and it has surely contributed to the decrease in the number of casualties, but the improvement of minor rivers lags behind and the development of slope land makes the situation worse for the vulnerability to sediment hazards. Namely, among 5,666 casualties due to water related hazards in the 31 years from 1967 to 1997, 29% are due to cliff failures and 25% to debris flows or comparatively large-scale landslides. The tendency that the majority of casualties are due to sediment hazards has become especially conspicuous recently. For example, in the case of the 'Nagasaki disaster' in 1982, 75% of the total 299 casualties and in the 'San-in disaster' in 1983, 90% of the total of 121 casualties were due to sediment hazards. More recently, almost all the remarkable disasters such as the 'Gamaharazawa debris flow disaster' in 1996 (14 people killed); the 'Harihara River debris flow disaster' in 1997 (21 people killed); the 'Hiroshima disaster' in 1999 (24 people killed); and the 'Minamata debris flow disaster' in 2003 (19 people killed) were sediment hazards.

This situation might be considered as inevitable under Japanese natural and social conditions. At present, 289,739 cliff-failure-prone steep slopes, 11,288 slowly and largely moving landslide-prone slopes, and 103,863 debris flow-prone ravines are designated. These designations are under the criterion that more than five families or some kinds of public facilities such as a school and administrative institution exist inside the possible hazardous area. Taking the debris flow-prone areas as example, the number of designated areas increased year by year: in 1977 (the first designation) there were 62,272; in 1986, 70,434; in 1993, 79,318; and in 2002, 89,518. This increment is partly due to the reassessment of the survey, but it is mainly due to the encroachment of inhabitants into potential hazardous areas.

The necessity of development regulation in potentially hazardous areas had been pointed out long before, but no laws to put the regulation into effect existed, so that it had been free to develop. With the 'Hiroshima disaster' as a turning point, the 'sediment disaster prevention act' has been enforced since May, 2000. In the possibly extremely hazardous areas designated by this act, all the organized housing land development and personal house buildings that do not have adequate strength to resist sediment hazards are prohibited, and the local government can advise the inhabitants already inside that area to move to a safer area.

In parallel with the implementation of this act, the designation of hazardous areas has been reassessed. Namely, areas designated so far are defined as category I, and the newly designated hazardous areas of category II and III are added. In category II areas fewer than five houses exist, and in category III areas no houses exist at present, but if houses were constructed, they may well meet with disasters. The total of category II and III ravines with possible debris flows was 94,345 in 2002. These numbers of Categories I to III are of course too many to be adequately treated by the structural countermeasures, hence, soft countermeasures such as the control of development and evacuation in advance are important. This is the reason why the sediment disaster prevention act is enforced.

Because the execution of the sediment disaster prevention act attaints the usage of personal properties, the delineation of hazardous areas must be accurate and with substantial reasons. The state of the art of debris flow engineering may partly meet the demand, but it is still far from reliable. Researchers are responsible for the improvement of reliability.

The importance of sediment hazards is not limited to Japan. There are many countries exposed to even worse conditions. In China, more than a hundred prefectures and cities suffer from debris flow disasters every year, and more than a hundred people are killed and more than two billion yuan are lost. As an example, in the disasters of 2002, 1,795 people died due to water-related hazards and among them 921 were due to sediment hazards like debris flows. In Colombia in 1985, more than twenty thousand people were killed by the lahars generated by the eruption of Nevado del

Ruiz volcano and Armero City disappeared. In Venezuela in 1999, the Caribbean cities were attacked by large-scale debris flows and more than twenty thousand people were killed. In Taiwan in 2001, a rainstorm associated with a typhoon generated many debris flows and killed 214 people. Many debris flow disasters also occur in other countries such as Nepal, Indonesia, the Philippines, Italy, Switzerland and France. Therefore, the improvement of debris flow research can contribute globally.

This book, at first, describes the various characteristics of debris flows and explains that debris flows can be classified into several types. Then, the mechanics of the respective debris flows are explained and in the process of discussion the advantages and drawbacks of various previous theories will be made clear. After that, the processes of occurrence and development, the characteristics of fully developed flow and the processes of deposition are explained from the mechanical point of view. Up to Chapter 5, discussions are concerned with fundamental mechanical theories; the rest of the book is concerned with the application of these theories. Namely, in chapter 7, there are computer simulations of some actual disasters for which we participated in the field investigations. The last chapter selects some debris flow controlling structures and discusses their effectiveness and performance designs, and it also discusses the soft countermeasure problems such as the identification of debris flow prone ravines and the prediction of occurrence by the concept of precipitation threshold.

My standpoint throughout the book is the fundamental quantitative explanation of the phenomena, and the design of structural countermeasures. The performances are discussed based on the fundamental mechanics, putting the empirical method out of the way as far as possible. As a consequence, many mathematical formulae appear, and I am afraid this makes the book daunting. But, even if readers ignore the complicated formulae, they will be able to understand the physical phenomena of debris flows as well as the current status and future needs of research.

This book does not aim to review the existing investigations thoroughly, but it intends to offer an overview of my works. Therefore, previous very important contributions might not be referred to. It is not due to disrespect but to make my thought clearer. Pyroclastic flows and snow avalanches are referred to as the phenomena similar to debris flow in Chapter 1, but in this book, due to the limitation of pages and in view of coordination as a book, I omitted detailed descriptions of these phenomena. I would like to discuss these problems on another occasion.

This book is published under my sole name, but the majority of the investigations referred to in this book were not, of course, done unassisted. I owe thanks to many colleagues some of whose names appear in the reference papers but others I was not able to reference. I would like to take this opportunity to thank them all.

Tamotsu Takahashi

Preface to the first English edition

The Japanese edition was published in September 2004 by Kinmiraisha in Nagoya, Japan. The book was acclaimed by Japanese engineers and researchers and was awarded the Japanese Association for Civil Engineers' 'Publishing Culture Prize' for the most valuable publication of 2004. Many colleagues expressed their interest in an English edition of the book, which would make it more broadly accessible to the international academic community. One of my motivations in undertaking the translation was the previous success of the English version of my earlier book Debris Flow, published in 1991, in the IAHR monograph series. This well-known publication continues to be cited frequently in international literature. In the preface of the original work, I wrote as follows:

> The important engineering subjects of the research; identification of the hazardous ravines, prediction of occurrence time, estimation of hazardous area and the risk, functions of various structural countermeasures, design criteria of structures, warning and evacuation systems, etc., have also made considerable progress. I would like to review those problems on another occasion.

A long time has elapsed since the publication of this original book, a period in which significant progress has been achieved in Fundamental Mechanics and in the field of Engineering. The 2004 Japanese edition reflects the intention expressed above and contains improved theories. This current English edition also incorporates part of another book I have published in Japanese, entitled *Mechanisms of Sediment Runoff and Countermeasures for Sediment Hazards*, which was published in 2006.

I hope that readers all over the world will enjoy reading this book and that they will find it an interesting and helpful contribution to the literature.

January 2007
Tamotsu Takahashi

Preface to the second English edition

The main themes in the first edition were focused on a single debris flow event and the following topics were discussed: the classification from the mechanical point of view; the characteristic behaviors of flow; the analyses and predictions of the processes during its initiation, flowing down and deposition; the reproductions of some representative debris-flow disasters using numerical simulations; and the methods for preventing or mitigating the debris-flow disasters. These themes should, of course, constitute the main parts of the second edition as well. But, from the countermeasure planning point of view, the preventing or mitigating functions of the structures should continually be shown for a long period during which many debris flows and highly sediment-laden floods occur. In this context, the analyses of sediment runoff over a long period are also important. Therefore, in the second edition, a sediment runoff model that is named 'SERMOW' and its improved version 'SERMOW ver.2' are introduced in the new chapter 6, where the sediment runoff volumes during a long period as well as the hydrograph, sediment graphs divided into several particle size groups for a single debris flow event can be predicted.

Recently, after the publication of the first edition, the large-scale landslides with the formation of landslide-dams occurred such as that induced by the Wenchuan earthquake of May 2008 in China, that induced by severe rainfall brought about by the typhoon Morakot in August 2009 at Shiaolin in Taiwan, and that induced by typhoon No. 12, in September 2011 at the Nara and Wakayama districts in Japan, among others. In this context, some considerations on the mechanics of the high mobility of large-scale deep-seated landslides and debris avalanches are added in the second edition.

Several other topics that may contribute to a deeper and more comprehensive understanding of debris flow mechanics are also added.

Fortunately, the first edition of this book was acclaimed by many readers, but a considerable time has elapsed since the publication, so that Taylor & Francis suggested me to revise. In response to the suggestion I decided to publish the second edition, renovating under the above-mentioned viewpoints.

July 2013
Tamotsu Takahashi

About the author

Tamotsu Takahashi (Kyoto, 1939) graduated as a Master in Civil Engineering at Kyoto University in 1965. From 1965 to 1967, he then worked as a research assistant at the Disaster Prevention Research Institute (DPRI) of the same university and, after a year in the Civil Engineering Department as a lecturer, he returned in 1968 to the DPRI as an associate professor. With his research on flood flow dynamics in river channels, he obtained the doctoral degree in 1972. After this, he worked as a post-doctoral fellow at Lincoln College, New Zealand, where he investigated miscellaneous problems that were associated with braided rivers. Upon returning to DPRI in Japan, he put importance on the study of sediment runoff problems that were involved with debris flow and bed load on very steep slope channels. Consequently, in 1982, he was awarded a full professorship for a newly founded research section on the investigation of anti-flood hazards systems. He then added slightly more themes to his portfolio and extended his research to the water flooding and sediment problems in urban areas. From 1992 he moved to the research section on the investigation of sedimentation problems.

After his retirement in 2003, he continued working on debris flow and sediment runoff problems as a professor emeritus at Kyoto University. From 1995 to 1997 he served as the director of DPRI and during this appointment, he has reorganized the entire DPRI and has thoroughly promoted the scientific investigation of the Great Hanshin Earthquake which took place in Kobe in 1995 as the director of DPRI and the head of the Japanese Group for the Study of Natural Disaster Science. He is now working for the foundation 'Association for Disaster Prevention Research' as the chief director.

Professor Takahashi has authored numerous papers and held many invited keynote lectures. He also received several awards for his outstanding work from the Japan Society of Civil Engineers and from the Japan Society of Erosion Control Engineering. His successful book 'Debris Flow', published in 1991 by A.A. Balkema Publishers in the IAHR monograph series was the first systematic approach to the subject and is still frequently referred to. The original Japanese language version of this current new and extended edition was received very well and the author was awarded the Publishing Culture Prize from the Japan Society of Civil Engineers in 2004 for it. He was also awarded the Akagi Prize in 2008 for his outstanding contributions to the prevention and mitigation of debris flow disasters.

Chapter 1

What is debris flow?

The photograph shows the front part of the viscous debris flow at Jiangjia gully, Yunnan, China. It is flowing from upper right to lower left. This flow contains solids fraction more than 60% by volume. Notwithstanding such a high solids concentration, it flows fast on slopes as mild as 3 degrees.

INTRODUCTION

Since ancient times there have been witness records of debris flows under way, stopping and depositing (e.g. Schlumberger 1882; Blackwelder 1928; their Japanese translation appears in Takahashi 1983). Thus, the nature of debris flows should have been known, at least fragmentarily, by some geographers, geologists and engineers who are concerned with works in mountain rivers. Residents in steep mountain areas in Japan have called debris flow by terms such as 'Ja-nuke', 'Yama-tsunami', and 'Yama-shio', and they have handed these terms on to the next generation to warn them of these phenomena. The meanings of these Japanese terms are 'the run off of king snake', 'the tsunami at mountain' and 'the mountain tide', respectively. But, as they have occurred in remote steep mountain and moreover in adverse conditions of severe rainstorms, there have been few scientific observational records. Therefore, no one has been able to understand why such huge boulders were easily transported down to the village on alluvial cones to cause disasters. It was a 'phantasmal disaster'.

In the last half of the 1960s scientists began to investigate the mechanisms of debris flow. Particularly in Japan, adding to the theoretical and experimental investigations, some deliberate observation systems were established in the basins where debris flow often occurs, and some clear photographic images of debris flow under way were taken for the first time in the world. Since then, some debris flow images have been shown on TV, and now the idea of debris flow phenomena is understood not only by scientists but also by the general public. When broadcasting warnings of heavy rain, the term 'debris flow' (dosekiryu in Japanese: doseki means earth and stone; and ryu means flow) is used as a widely understood phenomenon.

In this chapter, at first, various subaerial sediment moving phenomena are analyzed from a mechanical point of view, and debris flow is defined as a significant category among them. Next, referring to some observational records, debris flows are classified into three types. The characteristic differences between these three types of debris flows are explained by the physical mechanisms of the particle dispersion in the flow. Finally, the classic classifications of debris flows by the causes of occurrence and the relationship between occurrence and rainfall characteristics are explained.

1.1 VARIOUS SEDIMENT MOVING PHENOMENA

Natural sediment motion/transportation on a subaerial slope is divided into two types; one is motivated by gravity and moves *en masse* and the other is driven by fluid dynamic forces and moves as an individual particle motion. In this book, phenomena are exclusively considered as subaerial even though they occur subaqueously as well. Herein, the term sediment means all the particulate substances from clay to huge boulders. The sediment hazards in this book mean the hazards caused by the motion of sediment. The individual particle motions include bed load, suspended load and wash load in river flows; blown sand in the desert or at the seacoast; and yellow sand that reaches Japan from the distant Gobi Desert. The representative geophysical sediment motions/transports in a mass include:

1   landslides and landslips (cliff failures);
2   debris avalanches;
3   pyroclastic flows; and
4   debris flows and immature debris flows.

Some supplementary explanations may be necessary for the terms 'landslide' and 'landslip' or 'cliff failure'. For many years in Japan the gravitational mass movements of more or less dry rock, debris or earth on slopes have been classified into two categories. The specific Japanese names are given to the respective phenomena. One is 'Ji-suberi' and the other is 'Houkai'. The translation into English may be 'landslide' for the former term and 'collapse' for the latter. Ji-suberi specifically means the slow and persistent motion of a large earth block that often occurs on a rather gentle slope as a recurrence motion of a once moved block. Houkai specifically means a juvenile landslide that often occurs on a steep slope as a translational motion of a shallow earth block or fragmented rocks. In 1996, the World Landslide Inventory Commission defined the term 'landslide' as the gravitational mass down slope motions of rock, debris or earth (Sassa 2003). By this definition, it covers all the phenomena of Ji-suberi and Houkai, and moreover, it covers debris avalanches and even debris flows. Because the specific phenomena included in this broad definition are mechanically different, they need to be investigated by different methods. Japanese statistics of disasters have discriminated between Ji-suberi and Houkai, so it is not convenient to put them together. Therefore, in this book, the term 'landslide' is applied to 'Ji-suberi', i.e. the deep-seated earth block motion, and 'landslip' or 'cliff failure' is used for 'Houkai', i.e. a shallow surface soil motion. But, sometimes, 'landslide' is used as generic term of these two phenomena, but not for debris avalanche and debris flow.

Some comments on the term 'debris avalanche' may also be necessary. In some papers written in English, small-scale landslips are also called debris avalanches. But, in this book, I hold this term to mean the very large-scale collapse of a mountain body that reaches distant flat areas of low slope. This phenomenon is sometimes called 'Sturzström'. This is named by Heim (1882). 'Sturz' means collapse and 'Ström' means flow. Heim believed that the phenomenon he experienced was a flow of debris as if it were a kind of liquid. Fluid mechanically, flow is defined as a phenomenon in which irreversible shearing deformation prevails everywhere inside. In the case of a debris avalanche, however, motion starts as a block and it slips on the slip surface. Therefore, the processes from start to stopping would be better called an avalanche rather than a flow.

The essential mechanisms of the four geophysical massive sediment motions mentioned above are outlined as follows:

1   In the cases of landslides and landslips, the earth block slips on a slip surface, and the deformation inside the block is small. Therefore, it is essentially the motion of rigid body. Except for the case in which motion is transformed into a debris avalanche or debris flow, the distance of motion is short. When it reaches an almost horizontal area, it can reach a distance at most twice the height to the scar.

2   Debris avalanches are mainly caused by a very large landslide of a few million cubic meters or more. The rapid demolition of the slid block occurs early in the initial slipping stage and it reaches much further than a landslide or landslip. Many theories for its large mobility have been proposed, but still no conclusive theory exists. A slipping rigid body will decelerate and finally stop when the kinetic friction force becomes larger than the gravitational driving force. The friction force is the product of the load acting perpendicular to the slip surface and the kinetic friction coefficient. Because the kinetic friction coefficient has a nearly constant value, if the mass does not change, the friction force operating on the body becomes larger as the slope angle becomes flatter; contrary, the driving force which becomes smaller with a decreasing slope gradient. Hence, with a decreasing slope gradient motion stops. In general, the kinetic friction coefficient is of the order of 0.6. This means that the motion of a slipping body will stop on a slope flatter than 30°. In the case of a debris avalanche, however, motion sometimes continues even on a slope as flat as 3°. So there must be some kind of lubrication mechanism to lessen the apparent kinetic friction coefficient.

3   Numerous pyroclastic flows have occurred at the Unzen volcano since the beginning of the last eruption in 1990, and thanks to the TV shows of actual pyroclastic flows, it became a well-known phenomenon in Japan. The pyroclastic flow at Unzen was produced by the collapse of the lava dome; during the process of falling down, a huge rock block was crushed into fine material and it flowed down to a distant flat area. This is one type of pyroclastic flow; there are several types based on the classification by the processes of occurrence, material composition and magnitude. The mechanics of flow in any type are controlled by the gas ejection from the material itself. The volcanic gas ejected from the material can form an upward gas flow that is fast enough to sustain the particle's weight; the gas flow produces a fluidized layer within the flow. Inter-particle friction force in the fluidized layer is minimal and thus large mobility is attained. More detailed discussion appears in Takahashi and Tsujimoto (2000).

4   Particles dispersed in water or slurry move within the debris flow. The buoyancy acting on particles must have some effects on the large mobility as it can flow even on a gentle slope as flat as 3°. But buoyancy by itself is not enough to sustain the entire weight of heavy particles, hence, some other mechanisms to sustain particles and to keep the distances between particles wide enough to make the motion of particles easy must operate.

Various classifications of sediment motions have been proposed. Among them that proposed by Varnes (1978) is the most popular one. In this the existence domains and the schematic illustrations of various moving patterns are demonstrated on a plane whose orthogonal two axes are the types of motion (falling, tilting and toppling, slide, lateral spreading, and flow) and the material types before the initiation of motion (bedrock, coarse rock debris, rock debris, sand, and fine particles), respectively. Furuya (1980) slightly modified Varnes' diagram so as to include surface landslips and to improve some defects of Varnes' diagram. Whereas Varnes demonstrated the entire phenomena on one plane, Ohyagi (1985) split the Varnes' plane into multi-layered planes with respect to the velocities of motion and changed it to a three-dimensional expression by adding the vertical axis representing velocity. Further classifications include those by, among others, Pierson and Costa (1987) who focused on velocity and solids concentration; and Coussot and Meunier (1996) who focused on velocity, solids concentration and material's cohesion. These classifications are empirical and qualitative; although easy to understand, they do not consider the physical mechanisms of motion and transition process from rigid body to flow.

Takahashi (2001) classified the subaerial mass flows consisting of granular materials (particle sizes may vary from powder to mountain blocks) focusing attention on the essential mechanism to control the phenomena. That classification was lately modified as shown in Figure 1.1 (Takahashi 2006).

Subaerial massive sediment motions, closely related to sediment hazards, are basically divided into fall, flow and slip. The eight blocks in Figure 1.1 represent these

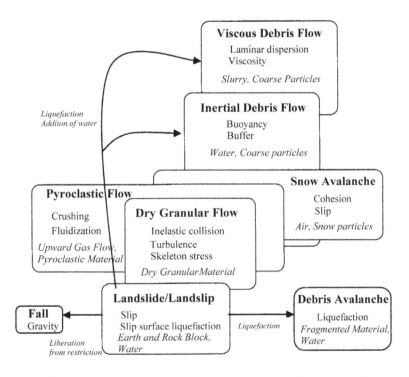

Figure 1.1 Subaerial mass movements; their mechanical resemblances and differences.

phenomena based on the specific aspects. The upper five blocks represent the phenomena in which particles are dispersed in the flowing body, and the lower three blocks represent the phenomena in which the moving bodies are mostly the agglomerates of soil and rocks. If the slipping rigid body (it is not necessarily one piece) is liberated from the constraint of the ground surface on which it slips, it becomes a free falling body, and if liquefaction at the lower part of the slipping body proceeds during motion, it acquires high mobility and can be called a debris avalanche. If the entire body is liquefied, it is then a debris flow. Thus, the arrows starting from the landslide/landslip block towards the debris avalanche block or debris flow blocks represent the processes from initiation to full development. High water content is necessary for the development of liquefaction, so that, for a large-scale debris avalanche, the mountain body that starts moving should have been saturated with water at least at the lower part. But, in the case of debris flow, if the volume of slid body is small, even if it is not saturated with water, the addition of water from outside may be enough to transform it into a debris flow. The processes of liquefaction and the acquirement of high mobility will become clear in chapter 3, but, a supplementary explanation for the mechanism of debris avalanche may be necessary here. Namely, there have been some theories claiming that the mere crashing of an aggregate into fragments while in motion is enough to acquire its high mobility; e.g. Davies (1982). But, as it will become clear later in this book, dry granular flow is not able to keep going on a slope flatter than the angle of repose of the material.

The four blocks representing dry granular flow, pyroclastic flow, snow avalanche and inertial debris flow overlap a great deal with each other. The overlapping parts are the commonly operating mechanism (written by roman letters on the blocks); i.e. the stresses due to inelastic collision, turbulence and quasi-static skeletal stress, although these are not necessarily equally dominant in the respective motions. The terms written in italic letters on the respective blocks are the principal material or the medium existing in the respective motions. The protruding part of each block is the peculiarity of the mechanism in each phenomenon. Namely, to explain the mechanism of pyroclastic flow, besides the commonly operating stresses written on the block of dry granular flow, the effects of particle crushing (especially for the case of the Merapi-type pyroclastic flow which is defined as the one that starts from collapse of a lava dome) and the generation of fluidized layer due to gas emission from the material itself should be taken into account. For a snow avalanche, besides the commonly operating stresses, the effects of cohesion that give rise to the formation of snowballs and the slip on the snow surface must be crucial (Takahashi and Tsujimoto 1999; Takahashi 2001). For an inertial debris flow, besides the commonly operating stresses, the buoyancy effect and possibly the buffer effect of interstitial fluid to moderate particle collisions and the additional mass effect to accelerate the surrounding fluid would be important. If a debris flow contains much fine sediment, such as clay and silt, the viscosity in the interstitial fluid (slurry made up by the mixing of clay and silt with water) becomes very large and turbulence is necessarily suppressed. Then, the effect of particle collision becomes minimal and instead the laminar dispersion mechanism (Phillips *et al.* 1992) should operate. Therefore, the common mechanism between viscous debris flow and inertial debris flow is only the buoyancy. Viscous debris flow also has little in common with debris avalanche, pyroclastic flow and snow avalanche as shown by no overlap between the blocks representing these phenomena.

Table 1.1 Magnitude and mobility of mass movements.

Phenomena	Magnitude (m^3)	Velocity (m/s)	Arrival distance (km)	Equivalent friction coefficient
Landslide	$10\sim10^6$	$10^{-6}\sim10$	$<0.3^*$	$>0.5^{**}$
Landslip	$2\times10^{5**}$	–	0.7^{**}	0.25^{**}
Debris flow	$10^3\sim10^6$	$20\sim0.5$	$0.2\sim10$	$0.3\sim0.05$
Debris valanche	$10^7\sim10^{10}$	$10\sim10^2$	<30	$0.3\sim0.05$
Pyroclastic flow	$10^4\sim10^{11}$	$10\sim50$	<60	$0.4\sim0.2$

*Landslip smaller than $1.5\times10^4\,m^3$.
**Las Colinas, El Salvador, 2001.

Sediment mass movements sometimes trigger serious disasters. The knowledge of the respective phenomena, such as with what magnitude and destructive power they will strike us and how large an area they will affect, is important to mitigate disasters. Table 1.1 summarizes and classifies the phenomena from such a viewpoint. The numerical values in the table show a rough standard deduced from past records. In comparison to the obvious fluidity of debris flows, debris avalanches and pyroclastic flows, landslides and landslips or cliff failures cannot reach far; the comparatively dry earth block that failed can reach, at most, the distance of only twice the height to the top of the scar if it moves on a horizontal plane at the foot of the slope; i.e. the equivalent friction coefficient is larger than 0.5, where the definition of the equivalent friction coefficient is given in Figure 1.2.

The fact that the equivalent friction coefficient for a dry earth block is larger than 0.5, is quite reasonable, if it should not be much different from the kinetic friction coefficient of a rigid body. As is explained later, in a dry granular flow, the particle dispersive pressure that is produced by the particle encounters is vertically transferred via the particles moving in contact with each other and that pressure at the bottom of the flow is equal to the total load of particles above the bottom. The apparent friction angle between the particles and bottom is equal to the angle of repose and this is not much different from the kinetic friction angle of a rigid body. There are many experimental data that confirm this under comparatively large shearing velocity (e.g. Hungr and Morgenstern 1984). Namely, the equivalent friction coefficient for a dry granular flow is nearly equal to that of a landslide and of a landslip. What, then, is the reason for the high fluidity of debris flows, debris avalanches and pyroclastic flows? Hints as to the reasons are given in italics in the respective blocks in Figure 1.1.

1.2 DEFINITION OF DEBRIS FLOW

The definition of debris flows has already been mentioned in the physical explanation of differences between debris flows and other mass movements. The clearer expression may be as follows: Debris flow is a flow of sediment and water mixture in a manner as if it was a flow of continuous fluid driven by gravity, and it attains large mobility from the enlarged void space saturated with water or slurry.

The strong mobility of debris flow is evident by comparing the equivalent friction coefficients with other mass movements shown in Figure 1.2. Although it may have little physical meaning (Campbell *et al.* 1995; Takahashi 2006), this concept has been widely used in the discussion on the mobility of massive motions. As illustrated in the definition sketch in Figure 1.2, the equivalent friction coefficient is defined by the tangent of the angle of elevation from the distal end of the deposit to the top of the slide scar. In the entire processes from initiation to deposition, the potential energy of the slid earth block is converted into kinetic energy and it is consumed by the frictional resistances between the earth block and the ground surface and by the internal shearing deformations, hence, strictly speaking, the elevation angle should be measured from the center of gravity of the deposit to the center of gravity of the earth mass before sliding. Because the identification of these points is very difficult, for the sake of convenience, the mentioned angle of elevation is adopted.

Many large-scale landslides/debris avalanches plot their equivalent friction coefficients as they decrease with increasing volumes; when the volumes are less than a million cubic meters, the equivalent friction coefficients are larger than 0.6, similar to that of a rigid body on a rigid floor, but when the volumes become a hundred million cubic meters or more, it becomes very small meaning that the motion reaches a distant flat area. The wide scatter of these plotted points is partly due to the different mobility associated with the difference in water content from dry to nearly saturated.

Of the data of pyroclastic flows in Figure 1.2, the four points plotted closest to the left end are for those produced by the collapse of lava domes at the Unzen volcano and other points are for the ones due to the collapse of the eruption plume or those due to lateral blowout. In the case of dome collapse, the top of the perspective line indicating the elevation angle is the summit of the dome, but in the case of column collapse, it is the highest point of the plume. In any event, pyroclastic flows have high mobility even if the volume is not large. The mechanism to gain such a high mobility was very briefly mentioned in the last section.

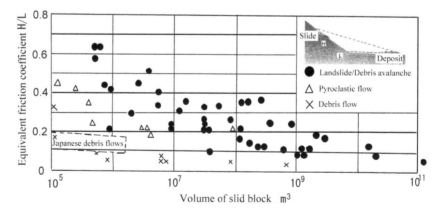

Figure 1.2 Equivalent friction coefficients in debris flow, landslide/debris avalanche and pyroclastic flow. Data of Japanese debris flow, landslide/debris avalanche; Chigira (2001), other debris flows; Iverson (1997), other landslide/debris avalanche; Hsü (1975), pyroclastic flow; Kaneko and Kamata (1992).

As is clear in Figure 1.2, debris flows have a much higher mobility than any other pyroclatic flow and landslide. This fact suggests that the larger water content than any other mass movements makes the mobility very large. The mechanism to have the largest mobility is the key issues to be discussed in the next chapter.

1.3 CLASSIFICATION AND CHARACTERISTICS OF DEBRIS FLOWS

The definition of debris flows provided in the last section is applicable regardless of their constitutive material's sizes and their distributions, the sediment concentrations in flow, the properties of interstitial fluid, and the hydraulic conditions of flow such as velocity, depth and channel slope gradient. The behaviors and the destructive powers of debris flows differ depending on these factors, however. Each researcher likely has an interpretation of debris flow that is familiar to him or her and talks about such a debris flow, but others may have another interpretation. This can often cause unnecessary confusion. One must clarify which kind of debris flow is the subject of discussion.

In this section, the classification of debris flows mainly based on appearance is described.

1.3.1 Stony-type debris flow

A debris flow occurred in the Shiramizudani experimental watershed of the Hodaka Sedimentation Observatory, DPRI, in 1968. It left behind curious traces inside and in the margins of the channel within which it flowed; the channel was about 10 m in width and a few tens of centimeters in depth. Although the central part of bed was covered by comparatively small particles, both hems were comprised of boulders larger than 1 meter in diameter as if they were raised from the bottom and aligned artificially (Figure 1.3). Photo 1.1 shows another debris flow, but it shows similar sight.

The nature of the debris flow described above was enough to give me an incentive to investigate the mechanics of flow, but, at that time, debris flow was considered a 'phantasmal disaster' that had unknown origins. Several years past without conspicuous advances until Okuda and his colleagues in DPRI succeeded in taking photos of

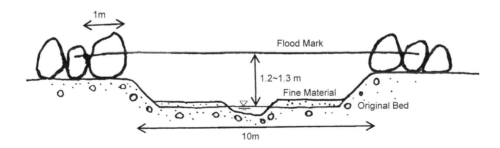

Figure 1.3 Cross-section of the Shiramizudani Ravine after the passage of a debris flow.

Photo 1.1   Boulders in line along both sides of the channel (the Dacho river in Yunnan, China).

debris flow on going at Kamikamihorizawa in 1976. Photo 1.2 is the time sequential photographs of that debris flow, with one picture taken every second. They analyzed these photographs in detail and obtained the following conclusions on the nature of debris flow (Okuda *et al.* 1977):

1   The forefront of the debris flow looks like a bore and the depth of flow suddenly becomes large from virtually no preceding flow.
2   The biggest stones accumulate at the front part and the forefront contains little water; it may be called stone flow.
3   The flow is elevated much along the right (outer) bank, presumably because the photographed section locates a little downstream of a slight bend in the stream channel.
4   The front part of the flow, where big boulders accumulate, lasts only a few seconds and the following part that lasts long looks like a mud flow with gradually decreasing discharge.
5   Estimated from the sharpness of photo images, the velocity seems to distribute laterally and the central part is something like a plug in Bingham fluid flow; the velocity at the central part is the largest and almost no lateral velocity gradient exists. (*Author's remark: It is wrong to conclude automatically that if no velocity gradient exists it is a plug. The turbulent water flow in a wide channel has almost uniform lateral velocity distribution.*)

Debris flow observations in Kamikamihorizawa have given a great many valuable outcomes. They are concerned with the whole picture of the processes of debris flow; i.e. the situations at the source area, behaviors while in the channel, and the processes of deposition. The following are the various natures of stony debris flows made clear in the Kamikamihorizawa ravine.

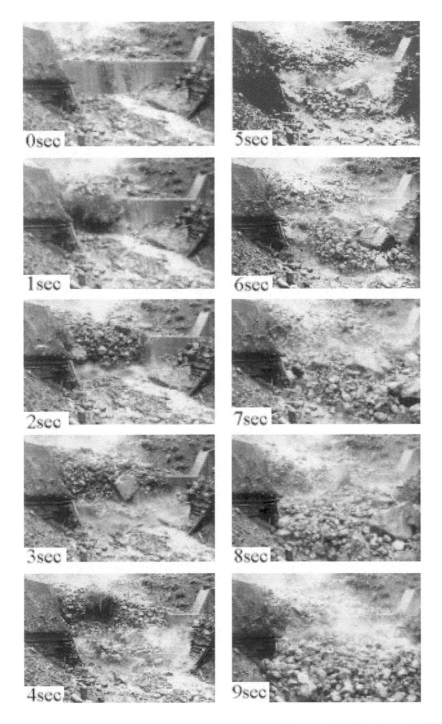

Photo 1.2 A stony-type debris flow in Kamikamihorizawa. 3 August 1976 (Okuda *et al.* 1977).

The nature of stony-type debris flow in initiation and flowing regions

Although the processes of a large-scale debris flow initiation have not been observed yet, a witness record of an occurrence of a small one is available (Okuda *et al.* 1978), they described it as follows:

> Subsequent to a thunder storm, turbid water flows loading particles of a few ten centimeters to about 1 meter appeared in rills and gullies. These were quasi-steady flows without bores and other waves, but they were suitable to be called debris flows. These flows were built up at the confluences and ran down as the masses having a velocity of about 5 m/s at an interval of five to ten seconds. The piling up of flow at the channel knickpoints seemed to give rise to repeating bores.

This description suggests that the debris flow was generated by the appearance of surface water flow; it was not due to the transformation of landslides.

A number of wire sensors were installed along the Kamikamihorizawa channel, and the times a debris flow front cut them in sequence were recorded to find out the translation velocity of the debris flow front. Figure 1.4 demonstrates the data thus obtained for 14 debris flows. The upper sequential line graph shows the longitudinal changes of the translation velocities of the debris flow fronts; the middle graph shows

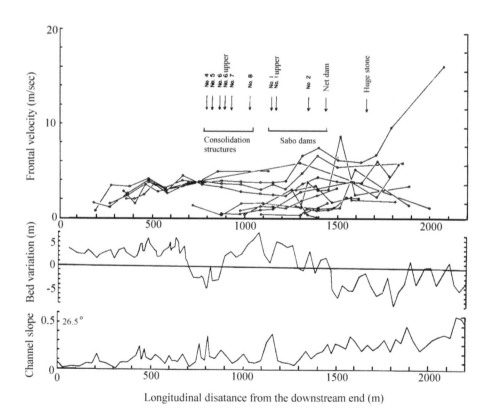

Figure 1.4 The longitudinal change in the debris flow translation velocities.

the change in the channel bed elevation during 15 years from October 1962 to October 1977, where positive values mean aggradation and negative values mean degradation; and the lower graph is the longitudinal channel gradients in October 1977. The origin of the abscissa axes is located at around the distal end of the debris flow fan. Yakedake Volcano erupted in June 1962. Just after the eruption, debris flows frequently occurred, but the frequency gradually diminished with time. The general tendency shown in this figure is as follows (Okuda *et al.* 1978):

Concerning the translation velocity:

1   The front velocity decelerates around 'huge stone' and then gradually accelerates in the reach of the channel constriction down to the 'dam no.2'.
2   Downstream of the dam no.2 it retards again because debris flow comes into the sedimentation pocket of 'dam no.1 upper'.
3   In the reach downstream of dam no.1, where the serial bed consolidation structures are set, the velocity is nearly constant.
4   The flow then comes out to the fan area and finally stops.
5   Even the velocities upstream of the fan top are different case by case, those downstream of the fan top become nearly the same, but the positions and velocities at stoppage are different case by case.

Concerning the channel bed variations:

1   The tendency towards gully bed degradation is evident in the upstream reach, but in the reach where dams and consolidation structures are constructed the bed aggrades.
2   Downstream from this reach to the fan top area, the tendency towards degradation reappears but soon the flow begins to unload the freighting sediment and it develops the fan.
3   The degradation of bed and the deceleration or nearly constant velocities upstream from the point 1,500 m to 1,900 m suggest that it is in this reach that the debris flow rapidly develops by entraining bed materials.

From these facts, we can understand the following characteristics in the development and declination processes. Namely, where channel slope is steeper than 16°; upstream of the point 1,500 m, debris flow develops by the erosion of bed, and downstream of that reach where the countermeasure structures are installed, the erosion of bed is restrained or, on the contrary, deposition takes place. The reason for this would be either by the bed consolidating and sediment storing effects of those structures or because the debris flow has fully developed in the upstream reach and it comes down to this reach with the concentration no more bed erosion is possible. Downstream of this reach between 700 m and 900 m, possibly by the deposition upstream and helped by the locally steepened channel gradient, the ability to erode the bed is rejuvenated. Thus, the redeveloped debris flow finally comes out to the fan area, and due to the decrease in slope gradient, it stops. Furthermore, we can understand that, similar to the case of water flow, the deeper the depth of flow and the steeper the channel gradient, the larger the velocity of debris flow becomes, but velocity also depends on solids concentration, the denser the concentration, the slower the velocity becomes. Sometimes,

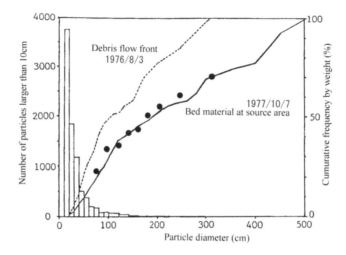

Figure 1.5 Distributions of particle diameters larger than 10 cm in a debris flow and the valley bed at the headwater area (Okuda *et al.* 1978, partly altered).

it is said that debris flow develops endlessly like a snowball rolling down a steep slope covered by snow, but, in fact, the growth is limited within the steep developing reach, and a fully developed debris flow has no more bed eroding ability and does not grow. We can see many debris flows pass through a slope or channel without erosion as an example described in the preface.

The surface velocity of the debris flow that occurred on 21 September 1979 was measured by the spatial filter speedometer. The measured velocities during four minutes including the instant of the passage of front were around 6 m/s and these values were larger than the velocity obtained by the cutting of wire sensors set along the channel, which was 3.3 m/s (Okuda *et al.* 1980). Under the condition of neither erosion nor deposition, the translation velocity of the front is equal to the cross-sectional mean velocity. The translation velocity is equal to that obtained by wire sensors. Therefore, the larger surface velocity than the translation velocity suggests the existence of vertical velocity distribution inside the flow; with the slower velocity in the lower layer and the faster velocity in the upper layer than the cross-sectional mean velocity.

The number of particles coarser than 10 cm was counted at a position in the headwater area where evident bed degradation took place, which is depicted by the bar graph in Figure 1.5. The cumulative size distribution calculated from this frequency analysis is also shown on the same figure by the solid line. The broken line on the same figure shows the cumulative size distribution at the front of debris flow measured by analyzing the photographs (Okuda *et al.* 1978). The largest particle size at the front in this debris flow was 310 cm, and the existence ratio of this size in the headwater area – about 70% – can be plotted as the most right-hand circle on the solid line, then we can reduce the scale of the size frequencies at the front (broken line) by the same ratio (70%) as illustrated by the other circles. We clearly see these circles plot near the solid

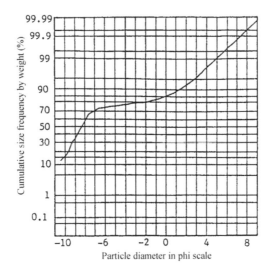

Figure 1.6 Particle size distribution at the front of a debris flow on 4 September 1978 (Okuda *et al.* 1979).

line. This means that the debris flow on 3 August 1976 was generated by the erosion of the bed in in which only the particles less than 310 cm were selectively entrained.

Figure 1.6 shows the size distribution in the front part of a debris flow in which all the comprising materials (from 2 μm to 200 cm) are taken into account (Okuda *et al.* 1979). The particle diameter in this figure is given by φ-scale, where $\varphi = -\log_2 d$ and d is the particle diameter in millimeters. We can see in this example, that the front part lacks particles whose size is between 4 mm and 128 mm. The bed material in the developing reach, however, contains many such particles, and therefore this must be the result of particle segregation within the flow.

As is evident in Photo 1.2, many big boulders accumulate at the front and they are scarce in the rear part. The size segregation effect seems to transport big boulders ahead and, by way of compensation, leave the particles of 4 mm ~ 128 mm in size behind. Finer particles, less than 1 mm, are contained in about 20% of the front part. This suggests that such fine particles comprise the interstitial muddy water which behaves as a liquid.

A sediment sampler (50 liters in capacity) set on the bed consolidation structure 'no.6 upper' trapped, on several occasions, a portion from the forefront of the debris flow. The sample analyses revealed the apparent specific densities of these water and solid mixtures were between 1.4 and 1.85 g/cm^3. A pressure gauge was also buried on top of this structure and measured the hydraulic pressure in the interstitial fluid at the bottom of the flow just upstream from the free drop. The pressures measured by this equipment corresponded to the equivalent densities of about 1 g/cm^3 and about 1.5 g/cm^3, respectively at the forefront where the flow depth and particle concentration were the maximum and at the rear part where the material was much liquid, in which hydrostatic pressure distribution was assumed. These equivalent densities were always less than the density of the hypothetical liquid that considers the entire mixture of water and particles comprising debris flow is a continual fluid. This pressure deficit was largest at the stony front part and became less towards the muddy rear part. This means that some parts of the load are not supported by the fluid pressure but are transmitted directly to the bottom possibly by the effect of particle collision.

The stony-type debris flows in the process of deposition in Kamikamihorizawa

The process of deposition was observed on 21 September 1979 (Okuda *et al.* 1980). The behavior of the flow nearby the fan top was similar to that in the gully upstream because of the restraint of flow within the incised channel. Even in the region further downstream where the incised channel disappears, the flow did not widen immediately so as to flow over the entire debris fan, but it flowed gradually enlarging its width. On a flat gradient of 4~5 degrees, a big boulder of about 4 m in its major axis was observed to roll down with the greater part of its gigantic body protruding. This debris flow flowed down the fan more than 500 m and made a comparatively flat deposit.

Debris flows are known to form elongate and slightly higher landforms on stopping. These deposits are called 'debris flow lobes'. The above mentioned debris flow formed 'lobe A' that arrived at an exceptionally distant place (see Fig.1.7). The height of it was small and it was comprised of mainly sand and gravels with a scattering of large stones.

The fan of Kamikamihorizawa consists of the imbrication of many lobes. These lobes can be classified into two types; one is the above mentioned flat-type; and the other is the 'swollen type', which has a half-pipe form as if the flow in a gully has been instantly frozen and it is mainly comprised of cobbles and large stones. The swollen-type lobes are located in the proximity of the fan top, whereas the flat-type ones are located in distal places. Sometimes, the swollen-type lobe obstructs succeeding flow and then the succeeding flow detours this lobe making a new channel. Photo 1.3(a) is an overhead view of lobe A near the distal end, and Photo 1.3(b) is a side view of the swollen-type lobe J (Suwa and Okuda 1982).

Photo 1.3 Debris flow lobes on the Kamikamihorizawa debris fan (Suwa and Okuda 1982) (a): The front part of lobe A. (b): The side view of lobe J (opposite side of tree).

Figure 1.7 shows the plane distribution of debris lobes deposited after 1978 (Suwa 1988). Lobe A can be found in the lower central part of the figure. Debris flows in 1970 flowed in the direction of arrows a, those between 1972 and 1975 flowed in the direction of b and c, those between 1978 and 1978 flowed in the b direction, in 1979 in the c direction, and from 1983 they flowed again in the direction of a. Thus, debris flows have changed their directions as a manner to bury hollows and the fan has grown creating many lobes. Though it is not evident in this figure, a kind of natural levee comprised of coarse rather uniform materials is often formed along both sides of debris flow routes by the pushing aside of frontal particles. If thinly sediment concentrated flood flow follows the debris flow and it passes through in between the thus formed levees, it erodes the way to form a channel. This peculiar incised channel form is the same type one that I wondered how this was formed seeing the trace of the Shiramizudani debris flow as described in the preface of this book and shown in Figure 1.3.

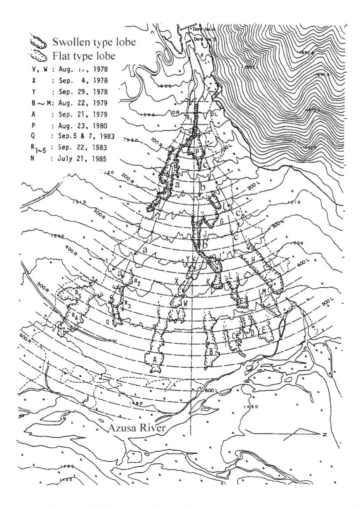

Figure 1.7 The distribution of debris flow lobes. Flow routes are shown by arrows (Suwa 1988).

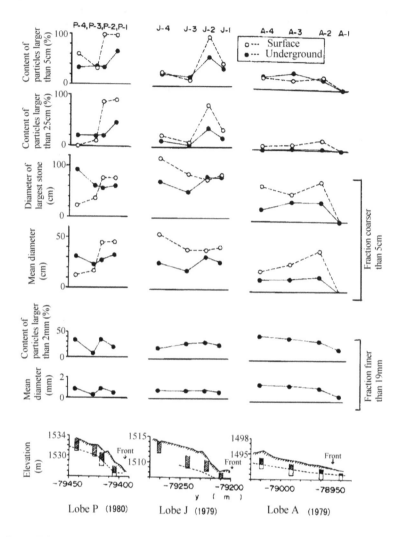

Figure 1.8 The structures of debris lobes (Suwa and Okuda 1982, partly altered).

Picking up the swollen-type lobes P and J and flat-type lobe A as representative ones, pits were dug on or in the neighborhood of each lobe to explore the structure of deposits (Suwa and Okuda 1982). The results of exploration are shown in Figure 1.8. The locations of the pits are indicated at the bottom of the figure, and the various characteristic values in each pit are plotted on the corresponding column position on the upper graphs. The solid lines and the broken lines on the graphs represent the values obtained at the deeper part and near the surface, respectively. It must be noted here that pits A-1 and J-1 are located outside the respective lobes and the locations of pits P-3 and P-4 have been washed by a water flood following the debris flow. Consequently, the representative pits which demonstrate the materials in the front part of each lobe are, P-1, P-2, J-2 and A-2. From the characteristic values of such representative pits, it is clear that materials on the surface are richer in cobbles and larger in the mean diameter than those in the deeper part, so that the structure of the sediment accumulation in debris lobe is inversely graded. Here the term 'inverse grading' means that, in contrast to the deposits due to normal water flood in which particles distribute from larger ones in the lower layer to the smaller ones in the upper layer (normal grading), particles distribute upwards from smaller to larger. Furthermore, the longitudinal particle size variation in each lobe shows the concentration of cobbles and the mean particle diameter increase towards the forefront. This corresponds to the tendency for bigger particles to concentrate towards the forefront and finer particles are left behind as debris flow proceeds downstream. If inverse grading in flow proceeds while in going down, faster velocity than the mean translation velocity in the upper part, as explained earlier, necessarily transports larger particles towards the forefront. Because the swollen-type lobe is formed by the sudden stopping of debris flow with little deformation from the shape in the upstream channel, the clear inverse grading within the swollen-type must be the reflection of the vertical structure in the flow in the upstream channel.

Comparison between the swollen-type and flat-type lobes shows that the mean particle diameter in the flat-type lobe is smaller and the flat-type lobe arrives at a more distant flatter place. This corresponds to the fact that the scale of debris flow that produces the flat-type lobe is larger than that of the swollen-type. The debris flow that formed lobe A was a big one and it transported large boulders whose diameters were up to 5 meters at the forefront when it passed through the fan top area, but at around the center part of the fan the biggest boulders recognized were about 2 m in diameter. Furthermore, in the deposit near the toe of the fan, the maximum diameter was less than 1 m. This meant that debris flow left the boulders putting them aside in its course in order from biggest to smallest as it proceeded downstream on the debris fan. Thus, the mechanism to cause inverse grading is essential in the phenomena of stony debris flows.

Large-scale flume experiments implemented at USGS

Large-scale flume experiments have been carried out at the US Geological Survey using a concrete flume; 95 m long, 2 m wide and 1.2 m deep (Major and Iverson 1999). The upstream 88 m reach has a constant slope of 31° and in the rest of the 7 m reach the channel gradient gradually flattens finally to adjoin a concrete runout plane that slopes 3 degrees. The experimental materials are mainly the mixture of 1–10% clay and silt, 10–60% gravels and 30–40% sand and the maximum particle size is generally 30 mm, but in some experiments it is as large as 150 mm. Prior to the experiment, the debris flow material of about 10 m^3 is prepared behind a steel gate at the head of the flume, soaked with water. On opening the gate, a debris flow arises which is separated in multiple surges whose heads have 10 to 30 cm depths and speeds of about 10 to 12 m/sec. The experimental flows decelerate rapidly and deposit sediment beyond the flume mouth on the concrete plane arriving as distant as 17 m, and the resulting deposits are elongated and less than 40 cm thick. The behaviors of flows and morphological and sedimentological features mimic the natural debris flows such as those that appear at Kamikamihorizawa. Namely, while the flow passes through a few tens of meters reach, gravels accumulate at the surge fronts and the rear parts become more liquid. The deposits form lobes and coarse particles concentrate along the margins.

Basal fluid pressure and basal total normal stress were measured in the flume at 67 m below the release gate and at several locations on the runout area. The total normal stresses obtained in the flume section were synchronous with the depths of flow, but the basal fluid pressures were somewhat asynchronous with the change in depths. Namely, at the instant of front passage the basal fluid pressure was almost zero and it gradually increased as much as the total normal stress in the succeeding flow, meaning that the material was fully liquefied in the succeeding flow part. These characteristics agree with the measurements in natural flow by Okuda *et al.* (1981) mentioned above. The measurements at the runout area also showed the fluid pressure at the front was small but at the following part, even in the process of deposition, the fluid pressure was sometimes kept large for a long time as much as was sufficient to liquefy the material. Namely, except for the frontal coarse part, debris flow deposit is not necessarily formed in the manner as if frozen at once, but it is formed either by the gradual sedimentation of material due to the backwater effect of a kind of natural dam built by the stopped front or by the gradual deposition of sediment while the succeeding flow is detouring or destructing the dam.

It must be noted that the experiments have been done in a very steep and rigid bed flume. Natural debris flows generally flow on an erodible channel, and if a run down debris flow contains too small a quantity of sediment, it will erode the channel bed and increase the sediment concentration, and on the contrary, if it contains too much sediment, it will deposit the sediment to decrease the concentration. Thus, debris flows on an erodible channel behave like an equilibrium state. In the experiments of the USGS, however, the sediment concentration is determined only by the condition of material arrangement before the experimental runs and the flow is not necessarily in equilibrium. The steep slope of 31° is sufficient to continue the motion even though the entire material is not liquefied; actually, the highest sediment concentration in the experiments were as much as 0.8 (Iverson 1997; Iverson *et al.* 2000), so that the experiments may include the case similar to the slippage of earth block with a partially liquefied layer in the lower part.

Many smaller scale debris flow experiments both in the rigid bed flumes and erodible bed flumes with various slopes have been done by our group and others in Japan. These results will be discussed in detail later.

1.3.2 Turbulent-muddy-type debris flow

The vigorous ejection of ash from an active volcano results in the thick mountain cover that is easily eroded even by a slight rainfall. Such ash cover erosion causes frequent occurrence of debris flows. This type of debris flows are, even though they contain many large boulders, mainly comprised of fine ash, and the behaviors of flow are different from those of stony-type.

Photo 1.4 shows the debris flows at the Nojiri River, Sakurajima. The flow is very much turbulent from the forefront to the rear end. We call this type of debris flow a '(turbulent) muddy-type'.

The sediment samples at the forefronts have been collected by the samplers set at the over flow section of a sabo dam (debris barrier) in the upstream area and in the

(a)   (b)

Photo 1.4 Debris flows in the Nojiri River, Sakurajima (Ohsumi Work Office 1988). (a) A debris flow
passing through a sabo dam. 18 December 1986. (b) A debris flow flowing down the channel
works.

channel works downstream in the Nojiri River. These data show the solids concentration at the sabo dam point is 35–42% by volume and that in the channel works is 54–72%, the median diameter of solids is 0.3–1 mm, and the fraction less than 0.1 mm is 10–30%. Because the flow is violently turbulent, we may apply Manning's formula for resistance law; $U = (1/n_m)H_a^{2/3}I^{1/2}$, where U is the sectional mean velocity in m/s unit, H_a is the depth of flow in m and I is the slope gradient of the channel which is equal to 1/18 in the channel works of the Nojiri River. The calculated resistance coefficient n_m for the parts before the peak, at the peak, after the peak and rear end are 0.030, 0.022, 0.027 and 0.020, respectively, and these values are almost the same to that of plain water flow (Ohsumi Work Office 1988). The resistance law is sometimes described by the velocity coefficient defined by $\psi = U/u_*$, where $u_*\,(=(gH_aI)^{1/2})$ is the shear velocity and g is the acceleration due to gravity. In the case of debris flows in the Nojiri River $\psi = 10$–20.

1.3.3 Viscous debris flow

We collaborated with the Institute of Mountain Hazards and Environment, Chinese Academy of Science and Ministry of Water Conservancy for eight years from 1991, concerning the mechanism of viscous-type debris flows. The Jiangjia Gully, Yunnan a tributary of the Xiaojiang River (one of the first class branches of the Jinsha River) is the experimental watershed of the Dongchuan Debris Flow Observation and Research Station of the Institute in which we observed several debris flows. The Jiangjia Gully has a watershed area of 46.8 km², is 13.9 km long in the trunk, 3,269 m at the highest altitude and 1,024 m at the confluence with the Xiaojiang River. It is a very steep watershed, of which 60% is occupied by landslides and more than one billion cubic meters of debris are accumulated ready to be flushed out as debris flows (Wu *et al.* 1990). Therefore, large-scale debris flows occur more than ten times a year. The observations of debris flow during flowage and the sampling of debris flow materials are accomplished in a straight channel reach, about 200 m long, 20–40 m wide, 5–6 m deep and with about a 6% slope.

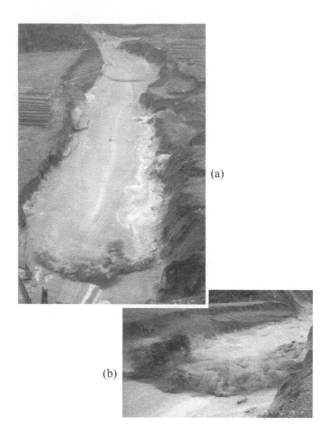

Photo 1.5   Debris flows in the Jiangjia Gully. (a) The bird's-eye view of two successive surges. (b) Violent turbulence at the forefront. Compare with the laminar flow in the rear part seen in (a).

The debris flows in the Jiangjia Gully are famous as the typical ones of viscous-type. The most conspicuous character of these flows is, as shown in Photo 1.5, its intermittency. Tens and hundreds of surges (intermittent bore-like flow) come out repeatedly with the time interval of from a few tens of seconds to a few minutes. Each surge rolls on and roars as if it were a breaking wave at the sea coast. The general time sequential aspects of an event in the observing channel are as follows:

Accompanying rainfall, at first, turbulent sediment loading flood flow of apparent density 1–1.5 t/m³ is generated. It is succeeded by a continuous debris flow of 1.8 t/m³. In due time, the first surge comes out. This surge decelerates within a distance of a few tens of meters up to two hundred meters and stops forming a deposit. The channel bed before the surge is a rugged surface but after the stopping of the surge the channel bed upstream of the stopped front becomes smooth. The next surge, helped by the just smoothened bed, goes further downstream than before and it elongates the deposit as well as thickening the depth of deposit by piling up on the deposit from the last surge.

Repetition of these processes results in complete smoothing the bed and the thickness of the deposit reaches an equilibrium stage (it seems to depend on the apparent density and viscosity but it is around 1 meter). Then, the numbers of surges repeatedly pass. The apparent density of debris flow in this stage is 1.8–2.3 t/m³. Occasionally, intermittency suddenly ceases in this stage and a continuous turbulent flow appears for a while, but it is followed again by the intermittent flows. This phenomenon may be caused by the supply condition of the debris flow material in the source area, but it cannot be said definitely. In general, after the repetition of intermittent surges, in a reverse of the developing process, it appears a continuous turbulent debris flow followed by the hyper-concentrated flood flow. The flood flow erodes the smoothened bed. Therefore, after one event, the channel bed returns to be as rugged as before. But, the elevation of the bed, the width of the channel and the longitudinal shapes of the side banks are not necessarily the same as those before the event.

The existing statistics show that the duration of one event is from a few tens of minutes to several tens of hours and the most frequent one is about two hours. About 70% of the duration is the continuation of intermittent surges (Kan 1994).

Figure 1.9 shows the temporal changes in depth, frontal velocity, time interval between successive surges and the volume of each surge for an event we observed (Suwa et al. 1997). It is clear that the deeper the depth of flow, the faster the frontal velocity, and this suggests that larger waves tend to overtake and cannibalize smaller waves. Actually these phenomena were confirmed in the observation. These phenomena also appear in the experimental flume (Iverson 1997).

The length of a surge tends to become shorter the larger the apparent density of material; a small surge is 30–100 m long and large one is 200–500 m long. Figure 1.10 shows an example of the simultaneous measurement of depths and velocities using the ultrasonic water level gauge and the radar speedometer (Kan 1994). This is the result of measurement at a position, where the horizontal axis is time, but the length of each surge (the 4th, 9th and 19th surges are shown and designated by the numbers in parentheses), that is obtained by converting time into length using the measured velocity, is also given on the same figure with the length-scale axis. The double illustrations of depth for the 4th surge by solid and broken lines show the envelope curve of turbulent wave heights detected by the ultrasonic meter and the substantial flow surface, respectively. It must be noted that the records from the ultrasonic meter indicate, at the forefront, the depth of flow is almost double.

The general characteristics of a surge can be outlined as follows:

- The front of a surge has a length of 3–5 m, where the depth and velocity are the maximum.
- The frontal velocity is usually 6–8 m/s, sometimes as large as 10 m/s.
- After the violently turbulent front part has passed, the 10–20 m long less turbulent part follows, in which velocity is still not much different from the frontal part.
- After this part, the wavy 30–50 m long part comes, in which depth and velocity gradually decrease.
- Then, the 50–80 m long laminar flow part continues.

The uniform stripe pattern in the center of flow in Photo 1.5(a) shows that no turbulence exists in this part. Until this time the flow takes place in the full width of the cross-section, but after this time the flow begins to gradually concentrate in the central part of the section with decreasing velocity, and finally the entire flow stops.

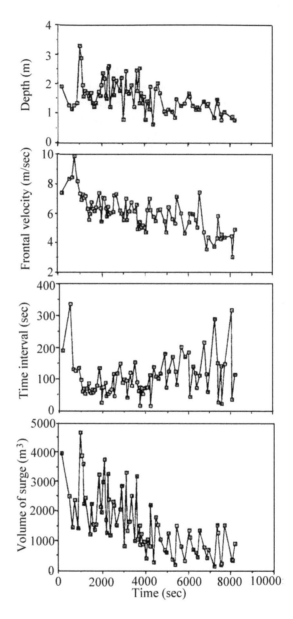

Figure 1.9 Depth, frontal velocity, time interval and volume of surges versus time for the debris flow of 13 August 1991. The time 0 is 11:37 a.m.

The white stripe on the center of the static bed just downstream of the approaching surge-front in Photo 1.5(a) is the trace of water flow joining from the left-hand side tributary that can be seen in the photo behind the front as a white wavy band adjacent to the left-hand side bank. It is gradually concentrated to the central part accompanied by the concentrating rear end part of surge. Just before the end of a surge, the surface

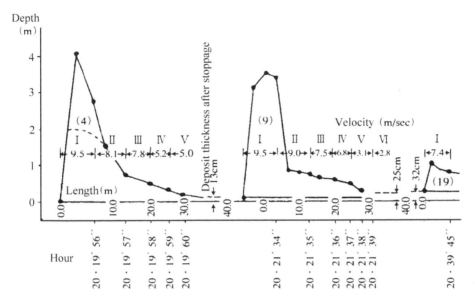

Figure 1.10 Depths and velocities of surges (Kan 1994).

of the flow becomes laterally concave and a thin concentrating flow appears towards the central part where the flow surface is the lowest within the cross-section. This means that no appreciable plug exists on the surface of flow, yet a thick deposit is left after the passage of a surge. Note that the existences of the plug and the deposit whose thickness is equal to the thickness of plug are the necessary conditions for modeling the flow as a Bingham fluid.

Figure 1.11 shows the particle size distributions in the samples obtained from normal flood, debris flows, slope surface in source area and debris flow deposit (Wu *et al.* 1990). As is clear from Photo 1.5, the viscous debris flow is different from the stony-type debris flow, as it has no accumulation of big boulders at the forefront. The materials in viscous debris flows are almost identical to that in the source area. The deposit downstream lacks the fine fractions a little, but, that is from the washout after deposition, so we can consider that the particle size distribution in the flow just before deposition is almost identical to that in the debris flow. Therefore, we consider that the viscous debris flow has no particle segregation effect in the process of motion. The determination of the maximum size particles in the slurry that behave as one integral fluid within the slurry is difficult problem. Nevertheless, seeing that the fractions of 3–5 in φ-scale (0.125 mm–0.031 mm) are scarce in the materials in the source area and debris flows, and considering that only particles less than 4 in φ-scale (0.063 mm) are contained in the normal flood, I assume that only the fractions less than 0.1 mm can behave like a continuum fluid in the slurry. Then, the debris flow can be considered as the mixture of slurry and coarse particles larger than 0.1 mm. Considering the apparent density of the entire debris flow and that 20–30% by weight of the entire material is fine particles less than 0.1 mm in size, we can calculate the apparent density of slurry as 1.3–1.5 g/cm^3, corresponding to a volume concentration of 20–30%. In general,

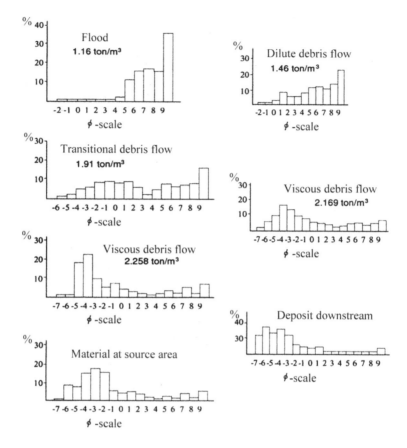

Figure 1.11 Particle size frequency in the materials of Jiangjia Gully (Wu et al. 1990).

viscous debris flow is considered as the flow of the dispersion of coarse particles in such dense slurry, the concentration of coarse particles in the slurry is more than 50% by volume.

1.4 THE SIGNIFICANCE OF THE MECHANICAL CLASSIFICATION OF DEBRIS FLOWS

The analysis of debris flow mechanics requires the estimation of stresses within the flow. Because debris flow consists of highly concentrated coarse particles and water or slurry, the candidate stresses are:

1   due to the collision of coarse particles;
2   due to the macro turbulent mixing of the fluid body comprised of particles and slurry;
3   due to the enduring particle friction between particles when coarse particles are contained denser than a boundary value;

4   due to the deformation of interstitial fluid or of the apparent viscous fluid
     consisting of the mixture of particles and slurry; and
5   that operating between particles and fluid resulting from the relative motion of
     the solid and fluid constituents.

The characteristics of flow are postulated to depend on the shearing rate du/dz, the representative particle diameter d_p, the depth of flow h, the particle density σ, the interstitial fluid density ρ, gravity acceleration g, the apparent viscosity of interstitial fluid or of an apparent viscous fluid comprised of particles and slurry μ, the coarse particle concentration by volume C, the restitution coefficient of particles e, and the friction coefficient between particles $\tan\varphi$, where u is the velocity of flow at height z measured perpendicularly from the bed surface. The stresses in the flow are the shearing stress and pressure, and these stresses have the dimension $[ML^{-1}T^{-2}]$, in which $[L]$ is the length, $[T]$ the time, and $[M]$ the mass dimensions.

The exchange of momentum on the occasion of particle collision gives rise to the collision stresses. Therefore, they should depend on the collision frequency, the particle mass and the restitution coefficient. Because the collision frequency is the frequency of particle encounter between particles embedded in the vertically adjacent two layers, it is postulated to be a function of the shear rate and the particle concentration. Through these considerations and the dimensional analysis, the shear stress due to particle collision T_c would be written as:

$$T_c \sim f(C, e)\sigma d_p^2 (du/dz)^2 \tag{1.1}$$

where $f(C, e)$ means a function of C and e, and this function is thought to become large with increasing C and e.

The macro turbulent mixing stress T_t is, by the application of fluid mechanics, written as:

$$T_t \sim \rho_m l^2 (du/dz)^2 \tag{1.2}$$

where ρ_m is the apparent density of interstitial fluid or entire mixture body. The mixing length l in this expression is important. When the comprising particles are large and the particle concentration is also large, it would have the scale of distance between particles, and as Iverson (1997) postulated, it would be estimated as $l \sim d_p$. In this case ρ_m is equal to the density of interstitial fluid ρ. But, as in the case of the turbulent-muddy-type debris flow, when the entire mixture body is violently turbulent, it would be estimated as $l \sim h$. In this case ρ_m is the density of the entire mixture body.

The quasi-static Coulomb friction stress at height z that is caused by the enduring grain contact T_{sq} is affected by the submerged weight of total particles existing above the height z, and it is written as:

$$T_{sq} \sim C(\sigma - \rho_m)g(h - z)\tan\varphi \tag{1.3}$$

The shearing stress due to the deformation of fluid is, for a Newtonian fluid:

$$T_{fq} \sim \mu(du/dz) \tag{1.4}$$

Concerning the solid-fluid interaction stresses, Iverson (1997) took notice of the permeability of the inter-particle void space. These stresses are related to the buffering effect to the direct inter-particle action. Herein, these stresses are not taken into account, because the dispersion of particles in viscous fluid is the result of this buffering effect as is explained later in section 2.8. Accordingly the expression like (1.4) is made possible.

The ratio of inertial grain stress T_c and viscous shear stress T_{fq} that is given by:

$$N_{\text{Bag}} = \frac{f(C, e)\sigma d_p^2 (du/dz)}{\mu} \tag{1.5}$$

represents the relative predominance of the respective stresses, and the value of this ratio is called the Bagnold number. In inertial debris flow the Bagnold number is large.

The ratio of turbulent mixing stress T_t and inertial grain stress T_c does not have such an important meaning when $l \sim d_p$, but, when l is assessed by h (large-scale mixing case), the ratio is given by:

$$N_{mud} = \frac{1}{f(C, e)} \frac{\rho_m}{\sigma} \left(\frac{h}{d_p}\right)^2 \tag{1.6}$$

and this means the relative depth defined by (h/d_p), can be the index to determine whether the debris flow is the inertial stony-type or the turbulent-muddy-type.

When l is assessed by h, the ratio of T_t and T_{fq} is given by:

$$N_{\text{Rey}} = \frac{\rho_m h^2}{\mu}\left(\frac{du}{dz}\right) \sim \frac{hU}{(\mu/\rho_m)} \tag{1.7}$$

where U is the cross-sectional mean velocity. This is nothing more than the well-known Reynolds number. The Reynolds number is the index to classify whether the flow is turbulent or laminar.

The ratio of T_c and T_{sq} indicates the relative importance between the inertial grain collision stress and quasi-static Coulomb friction stress. To transmit the quasi-static Coulomb friction stress, particles must always be in contact even though their relative position continuously changes. This condition requires that the solids concentration should be larger than a threshold value C_3. Bagnold (1966) says this condition is fulfilled when C is larger than 0.51 for natural beach sand, but it would depend on the composition of particle sizes; for widely distributed material, the threshold concentration would be larger because small particles will be stored in the void between large particles. Under such a densely concentrated condition, the other stresses except for T_{sq} become small and the motion would be a quasi-static one. Iverson (1997) gave the ratio of T_c and T_{sq} as follows:

$$N_{\text{Sav}} = \frac{\sigma d_p (du/dz)^2}{N(\sigma - \rho)g \tan\varphi} \tag{1.8}$$

where N is the number of particles above the height z. He claims that the grain collision stress is far smaller than the Coulomb friction stress if N_{Sav} (Savage number) is less than 0.1, and because in many debris flows N_{Sav} is less than 0.1, most debris flows are those in which Coulomb friction stresses predominate. If one multiplies the denominator by d_p, this term becomes equal to the Coulomb friction stress operating on the plane at height z. But, multiplying the numerator by d_p does not result in the grain collision stress. It is, furthermore, necessary to multiply by $f(C, e)$ to represent the collision stress. The coefficient $f(C, e)$ can be large (Campbell 1990), and more importantly, to give rise to the Coulomb friction stress, solids concentration should be larger than the threshold. Therefore, the quasi-static debris flow in which Coulomb friction stress predominate can only occur when its solids concentration is more than the threshold value of about 0.5.

The above discussion comes to a conclusion that there are two kinds of debris flows in a wider sense; one is the quasi-static debris flow in which Coulomb friction stress dominates and the other is the dynamic debris flow. Moreover, in the dynamic debris flow cases, there are three kinds; when the grain collision stress dominates, the debris flow becomes the stony-type, when the turbulent mixing stress dominates, it becomes the turbulent-muddy-type, and when the viscous stress dominates, it becomes the viscous-type. If the concentration is larger than the threshold value; C_2, neither dynamic debris flow nor quasi-static debris flow is possible, the dislocation of particles within the body cannot take place and the material becomes rigid. This threshold value C_2 is given by Bagnold (1966) for natural beach sand as 0.56.

In a fully developed debris flow, particles disperse densely in the entire depth, but, to be dispersed in the entire depth, the particle concentration should be larger than a limit. This limit value cannot be determined by only the static geometrical nature of a constituent like C_2 and C_3, but is determined dynamically as explained later. Below this limit, particles do not disperse in the entire depth, but they flow concentrated in a lower layer. This kind of debris flow is called an immature debris flow (Takahashi 1982).

Based on these considerations, a classification scheme and existence criteria of debris flows are given in Figure 1.12. The vertical axis represents the mean coarse particle concentration in the flow. If coarse particles are not contained in the flow (at the lowest point on the vertical axis), the flow is mere water or slurry flow. In a water flow, the shear stress is shared by the turbulent Reynolds stress and the viscous stress and the larger the viscosity the larger the relative importance of viscous stress becomes. Therefore, the flow regime changes from laminar to turbulent or vice versa along the lowest horizontal axis. The ratio of the end members of this axis, T_{fq}/T_t, is simply the Reynolds number. Therefore, the Reynolds number changes along this axis. The T that appears at the end member of the figure represents the total of the inner stresses.

When the solids concentration becomes larger but still less than about 0.02 (Takahashi 1991), the flow contains bed load or suspended load depending on the turbulence and viscosity. Particle collision stress appears but it is small.

When the solids concentration becomes larger but less than about 0.2, the flow becomes an immature debris flow, the collision stress dominates only in the lower particle mixture layer.

As the mean coarse particle concentration becomes larger but is smaller than C_3, the flow becomes a dynamic debris flow. In this case possible dominant stresses are particle collision stress, turbulent mixing stress and the viscous stress, because under such a concentration the quasi-static stress cannot be dominant. Therefore, the

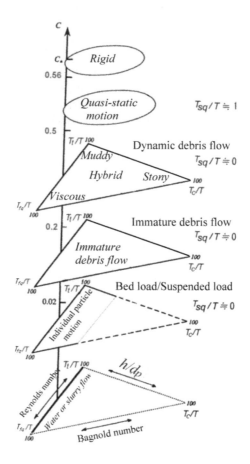

Figure 1.12 The existence criteria of various motions of the mixture of solids and fluid.

sub-classification of debris flow is possible on the ternary phase diagram as shown in Figure 1.12. The three apexes of the triangle mean respective T_c, T_t and T_{fq} occupy a hundred percent of the total inner shearing stress T. One of three axes of the ternary diagram indicates the Reynolds number as mentioned earlier. The ratio of the two end members at the apexes T_t/T and T_c/T; i.e., T_t/T_c, can be approximated as $f(C, e)(h/d_p)^2$, and the ratio of the two end members at the apexes T_c/T and T_{fq}/T; i.e., T_c/T_{fq}, represents the Bagnold number. Therefore, the three axes of the ternary diagram represent relative depth h/d_p, the Bagnold number and the Reynolds number, respectively. The region where the Bagnold number is large and the relative depth is small is where stony debris flow occurs. In the region where the Bagnold number and the Reynolds number are small, viscous debris flow occurs; and in the region where the relative depth and the Reynolds number are large, turbulent-muddy debris flow occurs. Thus, the area close to the three apexes are occupied by areas of stony, viscous and muddy debris flows, respectively, and the rest of the area in the triangle occupied by a hybrid-type of those three typical ones. The area occupied by each of the three typical debris flows should change depending on C.

As the concentration becomes larger, it exceeds C_3. In this region, collision, turbulent and viscous stresses become small and instead the quasi-static Coulomb stress becomes dominant so the flow becomes a quasi-static motion. And if the concentration becomes larger than C_2, the material becomes rigid.

Because grain collision stress and turbulent mixing stress are caused by inertial motion, debris flows existing nearby the relative depth axis; stony, hybrid and turbulent-muddy debris flows, can be called, as a whole, the inertial debris flow. The inertial debris flow has very different characteristics from the viscous debris flow. Flows existing near the axis representing the Reynolds number have very different characteristics from the stony-type debris flow; the flow consists of mainly fine particles. These flows are often called 'hyper-concentrated flow', of which a typical one in the small Reynolds number area is the mudflow in the Ocher Plateau, China and that in a large Reynolds number area is the flood flow in the Yellow River. Flows existing near the axis representing the Bagnold number have no general name, but both the viscous debris flow and stony debris flow move in a laminar fashion. Note that the more or less laminar motion of stony debris flow is the result of its difficulty in changing the particle positions up and down or back and forth in so highly concentrated a state, nevertheless, the interstitial fluid is often turbulent and particles make small oscillations themselves. The oscillation of coarse particles is essential for the origin of grain collision stresses that will be discussed later in detail.

The specific numerical values on the three axes of the ternary diagram that identify the existence domains of the respective dynamic debris flows within the triangular area, however, cannot be determined before the mathematical descriptions of the respective governing stresses are made clear in the next chapter. Nevertheless, we can compare the characteristics of those three kinds of natural debris flows using the physically significant parameters as follows:

The apparent density of material in the stony-type debris flow at Kamikamihorizawa is 1.5–1.85 t/m^3. This is equivalent to the solids concentration of 0.35–0.62 by volume. In the viscous debris flow at the Jiangjia gully, it is 1.7–2.3 t/m^3. This is equivalent to the solids concentration of 0.45–0.75. Though the observation point in the Jiangjia gully is located at a flatter place (the longitudinal channel slope is about $3°$) than in the Kamikamihorizawa (about $8°$), the viscous debris flow in the Jiangjia gully flows with larger solids concentration than the stony debris flow in the Kamikamihorizawa. The turbulent-muddy debris flow at the Nojiri River, Sakurajima having a solids concentration of 0.61 by volume was measured in the channel works of $3°$ in a longitudinal gradient. The solids concentration of 0.61 is equivalent to 1.98 t/m^3 in apparent density. I have some doubts about the accuracy of this measurement, anyway, the turbulent-muddy-type debris flow goes down in a flat channel having large mobility even in very high solids concentration.

Figure 1.13 compares the particle size distributions in the constituent of debris flows. The debris flows in Kamikamihorizawa contain many big boulders larger than a few tens of centimeters, whereas, the largest stones in the flows at Jiangjia gully are fist sized cobbles. The fraction of constituent less than 0.1 mm in stony debris flow is far less than in viscous debris flow. This figure also shows the particle size distributions in the samples collected while in motion and from the deposit in the Nojiri River, and that collected from the deposit in the Mizunashi River, Unzen Volcano (Hirano and Hashimoto 1993), where the sample in the Mizunashi River was collected

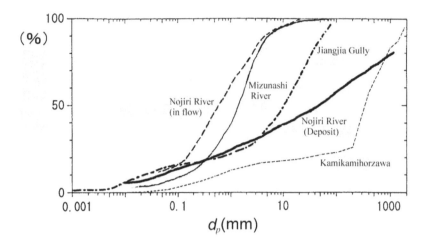

Figure 1.13 Distributions of particle sizes in the stony, viscous and turbulent-muddy-type debris flows.

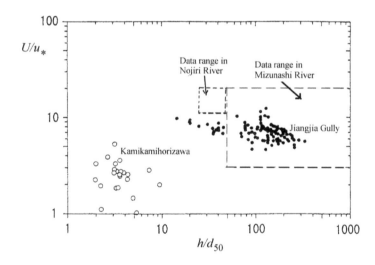

Figure 1.14 Velocity coefficients versus relative depths in various debris flows.

only the fraction of constituent less than 20 mm. Muddy-type debris flows, sometimes when their scales are large, contain many large boulders as well, but they mainly contain the fine component of 0.1 mm or less as plenty as in viscous debris flows at the Jiangjia gully. As mentioned earlier, continuous turbulent debris flow sometimes appears in the Jiangjia gully, so that we cannot judge whether the flow is viscous-type or turbulent-muddy-type based only on the percentage of fine particles less than 0.1 mm.

Figure 1.14 shows the relationship between the velocity coefficients (U/u_*, where U is the velocity of flow and u_* is the shear velocity, the definition of which is given in section 1.3.2) and the relative depths in respective kinds of debris flows. The smaller the velocity coefficient, the greater the resistance to flow is. The velocity U in Figure 1.14 is the translation velocity of the surge front in the cases of Kamikamihorizawa, Jiangjia gully and Nojiri River, and in the case of Mizunashi River, it is the surface velocity of flow measured by the radar speedometer (Hirano and Hashimoto 1993). The relative depth in the figure was obtained using the median size (d_{50}) in the respective samples; these were about 10 mm, 300 mm, 40 mm and 1 mm in the Jiangjia gully, Kamikamihorizawa ravine, Nojiri River and Mizunashi River, respectively. For the data from Nojiri and Mizunashi only the distributing areas are illustrated. The debris flow in the Nojiri River has, as mentioned earlier, mobility as large as a plain water flow, and the area of data distribution in Figure 1.14 is narrow, but, in the Mizunashi River they distribute in a very wide area. This may be caused by the wide scatter in measured values of depth and velocity. The velocity coefficients in the Jiangjia gully and in the Kamikamihorizawa are evidently different from each other. The stony-type debris flows have larger resistance to flow than the viscous-type. Note that the distribution ranges of these two types with respect to the relative depths are different; the stony-type flows can only occur in the range where the relative depth is less than about 20–30. However, if hypothetically large-scale debris flow having a relative depth larger than 20–30 occurred in the Kamikamihorizawa, it would be the muddy-type debris flow even though it contained as many large particles as in the actual stony-type debris flows. Comparing the velocity coefficients of viscous debris flow with that of the muddy debris flow, we can see that the viscous debris flow has a slightly smaller mobility than the muddy debris flow and the plain water flow.

1.5 CLASSIFICATIONS BASED ON OTHER VIEW POINTS

The classification of debris flow from other than the mechanical point of view is of course possible.

In China, they classify debris flows into mud-stone flow and water-stone flow depending on whether the interstitial fluid is slurry or water. The mud-stone flow is further classified based on the apparent density of material; 1.3–1.8 t/m³: fluid debris flow, 1.8–2.0 t/m³: quasi-viscous debris flow, 2.0–2.3 t/m³: viscous debris flow, and over 2.3 t/m³: highly viscous debris flow (Kan 1996). The apparent density of material is one of the important factors to determine the characteristics of flow, but they cannot be explained only by this factor as mentioned earlier.

The classifications by causes and processes of occurrence are often used. The causes of generation are: (1) rainfall; (2) melting of ice and snow; (3) earthquake; (4) volcanic eruption; and (5) destruction of lake. It is not necessary to explain cause (1).

Abrupt atmospheric temperature rise sometimes causes the bursting melt of snow and glaciers and induces debris flow (Sharp and Nobles 1953). The melting of snow or glaciers is also caused by volcanic eruption (Murano 1965; Pierson and Scott 1983; Takahashi and Egashira 1986), and in such cases the induced debris flow may be classified as having both causes (2) and (4). On the slope of Mount Fuji, in early spring, a kind of snow avalanche that is called 'Yukishiro' in Japanese which may be translated in English as 'slush-flow' occurs. This is caused by both atmospheric temperature rise and rainfall, and the slush goes down entraining basal solid grains (Kobayashi 1995).

Strong earthquakes generate landslides, and these landslides sometimes transform into debris flows. The western Nagano earthquake in 1984 caused a large-scale landslide of forty million cubic meters on the slope of Mount Ontake. The slid earth flowed down about 10 km and choked the Otaki River. Soon after the emplacement of deposit, debris flow passed on the deposit surface. A stratovolcano such as Mount Ontake is very rich in voids in the mountain body, and if these void spaces are nearly saturated by water, at least some parts of slid earth body can be transformed into debris flow without the supply of water from outside. This mechanism will be discussed later.

Debris flows on the slope of a volcano are induced not only by the effects of melt water, but also by the overflow of a crater-lake (Mugiono 1980) or by the flood runoff of severe rainfall. Debris flows on volcanic mountains are sometimes called 'lahar', that is an Indonesian genetic term. The high temperature one is a 'hot lahar' that is induced on the occasion of eruption, and that induced by rainfall having no close connection with volcanic activity is a 'cold lahar'. The lahars that have a close relationship with volcanic activity are often called 'volcanic mud-flow'. The volcanic mudflow is sometimes a turbulent-muddy debris flow and other times a hyper-concentrated flood flow irrespective of temperature, in which heavily ejected fine ash or pyroclastic materials are the main constituent. Mount Yakedake on which the Kamikamihorizawa is located is a volcano, but considerable time has elapsed since the last eruption in 1962 and the ash cover is now removed to cause stony debris flows. A large-scale landslide that accompanies a phreatic explosion can sometimes induce debris flow by the same mechanism as the case due to an earthquake induced landslide (Takahashi 1981; Hoshino *et al.* 1998).

The destruction of a lake typically occurs due to natural dam destruction and moraine destruction. The contributing factors for destruction include rainfall, melting of glaciers, and plunging of huge blocks.

The above mentioned various causes of debris flows can be mechanically sorted out as the cases in which:

1   the deposit on the gully bed is eroded by the supply of water from outside;
2   a moving landslide block transforms into debris flow by the effects of water already contained before the slide or by the supply of water from outside; and
3   the dam naturally built by landslide collapses.

The detailed mechanisms will be discussed later.

The key factors to be known beforehand in controlling the generation of debris flow are 'from where', 'in what way' and 'how much' water and sediment are given. Before concluding this chapter, we will see the rainfall conditions to generate debris flows in Kamikamihorizawa, Jiangjia gully and Nojiri River.

The rainfall intensity every five minutes, the surface runoff discharge of water, and the level of groundwater that is forced to pool at a shallow depth beneath the ground surface by the insertion of impermeable sheet are measured at the source area of the Kamikamihorizawa as illustrated in Figure 1.15 (Suwa and Sawada 1994). At the moments shown by arrows debris flows occurred, and they have intimate relations with the short time rainfall intensities. Namely, for the occurrence of debris flow, about 10 mm of antecedent rainfall is necessary, and once this condition is satisfied even if

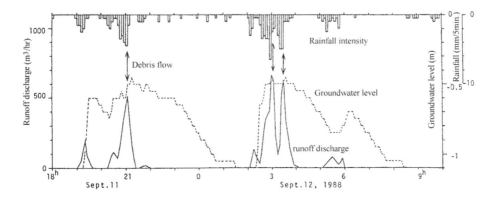

Figure 1.15 Relationship between debris flow occurrence and rainfall and other factors.

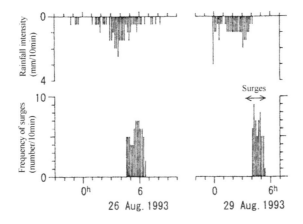

Figure 1.16 The relationship between debris flow surges and rainfall intensities in the Jiangjia gully.

the soil layer is not saturated with water, debris flow occurs when 4–18 mm rain falls in ten minutes.

Figure 1.16 shows the frequency of debris flow surges and rainfall intensities in ten minutes in the Jiangjia gully watershed (Suwa and Sawada 1994). As the rain-gauge (Daaozi rain-gauge) is located comparatively far from the source area of debris flow, it cannot be said conclusively, but the rainfall intensities and occurrence of debris flow have no obvious relationship. The rainfall amount within a long period of one or more hours seems to be effective. In the Nojiri River, debris flow occurs under the condition of more than 10 mm of antecedent rainfall and 5–10 mm rainfall in ten minutes (Ohsumi Work Office 1988). This is similar to the condition in the Kamikami-horizawa. The fast concentration of runoff water and the erosion of the bed by such surface water can easily generate debris flows in the Kamikamihorizawa and in the Nojiri River basins. However, in the Jiangjia gully, although debris flow is easily gen-erated, long lasting rainfall is necessary. This would suggest that the debris flow in the Jiangjia gully is generated from landslides which need a long time to be soaked with water. In this context, the intermittency of debris flow in the Jaingjia gully may correspond to the difference in the initiation time of each landslide that generates each surge at the source area.

The debris flows in these three basins occur extraordinarily easily. The existing data collected in Shodo Island (Ashida *et al.* 1977) show that the approximate threshold to generated debris flow is 300–400 mm in antecedent rainfall and an intensity of 40–50 mm in one hour. It suggests that, except for the water supply condition, the soil properties and quantities on slopes and stream beds in the basin very much affect the onset of debris flow.

Chapter 2

Models for mechanics of flow

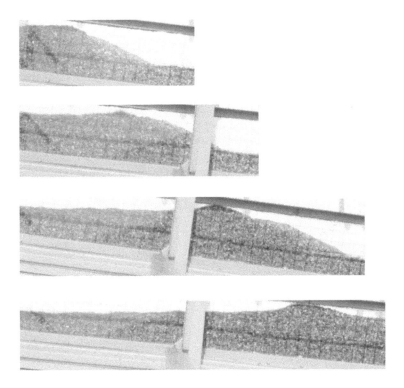

The photographs show the time interval shots of an experimental debris flow. The material laid uniformly to the height of the thick black line drawn parallel to the bed was soaked with water prior to the experimental run, and water flow was suddenly started from the upstream end to generate a debris flow. The forefront has an obvious snout shape and the rear part flows with a flatter slope than the original bed slope. The flow proceeds from left to right and time elapses from the upper to lower photographs.

INTRODUCTION

Debris flows are the flow of water and solids mixture, in which solids are the assemblage of widely distributed grains. Therefore, the essential model for the mechanics of flow should consider the constituent as the mixture of continuous fluid (water or slurry) and densely assembled discrete particles, and it should explain both the macro behaviors such as mean velocity of a bulk body and flooding limit on an inundation area and the micro individual particles' motions, simultaneously. However, the existing discrete element models that are suitable for discussing individual particle motion cannot treat solid-fluid interaction well, and more importantly, they cannot predict the macro behaviors practically required. At present, the continuum mixture theory that divides the constituent into the continuum solid phase and the continuum fluid phase taking the interaction between the phases strongly into account is the most influential.

This chapter first outlines the physical methods to discuss the motion of a solid and fluid mixture. Namely, after a brief review of the discrete element method that may become a viable method in future, the classical single-phase hydraulic models; the viscoplastic fluid model and dilatant fluid model, are explained to make clear the historical progress and their advantages and disadvantages. Especially, the dilatant fluid model is proved to be suitable for modeling the stony-type debris flow by comparison with the laboratory experiments. Next, the constitutive relationships between stress and strain for dry granular flow, developed by us, are introduced. Applying this granular flow theory to the hypothetical solid phase in inertial debris flows, the mechanics and prediction of inertial debris flows are discussed as a continuum mixture theory. Also for viscous debris flow, considering the mechanics to disperse heavy grains in a viscous fluid, it is proved that the mixture of fluid and grains in this case can be modeled as single-phase Newtonian fluid.

2.1 MODELS FOR SOLIDS AND FLUID MIXTURE AS THE MULTI-PHASE FLOW

The existing simulation models for granular flow with the negligible effects of interstitial fluid (flow in vacuum or under atmosphere) can be categorized into the continuum constituent model and the discrete particle model. The former model describes the granular assembly as a continuum material and its characteristics in flow are analyzed by the Eulerian forms of continuity and momentum equations. The discrete characteristics of grain motion are necessarily neglected. This method is useful for knowing the macro behaviors of motion such as the depth of flow and velocity as a bulk. The latter model, on the other hand, traces the motion of individual grain by the Lagrangian equation.

To describe the phenomena by the former model, the following equations are necessary:

1   the stress balance equation between the stresses operating from outside and the inner stresses that arise within the hypothetical continuum fluid;
2   the continuity equation that describes the mass conservation condition; and
3   the constitutive equation that describes the relationship between stress and deformation velocity in the hypothetical continuum.

Equations 1) and 2) can be borrowed from classic fluid mechanics. With the constitutive equation a simple discussion using dimensional analysis has been done in 1.4, but it is a formidable task to obtain it for general cases with various grain concentrations, sizes and their distributions under various particle interactions. Furthermore, to consider the assembly of discrete particles as a continuum might be an unreasonable assumption. But, if an appropriate form of it under a particular flow condition is given, the former model can be a very useful one to understand the flow characteristics *en masse* under that condition.

Similar discussion would be applicable for the case of debris flow. To describe debris flows by the continuum constituent model, because the interstitial fluid is water or slurry and the density difference between fluid and grains is small, the effects of interstitial fluid may not be neglected. The essential problem in this case is how the constitutive equation is given, considering the interactions between particles and the interstitial fluid. In this context, the way of thinking branches into two. One is the single-phase model that considers the mixture of fluid and solids as a kind of continuous fluid whose properties implicitly reflect the fluid-particles interaction effects. In this model only one constitutive equation for this hypothetical fluid is given. The theory of debris flow started with this idea and to this day it is widely used. Another way of thinking is the two-phase model in which the solid phase and the fluid phase are thought as independent continuums and the constitutive equations for the respective phases are obtained taking the interaction between the two phases into account. Irrespective of which model is selected, once the necessary one or two constitutive equations are given, the macro characteristics of debris flow such as depth, velocity and solids concentration can be analyzed and the prediction of hazards or effective countermeasure planning become possible. The state-of-art for describing the interaction effects between the two phases is insufficient, but recent debris flow theories tend to select the two-phase model.

The discrete particle model for granular flow, in which the effect of interstitial fluid is negligibly small, describes the motion of arbitrary selected objective particle in flow by the application of Newton's second law equation; $\mathbf{F} = m\mathbf{a}$, where $\mathbf{F}$ is operating force, m is the mass of particle, and $\mathbf{a}$ is the acceleration. The operating force upon the objective particle is the resultant force produced by the contact with the neighboring particles. Two kinds of approach for treating the contact have been adopted; the hard-contact technique and the soft-contact technique. In the hard-contact technique, the particle is considered as a rigid body, and the contact arises only between the two adjacent particles in an instantaneous moment (Campbell and Brennenn 1985; Straub 2001). This technique has a shortcoming in that it cannot treat the simultaneous collision of multiple bodies.

The soft-contact technique is free from the shortcoming of the hard-contact technique (Campbell 2001). At soft-contacting, particles deform a little and the forces proportional to the stiffness of particles act on the contact points. The interaction between particles on a contact point is modeled by the deformation of an interface instrument that consists of a spring and a dashpot connected in parallel. A pair of such an instrument is mounted normally and tangentially on each contact point as shown in Figure 2.1, each instrument produces a normal repulsive force and a tangential

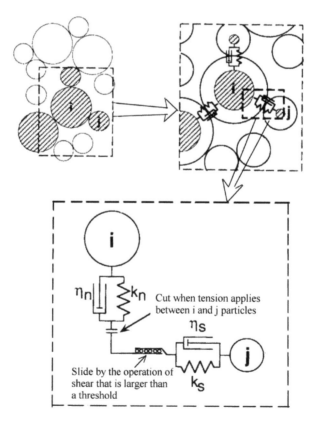

Figure 2.1 Particle contact in soft-contact model. k_n, k_s: spring constant, η_n, η_s: damping coefficient. (Gotoh and Sakai 1997, partly altered).

friction force, respectively. To handle a non-cohesive particle, a joint that is nonresistant to tension in a normal direction and a slipping joint that is effective under a tangential force more than a threshold should be serially connected to the respective instruments (Gotoh and Sakai 1997). The spring constant and damping coefficient for each instrument should change depending on the direction and particle diameter. Thus, many parameter values must be given beforehand. Moreover, the calculation should be carried out under optimum time steps (Harada 2001).

It is quite natural to apply the discrete particle model to a two-phase flow. The hard-contact technique may be applicable when grain concentration is small, but, for the flow on a movable bed, the particle exchange between that involved in the flow and that on the bed arises by a contact of particle in flow with some of the bed particles. As the multiple particle collision is not allowed in the hard-contact technique, the soft-contact technique becomes indispensable for the flow on a movable bed.

In the discrete particle model, the behaviors in particle motion can be microscopically as well as macroscopically traced in chronological order and the results can be given graphically even for the phenomena that may be difficult to observe in a physical model test. For example, it will be suitable to trace the phenomena of particle segregation in stony debris flow. But, if the number of particles becomes large, the load on the computer becomes enormous. The quantitative macro characteristics of flow such as depth and velocity can be obtained only after the completion of simulation.

In the analysis of a solids-liquid mixture flow by the discrete particle model, the assessment of interaction between particles and fluid is crucial. One example of such an assessment is the bed load transport in the large shear stress range. In this case, it gives rise to a highly concentrated sheet flow layer close to the bed. For the analysis of this phenomenon, an iterative operation is necessary, in which, at first stage, the flow field without particle motion is solved; in the second stage, the particle motion in the obtained flow field is solved; in the third stage, the existence of the particle transporting layer modifies the flow field by the reaction force; and in the fourth stage, the modified flow field changes the particle transportation. Thus, asymptotically the characteristics of the sheet flow layer can be obtained (Bakhtiary 1999). But, for the case of debris flow, the applicability of such a method is still unclear, because the drag force acting on particles due to the velocity difference between fluid and particle may have trivial effects.

Let us consider more about continuum constituent model for debris flow. If the solid phase is considered as a continuum and described by the Eulerian equations, because the liquid phase is naturally described by the Eulerian flow equation, mathematically this two-phase model becomes Euler-Euler coupling model. In this model, two momentum conservation equations for liquid and solid phases, respectively, are written as (Iverson 1997):

$$\sigma C\left(\frac{\partial \mathbf{v_s}}{\partial t} + \mathbf{v_s} \cdot \nabla \mathbf{v_s}\right) = \nabla \cdot \mathbf{T_s} + \sigma C\mathbf{g} + \mathbf{F_i} \tag{2.1}$$

$$\rho(1-C)\left(\frac{\partial \mathbf{v_f}}{\partial t} + \mathbf{v_f} \cdot \nabla \mathbf{v_f}\right) = \nabla \cdot \mathbf{T_f} + \rho(1-C)C\mathbf{g} - \mathbf{F_i} \tag{2.2}$$

where v_s and v_f are the solid and fluid velocities, respectively, T_s and T_f are the solid phase and liquid phase stress tensors, respectively, F_i is the interaction force per unit volume that results from momentum exchange between the solid and fluid constituents.

The addition of the respective right- and left-hand side terms of these two equations yields a momentum conservation equation that is applicable to the bulk mixture. This procedure eliminates the assessment of complicated interactions between fluid and solid phases. If nearly steady uniform flow in a Cartesian coordinate system is considered and the relative velocity between solid and fluid phases is neglected, the following two equations for pressure and shear stress, respectively, are finally obtained (Takahashi, 1991):

$$P_{ds} + P_f = (\sigma - \rho)g\cos\theta \int_z^h C\,dz \tag{2.3}$$

$$T_{ds} + T_f = g\sin\theta \int_z^h \rho_T\,dz \tag{2.4}$$

where P_{ds} is the pressure in solid phase, P_f is the pressure in excess over the hydrostatic one, T_{ds} is the shear stress in the solid phase, T_f is the shear stress in fluid phase, $\rho_T(=(\sigma - \rho)C + \rho)$ is the apparent density of the debris flow material, and θ is the slope of the flow surface.

These two equations comprise one of the fundamental equations 1), 2) and 3) that appeared earlier; i.e. the two-phase continuum model version describing the stress balance in the flow of mixture. The essential problem in this two-phase mixture theory is how the left-hand sides of Equations (2.3) and (2.4) (the constitutive equations) are described.

2.2 SINGLE-PHASE CONTINUUM MODELS

In the single-phase continuum models for debris flow, the mixture of particles and fluid is considered as a kind of continuous fluid which behaves as debris flow in various situations. The characteristics of the apparent fluid are determined by the relationship between the operating shear stress and the rate of strain that is called the constitutive law or the consistency. Some of these relationships are given by Figure 2.2. The constitutive equations of the respective curves are as follows:

$$\textit{Newtonian fluid}; \quad \tau = \mu(du/dz) \tag{2.5}$$

$$\textit{Bingham fluid}; \quad \tau = \tau_y + \eta(du/dz) \tag{2.6}$$

$$\textit{Herschel} - \textit{Bulkley fluid}; \quad \tau = \tau_y + K_1(du/dz)^n, n \le 1 \tag{2.7}$$

$$\textit{dilatant fluid}; \quad \tau = K_2(du/dz)^n, n > 1 \tag{2.8}$$

The Newtonian fluid is represented by the laminar flow of plain water in which the shear stress τ is linearly proportional to the rate of strain (du/dz), where the proportional coefficient μ is the dynamic viscosity. The other fluids in the figure are lumped

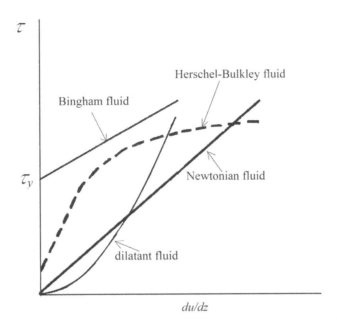

Figure 2.2 Consistency curves for some kinds of fluid.

together as non-Newtonian fluids. Among them, Bingham fluid does not deform if the operating shear stress is smaller than a threshold, but, if the operating stress is larger than the threshold it behaves as if it were a Newtonian fluid, where the threshold stress τ_y is called the yield stress (strength) and the proportional coefficient η is called the rigidity modulus or the viscosity of Bingham fluid. The Herschel-Bulkley fluid increases its mobility with increasing shear stress, and the Bingham fluid is a particular one in which the power n is equal to 1. Water saturated static clay forms some void rich structures and when such a skeletal structure is destroyed by the operation of shear stress larger than the yield strength, particles disperse within the mass decoupled each other diminishing the resistance to flow. This is the case of the Herschel-Bulkley fluid. The dilatant fluid, on the contrary, decreases its mobility with increasing shear stress due to the increasing frequency of getting over other particles with increasing shear rate.

Debris flow obviously behaves differently to plain water flow, so that it may be appropriate to consider it as the flow of a non-Newtonian fluid. But, as different models are available, one must find out what characteristic relation between the operating shear stress and the velocity of deformation (velocity gradient perpendicular to the bed) exists in the particular flow about which one is talking.

2.2.1 Visco-plastic fluid model

One whole event of debris flow, irrespective of its genetic types whether due to the transformation of a landslide earth block or due to the mobilization of a channel bed, commences from the gradual or abrupt motion of static material and ends by

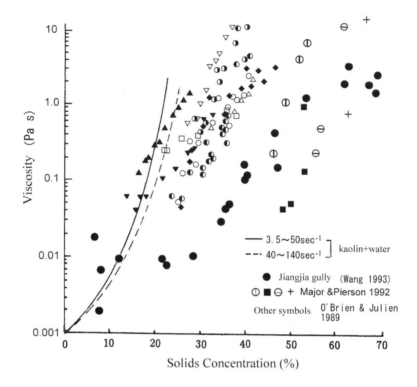

Figure 2.3 Viscosity of the interstitial fluid as Bingham fluid.

stopping as a deposit at flat area after a motion of considerable distance. Therefore, if one considers that the static material before the commencement of motion has a yield strength, one can explain the starting and the continuation of motion attributing to the operation of stress more than the yield strength weakened by the addition of water. Moreover, one can explain the stopping of flow attributing to the decline of operating shear stress less than the yield strength. The debris flow modeling as a kind of viscous fluid having strength is called 'visco-plastic fluid model'. The Bingham fluid model that assumes the relationship as Equation (2.6) for a planer simple shearing flow was firstly and independently proposed by Yano and Daido (1965) and Johnson (1970), since then, this model has been used to explain many actual debris flows.

The application of the Bingham fluid model requires the determination of parameters τ_y and η, respectively. Usually, these values are obtained by the rheometer test of samples collected from the field. Although a large-scale rheometer which enables the test of samples as much as $1\ m^3$ and containing quite large particles is used, it is impossible to test the bulk material containing many boulders larger than $1\ m$ in diameter. Therefore, many tests have been done only for the interstitial fluid. The large-scale experiments done by USGS explained in section 1.3.1 may be said to be the tests of debris flow materials in a near prototype scale.

Figure 2.3 shows the viscosities of a Bingham fluid measured in our laboratory flume experiments using kaolin clay and the data for samples collected from fields.

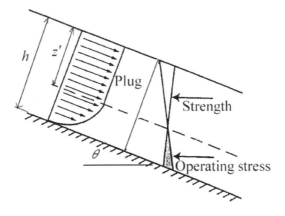

Figure 2.4 Velocity distribution in a Bingham fluid.

Because kaolin tends to have larger mobility with increasing shear stress, the measurements of viscosity are done under both a small shearing rate ($3.5–50\,\mathrm{sec}^{-1}$) and a large one ($40–140\,\mathrm{sec}^{-1}$). Viscosity in kaolin has almost one to one relationship with concentration. Similar relationships seem to exist for other samples as well, but these values differ very much depending on samples, suggesting that the prediction of viscosity is difficult. Herein, the values of yield strength are not referred to, but they also range from about 1 Pa to 100 Pa.

The shearing stress operating on the plane at height z in a planar flow of depth h and θ in the channel slope is given by:

$$\tau = \rho g(h - z)\sin\theta \tag{2.9}$$

If it is a Bingham fluid, Equation (2.6) is applicable. Therefore, the velocity distribution perpendicular to the bed is obtained by the integration of Equation (2.10) under the boundary condition; at $z = 0$, $u = 0$:

$$\tau_y + \eta(du/dz) = \rho g(h - z)\sin\theta \tag{2.10}$$

The result is:

$$\frac{u}{u_*} = \frac{\rho u_* h}{\eta}\left\{\left(1 - \frac{z'}{h}\right)Z - \frac{1}{2}Z^2\right\} \tag{2.11}$$

where $z'(= \tau_y/\rho g\sin\theta)$ is the depth measured from the flow surface to the plane where the operating shear stress is equal to τ_y, $Z = z/h$, u_* is the shear velocity defined by $(gh\sin\theta)^{1/2}$. When $z' = 0$, it becomes a parabolic curve applied to a Newtonian fluid. Equation (2.11) has a shape illustrated in Figure 2.4. The part shallower than z' behaves as a rigid board because the operating shear stress is smaller than τ_y. This part is called the 'plug'.

Let us try to apply a Bingham fluid model to the debris flow in the Jaingjia gully. Provided the depth is 1 m, the channel slope is 3° and the apparent density of material is

2.2 g/cm^3, the substitution of representative viscosity and yield strength obtained from the rheometer tests of field samples; 1 Pa · s and 100 Pa, respectively, into Equation (2.11) yields $z' = 9$ cm and a cross-sectional mean velocity of $U = 470$ m/sec, which is unreasonable in comparison to field observations. On the other hand, the substitution of field observed deposit thickness after the passage of a surge and the velocity of debris flow; 50 cm and 6 m/sec, respectively into Equation (2.11) yields $\tau_y = 5600$ Pa and $\eta = 230$ Pa · s. As mentioned earlier, the yield strength of slurry in the Jiangjia gully being 1–100 Pa and the viscosity being 0.01–10 Pa · s (see Figure 2.3), the yield strength as well as the viscosity of the bulk debris flow material that contains plenty of coarse particles are from a few tens to hundreds times those in the interstitial fluid. The physical reasons for this should be made clear and these characteristic values should be obtained theoretically, but now, only the inverse analysis of observational data can explain these values. The pile of data obtained by rheometer tests would be useless for predicting the velocity and the thickness of deposit of actual debris flows, even though the qualitative tendency to increase in both viscosity and strength with increasing solids concentration is known by these tests.

Johnson (1970) modified his own Bingham fluid model to take account of the far lower mobility of actual debris flow than that expected by the rheometer tests, as the following 'Coulomb-viscous model':

$$\tau = c + \sigma_n \tan \varphi + \eta(du/dz) \tag{2.12}$$

where c is the cohesive strength, σ_n is the internal normal stress, and φ is the internal friction angle. However, no physical reasoning to think that way is given.

Coussot et al. (1998) claim that the parameters of debris flow material τ_y, K_1 and n can be uniquely determined if the Herschel-Bulkley model is adopted. In reference to the sensitive reaction of τ_y and η to the slight change in material composition in a Bingham fluid model, this discussion would be valid only in narrow range.

A compromise model between the visco-plastic models and the inertial fluid models is the 'polynomial model' (Chen 1988; Julien and Lan 1991). This model adds the term $(du/dz)^2$ to the equation for the visco-plastic model. This model was introduced to explain the wide variety in characteristic behaviors of actual debris flow, but the physical background is suspicious. Moreover, the number of parameters to be determined is more than the simple visco-plastic model, though it is difficult to determine the parameter values even in the simple model.

The fact that the debris flow in the Jiangjia gully lacks the plug, as mentioned in section 1.3.3, means that the stoppage of flow at the rear end of a surge is not due to the strength of the fluid but due to the deposition of particles confronting the deficit in the particle sustaining mechanism. Therefore, neither the Bingham fluid model nor other visco-plastic models for viscous-type debris flow are suitable for modeling this behavior.

2.2.2   Dilatant fluid model

The stony-type debris flow, whose major solid component is coarse grains larger than fist-size can often reach to an area flatter than about 4° keeping its high mobility and ability to carry big boulders. The reason for this is the reduction of resistance to flow due to the dispersion of grains separated from one another (Takahashi 1978). Because the grains are heavier than the ambient fluid, they tend to be deposited. Therefore, to keep the flow going, some mechanisms to disperse the grains rather than them settling must be operating. When the grain concentration is comparably small (0.5 or less by volume), the excess pressure in the interstitial fluid, which is one of the possible grain sustaining mechanisms, would be very short-lived because of the large voids between the particles. One contributory mechanism to sustain grains in such a case would be the repulsive forces of grain collisions.

The constitutive equation given by Bagnold

Bagnold (1954) was the first to discover the significance of inter-particle collision as the cause of grain dispersion. He carried out experiments using the apparatus as shown in Figure 2.5. The apparatus consisted of double concentric drums whose inner one was stationary and the outer one rotated so as to create the prescribed shearing rates. A mixture of glycerin-water-alcohol solution and neutrally buoyant grains made from paraffin wax and lead stearate was sheared in the annular space between the two concentric drums. The torque spring and the manometer attached to the inner drum measured the shear stress and pressure, respectively, within the experimental mixture of grains and fluid. The experimental results showed that when the rotating velocity was small (a small shear stress region) pressure and shear stress varied linearly with the change in shearing rate (du/dz), but when the rotating velocity was large (a large stress region) both pressure and shear stress varied proportional to the square of (du/dz). In a small stress region (macro-viscous region) grains merely contributed to the increase in the viscosity of the fluid, whereas in a large stress region (inertial region) the effects of the fluid viscosity became negligibly small, and in between the two regions there was a transitional region. He hypothesized that the pressure and shear stress in the inertial region was produced by the inter-particle collision.

Bagnold hypothesized the pressure generating mechanism due to inter-particle collision on the assumption of a particle dispersion system as illustrated in Figure 2.6. The particles embedded in each layer move in equal spacing keeping the velocity of the layer. The oscillations are caused by successions of glancing collisions as the grains of one layer overtake those of the next slower layer. The velocity difference between the two adjacent layers is δu. The repulsive pressure arising from the particle collision is assumed to be proportional to the mean momentum change by one collision, the frequency of collision and the number of particles per unit area in one layer.

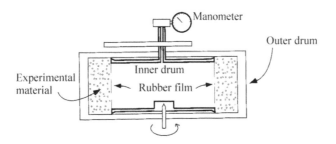

Figure 2.5 Schematic diagram of Bagnold's experimental apparatus.

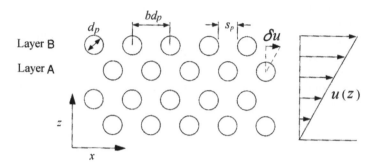

Figure 2.6 Grain arrangement that Bagnold assumed.

As defined in Figure 2.6, the center to center distance between adjacent particles can be written as:

$$b = (s_p/d_p) + 1 = (1/\lambda) + 1 \tag{2.13}$$

where s_p is the space between the two particles, and $\lambda(=d_p/s_p)$ is the 'linear concentration' defined by Bagnold which is an expression for grain concentration. The relationship between volume concentration C and the linear concentration λ is:

$$C = \frac{C_*}{b^3} = \frac{C_*}{\{(1/\lambda) + 1\}^3} \tag{2.14}$$

where C_* is the maximum concentration when $\lambda = \infty$ ($s_p = 0$) and is equal to about 0.65 for uniform natural grains.

The grains of layer B are sheared over those of layer A at a mean relative velocity δu and if the collision angle to z direction is α_i, and grain is elastic, each B grain experiences a change of momentum per one collision $m\,\delta u \cos \alpha_i$, where m is the mass of the grain. Each B grain make $f(\lambda)\delta u/s_p$ collisions with an A grain in unit time, where $f(\lambda)$ is an unknown function of λ. This is introduced here based on the conjecture that when λ is large the geometrical restraint will affect the collision frequency, and if λ is small it is equal to λ. The number of grains in a unit area in each layer is $1/(bd_p)^2$. Therefore, the repulsive pressure operating in the z direction is written as:

$$p \propto m\,\delta u \cos \alpha_i\, f(\lambda) \frac{\delta u}{s_p} \frac{1}{(bd_p)^2} \tag{2.15}$$

Substitution of $m = \sigma \pi d_p^3/6$, $\delta u = b d_p du/dz$ and $s_p = d_p/\lambda$ into this equation gives:

$$p_c \propto \frac{\pi}{6}\sigma\lambda f(\lambda)d_p^2\left(\frac{du}{dz}\right)^2 \cos \alpha_i \tag{2.16}$$

Similarly, we obtain the shear stress as:

$$\tau_c \propto \frac{\pi}{6}\sigma\lambda f(\lambda)d_p^2\left(\frac{du}{dz}\right)^2 \sin\alpha_i \tag{2.17}$$

The deduction of Equations (2.16) and (2.17) is under the bold assumptions that the grains are elastic, the motion of particles is only in the x-z plane and the arrangement of grains is uniform. Bagnold was well aware of these shortcomings and he finally determined the formula introducing the experimentally obtained coefficients as follows:

$$p_c = a_i \cos\alpha_i \sigma\lambda^2 d_p^2 (du/dz)^2 \tag{2.18}$$

$$\tau_c = \tan\alpha_i \, p_c \tag{2.19}$$

In his experiments, the average of $\tan\alpha_i$ is 0.32 ($\alpha_i = 17.8°$) and $a_i = 0.042$.

Savage and McKeown (1983) did similar experiments to Bagnold's except that the inner drum was rotated in their apparatus and the effects of wall roughness were also tested. The experiments were all in the inertial region and verified Bagnold's relationships. The coefficient for relating the stress to the term $(du/dz)^2$, however, was larger than the one appearing in Bagnold's equation.

Hanes and Inmann (1985) also experimented in the inertial range by using a differently structured apparatus; a ditch like Saturn's ring curved into a horizontal plane was filled with the experimental material and covered by a stationary cap, then, the material was sheared by rotating the horizontal plane. The torque operating on the cap was converted to the shear stress. The experimental material was mainly dry glass beads but a mixture of water and sand was also used. The grains were not neutrally buoyant and as in the actual debris flows the direction of flow was perpendicular to the direction of gravity acceleration. The results of the experiment agreed well with Equations (2.18) and (2.19).

In Chapter 1, Equation (1.1) for the particle collision stress was deduced using dimensional analysis, and this has the same form as Bagnold's formula (2.15). In Bagnold's formula, $f(C,e)$ is given as $a_i\lambda^2 \sin\alpha_i$ through the analysis of experimental data. Some investigations to obtain $f(C,e)$ theoretically have been done ; e.g. Jenkins and Savage (1983), Lun et al. (1984), and Shen and Ackermann (1982, 1984), and for the case of a comparatively large restitution coefficient ($e > 0.8$), $f(C,e)$ values are rather strictly deduced. In these discussions, similar to the discussions on discrete particle models, different techniques that consider either the binary particles collision or multiple particle collisions are used. In Japan Tsubaki et al. (1982), among others, considered multiple particle collisions in which the effects of repulsion and rubbing motion during the particles' contact were taken into account. Miyamoto (1985) discussed the energy loss due to inelastic particle collision. Takahashi and Tsujimoto (1997) developed their own theory for granular flow which is referred to in detail later.

The first approximation of stony debris flow

Bagnold in his experiments measured directly the excess pressure within the mixture in the annular space over that within the plain fluid. Therefore, the pressure p_c in Equation (2.18) is equivalent to P_{ds} in Equation (2.3). The grain drag T_{ds} could not be measured directly but the sum of T_{ds} and T_f was measured. He, however, postulated that the residual fluid drag contribution T_f was equal to $(T_{ds} + T_f)/(1 + \lambda)$. Thus, T_f was negligible at high grain concentrations. Therefore, τ_c in Equation (2.4) is almost equal to T_{ds} in Equation (2.4) and T_f in this equation can be neglected in the case of stony debris flow.

Takahashi (1977) applied Bagnold's constitutive equations to the steady uniform open channel flow of a grain-water mixture, in which the grain was heavier than water. As the first approximation, he assumed a uniform distribution of grains throughout the depth and described the stress balance equations for x and z directions, respectively, as follows:

$$a_i \cos \alpha_i \, \sigma \lambda^2 d_p^2 (du/dz)^2 = C(\sigma - \rho)g(h - z)\cos \theta \qquad (2.20)$$

$$a_i \sin \alpha_i \, \sigma \lambda^2 d_p^2 (du/dz)^2 = \{C(\sigma - \rho) + \rho\}g(h - z)\sin \theta \qquad (2.21)$$

The right-hand side term in Equation (2.20) means the pressure at height z due to the submerged weight of grains contained in the part from the surface $z = h$ to the height z. Therefore, Equation (2.20) means that all the submerged weight of grains is sustained by the repulsion pressure due to grain collisions. The right-hand side term in Equation (2.21) is the shear stress at height z operating in the hypothetical single-phase fluid of the grain-fluid mixture. In this context, the left-hand side is the constitutive expression of this hypothetical continuum fluid. A fluid having the relationship; $\tau \sim (du/dz)^2$, is a kind of dilatant fluid, so Takahashi's model is called the dilatant fluid model. Referring to the discussion above, this model is understood as an equivalent model to the two-phase mixture model in which the dynamic fluid effects, P_f and T_f, are negligibly small.

Integration of Equation (2.21) under the boundary condition; $u = 0$ at $z = 0$, gives:

$$u = \frac{2}{3d_p} \left[\frac{g \sin \theta}{a_i \sin \alpha_i} \left\{ C + (1 - C)\frac{\rho}{\sigma} \right\} \right]^{1/2} \frac{1}{\lambda} \{ h^{3/2} - (h - z)^{3/2} \} \qquad (2.22)$$

Similarly, integration of Equation (2.20) under the same boundary condition gives:

$$u = \frac{2}{3d_p} \left[\frac{g \cos \theta}{a_i \cos \alpha_i} C \left(1 - \frac{\rho}{\sigma} \right) \right]^{1/2} \frac{1}{\lambda} \{ h^{3/2} - (h - z)^{3/2} \} \qquad (2.23)$$

It seems unreasonable that two independent velocity formulae are obtained. This contradiction arises because of the assumption of a uniform distribution of grain concentration throughout the depth. But, these two formulae become physically equivalent when the following formula is satisfied:

$$C = \frac{\rho \tan \theta}{(\sigma - \rho)(\tan \alpha_i - \tan \theta)} \qquad (2.24)$$

In other words, only when Equation (2.24) is satisfied, can the grain concentration in the flow be assumed to be uniformly distributed.

Takahashi (1977) did the laboratory experiments using a steel flume, 20 cm wide, 40 cm deep, and 7 m long. Its slope was variable from 0° to 30°. It had a transparent glass wall on one side through which the behaviors of the flow were recorded by 35 mm motor-driven cameras and a 16 mm movie camera. Prior to the commencement of experiments, material some ten centimeters thick was laid uniformly on the flume bed and soaked by water. As soon as the sudden discharge of water was introduced at the upstream end of the bed, a bore of the mixture of gravel and water was formed by the erosion of the bed and traveled downstream. If the bed slope was steeper than about 15° but flatter than about 25°, the bore height developed rapidly at first but gradually approached an equilibrium height. The flow in this stage was dispersing particles throughout the depth, and it was suitable to be called a debris flow. The discharges and grain concentrations were measured by capturing flow at the downstream end of the flume. The experimental materials had the properties given in Table 2.1, in which d_m is the particle mean diameter, d_{84}/d_{16} is the ratio of 84% diameter and 16% diameter that represents the range of particle size distribution. In one experimental series, a slurry of water and silt mixture ($\rho = 1.35$ g/cm^3) was supplied from the upstream end instead of water.

In the cases on a rigid bed, the flume bed was roughened by pasting experimental material and the prescribed discharges of water and grains were independently supplied from the upstream end to attain the flow of known grain concentration. The properties of experimental materials used in this series are given in Table 2.2.

Because the debris flow developing processes are described in another chapter, here, only the experimental results concerning the equilibrium concentration attained on the erodible bed and the velocity distributions for the cases on erodible and rigid beds are explained.

Table 2.1 Materials used in erodible bed experiments.

Material	d_m (mm)	d_{90} (mm)	d_{84}/d_{16}	C_*	$\tan \phi$	σ (g/cm^3)
A	5.8	18.1	5.52	0.756	0.75	2.60
B	3.5	4.2	1.67	0.627	1.02	2.60
C	3.0	4.45	1.99	0.600	0.75	2.58

Table 2.2 Materials used in rigid bed experiments.

Material	d_{50} (mm)	d_{84}/d_{16}	C_*	$\tan \phi$	σ (g/cm^3)	C	θ (°)
Natural gravel	4.0	–	0.65	0.75	2.65	0.45	18
Artificial aggregate	1.48	1.61	0.55	0.75	1.74	0.30	8

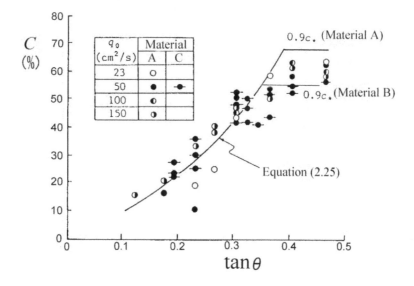

Figure 2.7 Equilibrium grain concentrations in stony debris flows.

Figure 2.7 shows the grain concentration in the forefront of the experimental debris flows on erodible bed. Experiments were carried out under various water supply discharges per unit width, q_0 (see legend of Figure 2.7). Although there is a wide scatter in the experimental data, the grain concentration seems not to depend on the discharge but only on the bed slope. The curve in Figure 2.7 is defined by Equation (2.25) in which $\tan \alpha_i$ in Equation (2.24) is substituted with $\tan \varphi$:

$$C = \frac{\rho \tan \theta}{(\sigma - \rho)(\tan \varphi - \tan \theta)}(\equiv C_\infty) \tag{2.25}$$

According to Bagnold (1966), $\tan \alpha_i$ and $\tan \varphi$ have similar values under the condition of equilibrium concentration.

Equation (2.25) indicates that if the channel slope satisfies:

$$\tan \theta \geq \frac{(\sigma - \rho)C_*}{(\sigma - \rho)C_* + \rho} \tan \varphi \tag{2.26}$$

C becomes larger than C_*. But, it is impossible to get the flow in such a high concentration, and according to the experiments, C is always less than about $0.9C_*$.

Figure 2.8 shows the results of the examination of whether Equation (2.25) is applicable to the case where the density of the interstitial fluid is larger than that of plain water. As is evident in Figure 2.8, this equation is surely applicable and the equilibrium grain concentration under heavier interstitial fluid that contains much silt and clay material becomes larger than that under plain water.

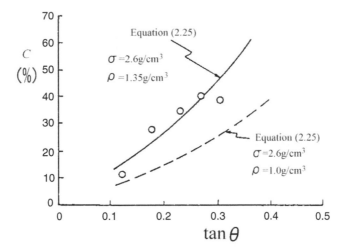

Figure 2.8 Equilibrium grain concentrations in stony debris flows generated by dense liquid.

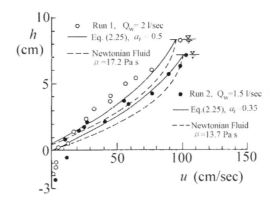

Figure 2.9 Velocity distribution in stony debris flow on erodible bed.

Figure 2.9 shows the theoretical and experimental velocity distributions on an erodible bed. Although Equation (2.22) fits each experimental velocity distribution rather well, there is some discrepancy at the lower part where it moves as if the bed were dragged by the upper rapidly moving flow. Moreover, in these two cases about a ten-fold larger a_i value than Bagnold's 0.042 is required to fit the theoretical curves to the experimental results. These velocity distribution forms and uncertainty in a_i value have been the main basis of refutation against the Takahashi's dilatant fluid model (e.g. Tsubaki *et al.* 1982; Chen 1987). Uncertainty in the a_i value, however, was not caused by the inappropriateness of the flow model but was brought about by the variability in the degree of bed saturation just before the generation of the debris flow. Namely, if the bed was dry, the water infiltration from the flow into the bed was effective in reducing the mean velocity (the increased C value brought about by the reduction of

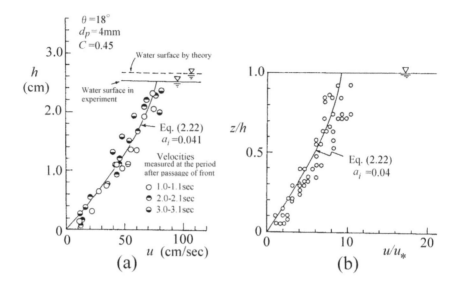

Figure 2.10 Velocity distribution in stony debris flow on rigid bed. (a) Material is sand grain, (b) Normalized velocity distribution. Material is light weight aggregate.

water quantity in the flow under a fixed a_i value gives a smaller u by Equation (2.22)) and in turn it was necessary to give a larger a_i value to fit the theory to the experimental velocities. This reasoning is verified by the results of rigid bed experiments shown in Figure 2.10 in which the dilatant fluid model is a good fit, giving $a_i = 0.04$. As for the discrepancy in the velocity distribution form, the following discussion on the second approximation will make the situation clear. The broken lines in Figure 2.9 show the velocity distributions of Newtonian fluid in which the apparent viscosities are obtained by the back analysis to fit the velocities. The velocity distribution of a dilatant fluid has a more uniform velocity gradient than that of a Newtonian fluid.

Figure 2.10 shows the results of rigid bed experiments and Equation (2.22). It shows how well this equation represents the experimental velocity distribution as well as the flow depth by substituting $\tan \alpha_i = 0.75$ and $a_i = 0.04$. These good agreements are perhaps due to the closeness of the grain concentrations to the equilibrium ones given by Equation (2.25).

The second approximation of stony debris flow

Although the first approximation can well explain the phenomena of stony debris flow, the assumption of a uniform grain distribution throughout the depth is, at least theoretically, too much simplification, and this causes the contradiction that there exists two independent Equations (2.22) and (2.23) for only one unknown velocity (Iverson and Denlinger 1987). Furthermore, as is evident in Figure 2.9, the velocity distribution on the erodible bed has an inflection point near the bottom and this peculiar velocity distribution cannot be explained by the first approximation. Although the motion below the inflection point may be qualitatively understood as the part of bed dragged by the flow above, it would be better to consider it as a part of flow.

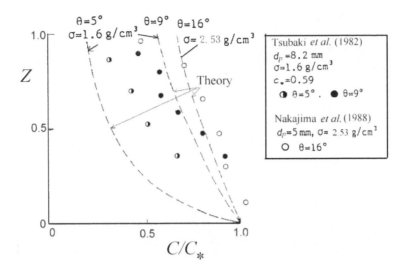

Figure 2.11 Theoretical grain concentrations and the experimental data.

To solve these disadvantages, although the characteristic form of $\tan \alpha_i$ as a function of C had not theoretically been discussed, Takahashi (1991) assumed the following formula referring to Savage and Sayed's (1984) experiment:

$$\tan \alpha_i = (1 + 1/\lambda)\tan \varphi = (C_*/C)^{1/3} \tan \varphi \qquad (2.27)$$

This formula is applicable in the range $\lambda > 3$ ($C > 0.3$), and contrary to that usually observed in quasi-static soil-mechanic testing, as $\tan \alpha_i$ becomes larger C becomes smaller. That mechanism can be understood by imaging that the particles embedded in the two adjacent upper and lower layers, respectively, can collide at a larger colliding angle to the z direction (a particle in a layer can cut deep into the other layer) the thinner the grain concentration. The validity of Equation (2.27) was confirmed by substituting this into Equations (2.18) and (2.19) and comparing these formulae with Bagnold's experimental results. The experiments by Straub (2001) also verify this.

The substitution of Equation (2.27) into Equations (2.20) and (2.21) gives the two independent formulae necessary and sufficient to obtain u and C at height z. The process of deduction can be found in Takahashi (1991). Figure 2.11 compares the theory with experiments concerning grain concentrations. Some data deviate from the theory largely, because the measurement of concentration distribution is very difficult. Nevertheless, as a whole, the tendency of distribution seems to be explained.

The normalized velocity distribution formula theoretically obtained is:

$$\frac{u}{u_s} = 1 - \frac{1}{2}(1 - Z)^{3/2}(2 + 3Z) \qquad (2.28)$$

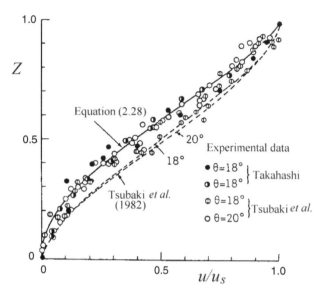

Figure 2.12 Theoretical velocity distributions and the experimental data.

where u_s is the surface velocity and $Z = z/h$. Equation (2.28) is compared with the experimental data in Figure 2.12. The peculiar form of the velocity distribution having inflection points near the bottom and near the surface is well explained.

Comparison with the field data

The cross-sectional mean velocity, U, of stony debris flow in a wide rectangular prismatic channel is given by the integration of Equation (2.22) as follows:

$$U = \frac{2}{5d_p} \left[\frac{g \sin \theta}{a_i \sin \alpha_i} \left\{ C + (1 - C)\frac{\rho}{\sigma} \right\} \right]^{1/2} \left\{ \left(\frac{C_*}{C} \right)^{1/3} - 1 \right\} h^{3/2} \tag{2.29}$$

Therefore, the unit width discharge q_T, is written as:

$$q_T \equiv Uh = \frac{2}{5} Rh^{5/2} \sin^{1/2} \theta \tag{2.30}$$

where R is a function of the solids concentration, particle diameter and others as is evident from Equation (2.29). Nevertheless, here, R is assumed to have a constant value of $5.4 \, \mathrm{m^{-1/2} \, s^{-1}}$. The shaded areas in Figure 2.13 show the variation ranges in flow depths against the unit width discharge in the extent of the channel slopes from $10°$ to $22°$ for Equation (2.30) and for the Manning equation that is applicable to a turbulent flow. The field data (Takahashi 1991; Crosta et al. 2003) are plotted inside or around the area representing Equation (2.30). This fact shows that the dilatant fluid model is an appropriate model for stony debris flows in which the effect of particle collisions dominate.

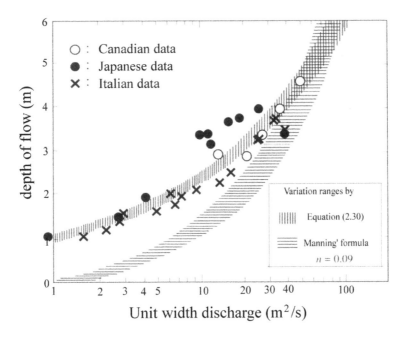

Figure 2.13 Comparison of the dilatant fluid model to the field data.

2.3 TWO-PHASE CONTINUUM MODELS (MIXTURE THEORY)

2.3.1 Stress equilibrium equations

The two-phase continuous fluid model considers the solid phase and the fluid phase as independent continuums and analyzes the flow characteristics by simultaneously solving the fundamental momentum and mass conservation equations for the respective phases under the interaction between the phases. The outlines of stress balance equations have already been mentioned in section 2.1, but, here follows a little more detailed discussion on these in the two-dimensional Cartesian coordinate system that is inclined to the direction of flow by θ.

The momentum conservation equations in the x-direction for the solid phase and fluid phase are, respectively:

$$\sigma C\left(\frac{\partial \hat{v}_x}{\partial t} + \hat{v}_x \frac{\partial \hat{v}_x}{\partial x} + \hat{v}_z \frac{\partial \hat{v}_x}{\partial z}\right) = \frac{\partial T_{ds}}{\partial x} + \frac{\partial T_{ds}}{\partial z} + \sigma Cg \sin\theta - C\frac{\partial p}{\partial x} + F_x \tag{2.31}$$

$$\rho(1-C)\left(\frac{\partial \hat{u}_x}{\partial t} + \hat{u}_x \frac{\partial \hat{u}_x}{\partial x} + \hat{u}_z \frac{\partial \hat{u}_x}{\partial z}\right) = \frac{\partial T_f}{\partial x} + \frac{\partial T_f}{\partial z} + \rho(1-C)g \sin\theta - (1-C)\frac{\partial p}{\partial x} - F_x \tag{2.32}$$

and in the z-direction (perpendicular to the bed) for the solid phase and fluid phase are, respectively:

$$\sigma C\left(\frac{\partial \hat{v}_z}{\partial t} + \hat{v}_x \frac{\partial \hat{v}_z}{\partial x} + \hat{v}_z \frac{\partial \hat{v}_z}{\partial z}\right) = \frac{\partial P_{ds}}{\partial x} + \frac{\partial P_{ds}}{\partial z} + \sigma C g \cos \theta - C\frac{\partial p}{\partial z} + F_z \tag{2.33}$$

$$\rho(1 - C)\left(\frac{\partial \hat{u}_z}{\partial t} + \hat{u}_x \frac{\partial \hat{u}_z}{\partial x} + \hat{u}_z \frac{\partial \hat{u}_z}{\partial z}\right) = \frac{\partial P_{ds}}{\partial x} + \frac{\partial P_f}{\partial z} + \rho(1 - C)g \sin \theta - (1 - C)\frac{\partial p}{\partial z} - F_z \tag{2.34}$$

where $\hat{v}_x$, $\hat{v}_z$ are the x and z components of the velocity of solid phase, $\hat{u}_x$, $\hat{u}_z$ are the x and z components of the velocity of fluid phase, respectively, T_{ds}, T_f, P_{ds}, P_f are the x and z components of the dynamic stresses within the solid phase and fluid phase, respectively, F_x and F_z are the interaction forces in the x and z directions, respectively, p is the static pressure in the fluid phase and t is time. The left-hand side terms of these equations indicate the accelerations of the respective phases.

Addition of the respective right- and left-hand side terms of Equations (2.31) and (2.32) gives:

$$\rho_T\left(\frac{\partial \tilde{u}_x}{\partial t} + \tilde{u}_x \frac{\partial \tilde{u}_x}{\partial x} + \tilde{u}_z \frac{\partial \tilde{u}_x}{\partial z}\right) = \frac{\partial T_{ds}}{\partial x} + \frac{\partial T_{ds}}{\partial z} + \frac{\partial T_f}{\partial x} + \frac{\partial T_f}{\partial z} + \rho_T g \sin \theta - C\frac{\partial p}{\partial x}$$
$$- \frac{1}{2}\left[\begin{array}{l} \sigma C\left\{\frac{\partial}{\partial x}(\hat{v}_x - \tilde{u}_x)^2 + \frac{\partial}{\partial z}(\hat{v}_x - \tilde{u}_x)^2\right\} \\ + \rho(1 - C)\left\{\frac{\partial}{\partial x}(\hat{u}_x - \tilde{u}_x)^2 + \frac{\partial}{\partial z}(\hat{u}_x - \tilde{u}_x)^2\right\} \end{array}\right] \tag{2.35}$$

and, similarly, addition of the respective right- and left-hand side terms of Equations (2.33) and (2.34) gives:

$$\rho_T\left(\frac{\partial \tilde{u}_z}{\partial t} + \tilde{u}_x \frac{\partial \tilde{u}_z}{\partial x} + \tilde{u}_z \frac{\partial \tilde{u}_z}{\partial z}\right) = \frac{\partial P_{ds}}{\partial x} + \frac{\partial P_{ds}}{\partial z} + \frac{\partial P_f}{\partial x} + \frac{\partial P_f}{\partial z} + \rho_T g \cos \theta - C\frac{\partial p}{\partial z}$$
$$- \frac{1}{2}\left[\begin{array}{l} \sigma C\left\{\frac{\partial}{\partial x}(\hat{v}_z - \tilde{u}_z)^2 + \frac{\partial}{\partial z}(\hat{v}_z - \tilde{u}_z)^2\right\} \\ + \rho(1 - C)\left\{\frac{\partial}{\partial x}(\hat{u}_z - \tilde{u}_z)^2 + \frac{\partial}{\partial z}(\hat{u}_z - \tilde{u}_z)^2\right\} \end{array}\right] \tag{2.36}$$

Equations (2.35) and (2.36) are the momentum conservation equations for the mixture in the x and z directions, respectively. In which ρ_T is the apparent density of the mixture, and $\tilde{u}_x$, $\tilde{u}_z$ are the x and z components of the velocity of the center of mass of a mixture volume element, and they are defined as follows:

$$\rho_T = \sigma C + (1 - C)\rho \tag{2.37}$$

$$\tilde{u}_x = \{\sigma C \hat{v}_x + \rho(1 - C)\hat{u}_x\}/\rho_T, \quad \tilde{u}_z = \{\sigma C \hat{v}_z + \rho(1 - C)\hat{u}_z\}/\rho_T \qquad (2.38)$$

The last terms in the right-hand sides of Equations (2.35) and (2.36) are the contribution to the stress in the mixture that results from the relative velocity between the solids and the fluid. Consequently, if that relative velocity is large, the phenomena become complicated. But, as in debris flow, if the relative velocity is small and if the flow is nearly steadily uniform, the left-hand side terms and the last terms of Equations (2.35) and (2.36) are approximately zero. Furthermore, T_{ds}, T_f, P_{ds}, P_f and p do not change in the x direction and so the derivative terms by x are all zero. Then, Equation (2.35) becomes:

$$\frac{\partial T_{ds}}{\partial z} + \frac{\partial T_f}{\partial z} + \rho_T g \sin \theta = 0 \qquad (2.39)$$

Similarly, Equation (2.36) becomes:

$$\frac{\partial P_{ds}}{\partial z} + \frac{\partial P_f}{\partial z} + \rho_T g \cos \theta - \frac{\partial p}{\partial z} = 0 \qquad (2.40)$$

If p distributes hydro-statically, then:

$$p = \rho g(h - z)\cos \theta \qquad (2.41)$$

Substituting Equation (2.41) into Equation (2.40) gives:

$$\frac{\partial P_{ds}}{\partial z} + \frac{\partial P_f}{\partial z} = -(\sigma - \rho)Cg \cos \theta \qquad (2.42)$$

At the surface of flow where $z = h$, T_{ds} and T_f are equal to zero, and the integration of Equation (2.42) from z to h results in Equation (2.3). This formula means that the total submerged weight of the solid phase above the layer z is in balance with the sum of the pressure within the solid phase and the dynamic pressure within the fluid phase at height z.

Similarly, noting that the shear stresses at the flow surface are zero, we obtain Equation (2.4). This formula means that the body force operating to drive the bulk mass above the height z in the x direction is in balance with the sum of the stresses within the solid phase and the fluid phase at height z operating in the x direction.

It must be noted that in the concept of dividing the mixture into a solid phase and a fluid phase, the fluid phase does not mean a pure fluid but one that contains fine particles dispersed in a fluid due to turbulence in the fluid or an electro-magnetic force between particles, and the solid phase is only the coarse particle fraction that is not suspended in fluid.

2.3.2 Coulomb mixture theory (Quasi-static debris flow)

Iverson and Denlinger (2001) evaluated the Savage numbers in various natural debris flows and in their own large-scale debris flow experiments and found that these values are always small. They concluded that in natural debris flows the Coulomb friction stress is the predominating factor rather than the grain collision stress and others. Herein, the Savage number was defined by the following expression:

$$N_s = \frac{\sigma(du/dz)^2 d_p^2}{(\sigma - \rho)gh} \tag{2.43}$$

This expression is not exactly the same as Equation (1.8) but has a similar concept.

The numerator of Equation (2.43) represents, as is evident by referring to Equation (2.18), the particle collision stress, and the denominator represents the load operating on the bed due to the submerged weight of particles. Iverson and Denlinger (2001) consider the denominator as the Coulomb friction stress, but one must note that the static particle contact stress can be transferred from the surface to the bottom only if the particles are all in contact with others and a kind of skeletal system is formed even though its pattern is ever changing due to the shift in contacting points during motion. This condition requires that the particle volume concentration is between C_3 and C_2, to which values are already referred in Chapter 1. Namely, if the particle concentration is less than C_3, the load represented by the denominator should be sustained by some other stresses.

The numerator of Equation (2.43) does not indicate the particle collision stress exactly. To indicate the exact collision stress, as seen in Equation (1.1), a function of the particle concentration must be multiplied. Iverson (1997) considers that this coefficient is equal to C. The denominator should also be multiplied by C to exactly demonstrate the load to the bed. As far as this context is concerned, the common C value is already deleted in Equation (2.43) and it represents the ratio of the collision stress to the Coulomb stress. However, as is evident in Equation (2.18), the coefficient to be multiplied to the numerator is not C but λ^2. The value of λ^2 becomes as large as 10 ($C_* = 0.65$ and $C \approx 0.3$) to 120 ($C_* = 0.65$ and $C \approx 0.5$) (Campbell 1990) in the range of large particle concentrations and $\sigma\lambda^2(du/dz)^2 d_p^2$ can be as large as $C(\sigma - \rho)gh$. Equation (2.20) is nothing but the stress balance equation at that situation. Therefore, the Coulomb mixture theory of Iverson and Denlinger (2001) is only applicable for the quasi-static flow in which the particle concentration is larger than C_3.

Iverson and Denlinger (2001) developed the depth-averaged flow theory for a complex three-dimensional field, but herein, to make the discussion clear-cut, the theory is simplified to a two-dimensional planer flow with constant slope. The stress balance equations are:

$$P_{ds} + P_f = (1/2)\rho_T gh \cos\theta \tag{2.44}$$

$$T_{ds}|_{z=0} = \{\rho_T gh(1 - \lambda_e)\cos\theta\}\tan\varphi_{bed} \tag{2.45}$$

$$P_f = (1/2)\lambda_e \rho_T gh \cos\theta = (1/2)p_{bed} \tag{2.46}$$

$$T_f = 3(1 - C)\mu(U/h) \tag{2.47}$$

where P_{ds}, P_f, and T_f are the depth-averaged values (not the values at height z as previously used) of the particle collision stress, dynamic fluid pressure and fluid shear stress, respectively, $T_{ds}|_{z=0}$ is T_{ds} at the bed which remains only on the bed after the depth-averaging operation, λ_e is the ratio of the effective pressure in the interstitial fluid to the total grain pressure and $\lambda_e = 1$ represents a case of complete liquefaction, $\tan \varphi_{bed}$ is the friction coefficient on the bed surface, p_{bed} is the fluid pressure on the bed, μ is the viscosity of the interstitial fluid, and U is the average flow velocity as a Newtonian laminar flow.

For a flow on a constant slope bed, the last term on the right-hand side of Equation (2.35) is neglected but the left-hand side term remains and the following momentum conservation equation is obtained:

$$\rho_T \frac{d(hU)}{dt} = -(\rho_T gh \cos\theta - p_{bed})\tan \varphi_{bed} - 3(1-C)\mu\frac{U}{h} + \rho_T gh \sin\theta \qquad (2.48)$$

If Θ, which is defined by:

$$\Theta = \tan\theta - (1-\lambda_e)\tan \varphi_{bed} \qquad (2.49)$$

is assumed constant, the solution of Equation (2.48) under the initial condition; $U = U_0$ at $t = 0$, is given as:

$$U = \frac{\rho_T gh^2 \cos\theta}{3(1-C)\mu}\Theta\left[1 - \exp\left\{-t\Big/\frac{\rho_T h^2}{3(1-C)\mu}\right\}\right] + U_0 \exp\left\{-t\Big/\frac{\rho_T h^2}{3(1-C)\mu}\right\} \qquad (2.50)$$

This result indicates that the velocity of the mixture gradually accelerates and approaches a steady state asymptotically, but the time necessary to approach a steady state can be so long that it is virtually unattainable. A similar result to Equation (2.50) can be obtained using Johnson's model (Johnson 1970) using Equation (2.12) by assuming cohesion $c = 0$. But, this applies only the simple planar flow case and not universally valid.

Denlinger and Iverson (2001) applied their own three-dimensional discussion to various three-dimensional terrains by carrying out the numerical calculations and checked their validity by experiments.

2.4 THEORY FOR SUBAERIAL RAPID GRANULAR FLOWS

The particles in flow to which the Coulomb mixture theory is applicable move in a laminar fashion as illustrated in Figure 2.14. The load of the particles is transmitted to the bed via a continuous skeletal system connecting the contacting particles from the surface to the bottom. Such a continuous system can only be maintained throughout the motion if the particle concentration is larger than a geometrically determined threshold C_3. The motion in such a case is slow.

On the other hand, in a rapid flow, the distance between the particles is large and particles frequently collide with each other. In the case of debris flow, the void space between particles is filled by water or slurry, but, as mentioned earlier, especially for the type in which particle collision stresses predominate, the dynamic stresses in the flow can be approximately written as the sum of the particle collision stresses and the fluid stresses in the interstitial fluid, in which the interaction between the particles and the fluid is neglected. In this section, as the basis for developing the mixture

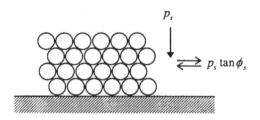

Figure 2.14 Skeletal stresses in a quasi-static grain flow.

theory of inertial debris flow, the stresses arising in subaerial granular flow are outlined and Takahashi and Tsujimoto's constitutive relations are introduced (Takahashi and Tsujimoto 1997).

Each particle in a rapid flow moves freely and independently and the velocity can be decomposed into the sum of the mean velocity and the fluctuation velocity. This concept resembles the kinetic-theory of molecules in gases. Because the mean-square value of the fluctuation velocities in the thermodynamics of gases represents the temperature, that of granules is referred to 'granular temperature'. The granular temperature represents the energy per unit mass and it produces pressure. However, granular and molecular systems have many essential differences. In particular, because the granules are inelastic, the granular temperature is dissipated by inter-particle collisions, and to maintain the motion energy must be continually added; e.g. gravity supplies energy into the flow on a slope.

Granular temperature can be generated by two mechanisms (Campbell 1990). The first is a by-product of inter-particle collision. When two particles collide, their resultant velocities will depend not only on their initial velocities, but also on the angle at which the two particles collide, the surface friction at the point of contact, and other factors, and therefore, they will contain random velocity components. The random velocities generated in this mode will be proportional to the relative velocities of the particles at the time of impact and hence must be proportional to the mean velocity gradient within the flow. Because the granular temperature is the mean-square value of random velocities, the granular temperature generated in this mode is expected to be proportional to the square of the mean velocity gradient. The second mode of temperature generation is due to the migration of particles in one layer to other layers. Following a random path, a particle moving with the velocity equal to its local velocity will pick up an apparently random velocity that is roughly equal to the difference in the mean velocity between its present location and the point of its last location. The granular temperature thus generated is also expected to be proportional to the square of the mean velocity gradient. Figure 2.15 is a schematic diagram of these granular temperature generation mechanisms in a granular flow that is comprised of intensively colliding particles. The stress produced by the first mode is called the 'collision stress' and that generated by the second mode is called the 'kinetic (streaming) stress'.

Many theories concerning rapid granular flows, based on the analogy between molecular motion and granular motion, have been proposed (Savage and Jeffrey 1981; Jenkins and Savage 1983; Lun *et al.* 1984; Johnson and Jackson 1987). However, the

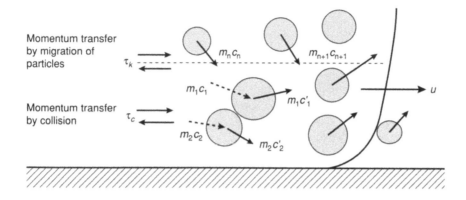

Figure 2.15 Dynamic stresses in a rapid granular flow.

majority of these investigations cover only the theoretical aspects of horizontal Couette flow, and few investigations have discussed the granular flow theory in comparison with the flows on a slope. The following discussion, is based mainly on Gidaspow's derivation (Gidaspow 1994), which focuses on the flows on a slope.

If the distribution function of the particle velocity is assumed to depend on the instantaneous velocity c, position r and time t, then, the number of particles whose velocity is in the range dc within the volume element dr is expressed as $f_p(c,r,t)\,dc\,dr$. Hence, the number of particles $n_p(t,r)$, velocity $v(t,r)$ in the mean flow field and granular temperature T are written as follows:

$$n_p = \int f \, d\mathbf{c} \tag{2.51}$$

$$\mathbf{v} = (1/n) \int \mathbf{c} f \, d\mathbf{c} \equiv \langle \mathbf{c} \rangle \tag{2.52}$$

$$T = (1/3n) \int \mathbf{C}_t^2 f \, d\mathbf{c} \equiv (1/3)\langle \mathbf{c}^2 \rangle \tag{2.53}$$

where $\mathbf{C}_t$ is the fluctuating velocity defined as:

$$\mathbf{C}_t = \mathbf{c} - \mathbf{v} \tag{2.54}$$

The external force of magnitude $m\mathbf{A}$ works on a particle, in which m is the mass of the particle and $\mathbf{A}$ is the acceleration; $\mathbf{A} = d\mathbf{c}/dt$. The change in the number of particles in a phase element is caused by inter-particle collisions and it is given by:

$$\{f(\mathbf{c} + \mathbf{A}\,dt, \mathbf{r} + \mathbf{c}\,dt, t + dt) - f(\mathbf{c}, \mathbf{r}, t)\}d\mathbf{c}\,d\mathbf{r} = \Delta f_c \, d\mathbf{c}\,d\mathbf{r}\,dt \tag{2.55}$$

in which Δf_c indicates the change in the distribution function due to inter-particle collisions. This formula can be rewritten at the limit $dt \to 0$ as:

$$\frac{\partial f}{\partial t} + \mathbf{c}\frac{\partial f}{\partial \mathbf{r}} + \mathbf{A}\frac{\partial f}{\partial \mathbf{c}} = \left(\frac{\partial f}{\partial t}\right)_{coll} \tag{2.56}$$

This is well-known Boltzmann equation.

Now multiply both sides of Equation (2.56) by a function of $\mathbf{c}$; $\psi(\mathbf{c})$, and the integration in the velocity space gives (Jenkins and Savage 1983):

$$\frac{\partial n\langle\psi\rangle}{\partial t} + \frac{\partial n\langle\psi\mathbf{c}\rangle}{\partial \mathbf{r}} - \mathbf{A}n_p\left\langle\frac{\partial\psi}{\partial \mathbf{c}}\right\rangle = \int \psi\left(\frac{\partial f}{\partial t}\right)_{coll} d\mathbf{c} \tag{2.57}$$

Equation (2.57), known as the Maxwellian transport equation, indicates that the change of mean value of a physical quantity $<\psi>$ in unit time per unit volume is given by the convection, velocity change and collision of particles. This is the fundamental equation in the discussion of granular flows.

2.4.1   Particle collision stress

If ψ in Equation (2.57) is momentum, the right-hand side represents the variation of momentum by collision, i.e. stress. For the derivation of stresses, the number of collisions in unit time should be counted exactly. Analogous to the single particle's frequency distribution function $f_p(\mathbf{c}, \mathbf{r}, t)$, a binary collision frequency distribution function $f_p^{(2)}(\mathbf{c}_1, \mathbf{r}_1, \mathbf{c}_2, \mathbf{r}_2)$ is introduced. Namely, $f_p^{(2)}(\mathbf{c}_1, \mathbf{r}_1, \mathbf{c}_2, \mathbf{r}_2)d\mathbf{c}_1\,d\mathbf{c}_2\,d\mathbf{r}_1\,d\mathbf{r}_2$ is the probability of finding a pair of particles in the volume $d\mathbf{r}_1\,d\mathbf{r}_2$ centered on points $\mathbf{r}_1$, $\mathbf{r}_2$ and having velocities within the ranges $\mathbf{c}_1$ and $\mathbf{c}_1 + d\mathbf{c}_1$ and $\mathbf{c}_2$ and $\mathbf{c}_2 + d\mathbf{c}_2$.

The binary collision space of particle P_1 (diameter d_1) and particle P_2 (diameter d_2) is as shown in Figure 2.16 in which the particle P_1 is fixed. The particle P_2 approaches

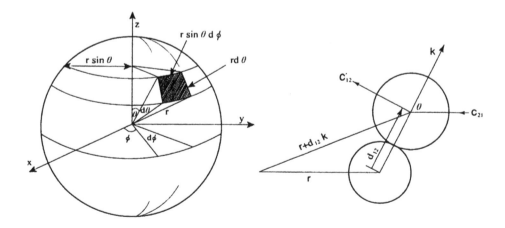

Figure 2.16 The space of binary collision of the spheres of diameters d_1 and d_2.

with the relative velocity c_{21} and collides with the particle P_1 at a point whose direction vector is k, where:

$$k = h\cos\theta + i\sin\theta\cos\phi + j\sin\theta\sin\phi \tag{2.58}$$

in which h, i and j are the unit vectors for the directions z, x and y, respectively. As the distance from the center of P_1 to the collision point is $d_{12} = (d_1 + d_2)/2$, the element of collision surface area that is shown by the shaded area in Figure 2.16, is:

$$dS = d_{12}^2 \sin\theta\, d\theta\, d\phi \tag{2.59}$$

The space within which particles approaching on this area are involved (collision cylinder) has the volume:

$$V = c_{21}\cos\theta \cdot d_{12}^2 \sin\theta\, d\theta\, d\phi \tag{2.60}$$

Consequently, the number of binary collisions per unit time per unit volume N_{12} is given as:

$$N_{12} = \iiint_{c_{12}\cdot k > 0} f_p^{(2)}(c_1, r_2 - d_{12}k, c_2, r_2)c_{12} \cdot k d_{12}^2\, dk\, dc_1\, dc_2 \tag{2.61}$$

where $dk = \sin\theta\, d\theta\, d\phi$ is the solid angle.

The mean value change of collisional property $<\psi_c>$ given by Equation (2.57) becomes, from Equation (2.61):

$$\langle\psi_c\rangle = \frac{1}{2}\iiint (\psi_2' + \psi_1' - \psi_2 - \psi_1)f_p^{(2)}d_p^2 c_{12} \cdot k\, dk\, dc_1\, dc_2 \tag{2.62}$$

where prime means the quantity after collision and for simplicity particles are assumed to have a uniform diameter, d_p.

The integrals of Equation (2.62) must be evaluated separately for P_1 and P_2. The value of $f^{(2)}$ at a distance of $d_p k$ is evaluated by means of a Taylor series as:

$$f^{(2)}(r - d_p k, r) = f_p^{(2)}(r, r + d_p k) - d_p k\nabla \cdot f^{(2)}(r, r + d_p k) \tag{2.63}$$

Consequently, the mean value of the collisional integral can be written as:

$$\langle\psi_c\rangle = -\nabla \cdot P_c + N_c \tag{2.64}$$

where

$$P_c = -\frac{d_p^3}{2}\int_{c_{12}\cdot k} (\psi_1' - \psi_1)(c_{12} \cdot k)k f_p^{(2)}\, dk\, dc_1\, dc_2 \tag{2.65}$$

and:

$$N_c = -\frac{d_p^3}{2}\int_{c_{12}\cdot k} (\psi_2' + \psi_1' - \psi_2 + \psi_1)(c_{12} \cdot k)k f_p^{(2)}\, dk\, dc_1\, dc_2 \tag{2.66}$$

Substituting:

$$\psi = m\mathbf{c} \tag{2.67}$$

into Equations (2.65) and (2.66), gives:

$$\mathbf{P}_c = -\frac{d_p^3}{2}\int_{\mathbf{c}_{12}\cdot\mathbf{k}} (m\mathbf{c}' - m\mathbf{c})(\mathbf{c}_{12}\cdot\mathbf{k})\mathbf{k}f_p^{(2)}\,d\mathbf{k}\,d\mathbf{c}_1\,d\mathbf{c}_2 \tag{2.68}$$

$$\mathbf{N}_c = 0 \tag{2.69}$$

In the case of the inelastic collision of equal mass particles, the change in velocity is given by the following equation, in which e is the coefficient of restitution:

$$\mathbf{c}_1' - \mathbf{c} = -\frac{1}{2}(1+e)(\mathbf{k}\cdot\mathbf{c}_{12})\mathbf{k} \tag{2.70}$$

The integration of Equation (2.68), considering Equation (2.70) and that the granular temperature T is $(1/3)\mathbf{C}_t^2$, finally yields:

$$\mathbf{P}_c = 2\sigma C^2 g_0(1+e)T\mathbf{I} - \frac{4\sigma C^2 g_0(1+e)}{3\sqrt{\pi}}\sqrt{T}\nabla\cdot\mathbf{v}\mathbf{I} - \frac{8\sigma C^2 g_0(1+e)}{5\sqrt{\pi}}\sqrt{T}\nabla^s\mathbf{v} \tag{2.71}$$

where $\mathbf{I}$ is the unit tensor, $\nabla^s\mathbf{v}$ is the deformation tensor, and g_0 is the radial distribution function defined as:

$$\lambda = g_0 - 1 \tag{2.72}$$

The first term on the right-hand side of Equation (2.71) is pressure, the second term is the viscous stress due to volume change and the third is the shearing stress.

2.4.2 Kinetic stress

Gidaspow (1994) discussed the kinetic stress at a dilute condition (g_0 is nearly equal to 1) by analogy with the mixing length theory in fluid mechanics. Takahashi and Tsujimoto (1997) extended his theory to a dense condition in a plain shear flow.

If an equilibrium state, where the velocity distribution function is independent of time and space is assumed, the velocity distribution function will be expressed by a Gaussian distribution as follows:

$$f_p = \frac{n}{(2\pi T)^{3/2}}\exp\left[-\frac{C_t^2}{2T}\right] \tag{2.73}$$

The mean value of the fluctuation velocity is:

$$\langle C_t \rangle = \frac{1}{(2\pi T)^{3/2}}\int_0^\infty C_t \exp\left[-\frac{C_t^2}{2T}\right]dC_t = \sqrt{\frac{8T}{\pi}} \tag{2.74}$$

In the case of a vertical, two-dimensional flow, the transport of a physical quantity in the z direction is given by:

$$Q = -l'\langle C_t \rangle \frac{db}{dz} \tag{2.75}$$

where b is the physical quantity, Q is the mean of the transportation rate, and l' is the distance of transportation.

The momentum σu is considered as the transported quantity, the rate of momentum transportation due to the fluctuation velocity in the z direction; i.e. the shear velocity, is given by:

$$\tau_k = \sigma l' \langle C_t \rangle \frac{du}{dz} \tag{2.76}$$

where τ_k is the kinetic stress.

The distance of transportation is the mean free travel distance before collision and it is related to the time before collision t_F as follows:

$$l' = \langle C_t \rangle t_F \tag{2.77}$$

The value of t_F is obtained by dividing the particle number in the unit volume by the collision frequency given in Equation (2.61). Consequently we obtain:

$$\tau_k = \frac{\sigma d_p}{3g_0} \sqrt{\frac{T}{\pi}} \frac{du}{dz} \tag{2.78}$$

Figure 2.17 shows the comparison of shearing stresses due to collision of particles calculated by the third term on the right-hand side of Equation (2.71) with that due to the random motion of particles calculated by Equation (2.78). The stress due to

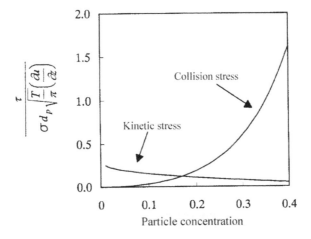

Figure 2.17 Collision and kinetic stresses versus particle concentrations in granular flow.

collision; τ_c, exceeds τ_k when the particle concentration is larger than about 15%, if $C_* = 0.65$ and $e = 0.85$ as is usual in the flow of sand in air.

2.4.3 Skeletal stresses

In the laminar flow as shown in Figure 2.14 the granular temperature is small and the enduring grain contact plays the dominant role in transmitting stresses in the flow; i.e. skeletal stresses. These stresses are not the dynamic stresses caused by the micro-behaviors of individual particles but the quasi-static stresses transmitted by a macro-structure of the constituents. The quasi-static pressure and shearing stress can operate only when the grain concentration is larger than a threshold value C_3 and less than another threshold value C_2, which are, as mentioned earlier, for natural beach sand 0.51 and 0.56, respectively (Bagnold 1966).

As explained in the discussion of the Coulomb mixture theory, the skeletal stresses are independent of the velocity gradient. In the case of a granular flow in air the effects of interstitial fluid are negligible, so that the skeletal stresses at height z in a steady uniform flow of depth h on a slope of gradient θ are written as:

$$p_s = \alpha_e \overline{C} \sigma g (h - z) \cos \theta \tag{2.79}$$

$$\tau_s = p_s \tan \phi \tag{2.80}$$

where $\overline{C}$ is the mean grain concentration above the height z and α_e is the ratio of the quasi-static skeletal pressure to the total pressure. This is assumed to be:

$$\alpha_e = \begin{cases} 0 & ; \quad \overline{C} \leq C_3 \\ \left(\dfrac{\overline{C} - C_3}{C_* - C_3} \right)^m & ; \quad C_3 \leq \overline{C} \leq C_* \\ 1 & ; \quad C_* \leq \overline{C} \end{cases} \tag{2.81}$$

The exponent m is unknown at present.

2.4.4 Constitutive relations

Based on the previous discussion, the constitutive relations for the planar flow as shown in Figure 2.18 are summarized as follows:

$$\tau = \tau_c + \tau_k + p_s \tan \varphi \tag{2.82}$$

$$p = p_c + p_s \tag{2.83}$$

where, from Equation (2.71):

$$\tau_c = \frac{4}{5} C^2 g_0 \sigma d_p (1 + e) \sqrt{\frac{T}{\pi}} \left(\frac{\partial u}{\partial z} \right) \tag{2.84}$$

$$p_c = 2 \sigma C^2 g_0 (1 + e) T \tag{2.85}$$

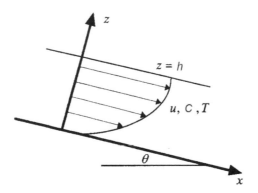

Figure 2.18 Planar steady uniform granular flow.

τ_k is given by Equation (2.78) and p_s is given by Equation (2.79). The second term on the right-hand side of Equation (2.71) is neglected because the flow is incompressible.

2.4.5 Application of the theory to dry granular flow

Consider a flow as shown in Figure 2.18, then the balance of stress equations are:

$$\frac{\partial \tau}{\partial z} + \rho_T g \sin \theta = 0 \tag{2.86}$$

$$\frac{\partial p}{\partial z} + \rho_T g \cos \theta = 0 \tag{2.87}$$

The non-dimensional expressions of these equations are:

$$\frac{\partial}{\partial Z}\left\{(f_2 + f_{22})\frac{\sqrt{T^*}}{\sqrt{AB}}\frac{\partial u^*}{\partial Z} + \frac{\alpha_e A \overline{C}}{B}(1 - Z)\tan \phi\right\} = -AC \tag{2.88}$$

$$\frac{\partial}{\partial Z}\{f_1 T^* + \alpha_e A \overline{C}(1 - Z)\} = -AC \tag{2.89}$$

where $Z = z/h$, $u^* = u/(gh \sin \theta)^{1/2}$, $T^* = T/gd_p \cos \theta$, $f_2 = 8\eta C^2 g_0/(5\sqrt{\pi})$, $f_{22} = 1/(3g_0\sqrt{\pi})$, $f_1 = 4\eta C^2 g_0$, $\eta = (1 + e)/2$, $A = h/d_p$, and $B = \tan \theta$.

The unknowns in Equations (2.88) and (2.89) are u^*, C and T^*, therefore, one equation is still missing. That is the equation giving granular temperature. Granular temperature represents the vibration energy of particles, therefore, to obtain it the particle's kinetic energy $(1/2)mc^2$ is substituted into ψ of Equation (2.57). Consequently, we obtain the following equation (Jenkins and Savage 1983; Gidaspow 1994):

$$\frac{\partial}{\partial z}\left(\kappa \frac{\partial T}{\partial z}\right) + (\tau_c + \tau_k)\frac{\partial u}{\partial z} - \gamma = 0 \tag{2.90}$$

where κ is the coefficient of thermal conductivity of the grain temperature and γ is the dissipation rate of the granular temperature due to inelastic collision. They are given by:

$$\kappa = 2\sigma C^2 d_p (1 + e) g_0 \sqrt{(T/\pi)} \tag{2.91}$$

$$\gamma = (12/d_p)(1 - e^2)C^2 g_0 \sigma T \sqrt{(T/\pi)} \tag{2.92}$$

The non-dimensional expression of Equation (2.90) is:

$$\frac{\partial}{\partial Z}\left(f_3 \sqrt{T^*}\frac{\partial T^*}{\partial Z}\right) + AB(f_2 + f_{22})\sqrt{T^*}\left(\frac{\partial u^*}{\partial Z}\right)^2 - A^2 f_5 T^{*3/2} \tag{2.93}$$

where $f_3 = 4\eta C^2 g_0/\sqrt{\pi}$ and $f_5 = 48\eta(1 - \eta)C^2 g_0/\sqrt{\pi}$.

The velocity, grain concentration and granular temperature of granular flow are obtained by solving Equations (2.88), (2.89) and (2.93) simultaneously, but because Equation (2.93) contains a diffusion term, this is not an easy task. Hence, as the first approximation, we assume that the first term on the left-hand side of Equation (2.90) is negligible and τ is equal to τ_c. Then, Equation (2.93) becomes:

$$T^* = \frac{Bf_2}{Af_5}\left(\frac{\partial u^*}{\partial Z}\right)^2 \tag{2.94}$$

Integration of Equation (2.88) from Z to the surface $Z = 1$ gives the following velocity distribution equation:

$$(f_2 + f_{22})\frac{\sqrt{T^*}}{\sqrt{AB}}\frac{\partial u^*}{\partial Z} + \frac{\alpha_e A \overline{C}}{B}(1 - Z)\tan\phi = A\int_Z^1 C dZ + \tau|_{Z=1} \tag{2.95}$$

where $\tau|_{Z=1}$ is the shear stress at the surface of flow.

Substituting Equation (2.95) into Equation (2.94) and approximating the grain concentration by the average concentration, we obtain the following granular temperature distribution:

$$T^* = \frac{B}{f_2 + f_{22}}\sqrt{\frac{f_2}{f_5}}\left\{A\overline{C}\left(1 - \frac{\alpha_e}{B}\tan\phi\right)(1 - Z) + \tau|_{Z=1}\right\} \tag{2.96}$$

The concentration distribution is obtained by transforming Equation (2.89) as:

$$\frac{\partial C}{\partial Z} = -\frac{1}{(\partial f_1/\partial Z)T^*}\left(CA + f_1\frac{\partial T^*}{\partial Z} - \alpha_e \overline{C}A\right) \tag{2.97}$$

The specific distribution curves of velocity, particle concentration and granular temperature, given by Equations (2.95), (2.96) and (2.97), require the boundary conditions at the surface and bottom of flow. Concerning the condition at the surface, $\tau|_{Z=1}$, we define the surface as the plane connecting the centers of particles at the uppermost layer of flow, and the particles are assumed to distribute as shown in Figure 2.6. Because one particle is contained in the area $b^2 d_p^2$, the body force operating on a unit area, i.e. $\tau|_{Z=1}$ is:

$$\tau|_{Z=1} = \frac{\pi}{6}\left(\frac{C_*}{C}\right)^{2/3} \tag{2.98}$$

According to our laboratory experiments, granular flows can be classified into the laminar, the laminar/dispersed and the dispersed types depending on channel gradient and conditions on the bottom. In the laminar-flow-type, the particle concentration exceeds 50% and the particles flow in an orderly manner. The laminar/dispersed-type appears on steeper slopes than the laminar-flow case, in which the randomness of particle velocity is conspicuous in the upper layer. On far steeper slopes particles vigorously exchange their positions due to collisions and the flow is the dispersed-type. In the case of the laminar/dispersed and the dispersed flows, slip arises on the bottom of flow.

The particle number N_b per unit area in the layer adjacent to the bottom is:

$$N_b = \frac{1}{d_p^2(C_*/C)^{2/3}} \tag{2.99}$$

and the frequency of collisions per unit time t_c, that is equal to $<c>/s_p$, is given using the definition of T given in Equation (2.53) by:

$$t_c = \frac{\sqrt{3T}}{d_p\{(C_*/C)^{1/3} - 1\}} \tag{2.100}$$

If the slip velocity on the bed is u_{sl}, the momentum change in the direction of flow per one collision is $\varphi' \pi \sigma d_p^3 u_{sl}/6$. Hence, considering that the shear stress is equal to the momentum change per unit area and per unit time, the following formula is obtained:

$$\frac{\varphi'\sqrt{3}\pi\sigma Cg_0\sqrt{T}u_{sl}}{6C_*} = \sigma f_1 T \tan\theta \tag{2.101}$$

The non-dimensional expression of this formula is:

$$u_{sl}^* = \frac{f_1 f_{88}}{\varphi'}\sqrt{\frac{B}{A}T^*} \tag{2.102}$$

where $f_{88} = 2\sqrt{3}C_*/(\pi Cg_0)$, and φ' is a coefficient representing the momentum change rate by the collision between the particle and the bed, $0 < \varphi' < 1$. The larger the φ' value, the larger the momentum loss by the collision and the smaller the slip velocity becomes. The φ' values in our experiments using polystyrene particles are 0.042 when the bottom is polyvinyl chloride board, 0.12 when the bottom is a rubber sheet and 0.50 when polyvinyl chloride cylinders of 2.9 mm in diameter are compactly stuck on the bottom perpendicular to the flow direction.

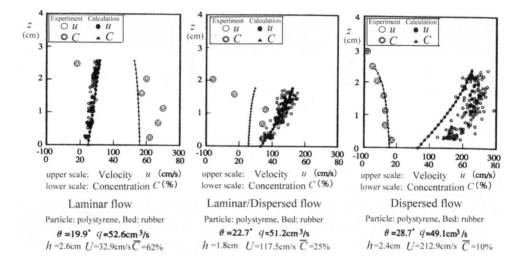

Figure 2.19 Theoretical and experimental velocity and particle concentration distributions.

Applying these boundary conditions we can now determine the velocity, concentration and granular temperature distribution forms. To verify the theory explained above, we carried out the experiments. A steel flume 10 cm in width, 5 m in length of which one side is a transparent glass wall was used. The bed condition was changed as described above. The experimental material was polystyrene particles whose mean diameter was 4.51 mm, density was 1.052 g/cm³, and the internal friction angle was 25.0°. The motion of particles was traced via the transparent wall using a high speed video (200 frames/s) camera. Analyzing the position of each particle in every 1/100 second using the video record, the velocities u in the main flow direction and w perpendicular to it were measured, and using these data the granular temperature $T = (1/3)\{(u - U)^2 + (w - W)^2\}^{1/2}$ was calculated, where U and W are the averages of u and w, respectively.

Figure 2.19 compares the velocity and particle concentration distributions by the experiments with those calculated by the theory in the respective cases of laminar, laminar/dispersed and dispersed flows produced by changing the slope of flume. The calculation was done under the conditions of $C_* = 0.65$, $C_3 = 0.51$ and m in Equation (2.81) is 0.5. For the laminar and laminar/dispersed-types, the theoretical velocities and their distributions coincide very well with the experimental results. For the dispersed-type, although the absolute values of velocities differed due to the too small estimation of slip velocity, the velocity distribution pattern is well explained. The characteristic velocity pattern in the laminar flow is linear and the velocity gradient is small, but in the laminar/dispersed and dispersed flows it is parabolic and the velocity gradient is large.

In the calculation of concentration distributions, the concentration at the bottom is determined by trial and error so as to fit to the measured average transport concentration. The calculated concentration distributions for all type flows are almost uniform. This tendency agrees with the experiments in which no conspicuous vertical changes in concentrations were observed except in the neighborhood of the surface.

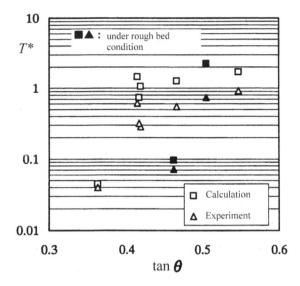

Figure 2.20 Non-dimensional granular temperature versus slope.

Especially in the dispersed flow, the theoretical concentration distribution fits very well to the experiment. For the laminar/dispersed and the laminar flow cases, the degree of agreement in the middle and lower layers is also rather good, but it is not good near the surface of the flow. This discrepancy is caused by defects in the determination of boundary condition at the surface of the flow and in the approximation of granular temperature.

Figure 2.20 shows the granular temperatures averaged within the entire depth. Although the calculated values are a little larger than the experimental values for higher granular temperature; i.e. in the dispersed and laminar/dispersed flows, the degree of agreement in the whole range is judged to be satisfactory. According to Equation (2.96), the granular temperature should linearly decrease from the bottom to the surface, but in the experiments it tends to be uniformly distributed or increases upward. This discrepancy may be caused by the neglect of the diffusion term in Equation (2.90).

To what extent particles can disperse by the repulsion of inter-particle collision is important not only in determining particle concentrations in subaerial granular flows such as pyrocalstic flows and snow avalanches, but also in classifying the stony debris flow and the immature debris flow.

If the slope is flatter than a threshold value θ_1, which is determined by the internal friction angle and the condition of the bed (smooth or rough), the flow becomes the laminar-type and the macro-concentration, $\overline{C}$, irrespective of the slope, has nearly a constant value of C_{ct}. The value of θ_1 is, for a smooth bed $\theta_1 \approx (\varphi - 5°)$, and for a rough bed $\theta_1 \approx \varphi$.

If the slope is steeper than θ_1, dynamic stresses predominate in the flow, and the shear stress is related to pressure as:

$$\tau = p \tan \alpha_i \tag{2.103}$$

Considering $\tau = \tau_c + \tau_k$ and $p = p_c$, and substituting Equations (2.78), (2.84) and (2.85), we obtain:

$$\tan\alpha_i = \sqrt{\frac{12(1-e)}{5\pi} + \frac{1-e}{\overline{C}^2 g_0(1+e)\pi}} \tag{2.104}$$

For granular temperature, different from the assumption we used to obtain Equation (2.94), τ_k is also taken into account, and from Equation (2.90) we obtain:

$$T = \frac{d_p^2(f_2 + f_{22})}{f_5}\left(\frac{\partial u}{\partial z}\right)^2 \tag{2.105}$$

The value $\tan\alpha_i$ corresponds to a kinetic friction coefficient and, for the steady uniform granular flow, it should be almost equal to $\tan\theta$. Therefore, modifying Equation (2.104) a little, we can give the relation between the slope and $\overline{C}$ as follows:

$$\begin{cases} \overline{C} = C_{ct} & ;\quad \theta \leq \theta_1 \\ \theta = \tan^{-1}\sqrt{\dfrac{12(1-e)}{5\pi} + \dfrac{1-e}{\overline{C}^2 g_0(1+e)\pi}} + \theta_1 - \varphi_{sp}\ ; & \theta \geq \theta_1 \end{cases} \tag{2.106}$$

where for a rough bed, $\theta_1 = \varphi$; for a smooth bed, $\theta_1 = (\varphi - 5°)$, and φ_{sp} is the angle of repose for spherical particles that is nearly equal to $26°$.

Figure 2.21 compares Equation (2.106) with experimentally obtained concentrations. We can conclude that Equation (2.106) rather well explains the relationship between slope gradients and particle concentrations. From Figure 2.21, we can see that on a rough bed flatter than $40°$, particle concentration cannot be thinner than 0.2.

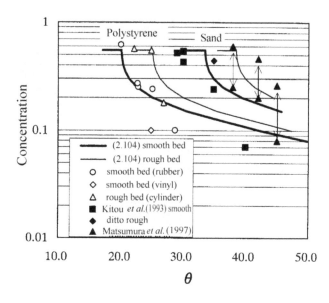

Figure 2.21 Particle concentration versus slope in subaerial granular flows.

2.4.6 Comparison with other constitutive relations for inertial range

Many constitutive relations for inertial granular flows under dry or wet conditions have been proposed based on experiments and theoretical considerations. Among them are those that obtained results from numerical experiments by computer simulations using the discrete element method. In Japan, the constitutive relations proposed by Tsubaki *et al.* (1982) and Miyamoto (1985) have been widely used.

Based on their own experiments on debris flows Tsubaki *et al.* (1982) proposed a model where the particle collision does not end instantly but a rubbing motion continues for a while when a particle overtakes the other particle after a collision. They also considered multiple particle collisions, in which a particle after colliding with the other particle moves together with the surrounding particles contacting with each other and the momentum change due to collision is transmitted to these surrounding particles. Moreover, an isotropic stress operates during the enduring contact motion that is related to a part of the submerged weight of particles in excess of the particle collision stress. The model finally yielded the following equations:

$$\tau_c = K_M \sigma d_p^2 \frac{(C/C_*)^2}{1 - C/C_*} \left(\frac{\partial u}{\partial z}\right)^2 \tag{2.107}$$

$$p_c = \frac{1}{\alpha_t}\tau + K_p \frac{C}{C_*}\frac{C - C_s}{C_s} \tag{2.108}$$

where $K_M = (\pi/6)(0.0762 + 0.102\mu_k)\{C_*/(\pi/6)\}^2 k_M$, $K_p = C_s\rho(\sigma/\rho - 1)gh\cos\theta\chi$, $\alpha_t = \gamma_t/\{1 + \rho/(2\sigma)\}$, $\gamma_t = (0.0762 + 0.102\mu_t)/(0.0898 - 0.067\mu_k)$, C_s is the particle concentration at the surface, μ_k is the kinetic friction coefficient between particles (≈ 0.1), k_M is an empirical coefficient (≈ 7.5), and χ is an empirical coefficient ($\approx 1/3$).

Miyamoto (1985) discussed the energy loss in a binary particle collision between the two particles respectively embedded in adjacent above and below layers. Here, the arrangement of particles are like shown in Figure 2.6 (in the case of Figure 2.6 the distance between the above and the below particles is not necessarily the same as that between the back and the front particles, but Miyamoto assumes the same distances.). By his model the energy loss per one collision is:

$$\phi_{di} = \frac{1}{2}(1 - e^2)\left(\frac{\pi}{6}\sigma d_p^3\right)(\delta u \sin \alpha_i)^2 \tag{2.109}$$

The frequency of collision in unit time is $\delta u/(b d_p)$ and the volume occupied by a particle is $(b d_p)^3$. Therefore, the energy loss by particle collisions in unit time per unit volume is:

$$\Phi e = \frac{\pi}{12}(1 - e^2)\sin^2\alpha_i\sigma\frac{d_p^2}{b}\left(\frac{\partial u}{\partial z}\right)^3 = k_d(1 - e^2)\sigma C^{1/3}d_p^2\left(\frac{\partial u}{\partial z}\right)^3 \tag{2.110}$$

where $k_d = (\pi/12)(\pi/6)^{-1/3}\sin^2\alpha_i$.

Because the relation between shear stress and energy loss can be written as:

$$\Phi e = \tau \frac{\partial u}{\partial z} \tag{2.111}$$

the shear stress due to inter-particle collision is written as:

$$\tau_c = k_d(1 - e^2)\sigma C^{1/3} d_p^2 \left(\frac{\partial u}{\partial z}\right)^2 \tag{2.112}$$

If the energy that is conserved throughout the process of collision is pressure, then, it is written as:

$$p_c = k_d e^2 \sigma C^{1/3} d_p^2 \left(\frac{\partial u}{\partial z}\right)^2 \tag{2.113}$$

From Equations (2.112) and (2.113) we obtain:

$$\tan \alpha_i = \frac{\tau_c}{p_c} = \frac{1 - e^2}{e^2} \tag{2.114}$$

Previous investigations agree in respect that both τ_c and p_c are proportional to $(\partial u/\partial z)^2$. The coefficients of proportion in respective investigations are, however, different. If we write τ_c as:

$$\tau_c = f_i \sigma d_p^2 \left(\frac{\partial u}{\partial z}\right)^2 \tan \varphi \tag{2.115}$$

The values of f_i from some previous investigations are shown in Figure 2.22. Close agreement of the curve of Tsubaki *et al.* (1982) to Bagnold's is due to their selection of numerical coefficients to fit to Bagnold's experimental results. In Miyamoto's (1985) calculation of Equation (2.112) $e = 0.8$ and $(\pi/6)^{-1/3}C^{1/3} \approx (C/C_*)^{1/3}$ are applied. In the case of Takahashi and Tsujimoto's (1997) the formula τ is calculated as the sum of τ_c and τ_k .

Figure 2.23 is the one prepared by Campbell (1990) arranging his own and others' results of computer simulations and the results of shear-cell experiments on glass spheres by Savage and Sayed (1984). To compare Figure 2.23 with Figure 2.22 one must multiply the scale of the abscissa by about 1.67 ($\approx 1/C_*$) and the scale of ordinate by about 3.1 ($\approx 1/\tan \alpha_i$). Note, here, that Bagnold's curve in Figure 2.22 represents his experimental results.

Comparing Figures 2.22 and 2.23, we can conclude that the respective constitutive relations except for Miyamoto's one represent the similar tendencies and agree rather well with experimental results. Especially, Takahashi and Tsujimoto's curve for $e = 0.85$ which has a similar tendency to the experiments shown in Figure 2.23 in respect of the rate of decrement of f_i, which slows down with decreasing concentration. This characteristic tendency is caused by the increase of the relative importance of the kinetic stress rather than the collision stress in the smaller concentration region. The large discrepancy in f_i between Miyamoto's and others is probably due to the

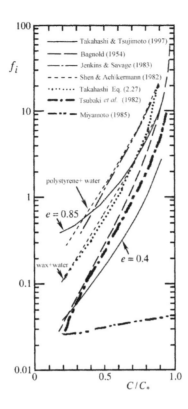

Figure 2.22 Comparison of f_i in various constitutive equations.

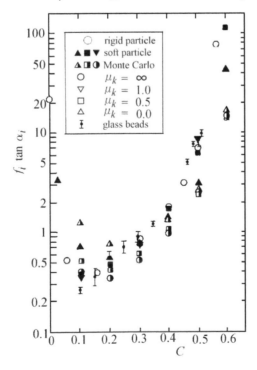

Figure 2.23 The f_i values in various experiments (Campbell 1990).

oversimplification of the energy calculation by Miyamoto. Bagnold developed a similar discussion but finally he introduced an empirical f_i value to match the experimental tendency that it was asymptotic to an infinitely large value at large concentration.

2.5 MECHANICAL CLASSIFICATION OF DEBRIS FLOWS REVISITED BASED ON THE THEORY FOR GRANULAR FLOWS

The high mobility of a debris flow is caused by the enlarged particle distances compared to those in the rigid state before mobilization. The space between particles in debris flow is filled with water or slurry, so the resistance to flow in that space originates from the small shearing resistance of the interstitial fluid. Therefore, as long as the particles are dispersed in the entire body, the resistance to flow should be far smaller than that in the quasi-static grain flow. The particles are, however, heavier than the ambient fluid, so they tend to settle down onto the bed; to maintain the flow, confronting the settling, some particle dispersing actions should work. This work consumes energy and the resistance corresponding to this is added to the resistance to flow. Because the sum of the shearing resistances of the interstitial fluid and to disperse the particles is less than the Coulomb friction resistance of sliding mass, a debris flow has a larger mobility than a landslide as shown in Figure 1.2.

The energy consumption to overcome the resistance to flow is compensated by the decrease in the potential energy along the ravine. If the flow is comprised of plain Newtonian fluid, no yield strength exists and flow continues as long as the channel is inclined, however mildly. However, in the case of debris flow, even if the mixture of particles and liquid is considered as a Newtonian fluid, sufficient energy to disperse heavy particles must be continuously supplied. Therefore, debris flow cannot arrive to a flatter area than a threshold inclination.

As described in section 1.4, there are four mechanisms to sustain particles within the flow. The first mechanism arises when the solids concentration is very dense, in which the particles in the shearing layer are moved in enduring contact with each other. Such motion is possible when the particle concentration in volume, C, is in between the two threshold values of C_2 and C_3. The shear stress at height z in this motion is given as:

$$\tau_s = (\sigma - \rho)gC(h - z)\cos\theta\tan\varphi \qquad (2.116)$$

This stress is independent of shearing velocity.

The second mechanism is that due to particle collisions which arise when C is less than C_3. According to Takahashi and Tsujimoto (1997), as described in section 2.4, the total shearing stress and the pressure are given as:

$$\tau_c + \tau_k = \sqrt{\frac{1}{15\pi}\frac{1}{\sqrt{1-e}}}\left\{\frac{4}{5}(1+e)C^2g_0 + \frac{1}{3g_0}\right\}\sigma d_p^2\left(\frac{du}{dz}\right)^2 \qquad (2.117)$$

$$p_c = \frac{2}{15}C^2g_0\frac{1+e}{1-e}\sigma d_p^2\left(\frac{du}{dz}\right)^2 \qquad (2.118)$$

The total submerged weight of particles above a shearing plane is supported by the repulsive pressure acting on that plane. Therefore, the total load of particles is transmitted to the bed through inter-particle collisions and the pressure within the interstitial fluid is almost equal to the hydrostatic pressure for the plain liquid flow of the same depth as the debris flow. The dispersive pressure due to inter-particle collisions decreases with decreasing particle concentration, and if the concentration becomes less than a threshold value particles can no longer be dispersed in the entire depth of flow. Here, this threshold value is described as C_4.

The third mechanism to sustain particles in flow is the suspension due to the large-scale turbulence in flow. This mechanism will arise when C is less than C_3 and the upward turbulent velocity in the flow is larger than the settling velocity of the particles. Because the settling velocity becomes greater the larger the particle diameter is, this mechanism will predominate when the comprising particle diameter is small. The action to suspend particles originates from the stress in the interstitial fluid and the suspended particles affect the stress and vice versa, so, in this case, we cannot separate the stress to suspend particles from the entire stress within the interstitial fluid. The existence of suspended particles works to increase the apparent density of the interstitial fluid and the suspended particle's load is sustained by the increased pressure in the fluid phase. If all the particles in the flow are suspended, the shearing stress in the flow is given as:

$$\tau_t = \rho_T l^2 \left(\frac{du}{dz}\right)^2 \tag{2.119}$$

where l is the turbulent mixing length and ρ_T is the apparent density of the entire debris flow material that is given as:

$$\rho_T = (\sigma - \rho)C + \rho \tag{2.120}$$

The fourth mechanism arises when the viscosity of the interstitial fluid is large. When the particles embedded in the adjacent shearing surfaces approach or recede, the ambient fluid is necessarily moved to cause a resistance to shearing that is additional to the viscose resistance of the fluid and it causes the action to disperse particles. Moreover, if the solids concentration is larger towards the bottom of the flow, the apparent viscosity of the virtual fluid (the mixture of liquid and particles) is larger towards the bottom, so the upward particle motion becomes easier than the motion towards the bottom that causes an equalization of concentration distribution. Thus, although the flow as a whole is laminar, a particle dispersing action works. As in the case of turbulent suspension, the particle dispersing stress and the fluid stress cannot be estimated separately. In this case, the shearing stress in the flow is given as:

$$\tau_\mu = \mu_a \frac{du}{dz} \tag{2.121}$$

where μ_a is the apparent viscosity of the virtual fluid comprised of the mixture of fluid and particles. According to Krieger (1972) it is given as:

$$\mu_a = \left(1 - \frac{C}{C_*}\right)^{-1.82} \mu \tag{2.122}$$

where μ is the viscosity of the plain fluid.

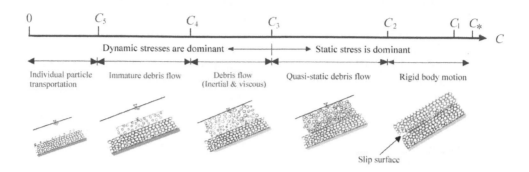

Figure 2.24   Classification of solids motions based on the concentration.

The above consideration reveals that the particle supporting mechanism governs the dominant stresses in flow and, consequently, they determine the characteristics of flow. Because the solids concentration in flow is one of the major controlling factors in the mechanism to sustain particles, the characteristic particle transporting phenomena can be classified by the value of C. Figure 2.24 illustrates the difference in the manners of solids motion for the different solids concentrations when the viscosity of the interstitial fluid is small. The horizontal axis in Figure 2.24 is the same as the vertical axis in Figure 1.12.

In Figure 2.24, C_* is the maximum packing concentration of the solids; C_1 is the concentration of solids in the sediment accumulation after shearing motion. According to Bagnold (1966) $C_* = 0.644$ and $C_1 = 0.604$ for natural beach sand. The values C_2 and C_3 for the same natural beach sand are given by Bagnold, as mentioned earlier, as 0.56 and 0.51, respectively. Takahashi (1982) gave the approximate values $C_4 = 0.2$ and $C_5 = 0.02$ based on his experiments. The immature debris flow that appears when C is in between C_4 and C_5 cannot disperse the particles in the entire depth of flow but the particle mixture layer appears only in the lower part of the flow. The upper part is the plain water flow or the turbid flow suspending only fine particles. The particle concentration in the lower mixture layer is approximately equal to C_4. The configuration of immature debris flow somewhat resembles the individual particle motion that appears when C is less than C_5, so it is sometimes called 'bed-load-like massive flow'. It must be noted that, in the individual particle transportation, the particles are moved by the dynamic drag force caused by the difference in the velocities between the particles and ambient fluid, in the immature debris flow the velocity difference is minimal and the particles and the fluid are altogether driven by gravity. No matter how it looks like bed load transportation, an immature debris flow belongs to the sediment gravity flow that is categorized as debris flow.

In the quasi-static debris flow, no dynamic stress operates but the static Coulomb friction stress operates that is given by Equation (2.116) at height z. Because the driving force acting on this plane is given as: $\rho_t g(h - z)\sin\theta$, the driving force becomes larger than the frictional resistance when the following formula is satisfied:

$$\tan\theta \geq \frac{(\sigma - \rho)C\tan\varphi}{(\sigma - \rho)C + \rho} \qquad (2.123)$$

The flow is possible only when Equation (2.123) is satisfied. If $\sigma = 2.65\,\mathrm{g/cm^3}$, $\rho = 1.0\,\mathrm{g/cm^3}$ and $\tan\varphi = 0.8$ are substituted into Equation (2.123), one understands that flow is possible when θ is steeper than 20–21° under the condition $C_2 > C > C_3$. The large-scale flume experiment at USGS are conducted under the condition $C = 0.6$, $\sigma = 2.7\,\mathrm{g/cm^3}$, $\rho = 1.1\,\mathrm{g/cm^3}$ and $\tan\varphi = 0.839$ (Iverson 1997). If this condition is substituted into Equation (2.123), the quasi-static debris flow is possible for $\theta > 21.4°$. The experimental flume has a rigid bed of 31° in slope gradient, so the condition for the occurrence of quasi-static debris flow is satisfied. Based on their experiments, Iverson and Denlinger (2001) developed a theory for quasi-static debris flow as previously explained in section 2.3.2. They claim that almost all debris flows are in the category of quasi-static debris flow, but, actually, many debris flows can continue their motion without deposition in the channel flatter than 10°. This means that there are many debris flows belonging to the category of dynamic stress dominating ones as shown in Figure 2.24.

There are three dynamic stresses in the dynamic debris flows; τ_c, τ_t and τ_μ. The relative predominance of the respective stresses will determine the characteristics of debris flow. For the sake of evaluating the relative predominance, the ratios of the respective stresses are described as follows:

$$\frac{\tau_c}{\tau_\mu} = \frac{4}{5}\sqrt{\frac{1}{15\pi}}\frac{1+e}{\sqrt{1-e}}C^2\left(1-\frac{C}{C_*}\right)^{1.82}\frac{\{(C_*/C)^{1/3}-1\}^{1/2}}{\{1-(C/C_*)^{1/3}\}}\mathbf{Ba} \tag{2.124}$$

$$\frac{\tau_t}{\tau_\mu} = \left\{\left(\frac{\sigma}{\rho}-1\right)C+1\right\}\left(1-\frac{C}{C_*}\right)^{1.82}\frac{l^2(du/dz)}{\mu/\rho} \tag{2.125}$$

$$\frac{\tau_t}{\tau_c} = \frac{5}{4}\sqrt{15\pi}\frac{\sqrt{1-e}}{1+e}\frac{\{(\sigma/\rho-1)C+1\}\{1-(C/C_*)^{1/3}\}}{C^2}\frac{\sigma}{\rho}\left(\frac{l}{d_p}\right)^2 \tag{2.126}$$

where $\mathbf{Ba}$ is the Bagnold number defined by:

$$\mathbf{Ba} \equiv \frac{\sigma d_p^2(du/dz)}{\{(C_*/C)^{1/3}-1\}^{1/2}\mu} \tag{2.127}$$

According to Bagnold (1954), when $\mathbf{Ba}$ is larger than 450, τ_c is far larger than τ_μ, so the flow is in the inertial range and when $\mathbf{Ba}$ is less than 40, τ_μ is dominant, so the flow is in the viscous range.

The term $l^2(du/dz)/(\mu/\rho)$ in Equation (2.125) can be evaluated as Uh/ν if l is represented by h and du/dz is represented by U/h. Uh/ν is the Reynolds number, Re, where U is the cross-sectional mean flow velocity and ν is the dynamic viscosity of the interstitial fluid. For the open channel water flow, it becomes turbulent flow with the Reynolds number larger than 1,500 and it becomes laminar flow with the Reynolds number less than 500.

Similarly in Equation (2.126), if l is represented by h, (l/d_p) becomes (h/d_p) that is defined as the relative depth of flow.

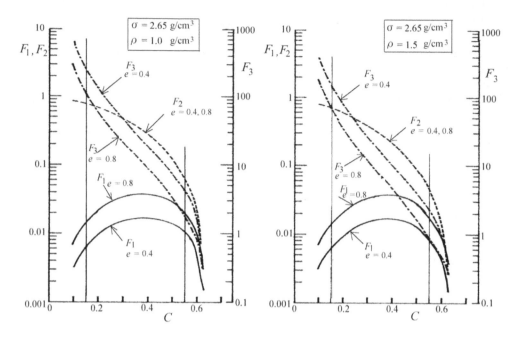

Figure 2.25 F_1, F_2 and F_3 variations versus particles' concentration, specific density, and the coefficient of restitution.

Writing the right-hand side terms of Equations (2.124), (2.125) and (2.126) as $F_1 \mathbf{Ba}$, $F_2 \mathbf{Re}$ and $F_3(h/d_p)^2$, respectively, the variations of F_1, F_2 and F_3 are examined in Figure 2.25. The maximum and minimum values of F_1, F_2 and F_3, respectively in the range of C from 0.15 to 0.55 are obtained from Figure 2.25. Eventually, Equation (2.124) suggests that, when $\mathbf{Ba}$ is larger than 1660~250, τ_c becomes larger than $10\tau_\mu$; meaning that the viscous stress becomes negligible in comparison to the particle collision stress. On the contrary, if $\mathbf{Ba}$ is less than 16~3, τ_c becomes smaller than $0.1\tau_\mu$ and the particle collision stress becomes negligible in comparison to the viscous stress. These two threshold values of the Bagnold number are about the same as the Bagnold results. Similarly, from Equation (2.125), when $\mathbf{Re}$ is larger than 12~167, τ_t becomes larger than $10\tau_\mu$, and when $\mathbf{Re}$ is less than 0.1~2, τ_t becomes smaller than $0.1\tau_\mu$. However, the applicability of Krieger's Equation (2.122) for the slurry comprised of cohesive fine particles should yet be examined carefully. If the apparent viscosity of the debris flow material can be directly estimated, it would be recommendable to use that value and the laminar to turbulent flow transition criterion of $\mathbf{Re}$ would be the same to the case of plain water flow. From Equation (2.126) we obtain that when h/d_p is larger than 0.2~3.5 τ_t becomes larger than $10\tau_c$ and when h/d_p is less than 0.02~0.35 τ_c becomes smaller than $0.1\tau_c$.

The discussion on the transition criterion from turbulent flow to collision stress dominated flow, however, needs other considerations whether the condition to suspend particles is satisfied. It is well known that a particle can be suspended by turbulence if $w_s/u_* < 1$, where w_s is the settling velocity of particle and $u_*(=\sqrt{gh \sin \theta})$ is the shear velocity on the bed. If the particle's diameter is large (more than 1 cm) and the settling velocity within the interstitial fluid is given by the following Newton's formula:

$$w_s = 1.82 \left\{ \left(\frac{\sigma - \rho}{\rho} \right) g d_p \right\}^{1/2} \tag{2.128}$$

The condition $w_s/u_* < 1$ is given as:

$$\frac{h}{d_p} > \frac{3.31(\sigma/\rho - 1)}{\sin \theta} \tag{2.129}$$

If we substitute $\theta = 10°$ and $(\sigma/\rho - 1) = 1.65$ into Equation (2.129), we obtain $h/d_p > 32$. This discussion ignores the effects of group settling and the suppression of turbulence in the densely particle loading flow, but Equation (2.129) indicates that, if the relative depth becomes large, particles will be suspended by turbulence. On the other hand, if Equation (2.129) is not satisfied, particles will not be suspended.

Newton's settling velocity formula (2.128) that is the premise of the above discussion is applicable in a strict sense only for the particles that are larger than 1 cm in diameter and it may be approximately applicable for the particles larger than 1 mm in diameter. However, for the particle less than 1 mm in diameter, the following Stokes' formula becomes more suitable:

$$w_s = \frac{g(\sigma - \rho)d_p^2}{18\mu} \tag{2.130}$$

where μ is the viscosity of the fluid phase. In this case the particle suspending condition that corresponds to Equation (2.129) is:

$$\frac{h}{d_p} > \frac{Re_*(\sigma/\rho - 1)}{18 \sin \theta} \tag{2.131}$$

where $Re_* (= u_* d_p/v)$ is the particle Reynolds number and $v (= \mu/\rho)$ is the dynamic viscosity of the fluid. The substitution of $d_p = 0.05$ cm, $\sigma/\rho = 2.65$ and $\theta = 20°$ into Equation (2.131) indicates that particles will be suspended only if the depth of the flow is larger than 2 cm; i.e. $h/d_p > 40$. Once particles are suspended, the apparent density of the fluid increases, so that the numerator on the right-hand side of Equation (2.131) becomes smaller and the particles become easier to be suspended. The turbulent-muddy-type debris flow arises by such a mechanism when the comprising particles are small.

The discussion in this section comes to a conclusion that the debris flows in which the dynamic stresses predominate can be classified within the area of the ternary diagram shown in Figure 2.26, whose three apexes mean that τ_c, τ_t and τ_μ occupy the hundred percent of the sum of these three stresses, respectively. Its three axes are the Bagnold number, the Reynolds number and the relative depth, respectively. For example, in the debris flow plotted at position A in Figure 2.26, τ_c, τ_μ and τ_t occupy 50%, 20% and 30%, respectively. If one defines the typical stony debris flow as the one in which the particle collision stress is more than 60% of the total of all the three stresses,

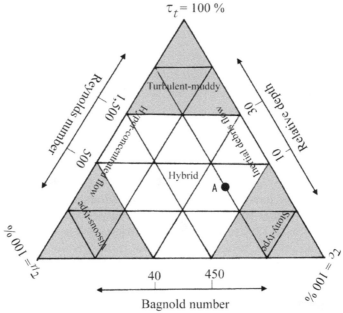

Figure 2.26 Classification of dynamic debris flows by three controlling factors.

the existence domain of the stony debris flow is given as the lower-right shaded area in Figure 2.26. Similarly, the turbulent-muddy-type debris flow exists in the shaded domain at the upper-center area and the viscous-type debris flow exists in the shaded domain at the lower-left shaded area in Figure 2.26. According to the discussion above, the stony-type debris flow appears when the Bagnold number is larger than 450 and the relative depth is less than 10. The turbulent-muddy-type debris flow appears when the relative depth is larger than 30 and the Reynolds number is larger than 1500. The viscous-type debris flow appears when the Reynolds number is less than 500 and the Bagnold number is less than 40. The rest area is the domain of the hybrid-type debris flow that has the characteristics intermediate of these typical three kinds of debris flows. Therefore, the debris flow plotted at position A is a hybrid-type debris flow whose characteristics somewhat resemble the stony-type debris flow.

The debris flows in the Jiangjia gully are comprised of particles representing less than fist sized cobbles. They usually come down to the observation point with many intermittent surges. The forefront of a surge is very much turbulent, but, at a little rear part, it changes to a laminar flow. The relative depth at the rear part distributes a wide range of 10 to 200. The backwards calculation of apparent viscosity by the field observation reveals that it is about 200 Pas. The flow velocity is about 4~8 m/s and the depth of flow is about 1~2 m. Therefore, the Reynolds number is of the order of 200. As shown in Figures 1.12 and 2.3, the mean diameter of the material is about 1 cm and the viscosity of the interstitial fluid is about 1 Pas, thereby the Bagnold number is in the order of 4. This fact confirms the debris flow in the Jiangjia gully is a typical viscous-type debris flow.

The debris flows in the Nojiri River are vigorously turbulent from the forefront to the end. The mean diameter of the comprising particles are around 40 mm and the relative depth of the flow is 30∼50. The fluidity is not much different from that of plain water flow. As the representative values of velocity, flow depth and the viscosity of the interstitial fluid, 10 m/s, 1 m and 0.01 Pas (that is 10 times that of plain water), respectively, may be adopted. Then, the Reynolds number is in the order of 10^6, that confirms the flow is turbulent-muddy.

The debris flows in Kamikamihorizawa have the characteristics to change the particle diameters from the forefront to the rear. At the more or less steady flow part that appears a little later from the passage of the forefront, the relative depth is 2∼10, mean particle diameter is about 30 cm, the flow depth is 60 cm∼3 m and the velocity is around 3∼4 m/s. Therefore, even we adopt the ten times larger viscosity than plain water as the interstitial fluid, the Bagnold number is in the order of 2×10^5. This value confirms the flow is the typical stony-type debris flow.

It was mentioned in section 1.4 that the debris flows whose existing domains are in the neighborhood of the relative depth axis in Figure 2.26; stony, hybrid and turbulent-muddy debris flows, can be called, as a whole, 'inertial debris flow' because the viscous stress in these domains is negligibly small (only the inertial stresses τ_c and τ_t dominate the flow).

It was also mentioned that the flows existing near the axis representing the Reynolds number are often called 'hyper-concentrated flows'.

2.6 THE MECHANISM OF INERTIAL DEBRIS FLOWS

The investigations on inertial debris flows, as will be guessed, started from typical stony debris flows. The dilatant fluid model earlier mentioned is such an example. Then, the models for immature debris flow and for turbulent-muddy debris flow were developed as the two-phase continuum models.

2.6.1 The supplementary explanation of mature stony debris flow

The mechanism of mature stony debris flow was explained in section 2.2.2 using the dilatant fluid model that is a kind of single-phase continuum model. This model is equivalent to the two-phase continuum models because the stresses in the interstitial fluid are negligibly small in the mature stony debris flow.

Similar discussions as before can be developed using the granular flow theory of Takahashi and Tsujimoto (1997) instead of Bagnold's dilatant fluid theory. Namely, from Equations (2.117) and (2.118) the balance of stress equations can be written as follows:

$$P_{ds} = \frac{2}{15}\frac{1+e}{1-e}Cg_0\sigma d_p^2\left(\frac{du}{dz}\right)^2 = (\sigma - \rho)g\cos\theta\int_z^h C\,dz \qquad (2.132)$$

$$T_{ds} = \sqrt{\frac{1}{15\pi} \frac{1}{\sqrt{(1-e)}} \left\{ \frac{4}{5}(1+e)C^2 g_0 + \frac{1}{3g_0} \right\} \sigma d_p^2 \left(\frac{du}{dz}\right)^2}$$

$$= g \sin\theta \int_z^b \rho_t \, dz \tag{2.133}$$

The difference between the combination of Equations (2.132) and (2.133) and the combination of Equations (2.20) and (2.21) is the difference of expression in the coefficients of the terms $\sigma d_p^2 (du/dz)^2$. In Equations (2.20) and (2.21), for the value of $a_i \sin\alpha_i$ the experimentally obtained 0.02 was used. Then, if e is set equal to 0.85 for Equations (2.132) and (2.133), those two combinations of equations become almost identical, meaning that by putting $e = 0.85$ we can get the velocity distribution and the mean cross-sectional velocity almost equivalent to those described by Equation (2.23) and (2.29), respectively. Another equivalent equation to Equation (2.19) is given as:

$$T_{ds} = \frac{1}{2}\sqrt{\frac{15}{\pi}} \frac{\sqrt{(1-e)}}{1-e} \left\{ \frac{4}{5}(1+e) + \frac{1}{3C^2 g_0^2} \right\} P_{ds} \tag{2.134}$$

Physically, the coefficient of P_{ds} in Equation (2.134) is equivalent to $\tan\alpha_i$ in Equation (2.24). However, substitution of $e = 0.85$ into Equation (2.134) yields 0.34~0.54 in the range of $C = 0.5$~0.2 that is somewhat less than the experimentally obtained $\tan\phi$. Some further discussions on the validities of the value of $a_i \sin\alpha_i$ and of the e value that neglects the buffering effect of the interstitial fluid seem to be necessary. However, if we adopt a lower e value than 0.85, the value of the coefficient of P_{ds} in Equation (2.134) that is equivalent to $\tan\alpha_i$ in Equation (2.24) becomes larger.

2.6.2 Immature debris flow

As explained in section 2.4.5, the lowest particle concentration with which particles are dispersed throughout the depth due to the action of inter-particle collision is about 0.2. On the other hand, the equilibrium concentration equation, Equation (2.25), gives a thinner concentration when the channel gradient is flat. In reality, however, under a concentration less than the limit value $C_l (\approx 0.2)$ the particles are no longer dispersed throughout the depth but are concentrated in the lower part of the flow. Consequently, above this particle mixture layer a water layer exists which may contain fine suspended sediment as illustrated in Figure 2.27. Takahashi (1982, 1987) named this kind of flow 'immature debris flow'.

Writing the thickness and the concentration of the particle mixture layer as h_l and C_l, respectively, the shear stress acting on the surface of the bed is given by:

$$T^o = (\sigma - \rho)C_l h_l g \sin\theta + \rho g h \sin\theta \tag{2.135}$$

In an equilibrium flow in which neither erosion nor deposition takes place, the operating shear stress should balance with the resistance stress produced by inter-particle encounters as well as the stress in the interstitial fluid caused by both turbulence and viscous deformation, $T^{\prime o}$. Therefore, the following equation must be satisfied:

$$T^o = (\sigma - \rho)C_l h_l g \cos\theta \tan\alpha_i + T^{\prime o} \tag{2.136}$$

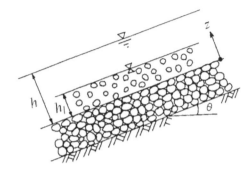

Figure 2.27 Schematic illustration of the immature debris flow.

On an erosive bed, from Equation (2.27), $\tan \alpha_i \approx \tan \varphi$ and, from the consideration in section 2.5, T'^o is negligible in comparison to particle collision stress. Thus, approximately:

$$(\sigma - \rho)C_l h_l g \sin \theta + \rho g h \sin \theta = (\sigma - \rho)C_l h_l g \cos \theta \tan \varphi \qquad (2.137)$$

This equation is also deduced by considering that the bed surface layer is in the critical stable state against massive failure under the static shear stress operated from the immature debris flow over it (see Chapter 3).

One obtains the following equation from Equations (2.137) and (2.25):

$$\frac{h_l}{h} = \frac{C_\infty}{C_l} \qquad (2.138)$$

Substituting the flattest slope angle to generate a fully developed stony debris flow (see Chapter 3) into Equation (2.25) and assuming that the thus obtained C is equal to C_l, the following relation can be obtained:

$$C_l \approx 0.4C_* \qquad (2.139)$$

Equation (2.139) was verified by the experiments using a nearly uniform bed material; $d_p = 1.14\,\text{mm}$, $\sigma = 2.61\,\text{g/cm}^3$, $C_* = 0.65$ and $\tan \varphi = 0.8$. In the experiments, material was laid on the flume bed of $10°–4°$ in slope, and water was supplied from upstream to create immature debris flows (Takahashi 1982).

In the grain mixture layer, under the assumption of a homogeneous concentration distribution within the layer, the shear stress balance equation (2.4) is written as follows:

$$a_i \sigma \lambda^2 d_p^2 \left(\frac{du}{dz}\right)^2 + \rho l^2 \left(\frac{du}{dz}\right)^2 = \{(\sigma - \rho)C_l(h_l - z) + \rho(h - z)\}g \sin \theta \qquad (2.140)$$

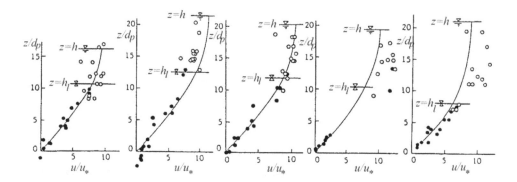

Figure 2.28 Velocity distribution in immature debris flow. Line; theory, circles; experiments.

in which Bagnold's dilatant fluid approximation is used for T_{ds}. In the upper layer the shear stress balance equation will be the same as ordinary turbulent water flow:

$$\rho l^2 \left(\frac{du}{dz}\right)^2 = \rho g (h - z) \sin \theta \tag{2.141}$$

Regarding the mixing length within the fluid phase, we assume:

$$l = \begin{cases} (\xi/\lambda)d_p & ; \quad z \le h_l \\ (\xi/\lambda)d_p + \kappa(z - h_l) ; & z > h_l \end{cases} \tag{2.142}$$

where ξ is a coefficient, and κ is the Kármán constant. Namely, we assume that l in the lower mixture layer has the scale of void space between particles and in the upper layer it has a scale proportional to the distance from the boundary between the upper and the lower layers plus the length scale of the mean void space in the lower mixture layer.

Integration of Equations (2.140), (2.141) and (2.142) under the boundary conditions; the velocity at the bottom is 0 and the velocities at the upper boundary of the lower layer and that at the lower boundary of the upper layer are the same, is not difficult but a little messy so that the mathematical procedure and resulting formulae are explained elsewhere (Takahashi 1982). Figure 2.28 compares the theoretical velocity distributions obtained under $\xi = 3$ and $\kappa = 0.4$ with experimental results. Although the velocities in the upper layer scatter rather widely, the theoretical velocity distributions generally reflect the experimental results.

We can obtain the cross-sectional mean velocity of flow by the integration of velocity distribution functions. The result is again complicated. Therefore, alternative to the theoretical results, which attach importance to the case of $h/d_p < 20$ in which particle suspension is negligible, an empirical resistance to flow formula is obtained as follows:

$$\frac{U}{u_*} = (0.4 \sim 0.7)\frac{h}{d_p} \tag{2.143}$$

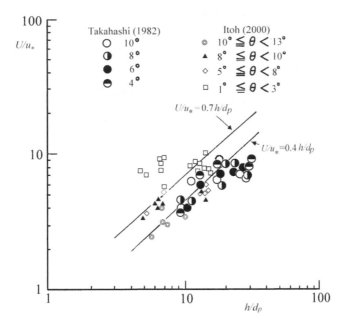

Figure 2.29 Velocity coefficients versus relative depths in immature debris flows.

Figure 2.29 compares Equation (2.143) with experimental data. Although a tendency for U/u_* to becomes small with increasing channel slope may be pointed out, Equation (2.143) does not contain the effects of the channel slope because the experimental data are very scattered.

The sediment discharge per unit width; q_s, is given by $C_l h_l u_l$, where C_l and u_l are the mean sediment concentration and the mean velocity of the mixture layer, respectively, and the total discharge of the sum of sediment and water per unit width q_t is given by Uh. Hence, the sediment transport concentration that is defined as q_s/q_t ($= C_{s\infty}$) can be theoretically obtained. The result is approximated as:

$$C_{s\infty} = 6.7 C_\infty^2 \qquad (2.144)$$

Figure 2.30 compares Equation (2.144) with experimental data. The empirical formula by Mizuyama (1981) that is given as:

$$C_{s\infty} = 5.5(\tan\theta)^2 \qquad (2.145)$$

is also shown. Both Equations (2.144) and (2.145) equally well explain the experimental results. Moreover in Figure 2.30 the curves of equilibrium concentrations for the mature stony debris flows given by Equation (2.25) and for bed load transportations they are given by the formula of Ashida *et al.* (1978) are also shown. These curves show the concentration switches from that of the bed load to that of the debris flow continuously corresponding to the channel gradient. Note that if the channel gradient becomes very steep, Equation (2.144) calculates the concentration larger than the equilibrium concentration for mature debris flow C_∞. Because the concentration in an

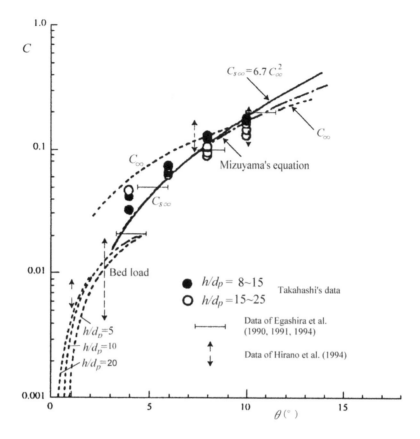

Figure 2.30 Sediment transport concentration by volume in immature debris flow.

immature debris flow cannot surpass C_∞, the applicable range of Equation (2.144) is limited within the range in which $C_{s\infty}$ is calculated less than C_∞ and larger than the concentration calculated as bed load; i.e. from Figure 2.30 about $4° < \theta < 10°$.

2.6.3 Turbulent-muddy debris flow

When the comprising particles are small but still larger than the size at which the electro-chemical adhesive force between particles becomes important, the turbulence in the flow will suspend particles and it becomes the turbulent-muddy-type debris flow. Although the contribution of particle collision stress becomes less important, Arai and Takahashi (1986) gave the shear stress balance equation as:

$$\rho_T g(h - z)\sin\theta = \rho_T l^2 \left(\frac{du}{dz}\right)^2 + a_i\sigma\lambda^2 d_p^2\left(\frac{du}{dz}\right)^2 \qquad (2.146)$$

where ρ_T is the apparent density of the debris-flow material and a uniform concentration distribution is assumed. This equation can easily be integrated by considering

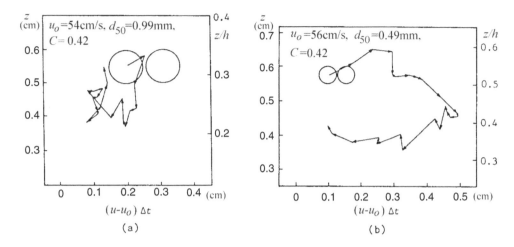

Figure 2.31 The moving path of a particle.

that under the condition $\lambda = 0$ (particle concentration is zero) the velocity distribution becomes the log-law distribution of plain water flow. Namely, for a flow on a smooth bed, we obtain:

$$\frac{u}{u_*} = \frac{1}{\kappa} \ln \left| \frac{z/h + \sqrt{(z/h)^2 + \Phi_1^2}}{a_o/R_* + \sqrt{(a_o/R_*)^2 + \Phi_1^2}} \right| \tag{2.147}$$

where κ is the Kármán constant; $a_o = 0.11$, $R_* = u_* h/\nu_o$, $\Phi_1 = \lambda^2 (a_i \sin \alpha_i/\kappa^2)(\sigma/\rho_T)$ $(d_p/h)^2$; and ν_o is the dynamic viscosity of water. The term Φ_1^2 is the contribution of particle collision stress and it becomes very small with decreasing particle concentration or decreasing (d_p/h), and Equation (2.147) without Φ_1^2 is nothing but the velocity distribution formula of plain water flow.

The experiments to verify the theory were conducted by Arai and Takahashi (1983, 1986). Figure 2.31 shows the path of an arbitrarily selected particle observed from a moving coordinate system having the velocity u_o in the direction of mean flow, in which the absolute mean velocity of the particle in the time interval of Δt ($= 0.04$ sec) is u. The arrow in Figure 2.31 indicates the direction and the distance traveled in Δt in the moving coordinate system and the two circles demonstrate the representative diameter of the particle and the mean relative distance between neighboring particles. The fluid mass incorporated with the particles moves in a large-scale eddy motion.

Figure 2.32 compares the theoretical velocity distribution forms with the experimental data, in which the velocity and the depth are normalized by the surface velocity and flow depth, respectively. Both theoretical and experimental data indicate that either with decreasing concentration or with increasing relative depth, the velocity near the bed increases and the velocity in the upper layer approaches a uniform state; the velocity distribution forms approach that of a plain water flow.

The absolute values of velocities by the theory were also confirmed to fit the experiments only if the Kármán constant in Equation (2.147) is correctly evaluated.

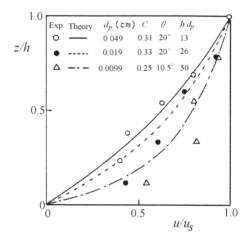

Figure 2.32 Theoretical and experimental velocity distributions.

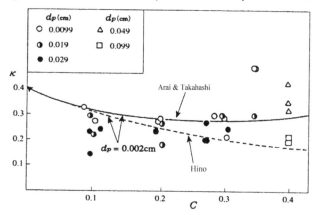

Figure 2.33 Variation in κ values with change in solids concentration.

The mixing length l in Equation (2.147) is evaluated as $l = \kappa z$. It is well known that the Kármán constant changes depending on the concentration of suspended sediment and Hino (1963) proposed a theory for that phenomenon in the range of concentration less than a few percent. Arai and Takahashi (1986) modified his theory to make it applicable to denser concentrations by modifying the acceleration diminishing function reflecting the action between fluid mass and particles. Figure 2.33 shows the theory and the experimental results, it seems to show a minimum κ value around $C = 0.2$–0.3.

The cross-sectional mean velocity is obtained by integrating Equation (2.147) as follows:

$$\frac{U}{u_*} = \frac{1}{\kappa}\left[\ln\left|\frac{1+\sqrt{1+\Phi_1^2}}{Z_o+\sqrt{Z_o^2+\Phi_1^2}}\right| - \sqrt{1+\Phi_1^2}+\Phi_1\right] \tag{2.148}$$

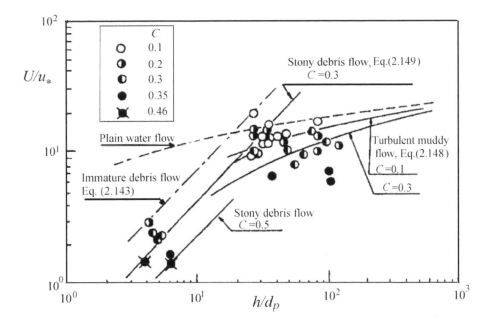

Figure 2.34 Resistance to flow in various types of debris flows.

where $Z_o = (a_o v_o)/(u_* h)$. Figure 2.34 shows the relationships between U/u_* and h/d_p in Equation (2.148) together with the experimental results. Figure 2.34 also shows the stony debris flow formula that is obtained by integrating Equation (2.22) as:

$$\frac{U}{u_*} = \frac{2}{5} \left[\frac{1}{a_i \sin \alpha_i} \left\{ C + (1 - C)\frac{\rho}{\sigma} \right\} \right]^{1/2} \left\{ \left(\frac{C_*}{C} \right)^{1/3} - 1 \right\} \frac{h}{d_p} \tag{2.149}$$

and the formulae for immature debris flow Equation (2.143) adopting the coefficient equal to 0.7 and for plain water flow are also demonstrated. If the relative depth is less than 10 and concentration is larger than 0.3, Equation (2.149) fits well to the experimental data, and if the relative depth is between 10 and about 30, Equation (2.148) fits the experimental data rather well. The resistance to flow in a turbulent-muddy debris flow is asymptotic to that of turbulent water flow with increasing relative depth.

A question is, sometimes, raised about the validity of Equation (2.20) for stony debris flow: for a neutrally buoyant particle σ is equal to ρ, so the right-hand side term of Equation (2.20) becomes zero. The left-hand side of Equation (2.20) is the dispersive pressure and Bagnold obtained a finite value for this by the experiments using a kind of neutrally buoyant particles. Therefore, they claim that Equation (2.20) cannot be applicable for the neutrally buoyant particle case nor for heavy particle cases because Equation (2.20) does not set restrictions for the specific gravity of particles. However, one must be aware that Equation (2.20) applies for the stony debris flow that is comprised of heavy coarse particles. The more strict description of stress balance

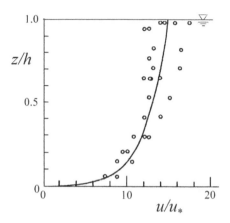

Figure 2.35 Vertical velocity distribution in the flow of highly concentrated neutrally buoyant particle.

equation in stony debris flow is given by Equation (2.146), and for mature stony debris flow that is comprised of heavy coarse particles, the first term in the right-hand side of Equation (2.146) is negligibly small, and so Equation (2.20) becomes applicable. For the case of neutrally buoyant particles the particle dispersive stress surely exists but it works merely to increase the pressure within the flow and the particles can easily be moved even though the particle concentration is large. Thus, in this case, the first term of the right-hand side of Equation (2.146) becomes dominant because l is far larger than d_p. Then, the stress balance equation for a flow of densely concentrated neutrally buoyant particles is approximately given by:

$$\rho_T l^2 \left(\frac{du}{dz}\right)^2 = \rho_T g (h - z)\sin\theta \tag{2.150}$$

where $\sigma = \rho_T = \rho$. This is equivalent to the equation for a turbulent water flow. Figure 2.35 shows the experimentally obtained velocity distribution of flow, in which the particles were the polystyrene beads of $\sigma = 1.03\,\text{g/cm}^3$; $d_p = 0.135\,\text{cm}$, the volume concentration of particles in water $C = 0.38$, and the channel slope was $\theta = 6°$. The curve in Figure 2.35 shows that the log law that is obtained from Equation (2.150) for a water flow on the rough bed described as:

$$\frac{u}{u_*} = 8.5 + 5.75 \log\left|\frac{z}{k_r}\right| \tag{2.151}$$

fits well to the experimental results.

In Equation (2.151), k_r is the equivalent roughness height that is, from the experiments, a function of C and d_p as shown in Figure 2.36, and for the case shown in Figure 2.35, $k_r/d_p = 0.45$.

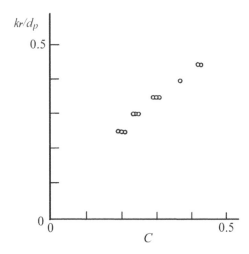

Figure 2.36 Variation in the equivalent roughness by different particle concentrations.

2.7 GENERALIZED THEORY FOR INERTIAL DEBRIS FLOWS

2.7.1 Theoretical considerations

The last section discussed the typical types of inertial debris flows individually. However, as is shown in Figure 2.26, there is a hybrid-type debris flow in between the typical stony and turbulent-muddy debris flows. The hybrid flow in the inertial range consists of the lower particle collision stress dominating layer and the upper turbulent mixing stress dominating layer. The relative thicknesses of these two layers in the flow depend on the relative depth h/d_p and the solids concentration C. If the particle collision layer occupies almost the entire depth it is a stony debris flow, and in the turbulent-muddy debris flow the turbulent suspension layer occupies almost the entire depth. Therefore, the general type of inertial debris flow is the hybrid one, in which both the particle collision layer and the turbulent suspension layer occupy considerable ratios. Takahashi and Satofuka (2002) gave a theory for such a general inertial debris flow. In this section, their theory is introduced and verified by the experimental data.

Figure 2.37 is a schematic illustration of a hybrid-type debris flow on an erodible bed. The upper layer of depth h_1 is the particle suspension layer, and in the lower depth h_2 layer particles move with frequent collision or enduring contact with each other. Because the flow is on a movable bed, the particle concentration on the bed is C_* and it decreases upward. Hence, at a distance above the bed, C first becomes less than C_3. Below this height particles are always in contact and the quasi-static stresses are transmitted through the skeletal structure within this layer, but above this height the skeletal structure does not exist and only the dynamic stresses are transmitted by means of inter-particle collisions. In the upper layer, where all the particles are suspended, the particles' load is transmitted to the lower layer as hydraulic pressure.

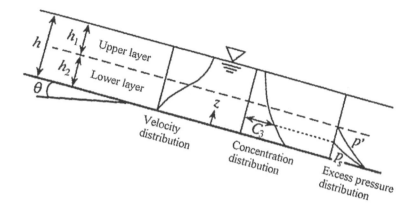

Figure 2.37 The general structure of inertial debris flow.

The stress balance equations in the lower layer to the flow direction and perpendicular to the bed, respectively, are written as:

$$\tau_c + \tau_k + \tau_s + \tau_t = g \sin\theta \int_z^h \{(\sigma - \rho)C + \rho\}dz \tag{2.152}$$

$$p_s + p_c = g \cos\theta \int_z^{h_2} (\sigma - \rho)C\,dz \equiv p' \tag{2.153}$$

where p' is the effective pressure that is equal to the submerged weight of total particles within the depth h_2 layer and transmitted by inter-particle contact.

At the bed where the particle concentration is C_* the velocity of flow u is equal to zero because in such a high concentration particles displacement is impossible. Hence, the dynamic pressure due to particle collision p_c is zero and all the excess pressure p' is sustained by p_s, which decreases upward and becomes zero at the height where C becomes equal to C_3. The vertical distribution function of p_s is unknown at present, but here, the following function is assumed:

$$p_s = \begin{cases} \dfrac{C - C_3}{C_* - C_3}p' \; ; & C > C_3 \\ 0 & ; \quad C \le C_3 \end{cases} \tag{2.154}$$

As for the relationship between p_s and τ_s, the Coulomb-type formula, Equation (2.80) is assumed.

The shear stress balance equation in the upper layer is written as:

$$\tau_c + \tau_t = \int_z^h \rho_T g \sin\theta\,dz \tag{2.155}$$

in which τ_k is already included in τ_t and τ_s is of course zero.

Provided that the granular temperature is given by Equation (2.105), τ_c, p_c and τ_k are written from (2.84), (2.85) and (2.86) as follows:

$$\tau_c = \frac{4}{5}\sqrt{\frac{1}{15\pi}}\frac{1+e}{\sqrt{1-e}}C^2 g_0 \sigma d_p^2 \left(\frac{du}{dz}\right)^2 \tag{2.156}$$

$$p_c = \frac{2}{15}C^2 g_0 \frac{1+e}{1-e}\sigma d_p^2 \left(\frac{du}{dz}\right)^2 \tag{2.157}$$

$$\tau_k = \frac{1}{3g_0}\sqrt{\frac{1}{15\pi(1-e)}}\sigma d_p^2 \left(\frac{du}{dz}\right)^2 \tag{2.158}$$

As for the turbulent mixing stress τ_t, following formula is used:

$$\tau_t = \rho_T l^2 \left(\frac{du}{dz}\right)^2 \tag{2.159}$$

where the mixing length is given by $l = \kappa z$. A question may arise as to whether, in the densely concentrated lower layer, the mixing length would have the scale of void between particles as already mentioned in the discussion of immature debris flow. There are two reasons for adopting this mixing length; 1) as discussed in section 2.5, no matter which mixing length is adopted in the lower layer τ_t is far smaller than τ_c or τ_s so that it hardly affects the velocity and solids concentration, 2) the interstitial fluid is widespread as a continuum in the entire depth so that the mixing length may have a larger scale than the void between particles. However, considering the intensity of turbulence, ρ_T, in the lower layer should be the density of interstitial fluid, whereas in the upper layer it should be the density of the mixture.

Substitution of these constitutive relations into (2.152) and (2.153) yields the following shear stress and pressure balance equations:

$$p_s \tan\phi + \rho_t l^2 \left(\frac{du}{dz}\right)^2 + \left\{\frac{4}{5}C^2 g_0(1+e) + \frac{1}{3g_0}\right\}\frac{1}{\sqrt{15\pi(1-e)}}\sigma d_p^2$$
$$= g\sin\theta \int_z^h \{(\sigma-\rho)C + \rho\}dz \tag{2.160}$$

$$p_s + \frac{2}{15}C^2 g_0 \frac{1+e}{1-e}\sigma d_p^2 \left(\frac{du}{dz}\right)^2 = g\cos\theta \int_z^{h_2}(\sigma-\rho)C\,dz \tag{2.161}$$

For the shear stress balance equation in the upper layer Equation (2.155) gives:

$$\left\{\rho_T l^2 + \frac{4}{5}C^2 g_0 \frac{1+e}{\sqrt{15\pi(1-e)}}\sigma d_p^2\right\}\left(\frac{du}{dz}\right)^2 = g\sin\theta \int_z^h \rho_T\,dz \tag{2.162}$$

The solids concentration distribution in the lower part of the lower layer where C is larger than C_3 is given from Equations (2.160) and (2.161) as the following formula:

$$\frac{dC}{dZ} = \frac{(sB^2C/\tan\theta) - B}{A(1 - Z + s\overline{C})} - K_4 \frac{1 - f(C)}{C^2 g_0} \frac{Z}{A} \tag{2.163}$$

and the velocity distribution in that layer is given by:

$$\left(\frac{du'}{dZ}\right)^2 = K_5 \frac{1 - f(C)}{C^2 g_0 B}(1 - Z + s\overline{C}) \tag{2.164}$$

where $Z = z/h$, $s = (\sigma/\rho) - 1$, $u' = u/\sqrt{gh}$, and $f(C) = (C - C_3)/(C_* - C_3)$:

$$\overline{C} = \int_{h_2/h}^{1} C\,dZ, \quad K_4 = 15\kappa^2 \frac{1-e}{1+e}\left(\frac{\rho}{\sigma}\right)\left(\frac{h}{d_p}\right)^2, \quad K_1 = 6\sqrt{\frac{1-e}{15\pi}},$$

$$K_2 = \frac{5}{2}\frac{1}{1+e}\sqrt{\frac{1-e}{15\pi}}, \quad K_5 = \frac{15}{2}\frac{1-e}{1+e}\left(\frac{\rho}{\sigma}\right)\left(\frac{h}{d_p}\right)^2 \sin\theta,$$

$$A = \left\{\tan\phi - \left(K_1 + K_2 \frac{1}{C^2 g_0^2} + \frac{K_4 Z^2}{2C^2 g_0}\right)\right\}\frac{1}{C_* - C_3}$$

$$- \frac{1 - f(C)}{C^3 g_0^2}\left\{\frac{2}{3}K_2(g_0 + 2) + \frac{1}{6}K_4 Z^2 g_0(g_0 + 5)\right\}$$

$$B = f(C)\tan\phi - \tan\theta + \{1 - f(C)\}\left(K_1 + K_2 \frac{1}{C^2 g_0^2} + \frac{K_4 Z^2}{2C^2 g_0}\right)$$

In the upper part of the lower layer where C is less than C_3, because $f(C) = 0$, the formulae of concentration and velocity distributions, respectively, are given as follows:

$$\frac{dC}{dZ} = \frac{(sB'^2 C/\tan\theta) - B'}{A'(1 - Z + s\overline{C})} - K_4 \frac{1}{C^2 g_0}\frac{Z}{A'} \tag{2.165}$$

$$\left(\frac{du'}{dZ}\right)^2 = K_5 \frac{1}{C^2 g_0 B'}(1 - Z + s\overline{C}) \tag{2.166}$$

where:

$$A' = \frac{1}{C^3 g_0^2}\left\{\frac{2}{3}K_2(g_0 + 2) + \frac{1}{6}K_4 Z^2 g_0(g_0 + 5)\right\},$$

$$B' = K_1 + K_2 \frac{1}{C^2 g_0^2} + \frac{K_4 Z^2}{2C^2 g_0} - \tan\theta$$

In the upper layer, if we treat it in a similar way to the case of ordinary suspended load transportation, the concentration distribution will be given as:

$$\frac{dC}{dZ} = -\frac{w'_s}{\beta \varepsilon_z} C \tag{2.167}$$

where β is a coefficient, ε_z is the turbulent diffusion coefficient, and w'_s is the non-dimensional particle settling velocity:

$$w'_s = \frac{w_s}{\sqrt{gh}} = \frac{F_R \cos\theta \sqrt{sgd_p}(1-C)^n}{\sqrt{gh}} = F_R \cos\theta \sqrt{\frac{sd_p}{h}}(1-C)^n \tag{2.168}$$

where w_s is the particle settling velocity, $(1-C)^n$ is a correction term to take the retardation in settling velocity due to the group settling of particles into account in which n is a constant and F_R is given by:

$$F_R = \sqrt{\frac{2}{3} + \frac{36}{d_*}} - \sqrt{\frac{36}{d_*}} \tag{2.169}$$

in which $d_* = (sgd_p^3)/v$ and v is the dynamic viscosity of pure water.

If we write the left-hand side term of Equation (2.162) as $\rho_T \varepsilon_Z (du/dz)$, ε_Z is given as:

$$\varepsilon_Z \equiv \frac{\varepsilon_z}{\sqrt{ghh}} = \left\{ \frac{4}{5} \frac{1+e}{\sqrt{15\pi(1-e)}} \left(\frac{\sigma}{\rho}\right) \left(\frac{d_p}{h}\right)^2 \frac{C^2 g_0}{sC+1} + l'^2_n \right\} \left(\frac{du'}{dZ}\right) \tag{2.170}$$

where $l'_n = \kappa Z$.

The velocity distribution in the upper layer is then given by:

$$\left(\frac{du'}{dZ}\right)^2 = \frac{\displaystyle\int_Z^1 (sC+1)\sin\theta \, dZ}{\frac{4}{5} C^2 g_0 \frac{1+e}{\sqrt{15\pi(1-e)}} \frac{\sigma}{\rho} \left(\frac{d_p}{h}\right)^2 + (sC+1) + l'^2_n} \tag{2.171}$$

At this stage we must consider the criteria of whether or not the upper layer exists, and the distance from the bottom of the boundary between the upper and the lower layers. For the existence of an upper layer, the particles must be suspended by turbulence and this condition is satisfied when $w_s/u_* < 1$. If this condition is not satisfied, the flow is either the stony-type debris flow or an immature debris flow. For the determination of the boundary height between the two layers we consider the following: The particle concentration in the upper layer is determined by the balance between the upward transportation due to diffusion and the downward transportation due to settling, where the boundary value of concentration is given at the boundary between the upper and lower layers. Hence, the mean concentration in the upper layer is less than the concentration at the upper boundary of the lower layer. The minimum particle concentration in the lower layer takes place at the upper boundary of the lower layer and the existing experimental data on immature debris flows show that the mean particle concentration within the mixture layer (lower layer) i.e. C_{lim} is about $0.5C_*$. Therefore, although it is somewhat expedient, the boundary height is set where the calculated particle concentration in the lower layer first becomes less than C_{lim} when the calculation proceeds from the lower to the upper layer; the concentration at the bed (lower boundary of the lower layer) is C_*. If the calculated concentration is larger than C_{lim} in the entire depth, we consider that the whole depth is composed of the lower layer regardless of whether the condition $w_s/u_* < 1$ is satisfied or not, so that $h_2 = h$. Note that even in such a case the effects of turbulent mixing are taken into account as can be understood by seeing that Equations (2.165) and (2.166) contain the terms relating to K_4 that reflects the turbulent mixing effect. If the concentration becomes less than C_{lim} at a certain height and the condition $w_s/u_* < 1$ is satisfied, flow is the hybrid-type, and if the concentration becomes less than C_{lim} at a certain height but the condition $w_s/u_* < 1$ is not satisfied, flow is the immature debris flow.

2.7.2 Verification by experimental data

Theoretical concentrations and velocity distributions were calculated for the combination of various channel slopes and particle diameters under the conditions listed in Table 2.3. In the calculations, flow depth was kept constant at 2 cm. The representative examples of the results are shown in Figure 2.38. If the grain diameter is 7 mm, even if the channel slope is as steep as 18°, the condition of particle suspension is not satisfied, and consequently, particles are densely dispersed due to the effects of inter-particle collision near the flow surface and velocity is small. This is the case of stony debris flow. If the channel slope becomes 6° for the same material, particles can no longer be dispersed in the entire depth. This is the case of immature debris flow. If the grain diameter is 1 mm and the channel slope is 6°, the condition of particle suspension is

Table 2.3 Parameter values used in the calculations.

$\sigma = 2.65\,\text{g/cm}^3$ $\rho = 1.0\,\text{g/cm}^3$ $\nu = 0.01\,\text{cm}^2/\text{s}$
$g = 980\,\text{cm/s}^2$ $\tan\phi = 0.7$ $e = 0.3$
$\kappa = 0.3$ $\beta = 1.0$ $n = 1.0$
$C_* = 0.65$ $C_3 = 0.5$ $C_{lim} = 0.5C_*$

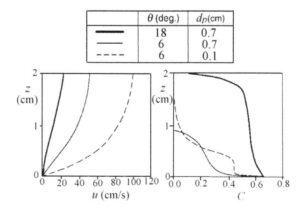

	θ (deg.)	d_p(cm)
——	18	0.7
—	6	0.7
- - - -	6	0.1

Figure 2.38 Velocity and particle concentration distributions by the theory.

satisfied, and although the concentration near the flow surface is very small, particles are dispersed by turbulence in the upper layer. This is the case of hybrid debris flow.

Figure 2.39 compares the theoretical velocity distributions with the experimental data of Hirano *et al.* (1992), where the velocity is normalized by the surface velocity. The debris flows in the experiments were generated by supplying water from upstream onto the water-saturated erodible bed. The bed slope was held constant at 14°, but the relative depth was changed by changing both the supplying water discharge and the particle size. The theoretical as well as experimental velocity distribution curves have a tendency to lower the break point on the curve; the boundary between the upper and the lower layers, and to increase the degree of concavity as the relative depth increases. Moreover, the theoretical curves well reflect the experimental results.

Figure 2.40 shows the relationship between the transport concentration C_{tr} $(= Q_s/Q_T, Q_s$; sediment discharge, Q_T; water plus sediment discharge) and the channel slope for various grain diameters. The experimental debris flows were generated by suddenly supplying water from upstream onto the water-saturated erodible bed in a flume 9.9 cm wide and 10 m long. We made every effort to hold the discharge Q_T constant at 531 cm³/s for all the experimental runs. Four kinds of materials 0.201, 0.066, 0.030 and 0.017 cm in the median diameters with densities between 2.64 and 2.66 g/cm³ and the internal friction angles between 38.5° and 39° were used for the bed. The debris flow samples were collected with a bucket at the outlet of the flume to measure Q_T and C_{tr}. As is evident in Figure 2.40, the theory can well explain the tendency that the smaller the particle diameter, the larger the equilibrium transport concentration becomes under a given channel slope.

In Figure 2.40, other than those corresponding to the experimental runs, the theoretical curves for $d_p = 1$ cm and 0.5 cm are also drawn. Under the same Q_T and grain

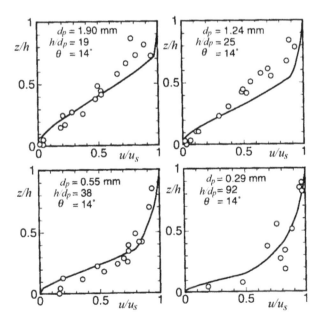

Figure 2.39 Theoretical and experimental velocity distribution forms.

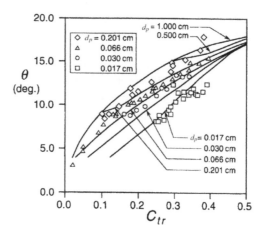

Figure 2.40 Theoretical and experimental equilibrium transport concentrations.

density with the experimental runs, particles larger than 1 cm cannot be suspended by turbulence even for a steep channel of 15°. The reason for the branching of curves corresponding to $d_p = 0.5$ cm and $d_p = 0.2$ cm respectively from the curve for $d_p = 1$ cm at the channel slopes of about 15° and 10°, respectively, is due to the beginning of suspension on slopes steeper than those corresponding to the branching points. The theoretical curve for $d_p = 1$ cm corresponds to the coupled equilibrium concentration curves for stony debris flow given by Equation (2.25) and for immature debris flow given by Equation (2.144), respectively. The theory, as well as experiments, indicates that the larger the relative thickness of the upper layer the larger the concentration becomes under the same channel slope, and the respective curves are asymptotic to the curve for stony debris flow at a steep channel slope. This tendency implies that the maximum concentration in turbulent-muddy debris flow will be about $0.9C_*$ because the maximum concentration in stony debris flow is about $0.9C_*$. Experiments for the particles less than 0.017 cm was not done, but we must take notice that the theory cannot apply for particles so small that adhesion between particles becomes important.

At this stage, I would like to clarify the difference between our theory and the theory of Egashira *et al.* (1997) that has been widely used in Japan. Their theory is concerned with stony debris flow which lacks an upper layer and its most distinguishing characteristic is the description of the static pressure ratio to the total pressures except for hydrostatic pressure as follows:

$$\frac{p_s}{p_s + p_c} = \left(\frac{C}{C_*}\right)^{1/5} \tag{2.172}$$

Namely, they claim that, even at the flow surface where the concentration is far less than C_3, p_s exists. They adopt Miyamoto's Equation (2.113) for p_c, then, the pressure balance equation for uniform flow corresponding to Equation (2.161) is given by:

$$p_s + p_c = p_s + k_d e^2 \sigma d_p^2 C^{1/3} \left(\frac{du}{dz}\right)^2 = g\cos\theta \int_z^h (\sigma - \rho)dz \tag{2.173}$$

Equation (2.172) was introduced to determine p_s in Equation (2.173), but Equation (2.172) can be rewritten as:

$$p_s = \frac{(C/C_*)^{1/5}}{\{1 - (C/C_*)^{1/5}\}} p_c = \frac{(C/C_*)^{1/5}}{\{1 - (C/C_*)^{1/5}\}} k_d e^2 \sigma d_p C^{1/3} \left(\frac{du}{dz}\right)^2$$

The static pressure p_s must be independent of velocity but this equation shows it depends on the velocity gradient. In this respect the theory is contradictory. Nevertheless, if the thus determined p_s is substituted into Equation (2.173), then, the left-hand side terms can be described only by the terms proportional to the square of velocity gradient. The reason why such a deduction is used can be conjectured as the result of adopting Miyamoto's constitutive relation Equation (2.113). As shown in Figure 2.22, Equation (2.113) gives exceptionally small p_c values under the relatively small value of velocity gradient as it appears in actual as well as in experimental debris flows. Hence, otherwise adopting a large p_s value, the left-hand side term of Equation (2.173) cannot balance with the right-hand side term. If Equation (2.172) is introduced, p_s becomes very large when C is large and by virtue of this formulation both sides of Equation (2.173) can balance even in the range of experimentally obtained small velocity gradient. Thus, Equation (2.172) was incautiously introduced only to synchronize the magnitudes of both sides of Equation (2.173) and has no physical basis. Equation (2.172) consequently changes Equation (2.173) to an expression similar to Equation (2.20), qualitatively and quantitatively. Therefore, we can also say that Equation (2.172) is merely introduced to modify Miyamoto's constitutive relations as it will have a similar magnitude to that obtained by Bagnold and others, and contrary to their claim, no static stress is considered. The idea that the static stress operates even if the concentration is less than C_3 has a difficulty in physics, because in such a concentration each particle is on average detached from other particles.

2.7.3 Approximate solutions for solids concentration and resistance to flow

Although the theory by Takahashi and Satofuka (2002) is verified by experimental data, it does not give the concentration and velocity in an explicit form so that it is inconvenient for practical use. In this section, by arranging the data of numerous numerical calculations, the practical explicit approximate formulae will be obtained.

As it is understood in the process of the previous discussion, the non-dimensional parameters that determine particle concentration are, mainly, the channel slope θ, the relative depth h/d_p and d_* that is related to the particle settling velocity. So, we calculated the particle concentrations by the theory under the extents of particle diameters of 0.01–10 cm, channel slopes of 4°–20°, and relative depths of 2–1024. Figure 2.41 shows the results of the calculation depicted as the relationships between channel slope and transport concentration in which particle diameter and relative depth are the varying parameters. These relationships for particles as small as $d_p = 0.01$–0.02 cm show that there are two slopes giving a concentration in the range of a large relative depth, but we do not know whether this phenomenon actually occurs because we have no experimental data under these conditions. Particles composing actual debris flows, however, are larger in size and in such cases the results of calculation seem to

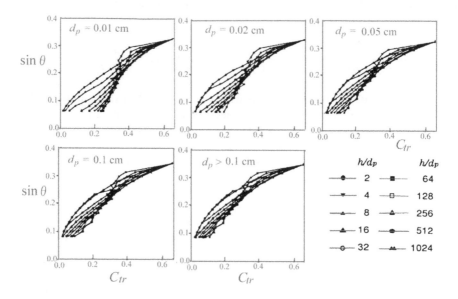

Figure 2.41 Theoretical transport concentration versus channel slope.

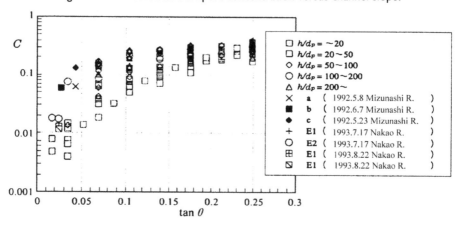

Figure 2.42 Relationship between channel slope and concentration by Hirano *et al.*

be applicable because the curves have a similar tendency to Figure 2.40; the curves for $h/d_p = 2$ (no upper layer exists) give the smallest concentrations corresponding to Equation (2.25) or Equation (2.144) and with increasing relative depth, especially in a flatter slope channel, the concentration becomes larger under a given channel slope.

Hirano *et al.* (1994) obtained the channel slope versus concentration relationships for the debris flows in the Mizunashi and the Nakao Rivers and for their experiments as shown in Figure 2.42. This figure also shows the tendency for the concentration to becomes larger with increasing relative depth and that the extent of variations in concentration is larger in the smaller channel slope range, furthermore, the absolute values of concentration are similar to those given in Figure 2.41.

Table 2.4 Formulae giving $b_1 = b_{12}$.

$$b_1 = -0.2030 + 0.2776\zeta - 0.1291\zeta^2 + 0.02498\zeta^3 - 0.001730\zeta^4$$
$$b_2 = 0.2215 - 0.3131\zeta + 0.1451\zeta^2 + 0.02763\zeta^3 + 0.001872\zeta^4$$
$$b_3 = -0.0473 + 0.0675\zeta - 0.0314\zeta^2 + 0.00588\zeta^3 - 0.000387\zeta^4$$
$$b_4 = 6.865 - 13.34\zeta + 7.309\zeta^2 - 1.561\zeta^3 + 0.1160\zeta^4$$
$$b_5 = 0.0394 + 8.779\zeta - 6.559\zeta^2 + 1.589\zeta^3 - 0.1267\zeta^4$$
$$b_6 = 0.9064 - 2.964\zeta + 2.048\zeta^2 - 0.4865\zeta^3 + 0.03851\zeta^4$$
$$b_7 = -28.70 + 76.44\zeta - 45.87\zeta^2 + 10.20\zeta^3 - 0.7758\zeta^4$$
$$b_8 = -9.935 - 39.41\zeta + 37.00\zeta^2 - 9.616\zeta^3 + 0.7964\zeta^4$$
$$b_9 = 1.306 + 6.001\zeta - 7.360\zeta^2 + 2.058\zeta^3 - 0.1755\zeta^4$$
$$b_{10} = 30.23 - 93.15\zeta + 63.48\zeta^2 - 14.78\zeta^3 + 1.153\zeta^4$$
$$b_{11} = 32.01 + 36.63\zeta - 50.40\zeta^2 + 14.17\zeta^3 - 1.217\zeta^4$$
$$b_{12} = -4.819 - 3.320\zeta + 9.628\zeta^2 - 3.008\zeta^3 + 0.2728\zeta^4$$

If the particle diameter becomes larger than 0.1 cm, the non-dimensional particle diameter d_* becomes ineffective in the transport concentration, although the particle diameter itself affects by the variation of relative depth. The reason for this is due to the representation of particle settling velocity by Equation (2.168); Equation (2.168) becomes, as a matter of practice, independent of d_* for large particle diameters.

The curves in Figure 2.41 can be approximately represented by:

$$C_{tr} = a_1 + a_2 x + a_3 x^2 + a_4 x^3 \tag{2.174}$$

where $x = \sin\theta$ and the respective coefficients are given by:

$$a_1 = b_1 + b_2 y + b_3 y^2, \quad a_2 = b_4 + b_5 y + b_6 y^2,$$
$$a_3 = b_7 + b_8 y + b_9 y^2, \quad a_4 = b_{10} + b_{11} y + b_{12} y^2$$

in which $y = \log_{10}(h/d_p)$, and by expressing $\zeta = \log_{10} d_*$, $b_1 - b_{12}$ are given as shown in Table 2.4.

Figure 2.43 shows Equation (2.174). Although the degree of fitness for fine particles less than 0.02 cm is not good, for larger particles the transport concentration is well given by Equation (2.174).

Figure 2.44 compares the theoretically obtained velocity coefficients with those obtained by the following approximate solution:

$$\frac{U}{u_*} = F_1\left[A_1 + A_2 \log_{10}\left(\frac{h}{d_p}\right) + A_3\left\{\log_{10}\left(\frac{h}{d_p}\right)\right\}^2\right] \tag{2.175}$$

where $A_1 = 1.1632 - 3.0374 C_{tr} + 1.0589 C_{tr}^2$, $A_2 = 2.6898 - 4.7747 C_{tr} + 3.9167 C_{tr}^2$, $A_3 = 0.8313 - 2.2134 C_{tr} + 0.6818 C_{tr}^2$, when $d_p > 0.1$ cm, $F_1 = 1$, when $d_p < 0.1$ cm, $F_1 = -13.72 B_1 (C_{tr} - 0.270)^2 + B_1 + 1.0$, in which $B_1 = A_4 + A_5 \log_{10}(h/d_p)$ $+ A_6\{\log_{10}(h/d_p)\}^2$, $A_4 = 0.50372 - 0.2773 \log_{10} d_* + 0.036998\{\log_{10} d_*\}^2$, $A_5 = 0.5530 - 0.24564 \log_{10} d_* + 0.030294\{\log_{10} d_*\}^2$, $A_6 = -0.20525 + 0.097021 \log_{10} d_* - 0.012564\{\log_{10} d_*\}^2$.

The approximate solutions almost perfectly fit to the theoretical solutions. Equation (2.175) indicates that the velocity coefficient, U/u_*, is a function of relative depth and particle diameter, but except for the case of very fine particles like $d_p = 0.01$ cm, d_* does not have much effect and the velocity coefficients are almost identical to that of $d_p > 0.1$ cm.

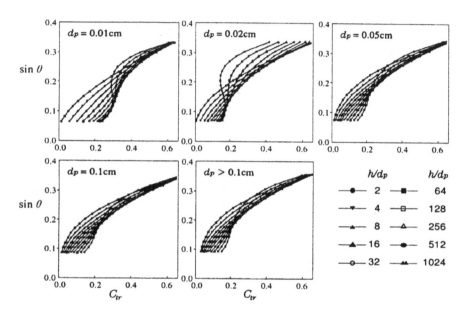

Figure 2.43 The approximate representation of transport concentration versus channel slope.

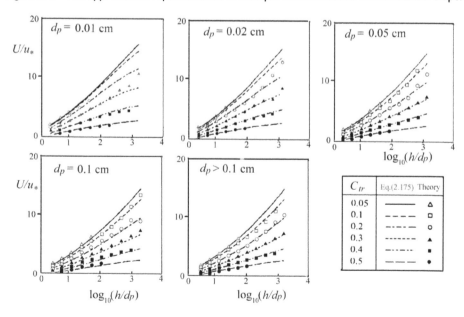

Figure 2.44 Approximate and theoretical velocity coefficients in debris flows.

According to Figure 2.34, the velocity coefficients for stony debris flows with relative depths at 3–6 are 1.4–3, and for turbulent-muddy flows with the relative depths at 30–100 are 5–10. The velocity coefficients for the actual debris flows shown in Figure 1.14 distribute around 1–4 in the Kamikamihori stony debris flows and 3–20 in the Mizunashi turbulent debris flows whose relative depths are 2–10 and 50 to 1000, respectively. And in the Nojiri turbulent debris flows the velocity coefficients are 10–20 with relative depths of 30–50. The tendency in Figure 2.44 is evaluated to fit rather well to actual as well as experimental debris flows data except for the cases in the Nojiri River. The velocity coefficients in the Nojiri River seem to be erroneously measured too large because these coefficients are almost identical with those of plain water flow. The accumulation of applied data will make clear whether the theory can practically be used.

2.8 NEWTONIAN FLUID MODEL FOR VISCOUS DEBRIS FLOW

2.8.1 Theoretical considerations

The inappropriateness of the Bingham fluid model for the typical viscous debris flows in the Jianjia gully was already discussed in Chapter 1, but, in this section, I will explain more about this, based on the data by field measurements (Arai *et al.* 2000). In our field measurement we used an ordinary video camera of 1/30 second per a frame and analyzed the surface velocity by the correlation method (Arai *et al.* 1998). Figure 2.45 shows the thus obtained temporal change in surface velocity at the center of a cross-section in a surge. As soon as the surge front arrives the surface velocity becomes the maximum, and after keeping that velocity awhile it gradually decelerates and finally stops. This process is already mentioned in Chapter 1.

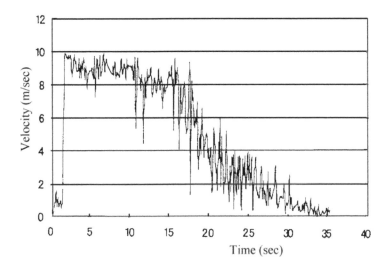

Figure 2.45 Temporal change in the surface velocity of a surge.

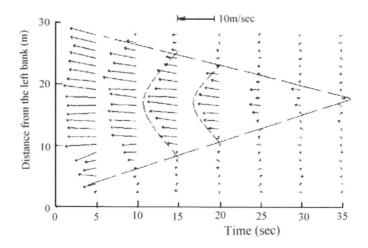

Figure 2.46 Temporal change in the lateral distributions of surface velocity in a section.

Figure 2.46 demonstrates the change in the lateral velocity distributions while passing through the measuring section. The big spearhead shape drawn by the two broken lines is the boundary outside of which the surface velocity is almost zero. Thus, the width of flow diminishes with time in parallel with the deceleration of velocity. This characteristic change clearly shows the stopping of flow is not due to the strength of the Bingham fluid. The lateral velocities near the forefront, where the flow is very much turbulent, distribute almost uniformly, whereas rearwards where flow is laminar the distribution is parabolic as the envelopes are drawn by the broken lines. If the depth of flow is laterally uniform, a parabolic surface velocity distribution is characteristic for the laminar flow of a Newtonian fluid.

The rear part of a debris flow surge in the Jiangjia gully is evidently laminar, but it contains a high density of coarse particles. In the case of inertial debris flow the causes of dense particle dispersion are the dispersive pressure due to inter-particle collisions and/or the suspension due to turbulent mixing. In a highly viscous flow, however, turbulence is minimal so that particles cannot be suspended by the fluid turbulence. If coarse particles are dispersed by inter-particle collisions, flow should behave as a dilatant fluid flow, but as mentioned above, the flow behaves as a Newtonian laminar flow. Therefore, the coarse particles in viscous debris flow are dispersed neither by turbulence nor by inter-particle collisions. Then, what is the mechanism to disperse coarse particles in a highly viscous laminar flow? Single phase fluid models, such as the Bingham fluid model and the Hershel-Bulkley fluid model are based on a uniform distribution of particles within the flow so that the mechanism to disperse particles is a black box. Iverson (1997) claims the following: the coarse particles in a viscous debris flow are all irreversibly settling towards the bottom, but, the time necessary to settle down all the coarse particles is much longer than the life span of the debris flow itself. Therefore, the particles are apparently dispersed as long as the debris flow continues. His reasoning has some quantitative basis, but it may fail to explain

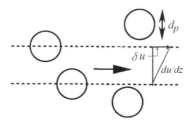

Figure 2.47 Particles in the adjacent shearing surfaces.

phenomena observed in the Jiangjia Gully. If this speculation is true, the particles near the bed are unilaterally deposited and the debris flow surge must diminish downstream. This phenomenon has not been observed in the Jiangjia Gully. Furthermore, it cannot explain the erosion process of deposit, which is left after a passage of surge, by a succession of new surges. Therefore, there should be a mechanism to keep the coarse particles dispersed; the mechanism to balance the settling motion and upward diffusive motion.

Takahashi (1993) considered that the approach of two particles embedded in adjacent shearing surfaces causes the perpendicular force to move the particles perpendicular to the general shearing plane and proposed formulae to evaluate pressure and shearing force.He later found that Phillips *et al.* (1992) had done the discussion based on a similar idea for the case of neutrally buoyant grains. Herein, the viscous debris flow is treated by modifying their theory to fit the case of heavy particles (Takahashi *et al.* 1998, 2000).

Consider the particles embedded in the adjacent shearing surfaces moving past one another as depicted in Figure 2.47. A particular particle in a shearing surface makes γC encounters in unit time with the particles in another surface, where γ is the magnitude of the local shear rate. The difference between the number of encounters at the lower side and the upper side of a particular particle is given by $d_p \nabla(\gamma C)$. This difference will cause a particle to be irreversibly moved from its original streamline to the area of the lower encounter frequency. If the particle migration velocity is assumed to be linearly proportional to the difference in encounter frequency and if each of these two-body interactions makes a displacement of $O(d_p)$ (order of scale d_p), the particle flux perpendicular to the main flow direction $\mathbf{N}_c$ is given by:

$$\mathbf{N}_c = -K_c d_p^2 (C^2 \nabla \gamma + C \gamma \nabla C) \tag{2.176}$$

where K_c is a constant.

If coarse particles are distributed anisotropically, the apparent viscosity varies spatially. Because the apparent viscosity is larger in the higher concentration region than in the smaller concentration region, the resistance to migration into the higher concentration region is larger than into the thinner concentration region. Assuming that the migration velocity is proportional to the variation of viscosity; $(d_p/\mu_a)\nabla\mu_a$, and that each two-body encounter produces the displacement of $O(d_p)$, the particle flux $\mathbf{N}_{\mu_a}$ due to the spatial variation in the viscosity is given by:

$$\mathbf{N}_{\mu_a} = -K_{\mu_a}\gamma C^2 \frac{d_p^2}{\mu_a}\frac{d\mu_a}{dC}\nabla C \tag{2.177}$$

where μ_a is the apparent viscosity of the mixture and K_{μ_a} is a constant.

The coarse particles in a debris flow are heavier than the interstitial fluid so that they have the tendency to settle down. The settling flux $\mathbf{N}_s$ due to gravity is given by Stokes law as:

$$\mathbf{N}_s = -\frac{2}{9}C\frac{d_p^2(\sigma - \rho)g\cos\theta}{\mu}g(C) \tag{2.178}$$

where μ is the viscosity of interstitial fluid and $g(C)$ is a hindrance function to account for the highly concentrated group settling; here, simply:

$$g(C) = \frac{1-C}{\eta_a} \tag{2.179}$$

is used, where η_a is the specific viscosity, and according to Krieger (1972) it is given as:

$$\eta_a = \frac{\mu_a}{\mu} = \left(1 - \frac{C}{C_*}\right)^{-1.82} \tag{2.180}$$

Therefore, the particle conservation equation can be written as:

$$\frac{\partial C}{\partial t} = -\nabla \cdot (\mathbf{N}_c + \mathbf{N}_{\mu_a} + \mathbf{N}_s) \tag{2.181}$$

In the steady state the left-hand side of Equation (2.181) is zero, then, considering Equation (2.180), Equation (2.181) becomes:

$$K_c\left(C^2\frac{d\gamma}{dz} + C\gamma\frac{dC}{dz}\right) + K_{\mu_a}\gamma C^2\frac{1}{\mu_a}\frac{d\mu_a}{dC}\frac{dC}{dz} + \frac{2}{9}C(\sigma - \rho)g\cos\theta\frac{1-C}{\mu_a} = 0 \tag{2.182}$$

It is noted that this particle conservation equation is independent of particle diameter.

If the shearing stress on the plane at height z is set equal to $\mu_a\gamma$ as is satisfied for a Newtonian fluid, the following velocity distribution equation is obtained:

$$\gamma = \frac{du}{dz} = \left(\frac{\rho u_*^2}{\mu_a}\right)\left\{\left(1 - \frac{z}{h}\right) + \frac{\varepsilon}{h}\int_z^h C\,dz\right\} \tag{2.183}$$

where:

$$\varepsilon = (\sigma - \rho)/\rho \tag{2.184}$$

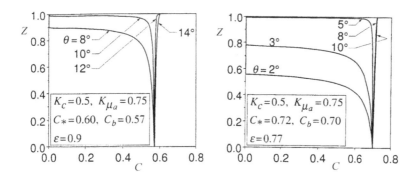

Figure 2.48 Distribution of coarse particle concentrations on various slopes.

From Equations (2.180), (2.182) and (2.183) the coarse particle concentration distribution function is given by:

$$\frac{dC}{dZ} = \frac{F_1}{F_2 F_3} \tag{2.185}$$

where $F_1 = -(2/9)\varepsilon(1 - C)/(K_c \tan \theta) + C(1 + \varepsilon C)$, $F_2 = 1 + 1.82(K_{\mu a}/K_c - 1)(C/C_*) \times (1 - C/C_*)^{-1}$,

$$F_3 = 1 - Z + \varepsilon \int_Z^1 C \, dZ$$

The calculated coarse particle concentration distributions under the combinations of parameters; $C_* = 0.6$, $C_b = 0.57$, $\rho = 1.38 \, \text{g/cm}^3$ $(\varepsilon \approx 0.9)$, and $C_* = 0.72, C_b = 0.70, \varepsilon = 0.77$ are shown in Figure 2.48. Respective combinations correspond to the conditions in the experiment which are referred to below and in the Jiangjia Gully. Both cases adopt $K_c = 0.5$ and $K_{\mu a} = 0.75$ obtained by Phillips *et al.* (1992). The concentration at the bottom of the flow C_b is assumed to be a little smaller than C_*, because substitution of $C = C_*$ into (2.185) does not make sense. As is clear in Figure 2.48, when the channel slope is steep, very high uniform concentration is maintained up to the surface of the flow, and for a relatively shallow slope channel, the high concentration suddenly decreases in the upper region. The coarse particle concentration in a certain slope channel becomes large and uniform as ε becomes small. Namely, the denser the interstitial fluid, the flatter the area debris flow of a certain solids concentration will reach.

The equilibrium coarse particle concentration for a given channel slope can be obtained by integrating Equation (2.185). The results corresponding to the conditions used in obtaining Figure 2.48 are shown in Figure 2.49.

The velocity distribution can be obtained by solving Equation (2.183) under the boundary condition; at $Z = 0, u' = 0$, where $u' = u/u_*$. The solution cannot be obtained analytically, but if the viscosity is approximated as a constant from the bottom to a height $Z = Z_2$, it is given by the following equation:

$$u' = \begin{cases} \dfrac{\rho u_* h}{\mu_a}(1 + \varepsilon \overline{C})(Z - \dfrac{1}{2}Z^2) & ; \quad 0 \le Z \le Z_2 \\ \dfrac{\rho u_* h}{\mu}\left\{Z - \dfrac{1}{2}Z^2 + X\left(Z_2 - \dfrac{1}{2}Z_2^2\right)\right\} & ; \quad Z_2 < Z \le 1 \end{cases} \tag{2.186}$$

where $X = (1 + \varepsilon \overline{C})\mu/\mu_a - 1$, $\overline{C}$ is the mean particle concentration below Z_2 and Z_2 is the non-dimensional height up to which the concentration is approximately constant. The velocity distribution by Equation (2.186) has a pattern like the one shown in Figure 2.50.

If one assumes that the concentration from the bottom to the surface is approximately uniform, $Z_2 = 1$ and the upper formula of Equation (2.186) applies throughout the depth. The integration of this upper formula gives the cross-sectional mean velocity and the result is written as:

$$\frac{U}{u_*} = \frac{1}{3}\frac{\rho u_*}{\mu_a}(1 + \varepsilon \overline{C})h \tag{2.187}$$

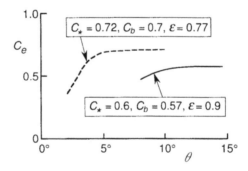

Figure 2.49 Equilibrium coarse particle concentrations versus channel slopes.

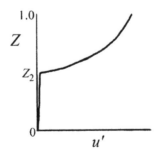

Figure 2.50 Schematic velocity distribution in an immature viscous debris flow.

2.8.2 Verification by experiments

Takahashi *et al.* (1997) carried out two kinds of flume experiments, in which a steel movable slope flume 10 m in length, 9 cm in width and 40 cm in depth was used. One side of the flume was a transparent glass wall through which the behavior of the flow could be observed. The bed was roughened by pasting gravel 3 mm in diameter to the bed.

For the experiments to measure the viscosity of the material, 13,000 cm^3 of the debris flow material was stored in the hopper set at 3.5 m upstream of the outlet of the flume. A high speed video set beside the flume 2 m downstream of the hopper recorded the flow structure at a rate of 200 frames per a second.

To measure the processes of flow, deposition and remobilization of the deposit, the hopper was set at 7.5 m upstream of the outlet. In these cases 25,000 cm^3 of material was prepared in the hopper and it was poured on the dry bed of the flume to generate a first surge. The discharge from the hopper gradually decreased with the volume reduction of material in the hopper, and shortly after all the material in the hopper was released a certain thickness of deposit remained on the bed of the flume nearly parallel to the bed surface. The same quantity and quality of the material as the first surge was prepared again in the hopper. Then, five to six minutes after the termination of the first surge, the second surge was introduced from the hopper to the now movable bed of the flume. In some cases, a third, fourth and fifth surge were produced with the same procedure. In the experiment, only the channel slope was changed to vary the conditions. The surface level of the flow was measured at two positions; 3 m and 6 m from the upstream end, respectively, using laser deflection sensors. The discharge variation from the hopper was calibrated by capturing the discharge directly with a calibration box. The outflow discharge from the flume was also measured using the calibration box. In this chapter only the characteristics during the first surge is explained and the results of sequential surge experiments will be used to verify the routing method in Chapter 4.

The well-mixed characteristics of the debris flow material in the Jiangjia gully was simulated (except for the component larger than 10 mm) by mixing silica sand and kaolin. The particle size distributions of the materials in the Jiangjia gully (the component less than 20 mm) and in the experimental one are compared in Figure 2.51. Although debris flows in the Jiangjia gully are said to flow with the solids volume

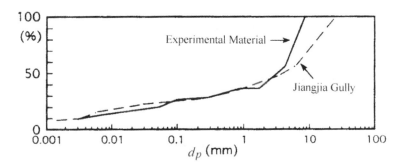

Figure 2.51 The particle size distributions of the materials in Jiangjia gully and in the experiments.

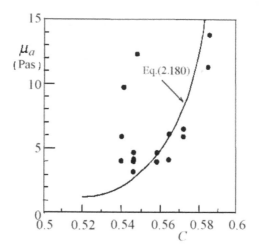

Figure 2.52 Apparent viscosity as a Newtonian fluid versus solids concentration.

concentration as much as 0.53 to 0.85 (Wu *et al.* 1990), the debris flow in the experiment could not reach the outlet of the flume if the solids concentration was higher than 0.58, and if it was lower than 0.55 the material easily segregated in the hopper before feeding into the flume. Therefore, many experimental runs were conducted under nearly constant solids concentration between 0.56 and 0.57. The debris flow material in the Jiangjia gully has a bimodal distribution and it lacks the fraction whose particle diameter is around 0.1 mm (see Figure 1.10). Therefore, in this material the particles less than 0.1 mm may be considered as material that composes the slurry. Similarly, if particles less than 0.1 mm in the experimental material are considered to compose the slurry, the slurry has an apparent density of 1.38 g/cm³ (volume concentration of the solids is 0.226). The slurry in the experimental material itself was well modeled as a Bingham fluid having a viscosity and yield strength of 0.036 Pas and 11.8 Pa, respectively.

The apparent viscosity of the experimental material as a Newtonian fluid was obtained as shown in Figure 2.52. At about a concentration of 0.54 the points widely scatter because this was the least concentration to be able to prepare as a nearly homogeneous material. Neglecting the scattered values at $C = 0.54$, the viscosity approximately follows Equation (2.180) under the conditions $\mu = 0.036$ Pas and $C_* = 0.6$ as shown by the curve in Figure 2.52.

The velocity distributions given by the upper formula of Equation (2.186) are compared with the experimental data in Figure 2.53, in which the experimentally obtained values of $\mu = 0.036$ Pas, $\overline{C} = 0.56$ and $C_* = 0.59$ are used. Figure 2.53 shows that not only the velocity distribution obeys the Newtonian laminar flow equation, but also the absolute velocity values are well explained by the theory introduced above.

Let us apply the theory to the Jiangjia gully debris flow. According to the observational data (Wu *et al.* 1990), the volume concentration of the fine particles in the slurry is about 30% and the volume concentration of the coarse particles in the debris flow

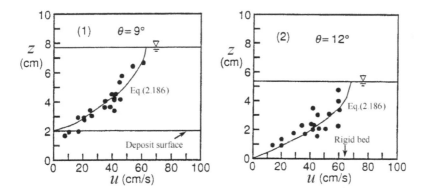

Figure 2.53 The calculated (curve) and experimental (circles) velocity distributions in the first (1) and second (2) surges.

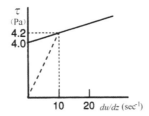

Figure 2.54 The stress-strain relationship of a slurry in viscous debris flow at the Jianjia gully (solid line), and the estimation of viscosity as a Newtonian fluid (broken line).

is about 55%. Thus, the densities of the slurry and the entire debris flow material are, respectively, $\rho = 1.5$ g/cm^3 and $\rho_T = 2.13$ g/cm^3; i.e. $\varepsilon = 0.77$. The total solids concentration in the debris flow material is 0.685. Estimation of C_* has a rather high sensitivity to the result of the calculation. Nevertheless, from the previous examinations (Wu *et al.* 1990), $C_* = 0.72$ is adopted. As for determining η_a, the deformable space not occupied by solids is important. Hence, in taking C_* and C values in Equation (2.180), the entire solids concentration including coarse particles and clay should be used. Thus, $\eta_a = 246$ is obtained. The viscosity of the slurry μ is affected by the electrical force between very fine particles and cannot be estimated by Krieger's formula. It must be found from experimental data at present. Previous rheometer tests for the slurry having solids concentration of about 30% regarded the slurry as a Bingham fluid, and they obtained a viscosity of $\mu_B = 0.02$ Pas and a yield strength of $\tau_B = 4$ Pa. The stress and strain relationship in the tested material was as shown in Figure 2.54. Note that the estimate of 4.0 Pa as the yield strength was obtained by linear extrapolation from the relationship at large strains. But, actually, the slurry has the characteristics of a Herschel-Bulkley fluid, and in the domain of a smaller strain rate of less than about 10/s; i.e. the order of strain rate in actual debris flow, it has a tendency to behave as a Newtonian fluid as drawn by broken line in Figure 2.54. Therefore, if the slurry is regarded as a Newtonian fluid and its viscosity is estimated from the slope of broken line in Figure 2.54, the viscosity has a value of about $\mu = 0.42$ Pas. Consequently, the

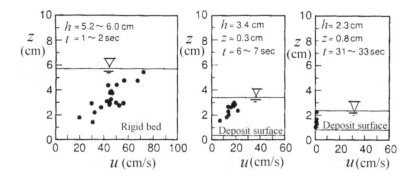

Figure 2.55 Temporal change in velocity, depth and deposit thickness ($\theta = 11°$).

apparent viscosity of the entire debris flow material is $\mu_a = 103$ Pas. Assuming $h = 100$ to 200 cm, $\theta = 3°$ and $\varepsilon = 0.77$ in Equation (2.187), we obtain $U/u_* = 5$ to 14. This value coincides with the observations shown in Figure 1.13.

Figure 2.55 shows the temporal changes in the velocity distribution, the flow depth and the deposit thickness in an experimental run. The deposit layer gradually thickens and the vertical velocity gradient becomes smaller with a decrease in the flow depth without making a solid plug near the surface of the flow. These characteristics cannot be explained by the widely accepted Bingham fluid model, in which stoppage of flow, and therefore the production of sediment accumulation of a certain thickness on the channel bed, occurs suddenly when the flow depth becomes shallower than the plug thickness. Armanini *et al.* (2003) verified the model of Takahashi *et al.* (199) with slight modifications based on their experiments.

2.9   EQUILIBRIUM SEDIMENT DISCHARGE IN INERTIAL DEBRIS FLOWS

Figure 2.56 compares the relationships between the non-dimensional equilibrium sediment discharge Φ, and the non-dimensional shear stress τ_*, in the formulae for the mature stony debris flow, the immature debris flow, and the generalized theory of inertial debris flow with the experimental data, where $\Phi = q_B/\sqrt{(\sigma/\rho - 1)gd_p^3}$, $\tau_* = u_*^2/[(\sigma/\varrho - 1)gd_p]$ and q_B is the unit width discharge of the bed material. The I value in Figure 2.56 indicates the channel slope. The data of Mizuyama, Smart and Takahashi are for the coarse particles transportation with the negligible amount of suspended sediment and the data of Arai and Takahashi are for the fine suspended particles. As shown in Figure 2.30 the immature debris flow for coarse particles appears in the channel slope range $4° < \theta < 10°$. Therefore, the data for the channel slope of $I = 0.2$ and $I = 0.3$ should be the one for the mature debris flow and the data for $I = 0.1$ should be for the immature debris flow.

The theoretical curve for the immature debris flow, for $I = 0.1$, well fits to the experimental results. The experimental data are scattered but the tendency that the

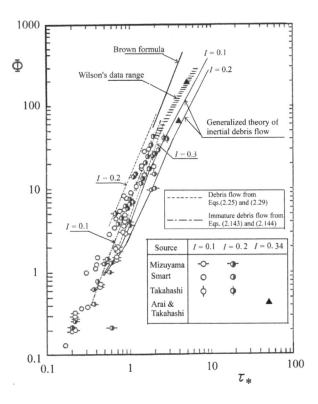

Figure 2.56 Equilibrium sediment discharges of the inertial debris flows.
Data sources: Mizuyama (1980); Smart (1984); Takahashi (1982); Arai and Takahashi (1986).

flatter the channel slope the larger the Φ value becomes can be seen and the theoretical curves for the mature debris flows whose channel slopes are $I = 0.2$ and $I = 0.3$ show a similar tendency. The generalized theory for the inertial debris flow seems to estimate a lesser sediment discharge than the specific stony and immature debris flow theories as long as those flows are comprised of coarse material.

It must be noted that particles are suspended in flow if the shear velocity of flow exceeds the particle's settling velocity. For the particles larger than about 1 mm for which Newton's settling velocity formula is applicable, the threshold τ_* value to suspend these particles is about 3.3, and the particles less than 1 mm for which Stokes' settling velocity formula is applicable are suspended if τ_* is larger than 1. Therefore, the curves for the generalized theory in Figure 2.56 in the large τ_* range indicate the cases in which majority of particles are suspended. However, the number of data for the fine particles on the steep slope bed is insufficient, so the evaluation of the generalized theory should be put off until the data for fine sediment become abundant.

It must also be noted that the bed load formula for plane bed on a shallow slope channel is given as (Takahashi 1987):

$$\Phi = 12\tau_*^{3/2}\left(1 - 0.85\frac{\tau_{*c}}{\tau_*}\right)\left(1 - 0.92\sqrt{\frac{\tau_{*c}}{\tau_*}}\right) \tag{2.188}$$

where τ_{*c} is the critical shear stress to initiate particle motion and this is equal to 0.04. This formula is known in Japan as Ashida, Takahashi and Mizuyama's bed load formula. Although the curve for Equation (2.188) is not shown in Figure 2.56, it almost coincides with the curve for $I = 0.1$ for immature debris flow in Figure 2.56 and moreover it almost coincides with Wilson's data (Wilson 1966) up to $\tau_* = 10$. Takahashi (1987) further modified Equation (2.188) to make the formula applicable to a steeper channel. The latter formula for $I = 0.2$ almost coincides with the curve of the generalized theory of inertial debris flow for $I = 0.1$, so these bed load formulae also show the tendency the steeper the channel slope the smaller the Φ value becomes. Anyway, the existing bed load formulae including the famous Meyer-Peter and Müller formula and the Engelund and Hansen formula estimate similar Φ values to stony mature and immature debris flows in the range $1 < \tau_* < 3$. Thus, the existing bed load formulae would be applicable for the cases of stony mature and immature debris flows even though they do not intend to explain the sediment discharge in debris flows.

In Figure 2.56 the Brown formula, $\Phi = 10\tau_*^{5/2}$ is shown. This formula is almost identical to the Graf and Acaroglu (1968) regression formula: $\Phi = 10.39\tau_*^{2.52}$. These formulae were proposed to explain the total sediment discharge of bed load plus suspended load on a rather flat slope bed. These formulae have the tendency to estimate a very large Φ value at a large τ_* that is larger than 3. There would be two reasons for this tendency: the small resistance to flow due to comparatively thin sediment concentration in the flow on a gentle slope bed seems to transport eventually more sediment than densely concentrated and highly resistant debris flow on a steep slope bed, and the effective shear velocity to transport particles in a highly sediment concentrated flow would be less than that estimated by $u_* = \sqrt{(ghI)}$ (see the discussion in section 3.1.2). The former reasoning can be proved by the following examination: If the channel slope gradient $I = 0.01$, the particle diameter $d_p = 0.2$ mm, the flow resistance law $U/u_* = 20$, and $\tau_* = 10$ are assumed, $h = 0.33$ m, $q = Uh = 1.187$ m^2/s and from Brown formula $\Phi = 3162$ are obtained. Then, the total sediment discharge, almost all in this case is suspended load, is obtained to be equal to 0.036 m^2/s. Thus, the sediment concentration in the flow is only about 3%. This means that even if the sediment discharge is very large the sediment concentration in the flow is far less than that in the debris flow. More experimental examinations are needed to arrive at reliable conclusions.

Chapter 3

Initiation, development and declination of debris flow

The photograph shows aspects of the debris flow source area in the Jiangjia Gully, Yunnan, China. It suggests that many small-scale debris flows (incipient debris flows) link up to generate a large-scale debris flow.

INTRODUCTION

We often see aerial photographs, taken after conspicuous sediment hazards, showing the long white trace along a ravine that widens at the outlet fan area, where houses on it are destroyed or buried. The elongated trace is the track of debris flow. In some cases there is an obvious landslide scar at the origin; but in other cases there is no such scar. The debris flow in the former case was obviously generated by the landslide, but the latter case shows that debris flow can be initiated without a landslide.

As stated in section 1.5, the mechanical causes of debris flow initiation can be classified into three types:

1   the deposit on gully bed is eroded by the supply of water from outside and the concentration of solids in the surface water flow becomes as dense as it can be and is called debris flow;
2   the landslide block transforms into debris flow while in motion by the effects of stored water in the slid earth block or by the supply of water from outside;
3   the collapse of a debris dam suddenly makes up debris flow.

In this chapter the processes of debris flow initiation, development and declination due to these three causes are discussed in detail.

A very large-scale landslide that is often called 'debris avalanche' reaches a great distance even if it does not transform into debris flow while in motion. Many discussions on the mechanisms of debris avalanches have introduced special reasonings other than those for debris flow. However, in this chapter a theory that is common to the transformation of a landslide into debris flow is developed. Thus, many debris avalanches are considered as the phenomena on the way to the transformation into gigantic debris flows.

3.1 INITIATION AND DEVELOPMENT OF DEBRIS FLOW DUE TO GULLY BED EROSION

3.1.1 The formation of incipient debris flow by the effects of surface water runoff

On a steep ephemeral gully bed; no water stream exists in ordinary time, sediment accumulates gradually due to the supply from side walls as rock-falls and landslips. Because a gully is the path of concentrated water, the gully bed will likely be washed out by a large-scale water flow when it rains severely. Sometimes, a small-scale landslip occurs by the effects of enhanced water content and liquefies while in motion on the slope. Less frequently, the rapid melting of snow or ice cover caused by the effects of an abrupt atmospheric temperature rise will supply an abnormally large amount of water.

Consider an infinitely long, uniform and void-rich sediment layer that is saturated with water. It has a thickness D and a slope θ, on that parallel surface, water of depth h_o is flowing. If both the hypothetical shearing stress, τ, which acts to drag the sediment block downstream and the resisting stress, τ_r, are assumed to distribute in straight lines, there should exist six cases corresponding to the relative arrangement of those two lines as shown in Figure 3.1 (Takahashi 1977; Takahashi 1991). In cases (1), (2) and (4) τ exceeds τ_r at the rigid bed surface and so the whole sediment layers are unstable; and in cases (5) and (6) τ_r is larger than τ and the whole sediment layers are stable. In case (3) the sediment layer itself is stable but the upper portion of thickness a_c is unstable. If, as assumed herein, the motion of the sediment layer starts under a sufficient quantity of water, i.e. enough to fill up the enlarged void space between particles, a massive flow of sediment and water mixture, i.e. debris flow will be generated. Therefore, in cases (1), (2), (3) and (4) debris flow may arise.

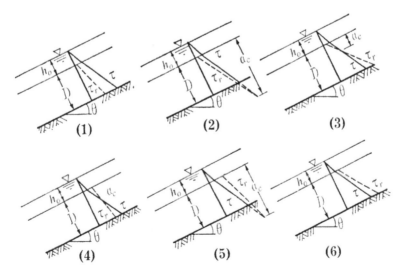

Figure 3.1 Stress distributions in a sediment layer.

The shear stress at depth a measured from the surface of the sediment layer is written as follows:

$$\tau = g \sin \theta \{C_*(\sigma - \rho)a + \rho(a + h_o)\} \tag{3.1}$$

and the resisting stress at depth a is written as:

$$\tau_r = g \cos \theta \{C_*(\sigma - \rho)a\} \tan \varphi + c \tag{3.2}$$

where c is the apparent cohesive strength of the sediment layer.

As is evident in Figure 3.1, one of cases (1), (4) and (5) occurs when $d\tau/da \geq d\tau_r/da$ is satisfied and when $d\tau/da < d\tau_r/da$, one of cases (2), (3) and (6) occurs. When τ $(= \rho g h_o \sin \theta)$ at the surface of sediment layer $(a = 0)$ is larger than c, cases (1), (2) and (3) will occur, otherwise cases (4), (5) and (6) will occur. Furthermore, cases (2) and (5) occur when $a_c \geq D$ and cases (3) and (4) occur when $a_c < D$.

In case (3) the part in which a is smaller than a_c should be unstable, but if a_c is less than one representative particle diameter, the sediment layer itself is stable and only individual particles existing on the surface of the sediment layer will be picked up by the fluid dynamic force in the surface flow. This is not the sediment mass flow (debris flow and immature debris flow) but the individual particle transport (bed load and suspended load). Therefore, if we write $a_c = nd_p$, n should be larger than 1 to generate sediment mass flow. Even when the condition $n \geq 1$ is satisfied, if a_c is far less than h_o, the coarse particles cannot be dispersed in the entire flow depth but move concentrating only in the lower part forming an immature debris flow. This means that to generate a (mature) stony debris flow a_c should be larger than kh_o, where k is a numerical coefficient near unity. This condition is not necessarily applicable to the initiation of a turbulent muddy debris flow, it may occur even on a flatter slope if h_o/d_p is larger than about 40 as discussed in Chapter 2.

Based on the above considerations, the criteria for the existence of cases (1)–(6) are obtained as depicted in Figure 3.2, in which the domains of stony debris flow occurrence are hatched and the domains of cases (1)–(6) are indicated by the number in parentheses. The domains labeled (3′) and (3″) correspond to the immature debris flow and the individual particle transport, respectively.

The curves which give the domain boundaries are deduced from Equations (3.1) and (3.2) as follows:

$$\sin \theta = \frac{c}{\rho g h_o} \tag{3.3}$$

$$\tan \theta_2 = \frac{C_*(\sigma - \rho)}{C_*(\sigma - \rho) + \rho} \tan \varphi \tag{3.4}$$

$$\tan \theta_1 = \frac{F_0}{F_1} \left\{ 1 + \frac{c}{kF_0 g h_o} \frac{(1 - c^2 k^{-2} F_1^{-2} g^{-2} h_o^{-2} + F_0^2 F_1^{-2})^{0.5}}{(1 - c^2 k^{-2} F_1^{-2} g^{-2} h_o^{-2})} \right\} \tag{3.5}$$

$$\tan \theta_2 = \frac{F_0}{F_2} \left\{ 1 + \frac{c}{F_0 g a_c} \frac{(1 - c^2 F_2^{-2} g^{-2} a_c^{-2} + F_0^2 F_2^{-2})^{0.5}}{(1 - c^2 F_2^{-2} g^{-2} a_c^{-2})} \right\} \tag{3.6}$$

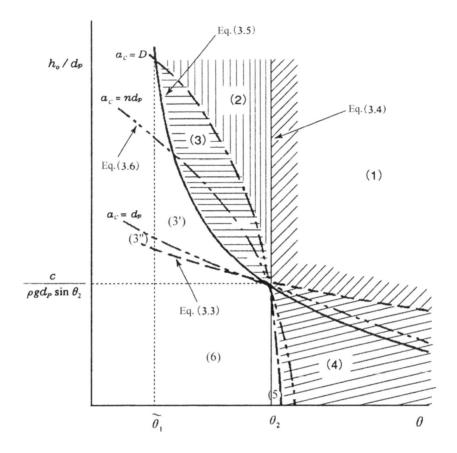

Figure 3.2 Occurrence criteria of various type flows on coarse cohesive sediment bed.

where:

$$\left.\begin{array}{l} F_0 = C_*(\sigma - \rho)\tan\varphi \\ F_1 = C_*(\sigma - \rho) + \rho(1 + k^{-1}) \\ F_2 = C_*(\sigma - \rho) + \rho(1 + h_o a_c^{-1}) \end{array}\right\} \tag{3.7}$$

Equation (3.4) is satisfied when $d\tau/da$ is equal to $d\tau_r/da$, which is the condition for a sediment layer to slide at the moment when the ascending ground water level coincides with the surface of the sediment bed. Therefore, if a sediment accumulation layer steeper than θ_2 has a cohesive strength smaller than $\rho g h_o \sin\theta$, the layer is in the domain of case (1). It will slide in the process of ascending ground water before it is thoroughly saturated by seepage water. If a sediment layer steeper than θ_2 has a cohesive strength larger than $\rho g h_o \sin\theta$, cases (4) or (5) will occur. Case (5), however, can only occur in the narrow domain shown in Figure 3.2 when cohesive strength is very large, the layer itself is stable and the erosion by the surface water flow can take place only as the individual particle motion. In case (4) which arises when the cohesive strength is not so large, a landslide occurs similar to case (1) before the appearance of surface water flow. Consequently, when the cohesive strength is not large, if a sediment layer is steeper than θ_2, the possibility of sliding occurring before the appearance of surface water flow is large. In such a case, unless the sediment layer has a very large void ratio structure and the quantity of seepage water contained at the moment of slide is sufficient to fill the reduced void after the destruction of such a skeletal structure, it is difficult to transform into debris flow. Nevertheless, there is a witness record of the occurrence of a debris-flow-like motion at the source area of a basin before the appearance of surface water flow (Imaizumi et al. 2003). Sometimes, surface water flow can appear even if the deeper part of sediment layer is not saturated, as an example is shown in Figure 1.15. The surface water flow in such a case can erode the sediment layer by the action of fluid dynamic force and the eroded material is entrained into the flow as it becomes debris flow, if the channel slope is steep and long. The transformation process of landslide into debris flow will be discussed later in another section.

Cases (2) and (3) are essentially the cases of a stable sediment layer. Case (2) corresponds to the situation in which a very deep surface water flow is suddenly supplied so that a_c becomes larger than D. Therefore, if the depth of surface water flow is not large, a_c becomes smaller than D and it becomes case (3). The domains for cases (2) and (3) (upper part of the curve indicating Equation (3.5) in Figure 3.2) are the domains of debris flow from the instant of initiation. In Figure 3.2 the curves representing Equation (3.6) are depicted for the cases corresponding to $a_c = D$, $a_c = nd_p$ and $a_c = d_p$, respectively. The domain between the curve for Equation (3.5) and the curve for Equation (3.6) that corresponds to $a_c = d_p$ is the domain of immature debris flow initiation (domain 3′).

The actual sediment bed may have little cohesion because it is often a pile of once mobilized sediment supplied from the side walls. When cohesion is negligible, Equations (3.5) and (3.6) are simplified as:

$$\tan\theta_1(\equiv \tan\tilde{\theta}_1) = \frac{C_*(\sigma - \rho)}{C_*(\sigma - \rho) + \rho(1 + k^{-1})} \tan\varphi \tag{3.8}$$

$$\tan\theta = \frac{C_*(\sigma - \rho)}{C_*(\sigma - \rho) + \rho(1 + h_o/a_c)} \tan\varphi \tag{3.9}$$

Equations (3.4), (3.8) and (3.9), the equation giving the critical slope steeper than which no sediment bed can exist:

$$\theta = \varphi \tag{3.10}$$

and the critical condition for initiation of bed load transportation (Ashida et al. 1977a):

$$\frac{\rho u_{*c}^2}{(\sigma - \rho)gd_p} = 0.034\cos\theta\left\{\tan\varphi - \frac{\sigma}{(\sigma - \rho)}\tan\theta\right\} \cdot 10^{0.32(d_p/h_c)} \tag{3.11}$$

divide the plane, whose two orthogonal axes are h_o/d_p and $\tan\theta/(\sigma/\rho - 1)$, into various characteristic domains in view of the non-cohesive sediment transport as shown in Figure 3.3. The boundary curves in Figure 3.3 are calculated under the conditions $C_* = 0.7$, $\sigma = 2.65\,\text{g/cm}^3$, $\rho = 1.0\,\text{g/cm}^3$, $k = 0.7$ and $\tan\varphi = 0.8$.

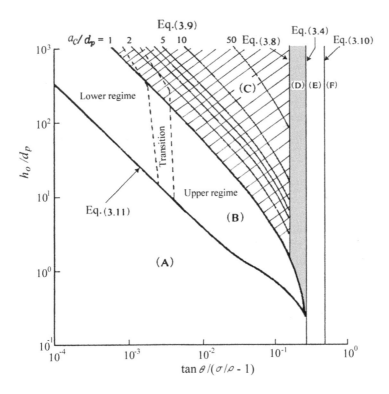

Figure 3.3 Occurrence criteria of various type flows on non-cohesive coarse sediment bed.

In Figure 3.3, the domain (A) is that of no particle motion; (B) is the domain of individual particle transport affected by fluid dynamic force, i.e. the bed load transport; and (C) (hatched domain) is the domain of immature debris flow. Numbers attached to the respective curves in this domain correspond to the thickness of the mobilized sediment layer and if a_c/d_p exceeds kh_o/d_p the flow becomes mature debris flow in the domain (D) (shaded domain). Note that the previously proposed domain of the 'upper regime' defined for the bed configuration under water flow (Ashida and Michiue 1972) contains both domains of individual particle transportation and immature debris flow. The domain (D) corresponds to the hatched domains (2) and (3) in Figure 3.2. The domain (E) is for the landslide caused by seepage flow and the domain (F) is for the unstable bed.

Equation (3.8) that defines the critical slope for the stony debris flow initiation, was verified by experiments (Takahashi 1977). Substituting the properties of the sediment bed and the k value used in the calculation of the boundary curves in Figure 3.3 ($k = 0.7$) into Equation (3.8), we obtain $\tilde{\theta}_1 = 14.5°$. This is in accord with the lower bound of channel slopes to generate actual debris flows.

The minimum value of h_o/d_p at the critical slope for debris flow initiation, $\tilde{\theta}_1$, is around 1.

Therefore, the process of debris flow initiation on a sediment bed whose slope is 14°–15° or more is speculated to be as follows: Associated with a continuous rainfall, the ground water level in the sediment bed gradually rises and in time surface water flow appears. As long as the depth of the surface water flow is small, only relatively small particles existing on the bed will be selectively transported as bed load and an irregularly-shaped, armor-coated bed will remain. Under such circumstances, when the rainfall intensity suddenly increases, the depth of surface flow rapidly increases. If the depth of surface flow becomes larger than the representative diameter of the bed surface particles (this idea is vague, but here it is considered as the mean diameter of the bed surface particles), the upper part of the sediment layer becomes unstable and will be entrained into the flow. The thickness of unstable layer is obtained from Equation (3.9) as:

$$\frac{a_c}{d_p} = \frac{C_\infty}{C_* - C_\infty} \frac{h_o}{d_p} \tag{3.12}$$

where C_∞ is the equilibrium particle concentration in the flow at the particular position that is being considered, that is given by Equation (2.25). If, as an example, $\theta = 20°$, $C_* = 0.7$, $\sigma/\rho = 2.65$ and $\tan \varphi = 0.8$ are substituted into Equation (3.12), $C_*/(C_* - C_\infty)$ is about 2.6. Namely, if the depth of surface water flow is of the order of the mean diameter of the bed, the upper part of the bed as thick as 2.6 times the mean diameter becomes unstable and starts to move. Thus, even a big particle whose diameter is as large as 2.6 times the mean diameter will be entrained into the flow. The mechanism of transportation of such big particles will be discussed in the next chapter. If $\theta = 15°$ is substituted into Equation (3.12), the coefficient $C_*/(C_* - C_\infty)$ is 0.8. This means that for $h_o/d_p = 1$, a_c becomes smaller than one particle diameter so that the premise for the initiation of debris flow is not satisfied. This contradicts with the conclusion that the critical slope to generate debris flow is about 14.5°. This nonconformity is brought by assuming $k = 0.7$. If $k = 1$ is used, $\tilde{\theta}_1$ becomes 16.3° and the nonconformity disappears. Although the value of k was obtained by experiments, it may be better to use $k = 1$.

Then, a rough estimate to judge the occurrence of stony debris flow with an obvious bore-like front is whether the surface flow deeper than the mean diameter of the bed surface particles suddenly appears on the channel reach steeper than 15 degree.

Even when the sediment bed is not saturated with water, the surface water flow can appear rapidly when severe rainfall begins suddenly. In such a case, even though the condition $h_o > d_p$ is satisfied, the sediment motion starts as individual particle motion. But, as long as steep sediment bed continues downstream, the particle concentration can develop downstream to form a debris flow.

The ordinate of Figure 3.3, which is the non-dimensional surface water depth, is not explicitly obtainable. It can be obtained by knowing the water discharge on the sediment layer and applying an appropriate resistance law. Therefore, in practice, the criteria is more easily found on the plane whose two orthogonal axes are $\sin \theta$ and the non-dimensional discharge $q_*(= q_o/g^{1/2}d_p^{3/2})$ as shown in Figure 3.4 (Takahashi 1987).

In Figure 3.4 immature debris flow occurs in the domain V and debris flow occurs in the domains VI and VII. The lines F and G represent Equations (3.8) and (3.4), respectively, and the curves A and E are equivalent to Equations (3.11) and (3.9), respectively. The curve B is obtained from the critical discharge equation for the initiation of particle motion presented by Bathurst et al. (1985).The curves C and D are,

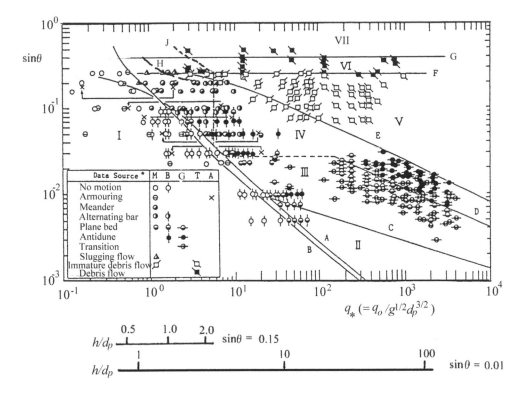

Figure 3.4 Criteria for flows and bed forms on an erodible steep slope channel composed of non-cohesive coarse particles.
* Data sources abbreviation: M: Mizuyama (1977), B: Bathurst *et al.* (1985), G: Gilbert (1914), T: Takahashi (1977, 1982), A: Ashida *et al.* (1984a).

respectively, the lower and the upper regime divides given by Garde and Ranga Raju (1963). The curve H is the experimentally obtained threshold for the debris flow initiation on a water saturated bed (Armanini and Gregoretti 2000) and the curve J is the threshold experimentally obtained on a non-saturated bed (Tognacca *et al.* 2000). The experimental data plotted in Figure 3.4 verify the division of flow types proposed herein. The boundary curve defined by Armanini and Gregoretti (2000) shows that the stony-type debris flow occurs on a bed steeper than about 12° when q_* is larger than 1 or 2.

The relationship between the discharge and the flow depth for water flow on the gully bed cannot be determined uniquely, but it depends on the slope. According to the experiments to find the resistance law for plain water flow on a very steep bed (Takahashi 1977), the following formula is approximately satisfied when h_o/d_p is nearly equal to 2:

$$\left(\frac{h_o}{d_p}\right)^3 = (0.14 - 0.125)\frac{q_o^2}{gd_p^3} \tag{3.13}$$

where q_o is the water discharge per unit width. For the flow on a gentle slope, the well-known logarithmic law is satisfied. To correlate Figure 3.3 with Figure 3.4, the two abscissa axes representing h/d_p for two different channel slopes are shown in Figure 3.4. The upper axis corresponds to Equation (3.13) and the lower one is for a gentle slope on which the logarithmic resistance law is applicable. As shown in Figure 3.4, the value $q_* = 2$ approximately corresponds to $h_o/d_p = 1$ on a steep slope that has been referred to as the critical value to generate debris flow. Therefore, we can conclude that a stony debris flow will occur on a gully bed steeper than $\tilde{\theta}_1$ when a surface water flow of $q_* \geq 2$ or $h_o/d_p \geq 1$ appears. Equation (3.13) and the discussion in the previous chapter indicate that if q_* is less than 10^3, stony debris flow will occur, and if q_* is larger than 10^4, turbulent muddy debris flow will occur. However, the critical slope to generate turbulent muddy debris flow should be a little flatter than $\tilde{\theta}_1$.

Pairs of points indicated by "x" marks are connected by arrows in Figure 3.4. This shows that in the early stage, just after water is supplied over a well-mixed sediment bed, all the particles available on the bed surface move and produce the bed form type in Domain IV, i.e. antidunes or meanders. Later a kind of armor coating develops and forms a stable step-pool bed form. The steps are formed by the conglomeration of coarse particles and on the back of them pools are formed. The experimental data in Figure 3.4 are only for slopes flatter than $\tilde{\theta}_1$, but in the field such a stair-like bed form can be also found on a steeper slope bed. On a saturated steeper slope bed, a small-scale surface water flow generates a peristaltic bed load transport and produces the steps. Once a stair-like bed form is produced, even if the general large-scale topographic slope of the channel is steeper than $\tilde{\theta}_1$, the local slope gradient at the back of step becomes flatter than $\tilde{\theta}_1$. In such a situation, unless the large-scale flood flow conceals the bed irregularity and the surface slope of the flow becomes as steep as the general topographic slope, debris flow by the destruction of the bed cannot be generated. Instead, if a large flood destroys the stair-like bed, erosion takes place quite vigorously and a large-scale debris flow will be generated. More descriptions concerning the configuration of a stair-like bed and its effect on debris flow generation appear in Chapter 6.

3.1.2 The development and decline of stony debris flow on sediment bed

The case of uniform bed saturated with water ($\tilde{\theta}_1 < \theta < \theta_2$)

Let us consider a water-saturated non-cohesive sediment bed that is a little flatter than θ_2. If a surface water flow of depth h_o is supplied, the stress distributions in the bed are given by Equations (3.1) and (3.2), and the part in which the operating stress τ is larger than the resisting stress τ_r starts to move with the surface water flow. If this sediment loading flow has a depth h and particle concentration C, then the shearing stress and the resistance stress operating at a depth a under the bed surface are described as follows:

$$\tau = g \sin \theta \{(Ch + C_* a)(\sigma - \rho) + (h + a)\rho\} \tag{3.14}$$

$$\tau_r = g \cos \theta \{(Ch + C_* a)(\sigma - \rho)\} \tan \varphi \tag{3.15}$$

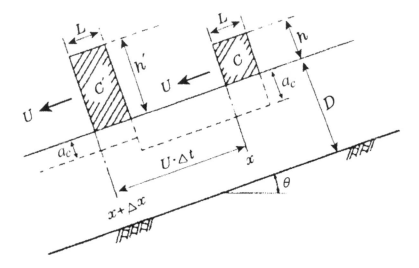

Figure 3.5 Debris flow developing process.

Hence, the depth a_c where τ is equal to τ_r is given by:

$$a_c = \frac{C_\infty - C}{C_* - C_\infty}h \tag{3.16}$$

The value of a_c is positive as long as C is less than the equilibrium concentration. This means that C and h will increase, helped by the longitudinally continuing bed erosion, and the flow asymptotically approach to an equilibrium state. If water and sediment added to the flow are stored in the part of length L as depicted in Figure 3.5, the water and particle conservation equations in this head part can be described respectively as:

$$Lh'(1 - C') = Lh(1 - C) + U\Delta t(1 - C_*)a_c \tag{3.17}$$

$$Lh'C' = LhC + U\Delta tC_*a_c \tag{3.18}$$

If the velocity U in a short time is assumed constant (the nearly constant translation velocity of debris flow in a developing stage was confirmed by experiments), the above equations at the limit of $\Delta t \to 0$ are rewritten as follows:

$$\frac{dh}{dt} = \frac{U}{L}\frac{C_\infty - C}{C_* - C_\infty}h \tag{3.19}$$

$$\frac{dC}{dt} = \frac{U}{L}(C_* - C)\frac{C_\infty - C}{C_* - C_\infty} \tag{3.20}$$

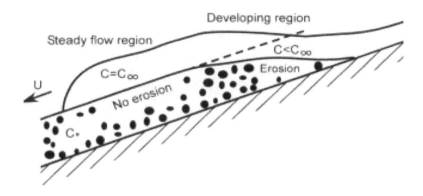

Figure 3.6 Longitudinal profile of debris flow on uniform bed by the constant rate water supply.

The solutions of these simultaneous equations under the initial condition; when $t = 0$, $C = 0$ and $h = (U_o/U)h_o$, are asymptotic at $t \to \infty$ to the following equations:

$$C = C_\infty \tag{3.21}$$

$$\frac{h}{h_o} = \frac{U_o}{U} \frac{C_*}{C_* - C_\infty} \tag{3.22}$$

where U_o is a hypothetical velocity of flow whose depth is h_o on the sediment bed. Equations (3.21) and (3.22) indicate that the water supply with a constant discharge over a long uniform sediment bed initiates a debris flow and it develops downstream and approaches an equilibrium state. If the water supply commences suddenly, a bulbous front, as illustrated in Figure 3.6, is formed. In the region within some upstream distance from the front no erosion takes place and this steady flow part elongates itself as it flows down.

Equation (3.22) can be rewritten as:

$$q_t = \frac{C_*}{C_* - C_\infty} q_o \tag{3.23}$$

where q_t is the discharge of debris flow per unit width and q_o is the discharge of the supplied water from upstream. The value of C_∞ approaches C_* as the channel slope becomes steeper. Therefore the steeper the channel slope, the far larger the discharge of debris flow becomes compared with the supplied water discharge. Figure 3.7 compares the results of experiments on a saturated bed to Equation (3.23) (Hashimoto *et al.* 1978).

Development and decline on a varying slope bed

Consider a channel bed with a continually changing slope as shown in Figure 3.8. As long as the surfaces of the seepage and the overland flows are as illustrated, the entire bed must be stable.

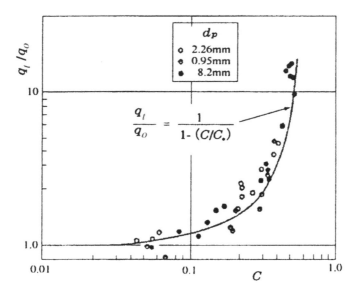

Figure 3.7 Debris flow discharge versus supplied water discharge.

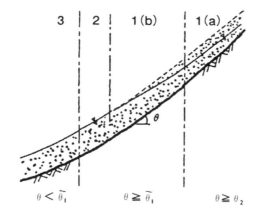

Figure 3.8 Varying slope bed and water seepage.

When a surface water flow suddenly enters the channel upstream, in the region 1(a) and 1(b), the bed will be eroded by the effects of the hydrodynamic force of the surface flow and the eroded sediment will be mixed with water, increasing the discharge and solids concentration downstream. Soon the flow reaches range 2, where the bed is saturated by seepage water. Then, the upper part of the bed layer becomes unstable due to the imbalance between the operating tangential and the internal resisting stresses. If there were no hindrance downstream, this layer would yield immediately, but the stable bed just downstream of the flow front hinders simultaneous movement. Therefore, only an upper part of the unstable layer begins to move and mixes with the flow. The debris flow developed progressively in this way proceeds to range 3. If the sediment concentration in the flow is too great in range 3, a part of the sediment will be deposited and the rest continues downstream.

The one-dimensional unsteady flow that proceeds in the manner written above can be described by the momentum and mass conservation equations (Takahashi *et al.* 1987). The mass conservation equation is:

$$\frac{\partial h}{\partial t} + \frac{\partial q_t}{\partial x} = i\{C_* + (1 - C_*)s_b\} + r \qquad (3.24)$$

and the momentum conservation equation is:

$$\frac{1}{gh}\frac{\partial q_t}{\partial t} + \frac{2q_t}{gh^2}\frac{\partial q_t}{\partial x} = \sin\theta - \left(\cos\theta - \frac{q_t^2}{gh^3}\right)\frac{\partial h}{\partial x} - \frac{q_t^2}{R^2 h^{2m}}$$

$$- \frac{q_t}{gh^2}i\{C_* + (1 - C_*)s_b\}\left\{(1 + k_c)\frac{\rho_{*b}}{\rho_T} - 1\right\}$$

$$- \frac{q_t}{gh^2}r\left(\frac{2\rho}{\rho_T} - 1\right) \qquad (3.25)$$

where i is erosion (>0) or deposition (<0) velocity, s_b is the degree of saturation in the sediment bed, r is the water inflow rate per unit length of channel, ρ_{*b} is the apparent density of the static bed $\{= C_* \sigma + (1 - C_*)\rho s_b\}$, and k_c contributes to the increase or decrease in the momentum of the flow and is approximated as 1 for erosion and 0 for deposition. R in Equation (3.25) is the resistance coefficient and the third term in the right-hand side of Equation (3.25) is the resistance term that depends on the constitutive equations for the flow.

In a stony debris flow, using the dilatant fluid model, m in Equation (3.25) is 5/2 and R is given as:

$$R = \frac{2}{5d_L}\left[\frac{g}{0.02}\left\{C_L + (1 - C_L)\frac{\rho_m}{\sigma}\right\}\right]^{1/2}\left\{\left(\frac{C_{*DL}}{C_L}\right)^{1/3} - 1\right\} \qquad (3.26)$$

where d_L is the mean diameter of the coarse particles in the debris flow which is supported by dispersive pressure, C_L is the volume concentration of the coarse fraction in the entire debris flow material, C_{*DL} is the volume concentration of the coarse fraction in the static bed when particles in flow are deposited, and ρ_m is the apparent density of the interstitial fluid.

In a plain water flow, the usual Manning-type resistance formula is applicable, so that m in Equation (3.25) is 5/3 and:

$$R = 1/n_m \qquad (3.27)$$

where n_m is Manning's roughness coefficient.

Near the upstream end of the channel, even if the relative depth is less than about 30, the quantity of sediment entrained in the flow by erosion would not yet be sufficient to be dispersed throughout the flowing layer; consequently an immature debris flow would appear, in which $m = 5/2$ and:

$$R = 0.7\sqrt{g/d_L} \tag{3.28}$$

If the flow is like a uniformly translating flow as illustrated in Figure 3.6 and i and r are less than the velocity of flow, the left-hand side terms and the second, fourth and fifth terms in the right-hand side of Equation (3.25) become small and it can be replaced by:

$$q_t = Rh^{2m} \sin^{1/2}\theta \tag{3.29}$$

This is nothing but the formula for steady uniform flow. The flow routing method that uses the steady uniform flow equation and the mass continuity equation simultaneously is the kinematic wave method that is often used in flood runoff analysis.

In this section, the solids component in the flow is divided into two fractions; a coarse particle fraction whose particles are sustained in the flow by the effect of inter-particle collisions and a fine particle fraction whose particles are suspended by turbulence in the interstitial fluid. The particle diameters for the two fractions may change with variations in the hydraulic condition of the flow, but, here it is assumed that they have fixed values. The continuity equations for coarse particle fraction and that for fine particle fraction are given by Equations (3.30) and (3.31), respectively:

$$\frac{\partial V_L}{\partial t} + \frac{\partial (q_t C_L)}{\partial x} = \begin{cases} iC_{*L} \; ; & i \geq 0 \\ iC_{*DL} \; ; & i < 0 \end{cases} \tag{3.30}$$

$$\frac{\partial V_F}{\partial t} + \frac{\partial \{q_t(1 - C_L)C_F\}}{\partial x} = \begin{cases} iC_{*F} & ; \quad i \geq 0 \\ i(1 - C_{*DL})C_F \; ; & i < 0 \end{cases} \tag{3.31}$$

where V_L and V_F are the volumes of coarse and fine particles, respectively, in a pillar-shaped space in the flow having a height h and a bottom area of unity, C_F is the volume concentration of the fine fraction in the interstitial fluid, and C_{*F} is the volume concentration of the fine fraction in the static bed.

To obtain the temporal changes in the thickness of sediment bed and the slope angle of the surface of sediment bed, the following two additional equations are necessary:

$$\frac{\partial D}{\partial t} + i = 0 \tag{3.32}$$

$$\theta = \theta_0 - \tan^{-1}(\partial D/\partial x) \tag{3.33}$$

where θ_0 is the initial bed slope.

The fundamental system of equations to analyze the process of debris flow development and decline by the erosion of and deposition on the sediment bed is now prepared, but to carry out the analysis the erosion and deposition velocities should be given appropriately. The erosion velocity is different depending on whether the bed is saturated with water or not.

At first, we consider the unsaturated bed. The erosion of the bed in this case is due to the picking up of individual particles from the bed surface by the effects of the shear stress created by the interstitial fluid, and that shear stress will become too small to pick up particles when the sediment concentration in the flow attains full growth. Therefore, the erosion of bed continues as long as the sediment concentration in flow is less than the equilibrium value. To represent such a character, analogous to the non-equilibrium bed load transportation formulae, the following erosion velocity equation is assumed:

$$i = K(\tau_{*f} - \tau_{*fc})\sqrt{\tau_f/\rho_m} \qquad (3.34)$$

where K is a coefficient, τ_f is the shear stress created by the inter-granular fluid, τ_{*f} $(=\tau_f/\{(\sigma - \rho_m)gd_p\})$ is the non-dimensional expression of τ_f, and τ_{*fc} is the non-dimensional critical shear stress in the interstitial fluid less than which particles on the bed can no longer be entrained into the flow. The shear stress τ_f would be equal to the difference between the operating shear stress and the shear resistance on the bed, so that it is given as:

$$\tau_f = \{(\sigma - \rho_m)C_L + \rho_m\}gh\sin\theta - (\sigma - \rho_m)ghC_L\cos\theta\tan\varphi \qquad (3.35)$$

This equation means if C_L attains full growth as:

$$C_L = C_{L\infty} \equiv \frac{\rho_m\tan\theta}{(\sigma - \rho_m)(\tan\varphi - \tan\theta)} \qquad (3.36)$$

τ_f becomes zero and particles can no longer be entrained. Substitution of Equation (3.35) into Equation (3.34) gives:

$$\frac{i}{\sqrt{gh}} = K\sin^{3/2}\theta\left\{1 - \frac{\sigma - \rho_m}{\rho_m}C_L\left(\frac{\tan\varphi}{\tan\theta} - 1\right)\right\}^{1/2}\left(\frac{\tan\varphi}{\tan\theta} - 1\right)$$
$$\times(C_{L\infty} - C_L)\frac{h}{d_L} \qquad (3.37)$$

Note that on a bed steeper than θ_2, the value of $C_{L\infty}$ obtained from Equation (3.36) exceeds C_* and it even surpasses the maximum possible compaction value. Because no flow is possible at such a high concentration and the experimental data reveal the maximum concentration is around $0.9C_*$. Therefore, on the bed steeper than θ_2, $C_{L\infty}$ and C_L in Equation (3.37) must be replaced by $0.9C_*$ and consequently $i = 0$.

On a bed flatter than $\tilde{\theta}_1$, the flow becomes an immature debris flow even if it attains full growth and τ_f in such a case is, from Equation (2.138):

$$\tau_f = \{(\sigma - \rho_m)C_l + \rho_m\}g\sin\theta - (\sigma - \rho_m)C_lh_l\cos\theta\tan\varphi \qquad (3.38)$$

Noticing $C_l h_l = C_L h$, and following the same procedure with the case of debris flow, the equation of erosion velocity for immature debris flow is obtained as:

$$\frac{i}{\sqrt{gh}} = K \sin^{3/2}\theta \left\{ 1 - \frac{\sigma - \rho_m}{\rho_m} C_L \left(\frac{\tan\varphi}{\tan\theta} - 1 \right) \right\}^{1/2} \left(\frac{\tan\varphi}{\tan\theta} - 1 \right)$$

$$\times (C_{s\infty} - C_L) \frac{h}{d_L} \tag{3.39}$$

where $C_{s\infty}$ is given by Equation (2.144).

Secondly, consider a debris flow moving on a bed saturated with water. In this case the stress distribution equations (3.14) and (3.15) have a slightly different form because the densities of the interstitial fluid in the flow and within the bed are different, i.e. ρ_m in the flow and ρ in the bed. Hence, the formula giving a_c becomes a little complicated (Takahashi 1991). Here, for the sake of simplicity, the stress distributions within the bed are assumed to satisfy Equations (3.14) and (3.15) by substituting ρ_m for ρ. Then, a_c is given by Equation (3.16), in which ρ_m is substituted for ρ. But, as mentioned earlier, the whole bed layer with a thickness a_c does not move as soon as the flow front arrives; there is a delay before the completion of erosion. Writing this delay as $(d_L/U)/\delta_e$, the erosion velocity is given by:

$$i = \delta_e \frac{a_c}{d_L} U = \delta_e \frac{C_{L\infty} - C_L}{C_* - C_{L\infty}} \frac{q_t}{d_L} \tag{3.40}$$

where δ_e is a coefficient. For the case of immature debris flow $C_{s\infty}$ is substituted for $C_{L\infty}$.

Experiments were carried out to verify Equation (3.40). The bed slopes in the experiments were varied between 12° and 15° and the entire bed layer was saturated with water prior to the experimental runs. Only the coarse particles, whose mean diameter was equal to $d_L = 1.88\,\text{mm}$, comprised the bed.

Figure 3.9 (a) shows the relationship between the substantial bed sediment erosion velocity (net volume of eroded solids in unit time), iC_{*b}, and (i/δ_e) by Equation (3.40). We can see from Figure 3.9 (a) that if $\delta_e = 0.003$ is substituted into Equation (3.40), the erosion velocity in the experiments is well reproduced.

It is difficult to know whether the bed is fully saturated or not before the onset of debris flow in the actual situation, so it is convenient if Equation (3.39) is applicable to the saturated case as well. In this context the relationship between iC_{*b} and i/K in Equation (3.39) is plotted in Figure 3.9 (b). Although the extent of the plotted point scattering is a little larger than those in Figure 3.9 (a), if $K = 2.3$ is substituted into Equation (3.39), the tendency of iC_{*b} variations is reasonably well explained. As is explained later $K = 0.06$ is suitable for the non-saturated bed, and this suggests that if the bed layer is saturated with water very rapid erosion will take place. However, in natural conditions, it is rare to be fully saturated as shown in Figure 1.15.

If C_L in Equation (3.37) or Equation (3.40) is equal to $C_{L\infty}$, i becomes zero and no erosion takes place. Because $C_{L\infty}$ becomes smaller in flatter reaches, if a debris flow that fully developed within the steep slope reach upstream comes down to flatter slope reach downstream, it carries an excess load of particles so it will deposit some coarse

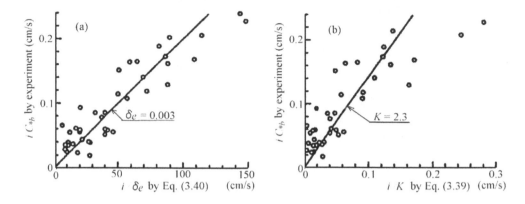

Figure 3.9 Adoptability of Equations (3.40) and (3.39).

particles and continue downstream, thereby reducing the concentration and the flow rate. The amount of excess coarse particles at that position is $h(C_L - C_{L\infty})$ per unit area. Describing the time necessary to deposit that amount as $(d_L/U)/\delta_d$, the depositing speed is given by:

$$i = \delta_d \left(1 - \frac{U}{p_i U_c}\right) \frac{C_{L\infty} - C_L}{C_{*DL}} \frac{q_t}{d_L} \tag{3.41}$$

The term $(1 - U/p_i U_c)$ is introduced to consider the inertial motion of debris flow. The details of why such term is necessary is explained in Chapter 5.

The idea of a delay represented by $(d_L/U)/\delta_e$ or $(d_L/U)/\delta_d$ is still vague. The erosion velocity may well be dependent on particle diameter so that Equation (3.40) may be reasonable, but for the deposition velocity the depth of flow may have more effect than the particle diameter. If the delay is evaluated by $(h/U)\delta_d$, the depositing velocity is given as:

$$i = \delta_d \left(1 - \frac{U}{p_i U_c}\right) \frac{C_{L\infty} - C_L}{C_{*DL}} U \tag{3.42}$$

The values of δ_d in Equations (3.41) and (3.42) are of course different. Which of these two equations should be adopted will be determined by the uniqueness of δ_d for many cases.

In the case of immature debris flow, the inertial motion becomes negligible and the term $(1 - U/p_i U_c)$ becomes approximately 1 and, of course, $C_{L\infty}$ should be changed to $C_{s\infty}$.

Side bank erosion velocity

The surface water flow at the source of a ravine may only be a thread-like stream on the gully bottom. If such a stream satisfies the condition for debris flow initiation, the flow will erode not only the channel bottom but also the side banks of the incised channel. The side bank erosion is possibly generated, similar to the bed erosion, by the shear stress created by the interstitial fluid but it may be considered to be half that stress operating on the bottom (Takahashi 1993a). Then, the velocity of side bank erosion is obtained from Equation (3.37) as:

$$\frac{i_s}{\sqrt{gh}} = \left(\frac{1}{2}\right)^{3/2} K_s \sin^{3/2}\theta \left\{1 - \frac{\sigma - \rho_m}{\rho_m} C_L \left(\frac{\tan\varphi}{\tan\theta} - 1\right)\right\}^{1/2} \left(\frac{\tan\varphi}{\tan\theta} - 1\right)$$

$$\times (C_{L\infty} - C_L)\frac{h}{d_L} \tag{3.43}$$

where K_s is a coefficient.

The recession of the side banks below the surface of flow gives rise to the instability of the channel walls above and soon those parts will fall into the flow. If the banks are vertical and the recession is parallel to the original walls, the mean recession velocity of the side walls is given as:

$$i_{sml} = \frac{i_s h}{l_l + h}, \quad i_{smr} = \frac{i_s h}{l_r + h} \tag{3.44}$$

where i_{sml} and i_{smr} are the mean wall recession velocities of the left-hand side bank and the right-hand side bank, respectively, and l_l and l_r are the height of left-hand and right-hand side walls, respectively, measured from the surface of flow.

3.1.3 Verification of the theory by experiments

Non-erodible side bank cases

To verify the theoretical model and to determine some coefficient in the formulae, laboratory experiments were carried out. The schematic arrangement of the varied slope channel is given in Figure 3.10. Water supply to the flume is either from the tap at the upstream end or from the side walls via a pair of rainfall generators. The channel width is 7 cm. The bed material A, whose size distribution is given in Figure 3.11, was laid on the bottom of the channel with a thickness of 10cm. Prior to starting the experiment, seepage flow was produced within the bed. The free surface of this flow appeared on the surface of the sediment bed at 150cm from the downstream end; the part of the bed upstream from this point was unsaturated. The degree of saturation s_b was assumed to be 0.8 for the unsaturated area. The boundary between the fine and coarse fractions of material was assumed to be at 0.3 mm; the size distribution of the coarse material is, then, given by line B in Figure 3.11, d_L equating to 1.8 mm.

A predetermined discharge of water was introduced suddenly from the upstream end or laterally from both walls, after which a debris flow was generated on the bed and developed downstream. The experimental data are given in Table 3.1.

Figures 3.12, 3.13, 3.14 and 3.15 compare the experimental results of Runs 1 and 2 with the numerically obtained values. After some trial and error computations, the numerical coefficients K and δ_e were set to 0.06 and 0.0007, respectively. Figures 3.12

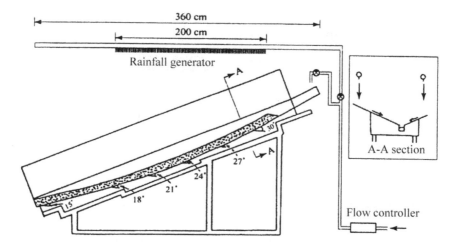

Figure 3.10 Experimental flume.

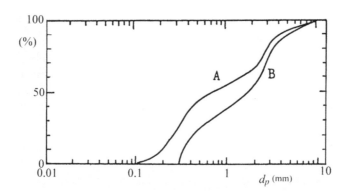

Figure 3.11 Size distributions in the experimental material.

and 3.14 suggest that both Equation (3.37) for the unsaturated region and Equation (3.40) for the saturated region predict well the erosion speed of the bed material.

Time variations in the flow depth including the front arrival time shown in Figures 3.13 and 3.15 are also well reproduced by the calculations. These, in turn, prove the validity of the resistance law, Equation (3.29), and the representative diameter of the bed d_L adopted in the calculation. The flow depth at each point in Figures 3.12, 3.13, 3.14 and 3.15 is asymptotic to a constant value. This is because of the removal of all the sediment and resulting exposure of the rigid channel bottom.

Figures 3.16 and 3.17 show the results of Run 3, in which the water supply from the side banks is longitudinally distributed. The theoretical erosion rate at 120 cm, where the bed was unsaturated, compares fairly well with the experimental value. At 220 cm the calculation fits the experimental results fairly well up to 24 seconds, after that time, deposition takes place in the experiment and the differences between the calculation and experiment increase. A similar tendency of small erosion rate at

Table 3.1 Experimental data.

Run	Bed length (cm)	Bed thickness (cm)	Water supply Position	Discharge (cm³/s)	Duration (s)
1	270	10	Upstream end	200	40
2	270	10	Upstream end	350	40
3	270	10	Side walls	200	40

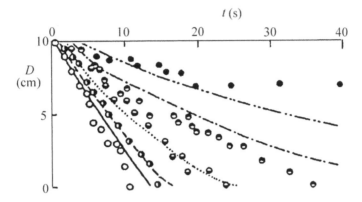

Figure 3.12 Bed erosion at various points in Run 1. Positions; see Figure 3.13.

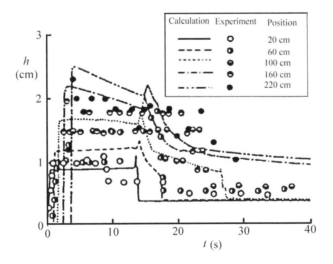

Figure 3.13 Depth versus time relationships in Run 1.

220 cm in the experimental values can also be seen in Figure 3.12. This is attributable to the existence of a bed girdle at the downstream end of the channel, which was installed to avoid the retrogressive erosion from the downstream end. The existence of the bed girdle was neglected in the computations.

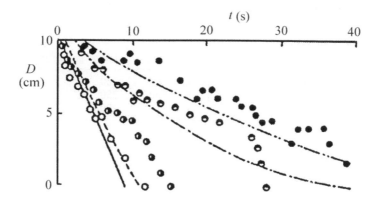

Figure 3.14 Bed erosion at various points in Run 2. Position; see Figure 3.15.

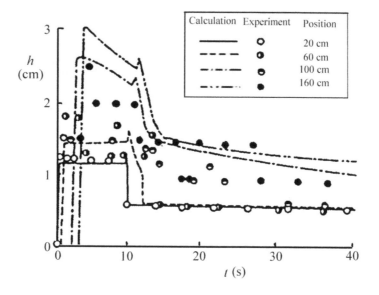

Figure 3.15 Depth versus time relationships in Run 2.

Figures 3.13, 3.15 and 3.17 reveal that, when water is supplied from upstream end, the debris flow develops downstream and the larger the supplied water discharge the larger the debris flow discharge becomes and the shape of the hydrograph becomes triangular. When water is supplied along the longitudinally extending side banks, the development of the hydrograph is small due to the scarcity of discharge upstream.

Erodible side bank case

An experiment to examine the process of debris flow development with side bank as well as bed erosions was carried out. The experimental flume was 3 m long, 40 cm wide and the slope was set at 3°. The mean diameter of the sediment bed particles was

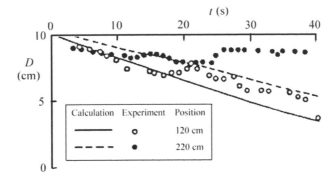

Figure 3.16 Bed erosion at various points in Run 3.

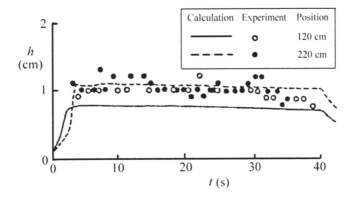

Figure 3.17 Depth versus time relationships in Run 3.

1.5 mm. A channel 5 cm wide, 2 cm deep and 3 m long was cut in the sediment bed adjacent to a transparent glass wall of the flume and a steady water flow of $150\,\text{cm}^3/\text{s}$ was supplied into this channel from the upstream end. The processes of erosion on the side bank as well as on the bed were recorded by video and the sediment discharge at the outlet of the flume was measured.

Figure 3.18 compares the temporal variations of the channel width and the bed level with the theoretical calculation. Figure 3.19 compares the experimentally obtained sediment discharge with the theoretical calculation. The side bank erosion in the experiment proceeded not uniformly but sporadic in time as well as in space. However in Figure 3.18, the longitudinally averaged width of the channel is given. In the calculation, $K = 0.06$, $K_s = 1.0$, $\delta_d = 1.0$, $p_i = 2/3$, $C_{*L} = 0.655$ and $C_{*F} = 0$ were used.

Figure 3.18 verifies the theory concerning both vertical and lateral erosions. Figure 3.19 clearly shows that the theory reproduces the rapid increase in discharge at the early stage and gradual decrease after, although the experimental values fluctuate largely possibly due to the sporadic side bank erosion.

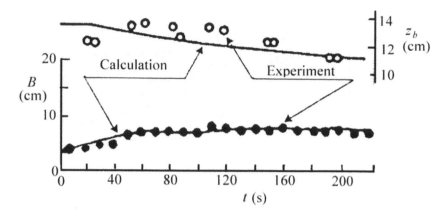

Figure 3.18 Vertical and lateral erosions of the stream channel.

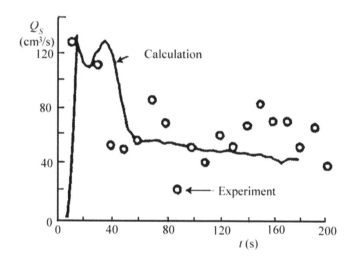

Figure 3.19 Sediment graph at the downstream end of the flume.

Division of particles into coarse and fine fractions

In the last section, the development and attenuation of stony-type debris flow were numerically analyzed under the assumption that the boundary between the fine and coarse particle fractions is 0.3 mm. The apparent density of the interstitial fluid, ρ_m, becomes heavy by entraining the fine particle fraction, thereby the mobility and the ability to load coarse particles are increased. The mechanism to suspend fine particles in the interstitial fluid must be as follows:

If the turbulent mixing velocity in the interstitial fluid exceeds the settling velocity of a fine particle and the void space among coarse particles that is filled by the fluid is larger than the diameter of that fine particle, the fine particle is suspended in the

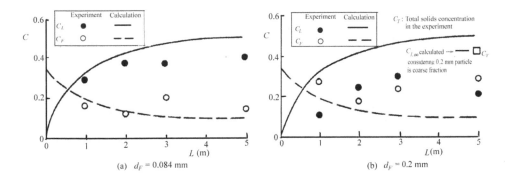

Figure 3.20 Comparisons of experimental results with the calculations under the assumption that the particles 0.084 mm and 0.2 mm in diameters, respectively, are the fine fractions in the interstitial fluid.

interstitial fluid. The turbulent velocity in the interstitial fluid would be approximately equal to the shear velocity within it. The shear stress allotted by the interstitial fluid in the total shear stress of the flow is given by Equation (3.35). Then, the shear velocity u_{*f} is given as:

$$u_{*f} = gh \sin \theta \left(1 - \frac{C_L}{C_{L\infty}}\right) \tag{3.45}$$

Two kinds of experiments were done to verify this conjecture. The first experiment used a rectangular channel whose slope was set at 18°, in which a water-saturated sediment bed comprised of uniform particles, $d_L = 1.07$ mm, was prepared. A debris flow was produced by suddenly giving a steady flow of $q = 53.3$ cm²/s from the upstream end of the flume. The supplied steady flow was a mudflow whose solids concentration was 35% by volume and the particle diameter was 0.084 mm. The second experiment was done under the same channel conditions as the first one. The supplied steady flow rate and solids concentration were also the same but the supplied mudflow was composed of particles 0.2 mm in diameter.

Figures 3.20 (a) and (b) compare the experimental results with the results obtained by the numerical calculations using the above-mentioned fundamental equations. Figure 3.20 (a) corresponds to the first experiment. The calculation was done under the assumption that the particle 0.084 mm in diameter was the composition that makes up the fine fraction in the supplied mud-flow and it contributes to make ρ_m large. The results of calculation fit the experimental results reasonably well. Therefore, the particles 0.084 mm in diameter were surely the composition of fine fractions suspending in the interstitial fluid.

Figure 3.20 (b) corresponds to the second experiment. The calculation considered the particles 0.2 mm in diameter as the composition of fine fraction, but in this case the results of calculation do not match the experimental results. This means the particles 0.2 mm in diameter cannot be considered as the composition to make up the fine fraction. Instead, if the particles 0.2 mm in diameter are considered as the composition of the coarse fraction together with $d = 1.07$ mm particles and ρ_m is made equal to 1, the calculated total solids concentration C_T at $L = 5$ m almost coincides with the equilibrium coarse particles concentration $C_{L\infty}$ calculated by Equation (2.25) as shown in Figure 3.20 (b). This means the particles 0.2 mm in diameter could not suspend in the interstitial fluid.

In fact, the settling velocity of the particle of 0.084 mm is significantly less than u_{*f} calculated by Equation (3.45), but the settling velocity of the particle of 0.2 mm is as large as u_{*f} calculated by Equation (3.45). Moreover, if the total solids concentration is equal to about 0.4, the void space between particles is about 0.2 mm. Then, the void space is too narrow to suspend 0.2 mm particles.

This result confirms that the diameter of the particle suspended in the interstitial fluid can be obtained by Equation (3.45). But, when the particles that compose a debris flow fall within a very wide size range, more detailed investigation is necessary to determine the boundary between fine and coarse fractions.

3.2   LANDSLIDE-INDUCED DEBRIS FLOW

There are two types of landslide that occurs on the occasion of severe rainstorms; the shallow landslide of about $1\,m \sim 2\,m$ thick and the deep-seated one of several tens of meters thick. The shallow landslide generally occurs with the strongest rainfall intensity, whereas, the deep-seated one often occurs after the termination of rainfall.

The shallow landslide contains plenty of water in itself and is also helped by a high flood runoff discharge. Therefore, under the severest rainfall it is easily transformed into a debris flow almost from the instant of the initiation of motion. Then, in the majority cases, the processes of debris flow initiation and development can be analyzed by the method introduced in section 3.1 giving a debris flow as the upstream boundary condition, whose hydrograph and other characteristics are pre-determined by the landslide characteristics. Because the shallow landslide occurs suddenly, it may be easier to understand the reason why a debris flow has a clear bore-like front rather than the case of the bed erosion type.

A deep-seated landslide needs a comparatively long time before the ground water level rises enough to make the earth block unstable. Hence, it often occurs later than the time of strongest rainfall and by that time flood runoff around the landslide may already have been reduced. Therefore, the mechanism for the transformation into debris flow would be completely different from the case of a shallow landslide. A large-scale landslide is also generated by snow melting, earthquake and volcanic eruption and sometimes in these cases it is accompanied by debris flow. A gigantic landslide can transform into a debris avalanche that has very large mobility and after the stoppage many disrupted earth-blocks originating from the mountain body (flow mounds) are scattered on the deposit. Sometimes, immediately after the stoppage of debris avalanche, debris flow is witnessed flowing on the surface of the deposit that threads its way through the flow mounds. Thus, some large-scale landslides transform completely into debris flow, some are partially transformed into debris flow that appears following the earth block's motion, and the others move as solid blocks from the initiation to the stoppage. A viable theory of landslide motion should be able to explain such a variety of motion comprehensively. The transformation into debris flow must

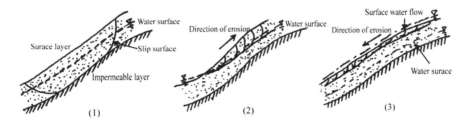

Figure 3.21 Three categories for the mechanism of surface landslide.

be brought about by the physical causes existing within the slid earth mass, but during the transformation, the process may be accelerated by the addition of water from outside. This kind of process would be the lemma of the comprehensive theory.

3.2.1 Mechanism of shallow landslides induced by severe rainfall

The mechanism of occurrence of shallow landslides can be classified into three categories as shown in Figure 3.21. The first category has an impermeable layer beneath the surface permeable layer. When it rains the water-saturated layer of a certain thickness appears within the permeable layer and the whole surface layer becomes unstable due to the effects of pore water pressure and the reduction of apparent cohesive strength in the saturated layer. In the second category the surface layer itself is stable even if the seepage flow that is parallel to the slope occupies the entire depth of the surface layer. At somewhere downslope the seepage flow oozes out of the surface so as to boil up the ambient particles or to generate severe erosion. Then, some upper parts of the surface layer lose their stability due to the elimination of support downslope. The local slides rapidly run up the slope. Many witness records indicate that a bursting out of water was the trigger of the landslide, which suggests many landslides occurred with the mechanism belonging to this category. In the third category the surface layer itself is stable but when the surface water flow appears it is violently eroded. This is the same mechanism as the debris flow occurrence in the stream channel discussed in section 3.1.

Generally, the length of a surface landslide is far greater than the depth, so the stability analysis for an infinitely long slope is applicable. Consider the situation where a uniform seepage flow in the surface layer takes place parallel to the surface of the slope as shown in Figure 3.22. The shearing force at the bottom of the surface layer that acts to drag down the surface layer is given by:

$$\tau = g \sin \theta [C_* \sigma D + (1 - C_*)\rho\{s(D - h) + h\}] \tag{3.46}$$

The stress resisting the driving stress is given by:

$$\tau_r = [g \cos \theta [C_* \sigma D + (1 - C_*)\rho\{s(D - h) + h\} - p_v]]\tan \varphi + c \tag{3.47}$$

where s is the degree of saturation in the void space above the surface of the saturated seepage flow, p_v is the pore water pressure, and c is the cohesive strength of the material.

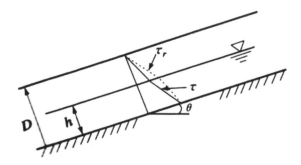

Figure 3.22 Stress distributions in the surface soil layer.

When $\tau > \tau_r$, a landslide occurs. As is evident in Equation (3.47), the pore water pressure acts to reduce the effective resisting stress and if an uplift pressure operates caused by confined groundwater the possibility of sliding is enhanced. If no uplift operates p_v is equal to the hydrostatic pressure which is given by:

$$p_v = \rho g \cos\theta \cdot h \tag{3.48}$$

The cohesive strength is effective to the stability of the layer particularly for the shallow topsoil layer. In the sandy soil layer the apparent cohesive strength is produced by the suction force of the void water and it disappears when the void is saturated with water. When $c = 0$, if $\sigma = 2.65$ g/cm^3, $\rho = 1.0$ g/cm^3, $\tan\varphi = 0.8$ and $D = 1$ m are assumed, Equations (3.46) and (3.47) result in the conclusion that, when the surface of the seepage flow coincides with the surface of the slope, landslides occur on the slopes of $\theta = 23°$ for the case $C_* = 0.7$ and on the slopes of $\theta = 20°$ for the case $C_* = 0.5$. If as the representative cohesive strength of sandy soil of weathered granite $c = 5 \times 10^3$ Pa is adopted (Yagi and Yatabe 1986), the critical slopes to give rise to landslides when the surface of the seepage flow coincides with the surface of the slope are $\theta = 36°$ and $\theta = 34°$ for $C_* = 0.7$ and $C_* = 0.5$, respectively. This means that even if the cohesive strength is small it is effective to stabilize the slope and landslides occur only when the entire surface layer is saturated with water. Particularly for the sandy soil, with the occurrence of landslides, the structure of the soil layer is easily shattered to small pieces and it will be transformed to debris flow.

Conditions to continue the downslope motion for a soil block

Consider a soil block is moving down the slope as shown in Figure 3.23. The lower part of the block is saturated with water and at the boundary between the surface of the slope and the block there are some liquefied parts produced by a large strain. The equation of momentum conservation for this soil block is given as follows:

$$\frac{d}{dt}(\rho_t LHU_e) = \rho_t g LH \sin\theta - L(\rho_t gH \cos\theta - \frac{L'}{L}p_v)\mu_k - (U_e - u_{li})h_{li}\rho_t u_{li}$$

$$- \frac{1}{2}\rho_t f u_{li}^2 L' + \rho_a(u_a - U_e)^2 h_a + \frac{1}{2}\rho_a \cos\theta \, g h_a^2 \tag{3.49}$$

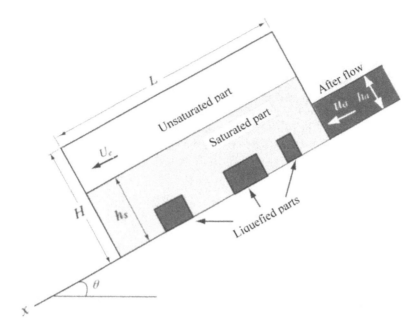

Figure 3.23 Moving block on a slope.

where U_e is the velocity of the earth block, u_{li} and h_{li} are the velocity and thickness of the liquefied layer, respectively, H and L are the thickness and length of the block, respectively, h_s is the thickness of the water-saturated part, h_a and u_a are the depth and velocity of the water flow behind the earth block, respectively, ρ_t, ρ_l and ρ_a are the apparent densities of the block, the liquefied layer and the following flow, respectively, μ_k is the kinematic friction coefficient between the slope surface and the block, f is the resistance coefficient of the liquefied layer to flow, and L' is the length of liquefied parts. The following formula is satisfied:

$$\rho_t = [C_* \sigma H + (1 - C_*)\rho\{s(H - h_s) + h_s\}]/H \tag{3.50}$$

The respective terms in Equation (3.49) mean: the left-hand side term is the momentum changing rate of the earth block, the first term of the right-hand side is the driving force due to gravity, the second term is the friction force due to the load of the earth block that is directly transmitted via the skeletal structure within the layer, the third term is the momentum left behind, the fourth term is the fluid dynamic resistance of the liquefied parts, the fifth term is the momentum supply from the after flow (if $u_a \leq U_e$, this term is zero), and the sixth term is the hydrostatic pressure operating on the earth block from the flow following the earth block.

When the total load of the block is supported by hydraulic pressure (the lowest part of the block is perfectly liquefied) $p_v = \rho_t g H \cos\theta$, and $L' = L$. When no liquefied layer exists, $p_v = \rho g\, h_s H \cos\theta$. In the latter case, the third and fourth terms in the right-hand side of Equation (3.49) are zero. When no after flow follows, the fifth and sixth terms are also zero.

If a completely liquefied layer appears and it is confined at the base of the block, this layer sustains the total weight of the block. In this case the motion of the block continues irrespective of the slope gradient. On the other hand, if no liquefied layer exists, the motion will stop except in the case that the downward slope gradient is steeper than a critical value θ_{ck}. This critical gradient can be obtained by substituting 0 for the left-hand side and the third and fourth terms of the right-hand side of Equation (3.49) and then solving Equation (3.49) to get the slope gradient that gives $U_e = 0$. The result is as follows:

$$\theta_{ck} = \tan^{-1}\left\{\left(1 - \frac{\rho}{\rho_t}\frac{h_s}{H}\right)\mu_k - \frac{\rho_a}{\rho_t}\frac{1}{\cos\theta}\frac{h_a^2}{LH}\left(F_{ra}^2 + \frac{1}{2}\right)\right\} \tag{3.51}$$

where F_{ra} is the Froude number in the after flow. Except for the case where a vast amount of water gushes out just after the sliding, the effects of after flow would be negligible. Then, the maximum value of θ_{ck} appears when the block is dry, and the value is $\theta_{ck} = \tan^{-1}\mu_k$. The minimum of θ_{ck} appears when the block is entirely saturated but not liquefied, and the value is $\theta_{ck} = \tan^{-1}[C_*(\sigma - \rho)\mu_k/\{C_*(\sigma - \rho)+\rho\}]$. If $\mu_k = 0.7$, $C_* = 0.7$, $\sigma = 2.65\,\text{g/cm}^3$, $\rho = 1.0\,\text{g/cm}^3$, θ_{ck} is between 35° and 20°. Noting that even a slight cohesive strength enhances the stability of the surface soil layer and in such a case sliding seldom occurs under unsaturated condition, but a void rich soil block will easily transform into debris flow under saturated condition, Equation (3.51) suggests that if the slope gradient ahead of a landslide is steeper than about 20°, the slid earth block will continue its motion downslope either as a rigid block or transforming into a debris flow.

The arrival distance of a slid earth block on a flat plain below a steep slope is important because it affects the delineation of hazardous area. Figure 3.24 shows the schematic diagram of the motion from initiation to stoppage. The height of a slide measured from the foot of the slope is H_p and the distal end of block after stoppage is at $x_L \cos\theta_d$ from the foot, here this is written as L_a. Japanese statistics of L_a is given in Table 3.2 (Japan Society of Erosion Control Engineering 2000), where the landslides are classified by the absolute arrival distance and that normalized by the height H_p; the number of the landslides falling in each category and its percentage to the total number are shown.

If the slid earth block is dry and no after flow exists, Equation (3.49) is written as:

$$\frac{dU_e}{dt} = g(\sin\theta - \mu_k\cos\theta) \tag{3.52}$$

From Equation (3.52) the velocity of the block at the foot of the slope, U_0, under the initial condition $U_e = 0$, is given as:

$$U_0 = \left\{2g\left(1 - \frac{\mu_k}{\tan\theta_u}\right)H_p\right\}^{1/2} \tag{3.53}$$

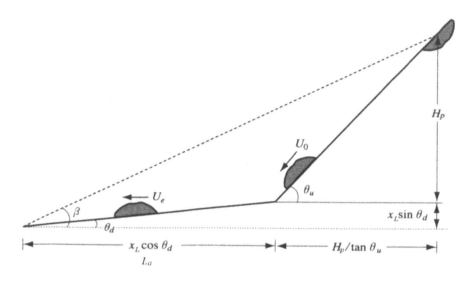

Figure 3.24 Schematic diagram of slid earth block motion.

Table 3.2 Statistics of Japanese landslides.

Arrival distance (m)	Relative arrival distance					
	$0 \leq L_a/H_P \leq 2$		$2 < L_a/H_P$		Total	
$0 \leq L_a \leq 50$	4,795	98.6%	20	0%	4,815	99%
$50 < L_a$	36	0.7%	12	0.2%	48	1.00%
Total	4,831	99.3%	32	0.7%	4,863	100.0%

The arrival distance of the block is, then, given as:

$$L_a = \frac{\cos\theta_d(1 - \mu_k/\tan\theta_u)H_p}{\mu_k \cos\theta_d - \sin\theta_d} \qquad (3.54)$$

where the velocity U_0 is assumed to change direction smoothly from that parallel to the slope to that parallel to the plain maintaining the same value.

Substitution of $L_a/H_p = 2$, $\theta_d = 0$ and $\theta_u = 30°$ to Equation (3.54) gives $\mu_k = 0.268$. The kinematic friction coefficient for a dry earth block may have a value around 0.7, so a slide that reaches a length of twice its height should have a far smaller friction coefficient than that of a rigid body.

When the entire earth block is saturated with water yet no liquefaction occurs, Equation (3.55) instead of Equation (3.52) is satisfied.

$$\frac{dU_e}{dt} = g\left\{\sin\theta - \frac{C_*(\sigma - \rho)}{C_*(\sigma - \rho) + \rho}\mu_k \cos\theta\right\} \qquad (3.55)$$

Namely, if $C_* = 0.7$, $\sigma = 2.65\,\mathrm{g/cm}^3$ and $\rho = 1.0\,\mathrm{g/cm}^3$, the apparent kinematic friction coefficient becomes 0.536 times to the case of dry block and it is about 0.38. Therefore, a landslide that reaches a length of twice its height should have more or less liquefied layer at the boundary at the ground surface and the block.

According to the original data on which Table 3.2 is based, the number of landslides that belongs to the class $L_a/H_p < 0.2$ is about 40% of all landslides and that belongs to the class $L_a/H_p < 0.4$ is about 70%. Under the same conditions as to obtain the apparent kinematic friction coefficient for $L_a/H_p = 2$, the apparent kinematic friction coefficients obtained for $L_a/H_p = 0.2$ and $L_a/H_p = 0.4$ are 0.518 and 0.469, respectively. These values are larger than 0.38, so these are the cases where, although at some parts some thicknesses of saturated layer exist in the block, no liquefaction takes place. Therefore, more than 70% of landslides move on the slope without liquefaction and stop and accumulate at the foot of the slope.

The $\tan\beta$ value as illustrated in Figure 3.24 is the so-called equivalent friction coefficient as mentioned earlier in section 1.2. The distal tip of the perspective line coincides with the arrival point of the slid block obtained by Equation (3.54) with the substitution of $\tan\beta$ into μ_k. The equivalent friction coefficient of a surface landslide is, at the smallest, about 0.27 as mentioned above, but for the debris avalanche it becomes smaller as discussed in the next section.

3.2.2   Debris avalanche

Debris avalanche at Mount Saint Helens

Mount Saint Helens in Washington State, USA erupted on 18 May 1980. This phreatic eruption was triggered by an earthquake that produced a gigantic landslide with a volume of about $2.3\,\mathrm{km}^3$. The deposited volume in the valley of the North Fork Toutle River swelled to about $2.8\,\mathrm{km}^3$ and buried the valley over a distance of about 25 km. The process and mechanism of landslide occurrence were discussed based on the time-lapse photographs taken in the process of sliding and it was concluded that the slide had occurred under the condition that the mountain body was nearly saturated with water. The time from the occurrence of landslide to the final stop of the avalanche was about 10 minutes or so (Voight et al. 1983).

Figure 3.25 shows the area of devastation due to the blast, debris avalanche and mud flow (Takahashi 1981). The fan-shaped part spread from the crater to the opposite mountain range is covered by the pumice flows that occurred after the debris avalanche. The thickness of the debris avalanche deposit is about 195 m near the Spirit Lake and decreases west-ward. The average thickness is about 45 m (Voight et al. 1983). I visited the area one month later the eruption and explored several places. Figure 3.26 is a result of cross-sectional survey at the section A-A demonstrated in Figure 3.25.

Upstream of the section A-A, the large fragments of mountain body (flow mound) were piled up as shown in Photo 3.1 (a), and in the downstream reach the number and scale of the flow mound decreased and the spaces between the flow mounds were flat. There were stripe patterns on the flat space as shown in Photo 3.1 (b) suggesting that the flat spaces were made by the mudflows that ran down after the debris avalanche stopped. On both sides of the cross-section there were the longitudinally continuous natural levees as seen in Photo 3.1 (c) which were as high as 30 m in the highest position.

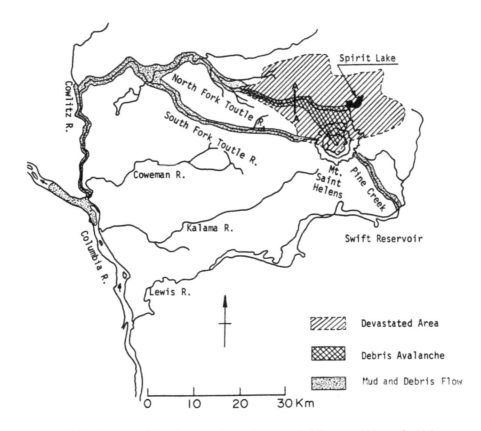

Figure 3.25 Mapping of the devastated area (except ash fall) around Mount St. Helens.

Photo 3.1 (c) also shows the scar marks high on the slope. There were holes a few tens of meters deep sporadically on the debris avalanche deposit as shown in Photo 3.1 (d). These may have been produced by the spouting of vapor because the deposit was very hot. The uppermost layer of about 2 m in the Photo 3.1 (d) is the mudflow deposit.

It was about 4 hours after the eruption when the mudflow front was witnessed to run down at around the section A-A. There are many points along the North Fork Toutle River where mudflow (lahar) was witnessed. The occurrence time of mudflow in the neighborhood of the Spirit Lake can be known by extrapolating the witness records and it is estimated at about 9 a.m. Because the time of eruption was 8:32 a.m. the mud flow must have been generated as soon as the landslide occurred. The mudflow continued for about 12 hours.

Debris avalanche at Mount Ontake

A huge landslide of 36 million cubic meters in estimated volume was caused by the Western Nagano Prefecture Earthquake that occurred in 1984, on the slope of Mount Ontake, Japan. The slid earth debris ran down along the Denjogawa Valley and the

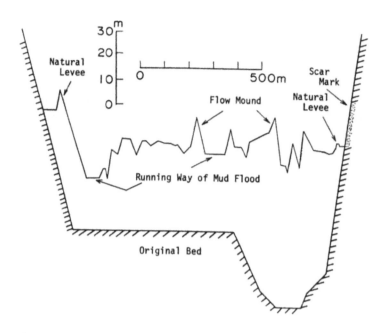

Figure 3.26 Cross-section of the debris avalanche deposit.

Otaki River about ten kilometers as a debris avalanche and it finally stopped, forming a debris dam choking the Otaki River.

The situation of flowage and deposition are summarized in Figures 3.27 and 3.28 (Okuda *et al.* 1985). Although some parts of the slid earth mass overflowed the confronting plateau of Mount Komikasa, the majority of it flowed along the Denjo River and eroded the riverbed and the side banks. Cross-section No. 8 was located at the bend and its outer bank was severely eroded. The depth of the Denjogawa Valley was about 100 m, but the debris avalanche flowed with a little larger depth so the material transported near the surface of flow overflowed the plateau around cross-section No. 7 and it was deposited in the valley of the Nigorizawa River. The main body of the debris avalanche continued down the Denjogawa Valley and at the junction of the Denjogawa and the Nigorizawa a part of debris flowed backward towards the upstream of the Nigorizawa. The strength of the flow decreased a little around Nigorigawa Spa and at cross-section No. 3 it left a deposit of about 50 m thick. The majority of the deposit here was composed of flow-mounds whose diameter was a few tens of meters. Near the outlet of the Nigorigawa a ridge of about 100 m high juts out from the right-hand side and the river kinks to the left. Here, the debris avalanche overflowed the ridge and poured out to the Otaki River. The flow collided with the right-hand side bank of the Otaki River, then, changed direction and continued another 2 km before stopping at the Korigase Gorge. According to some witness records, the mean velocity of the debris avalanche was around 26 m/s ∼ 20 m/s.

After the stoppage of the debris avalanche a mudflow flowed over the deposit. The mudflow continued even after about three hours later than the occurrence time of the huge landslide. The material that overflowed the Komikasa Plateau caused the debris

Photo 3.1 Various aspects of debris avalanche at Mt. St. Helens.

flow in the Suzugasawa River which originates from the Komikasa Plateau and pours into the Otaki River. This debris flow also continued intermittently until about three hours and twenty minutes later than the occurrence of the huge landslide (Suwa *et al.* 1985).

For the reason of high mobility of the Ontake debris avalanche Okuda *et al.* (1985) and Suwa *et al.* (1985) paid attention to the containment of plenty of water within the earth body. The source of the landslide was a fill of pyroclastic material over an old valley. There were many springs on the scar and small scale debris flows were generated from these springs. These facts suggested that the landslide material contained plenty of water before the slide. From the analyses of materials on the Komikasa Plateau, that overflowed around cross-section No. 7 and that was left on the side banks of the Denjo River, they made a hypothesis that the structure of the moving debris avalanche was as shown in Figure 3.29. They claim that the high mobility of the debris avalanche is caused by the existence of a highly water-containing lubricant layer beneath a rather dry layer. Layers I, II, III and IV in Figure 3.29 are scarce in water content in this order and layer III is saturated or nearly saturated with water and the material in IV is as liquid as the normal debris flow or mudflow. The material in layers I and II actually

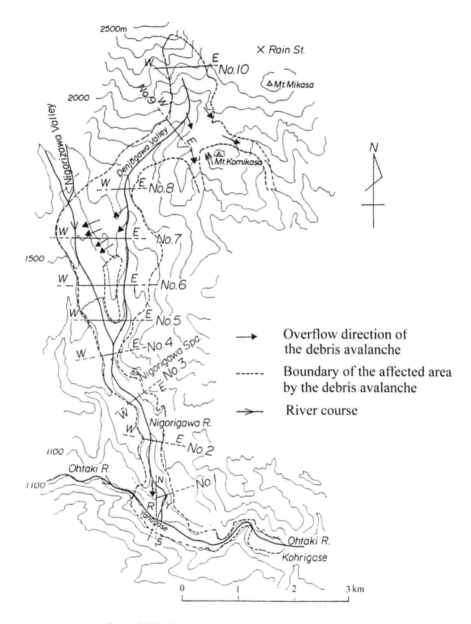

Figure 3.27 Flow area of the Ontake debris avalanche.

had the same quality of being composed of earth blocks and fractured debris except for the water content, and layer I diminished as it flowed down and was captured by and incorporated with layer II. The material in layer III was composed of fractured rock debris, soil and woody debris originally on the ground surface. They conjectured that the water content in this layer increased with the motion downstream.

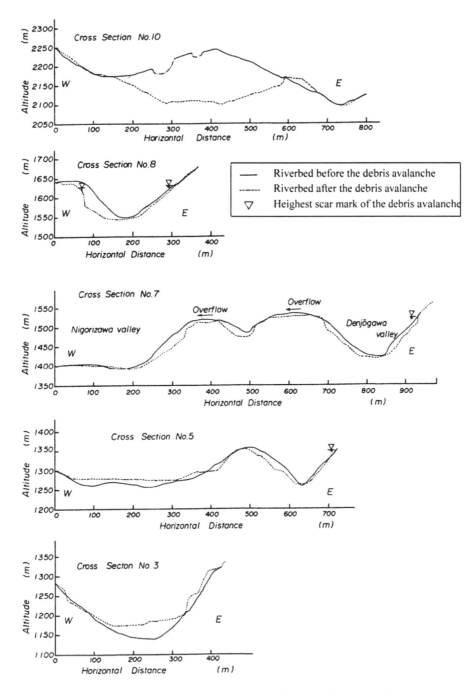

Figure 3.28 Ground surface variations after the debris avalanche.

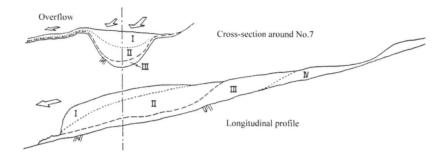

Figure 3.29 Schematic structure of the debris avalanche running down the Denjogawa Valley.

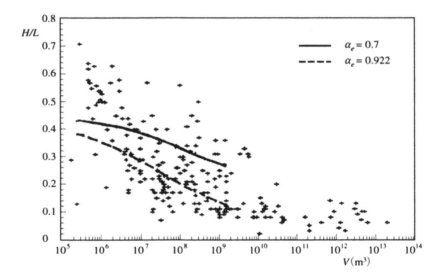

Figure 3.30 Equivalent friction coefficients for various scale landslides.

It is noted that they did not consider why and how the water content in layer III increased downstream and how layer IV was produced. These mechanisms should be the key to understand the mechanics of debris avalanche as explained in the following sections.

The hypotheses for the mechanics of debris avalanche

The equivalent friction coefficient for the debris avalanches of Mount Saint Helens and Mount Ontake are 0.09 and 0.13, respectively. A thorough collection of data was carried out as shown in Figure 3.30 (Campbell *et al.* 1995). The two lines in the figure are drawn based on Takahashi's theory that is explained later. In Figure 3.30 the plotted points scatter very much, but as a general tendency, the larger the landslide scale the smaller the equivalent friction coefficient becomes. This is already referred to in Chapter 1. The equivalent friction coefficient for a very large landslide is far smaller than that for a normal-scale landslide. Many scientists have been interested in this fact and many hypotheses have been advocated.

A. Heim was the first to recognize the high mobility of debris avalanche. He experienced a large rockfall at Elm, Switzerland. He concluded it was not a slide but a flow and the high mobility was maintained by the repeated collisions of fractured particles and he named such a kind of flow 'Sturzstrom' (Hsü 1975).

Kent (1966) postulated that the air entrained between the earth block and the slip surface induces upward air flow within the block and particles are fluidized (particles are suspended by the upward air flow). This hypothesis is introduced to explain pyroclastic flow (Wilson 1980) and turbidity current in the sea (Allen 1984) as well. But, according to the checking by Okuda *et al.* (1985) for the case of the Ontake debris avalanche, the upward air flow velocity can be only 2 cm/s and it cannot suspend soil particles larger than a few tens of microns in diameter.

Shreve (1966, 1968a, 1968b) considered that the sliding earth block jumps at the jutting point on the way and entrains air beneath the block. Then, the air layer becomes a cushion and a lubricant. Okuda *et al.* (1985) checked this hypothesis for the Ontake debris avalanche and concluded this hypothesis cannot be applied.

As the source of high pressure so as to be able to sustain earth block, the vapor pressure induced by the heat of friction was attributed by Habib (1975) and Goguel (1978). Erismann (1979) considered that the heat of friction melts rock and the thus produced thin liquid layer becomes a lubricant.

Melosh (1980) considered that the friction energy is transferred to the particle's vibration energy and the vibration causes the reduction of friction. Foda (1993) and Kobayashi (1994) attribute to the pressure wave produced by the Kelvin-Helmholts-type instability on the slip surface.

The hypotheses, that consider that air or water (vapor) play a crucial role for the high mobility of a debris avalanche, suffered a terrible blow when traces of debris avalanches were discovered on the moon's surface (Howard 1973). Making such a situation background, Hsü (1975) proposed to use the grain flow theory of Bagnold (1954). Because, as explained in the last paragraph of section 1.1, the higher mobility of granular flow than a sliding of a rigid-body cannot be obtained only by the production of shattered clast, he conjectured that the buoyancy effects in the apparent heavy gas densely suspending fine particles act to reduce the equivalent friction coefficient. Takahashi (1981) once used this idea to explain the debris avalanche of Mount St. Helens, where if the interstitial gas contains dust 50% by volume the apparent density of the interstitial gas becomes 1.325 g/cm^3 and the long arrival distance can be explained. Hsü considers the buoyancy effects can be active even in a vacuum like on the moon's surface but, since the particle diameter of dust is far larger than that of the gas molecule, question remains whether the buoyancy effects can arise.

Davies (1982) found the relationship between the end-to-end deposit length L^* and the deposit volume V as $L^* = 9.98 V^{0.32}$ using many previous data. Therefore, the centroid of the deposit is approximately at $0.5L^*$ upstream from the distal end of the deposit. Then, even if the tangent of the angle of the perspective line from the centroid of the deposit to the origin of landslide, that should be the physically reasonable definition of the equivalent friction coefficient, is 0.6 as in the case of the normal-scale landslides, the widely used definition of the equivalent friction coefficient that is the tangent of perspective line from the distal end of the deposit to the origin of landslide becomes smaller than 0.6 because the distal end is about $0.5L^*$ ahead. The larger the V the larger the L^* becomes. Therefore, the larger the volume of landslide the smaller the equivalent friction coefficient becomes. The essence of his discussion is based on the relationship between L^* and V that was found by himself and appeared above. This relationship is nearly fulfilled if the length, width and thickness of the deposit are L^*, $0.1L^*$ and $0.01L^*$, respectively. Then, why such an elongated deposit is formed must be explained. He attributed this elongated deposition to the mechanical fluidization based on Bagnold's experiment which seemed to have a tendency that the larger the shearing velocity the smaller the ratio of shear stress to pressure (the apparent friction coefficient) becomes. But, as explained in the last paragraph of section 1.1, the later experiments revealed that the ratio is almost constant irrespective of shearing velocity, so his argument lost ground.

Legros (2002) pointed out that the granular flow model cannot explain the elongated deposit; the large arrival distance of the center of gravity cannot be explained without reduced friction coefficient, and the extremely small equivalent friction coefficient found on the surface of the moon is an extreme case where the impact force of the meteoritic bombardment may be affected. He finally concluded that the role of the water-saturated layer at the lower part of the slid mass is the most important, although he did not develop the theory of motion.

The role of liquefied layer

When the existence of after flow is neglected and the velocities of the earth block and the liquefied part are assumed equal, the third, fifth and sixth terms of the right-hand side of Equation (3.49) are zero and Equation (3.49) can be written as:

$$\frac{dU_e}{dt} = g \sin\theta - g \cos\theta \,(1 - \alpha_e)\left(1 - \frac{\rho}{\rho_t}\frac{h_s}{H}\right)\mu_k - \frac{1}{2}\frac{\rho_l}{\rho_t}f\alpha_e\frac{U_e^2}{H} \tag{3.56}$$

where $\alpha_e \,(= L'/L)$ is the coefficient to describe the ratio of the length of liquefied layer to the entire length that represents the degree of liquefaction. If U_e, t and x are made dimensionless as $U' = U_e/(gH)^{1/2}$, $t' = t(gH)^{1/2}/H$ and $x' = x/H$, respectively, Equation (3.56) is rewritten as:

$$U'\frac{dU'}{dx'} = -aU'^2 + b \tag{3.57}$$

where $a = \rho_t f \alpha_e/(2\rho_t)$, $b = \sin\theta - \cos\theta(1 - \alpha_e)\{1 - \rho h_s/(\rho_t H)\}\mu_k$.

The solution of Equation (3.57) under the condition; at $x' = 0$, $U' = U_0'$ is given as:

$$U' = \left[U_0'^2\exp(-2ax') + \frac{b}{a}\{1 - \exp(-2ax')\}\right]^{1/2} \tag{3.58}$$

When no liquefied layer exists ($\alpha_e = 0$), $a = 0$ and, instead of Equation (3.58), the following equation is satisfied:

$$U' = (2bx' + U_0'^2)^{1/2} \tag{3.59}$$

The value of x' where U' becomes zero by Equation (3.59) is equal to L_a obtained by Equation (3.54).

The condition for an earth block to stop on the base plain of a mountain is $b \leq 0$, that is:

$$\tan \theta \leq (1 - \alpha_e)\left(1 - \frac{\rho}{\rho_t}\frac{h_s}{H}\right)\mu_k \qquad (3.60)$$

When $\alpha_e = 0$, Equation (3.60) is equivalent to Equation (3.51). This means when the liquefied parts exist the earth block can arrive at a flatter area than in the case of no liquefied layer.

When the gradient of the plain does not satisfy Equation (3.60), the earth block continues its motion. In this case, the velocity of the slid block is, from Equation (3.58), asymptotic to:

$$U_e = (gH)^{1/2}(b/a)^{1/2} \qquad (3.61)$$

This means the larger the value of b, i.e. the steeper the gradient of the plain and the larger the thickness of the slid block, the faster the velocity of the block approaches.

When the gradient of the mountain base plain satisfies Equation (3.60), b has a negative value and the arrival distance of a slid earth block is obtained by substituting $U' = 0$ in Equation (3.58) as:

$$x_L = -\frac{H}{2a}\ln\left\{-\frac{b}{a}\left(\frac{U_0^2}{gH} - \frac{b}{a}\right)^{-1}\right\} \qquad (3.62)$$

The values of a and b are not so sensitive to the volume of the block and $U_0^2/(gH)$ is equivalent to the Froude number, so it would not be so sensitive to the scale of phenomena. Therefore, Equation (3.62) indicates that the arrival distance of the slid earth block is controlled by the thickness of the block. Generally, the larger the volume of the slid earth block, the greater the thickness of the block. Thus, the arrival distance is controlled by the volume of the landslide. The tendency of the equivalent friction coefficient to become smaller with increasing landslide volume is due to the increase in the arrival distance on the mountain base plain with the increasing volume of the landslide.

Corresponding to Equation (3.53), the non-dimensional velocity of the slid earth block (Froude number) at the proximal end of the mountain base plain is obtained as:

$$U_0' = \left[\frac{b_u}{a_u}\left\{1 - \exp\left(-2a_u\frac{1}{\tan\theta_u}\frac{H_p}{H}\right)\right\}\right]^{1/2} \qquad (3.63)$$

where a_u, b_u and θ_u are a, b and the slope gradient at the base of the mountain slope, respectively.

Referring to Figure 3.24 the equivalent friction coefficient $\tan \beta$ is given by:

$$\tan \beta = (H_p + x_L \sin\theta_d)/(x_L \cos\theta_d + H_p/\tan\theta_u) \qquad (3.64)$$

Table 3.3 The equivalent friction factor and other characteristics of landslides as the function of thickness of earth block and degree of saturation.

H	α_e			
	0.7	0.8	0.9	0.922
30 m	1.62	1.62	1.62	1.62
	181 m	226 m	370 m	650 m
	0.433	0.438	0.422	0.394
100 m	1.62	1.62	1.62	1.62
	603 m	752 m	1,234 m	2,168 m
	0.398	0.384	0.347	0.292
300 m	1.55	1.57	1.58	1.58
	1,722 m	2,194 m	2,660 m	6,464 m
	0.316	0.291	0.237	0.178
500 m	1.42	1.46	1.49	1.49
	2,601 m	3,424 m	5,900 m	10,580 m
	0.273	0.244	0.187	0.135

Top row: u_0', Middle row: x_L, Bottom row: tan β

For the sake of grasping conceptually the values of the equivalent friction coefficient with the change of α_e and H, tan β is calculated under $\rho_t = 2$ g/cm^3, $\rho_l = 1.3$ g/cm^3, $\rho = 1.0$ g/cm^3, $f = 0.5$, $\theta_u = 25°$, $\theta_d = 2°$, $Hp = 1,500$ m, $h_s/H = 0.7$ and $\mu_k = 0.7$. The results are shown in Table 3.3. According to Equation (3.60) the motion of the earth block stops when $\alpha_e < 0.923$. In this range of α_e, the larger the value of α_e or the larger the thickness of the earth block the smaller the equivalent friction coefficient becomes. If one assumes the volume of landslide V is given as $10H^3$, when H is 30, 100, 300 and 500 m, V is 2.7×10^5, 10×10^6, 2.7×10^8 and 12.5×10^8 m^3, respectively. The relationships between these V values and tan β are drawn in Figure 3.30 for the two values of α_e. Incidentally, the Harihara River landslide was 18 m at its maximum depth and the volume was 1.6×10^5 m^3, the Ontake landslide was about 150 m in depth and the volume was 36×10^6 m^3 and the St. Helens landslide was about 600 m deep and the volume was about 28×10^8 m^3; these values prove $V = 10H^3$ is a reasonable estimate.

3.2.3   Model for the transformation into debris flow

From 6th to 10th July 1997, a severe rainfall amounting to 400 mm fell on the southern part of Kyushu Island, Japan. On 10th July, 4 hours after the stoppage of the rainfall, a large-scale landslide of 80 m at the maximum width, 190 m in length, and 30 m at the maximum depth occurred in the basin of the Harihara River (1.55 km^2). There was a dam to check debris flows (14 m high and 85 m wide) at about 300 m downstream from the foot of landslide, but the volume of the landslide (160,000 m^3) was so large that some parts of the mass overflowed the dam changing into debris flow, even though the dam stored about 50,000 m^3 of sediment. This debris flow destroyed 19 houses and killed 21 people. Because considerable time had elapsed after the stoppage of

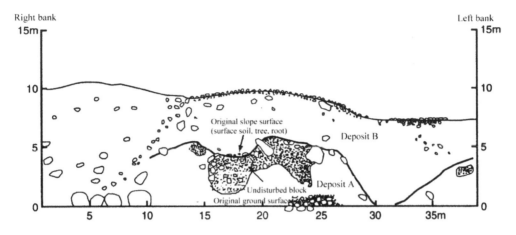

Figure 3.31 The cross-section of the trench upstream of the dam.

rainfall to when landslide occurred, the flood runoff discharge in the river should have been small in such a small basin. Therefore, the hypothesis that the slid earth mass choked the river to make a natural dam and then it was destroyed by the running water in the river should be rejected. In fact, no mark of natural dam formation was seen. Thus, although some parts of slid earth mass were checked by the dam, the rest part transformed into debris flow while it continued its motion practically without any water supply from outside.

After the disaster a trench cut inspection of deposit was done at about 170 m upstream of the dam (Yamada *et al.* 1998). It revealed that there were two different flow units as shown in Figure 3.31; one was the relatively undisturbed earth block in the lower part (deposit A) and the other was the disturbed debris flow deposit (deposit B) laid over the deposit A. This fact suggests that a considerable part of the slid earth mass reached at least to the trench cut position with little disturbance (the trees originally on the surface of the mountain slope could be found on the boundary between the deposit A and the deposit B), and after the stoppage of the earth blocks the debris flow passed over and deposited on the stopped earth blocks. As mentioned earlier, water supply from outside was scarce, so that the debris flow that arrived later than the earth blocks should be the part originally made up the lower part of the earth mass and the part was liquefied into debris flow while in motion and left behind the earth blocks.

Such a stratigraphy described above was also recognized in the deposit of debris flow that accompanied the landslide caused by the phreatic explosion at Sumikawa Spa, Akita Prefecture, Japan, in May, 1997 (Hoshino *et al.* 1998; Sassa *et al.* 1998). The debris flows flowing over the deposit of debris avalanches were observed at St. Helens Volcano in 1980 (Takahashi 1981) and at Ontake in 1984. These two cases were generated by earthquakes with no rainfall. Especially for the Ontake case, the role of the liquefied layer beneath the moving earth mass was emphasized for the high mobility (Suwa *et al.* 1985) as explained in section 3.2.2.

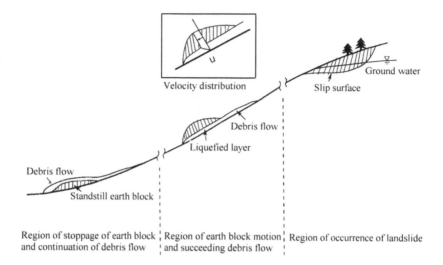

Velocity distribution

Ground water

Slip surface

Debris flow

Liquefied layer

Debris flow

Standstill earth block

Region of stoppage of earth block   Region of earth block motion   Region of occurrence of landslide
and continuation of debris flow   and succeeding debris flow

Figure 3.32 A process of transformation of landslide into debris flow.

Based on the discussion above, Takahashi (2001a) constructed a model for the transformation of landslide into debris flow as illustrated in Figure 3.32. If a deep-seated landslide occurs by the rise of the ground water level, at least at the neighborhood of the slip surface, the void rich soil structure will be destroyed and individual particles are dispersed in the ground water. Then, the weight of the earth mass that was previously sustained by the skeletal structure will be sustained by the pressure in the destroyed layer; the liquefied layer. If the earth mass is dense (void ratio is less than about 0.5), even if the skeletal structure is destroyed, particles cannot be detached from each other so that it is hardly liquefied (Fleming *et al.* 1989). But, if the earth mass is sparse, it is suddenly liquefied and the motion is accelerated (Iverson *et al.* 2000). Because the permeability of the earth mass is small, the high pressure within the liquefied layer will not easily be released and the earth block continues its motion supported by the liquefied layer. The velocity distribution in the liquefied layer must be as shown in Figure 3.32, i.e. zero at the bottom and the maximum at the boundary between the liquefied layer and the earth block, so that the earth block goes faster than the liquefied layer, then the liquefied layer is left behind as debris flow. The motion of the earth block continues as long as the slope gradient is steep enough and liquefaction at the boundary between the earth block and the ground continues. Thus, the earth-block proceeds cannibalizing its body to produce a new liquefied layer and elongates its tail as a debris flow. If the water saturated part of the earth block has been thoroughly consumed by producing a liquefied layer or the earth block arrives at a flatter area, it will stop, and the debris flow following the earth block will go over the stopped earth block or detour round it and continue to run down. Thus, the surficial part of the earth block from the original mountain slope remains underneath the deposit of debris flow. In the Harihara River debris flow case, the earth block was worn down from its maximum thickness of 30 m to about 5 m during the run down motion of about 200 m, and it stopped upstream of the dam.

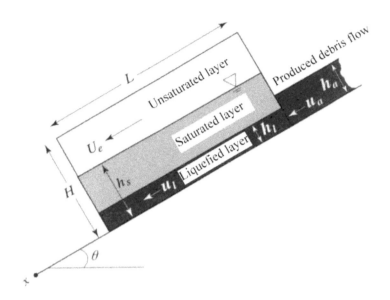

Figure 3.33 Moving earth block producing the liquefied layer and debris flow.

The importance of liquefaction within the saturated zone as a cause of the rapid motion of a landslide mass has been discussed by many authors. They are based on various concepts; e.g. Iverson *et al.* (1997), Okura *et al.* (2002), Wang and Sassa (2002) and Takahashi as explained in sections 3.2.1 and 3.2.2 consider that the liquefied layer is produced inside the moving earth block; Hutchinson (1988) and Sassa (1997) consider that the liquefied layer is produced inside the previously stable soil layer on which landslide mass suddenly rides, and Sassa (1997) considers that the fracture of coarse particles on the slip surface produces a thin liquefied layer. However, no previous model is able to analyze the simultaneous motions of the earth mass and the debris flow from the onset to stoppage.

3.2.4 Mathematical model for the one-dimensional motion of a deformable earth block with the liquefied layer

Let us assume that the moving earth block is a rectangular solid having a unit width like the one shown in Figure 3.33. This model is different from those of Ashida *et al.* (1985), Iverson *et al.* (1997) and Takahashi's that appeared in sections 3.2.1 and 3.2.2 in respect to the volume change of earth block; the earth block in the former theories is a non-deformable rigid body, but this model considers the earth block volume attenuation downstream.

The momentum conservation equation is rewritten from Equations (3.49) and (3.56) as:

$$\frac{d}{dt}(\rho_t LHU_e) = \rho_t gLH \sin\theta - \rho_t gLH \cos\theta(1 - \alpha_e)(1 - \rho h_s/\rho_t H)$$

$$-(U_e - u_l)h_l \rho_l u_l - (1/2)\rho_l f u_l^2 L + \rho_a(u_a - U_e)^2 h_a$$

$$+(1/2)\rho_a \cos\theta g h_a^2 \qquad (3.65)$$

For the perfectly liquefied layer α_e is 1 and the second term of the right-hand side of Equation (3.65) is eliminated, but, herein, this term is left to make possible the numerical reproduction of the experimental cases where the production of a perfectly saturated layer is rather difficult.

The volume conservation equation of the earth block is:

$$\frac{d(HL)}{dt} = -(U_e - u_l)h_l \tag{3.66}$$

and the equation for the changing rate of the liquefied layer thickness is:

$$\frac{d(h_l L)}{dt} = i_c L - (U_e - u_l)h_l \tag{3.67}$$

where i_c is the changing rate of earth block to liquefied layer at the boundary between the earth block and the liquefied layer, and $(U_e - u_l)h_l$ is the volume left behind as debris flow.

The shear stress τ operating on the boundary plane between the solid earth block and the liquefied layer will do as much work as τu_l per unit time. Because the earth block has strength τ_y, as much energy as $i_c \tau_y$ should be given to the earth block for it to be liquefied at the rate of i_c. Although the shear stress operating on the boundary plane does not entirely contribute to liquefy the earth block, herein, i_c is simply assumed to be determined by the relationship $i_c \tau_y = \tau u_l$. If $\tau/\tau_y = \beta$, the following formula is obtained:

$$i_c = \beta u_l \tag{3.68}$$

If the strength of the earth block is represented by the cohesion c and the liquefied layer becomes a Newtonian fluid, the following formulae are obtained:

$$\tau_y \approx c \tag{3.69}$$

$$\tau = \mu(du/dz) \approx \mu(u_l/h_l) \tag{3.70}$$

Therefore, i_c is given as:

$$i_c = \frac{\mu u_l^2}{ch_l} \tag{3.71}$$

where μ is the viscosity of the liquefied layer. As a rough standard of actual phenomena, $c = 35 \times 10^3$ Pa, $\mu = 100$ Pas, $h_l = 100$ cm and $u_l = 10$ m/s are substituted into Equation (3.71), i_c has a value of the order of 0.3 m/s.

If the velocity in the liquefied layer distributes linearly, U_e is given as:

$$U_e = \left(\frac{2H - h_l}{H}\right) u_l \tag{3.72}$$

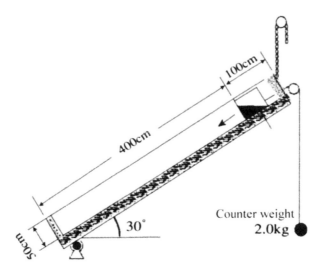

Figure 3.34 Experimental flume.

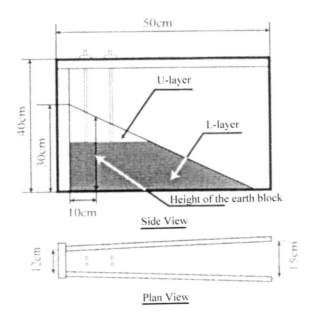

Figure 3.35 Box to install the test piece.

Verification by experiments

A steel movable slope flume 500 cm in length, 20 cm in width and 50 cm in depth with one side as a transparent glass wall was set at a slope of 30°, as shown in Figure 3.34. The test piece of earth block was installed in a box whose bottom and rear end are open as shown in Figure 3.35, and let the test piece together with the box slide on the

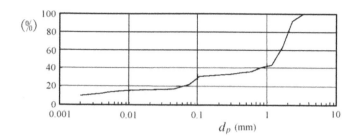

Figure 3.36 Experimental material.

bottom of flume. The reason why the test piece is installed in the box is to prevent the collapse of the block at the forefront and let the abrasion of the block proceeds only at the bottom while in motion. The weight of the box was cancelled by the counter weight. The four iron bars fixed to the box and penetrating into the earth block restrain the relative motion of the earth block to the box. The stopper at the downstream end of the flume prevents the box from dropping off the flume. The moving behavior of the earth block was recorded by a video camera.

The experimental material is a mixture of kaolin and silica sand whose size distribution is shown in Figure 3.36. The lower part of the test piece (L-layer) has a larger water content than the upper part (U-layer). The water content w (weight of water/total weight of water plus solids) in the L-layer was changed from 15.3% to 18.0% depending on the test cases and that in the U-layer was fixed at 13.0%. The substantial weight of the L-layer material was fixed at 2 kg for all the experimental cases.

The experiments revealed that the phenomena can be classified into three types depending on the degree of saturation within the L-layer. The first category arose with water contents of 15.3% and 15.6%, the earth block stopped after it had proceeded 2–2.5 m within 3–4 seconds. A deposit layer of about 5 cm was left behind and both the earth block and the deposit did not move any more. The second category arose with water contents of 16.0% and 16.7%. The earth block once stopped after traveling 2-3 m within about 3 seconds, but the deposit left behind the block gradually moved and pushed the block an additional 1 m or so. The third category appeared with water contents of 17.0%, 17.4% and 18.0%. The earth blocks in these cases arrived at the downstream end of the flume within about 3 seconds and stopped on the back of the stopper. The L-layer was thoroughly worn out within about 2 seconds, thereafter the earth block motion decelerated. The part left behind the block was completely liquefied and it flowed with about 2 cm in thickness running over the already stopped U-layer and was deposited by the blockade of the downstream-wall of the box.

Figure 3.37 shows the earth block's traveling velocity and abrasion rate versus time for the respective categories. In the first category ($w = 15.3\%$), both the traveling velocity and the abrasion rate were smaller than in the other categories, and the peak times of traveling velocity and abrasion rate coincided indicating the validity of representing the abrasion rate by Equation (3.68). In the second and third categories, both traveling velocity and abrasion rate were larger than those in the first category

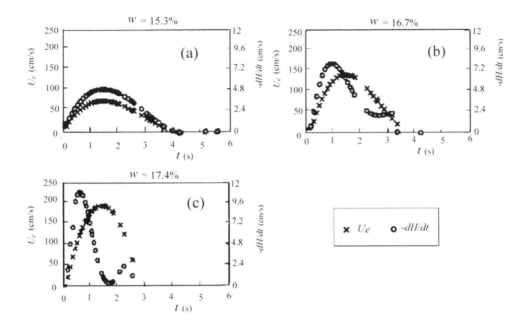

Figure 3.37 Traveling velocity and abrasion rate of the earth block in the experiment.

as one would expect, but the maximum abrasion rate preceded the maximum traveling velocity. The latter characteristics would presumably be the result of the vertically distributing water content within the L-layer; the larger value in the lower part. Such a distribution was brought by the vertical infiltration of water within the L-layer. The abrasion rate (liquefaction rate) will become smaller if the water content becomes smaller.

The fundamental Equations (3.65), (3.66) and (3.67) were numerically solved under the conditions that the initial thickness of the earth block was 24 cm and the thickness of the L-layer was 10 cm, where the adopted parameter values were $\theta = 30°$, $\rho_t = 2\,g/cm^3$, $\rho_l = 2.15\,g/cm^3$, $\mu_k = 0.75$, $f = 0.3$ (when saturated) and $f = 0$ (when unsaturated). The entire L-layer was assumed to be saturated, but to reflect a slightly smaller degree of saturation in the upper part of the L-layer, the β value in upper 3 cm part of the L-layer was gradually decreased. The appropriate values of α_e and β were determined through the sensitivity examinations of these respective values on the traveling velocity and the abrasion rate.

Figure 3.38 shows the results of calculation for the cases of higher degree of saturation. The adopted values of α_e and β are given in Figure 3.38. If one compares Figure 3.38(b) with Figure 3.37(c), one will understand that the fundamental equations well explain not only the magnitudes of the traveling velocity and abrasion rate but also the tendencies of variation in those values. However, the time lag between the occurrence of the maximum traveling velocity and the occurrence of the peak abrasion rate in the calculation was too short in comparison with the experiments. It suggests that the

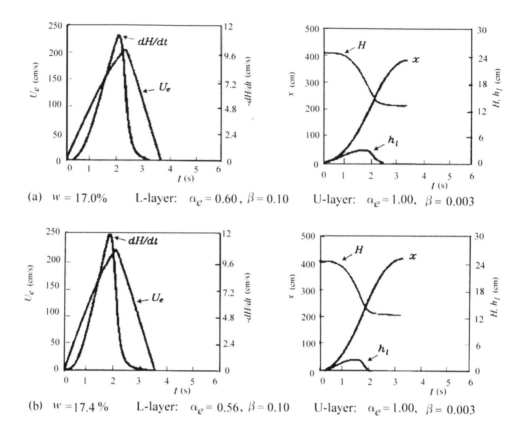

(a) $w = 17.0\%$ L-layer: $\alpha_e = 0.60$, $\beta = 0.10$ U-layer: $\alpha_e = 1.00$, $\beta = 0.003$

(b) $w = 17.4\%$ L-layer: $\alpha_e = 0.56$, $\beta = 0.10$ U-layer: $\alpha_e = 1.00$, $\beta = 0.003$

Figure 3.38 Results of numerical calculation for higher degree of saturation cases.

thickness of saturated part within the L-layer in the experiment was thinner than in the calculation.

3.2.5 Numerical simulation of earth block and debris flow motions across a three-dimensional terrain

The analytical model for the motion of an earth block mentioned above can trace the motion of an earth block only on a one-dimensional slope. It can neither trace the motion of debris flow following the block nor analyze the motions of the earth block and debris flow across a three-dimensional terrain. Here, a new simulation model that is able to reproduce the experimentally obtained moving characteristics of the gradually liquefying earth block and the debris flow produced by the liquefaction and following the earth block is introduced.

Behaviors of earth block and debris flow in the experiments

Photo 3.2 shows the situation of the Harihara River debris flow. A deep-seated landslide that occurred on the slope seen on the upper part of the photograph merged into

Photo 3.2 Debris flow at the Harihara River.

the river channel with a merging angle of about 45°. The earth mass ran up the opposite bank slope and then came back to the river channel to continue downstream in
the river channel having a trapezoidal cross-sectional shape. Some parts of the earth
mass were trapped by the check dam that is seen in the center of the photograph, but
the other parts overflowed the dam and flooded over the residential area at the lower
part of the photograph. The longitudinal gradient of the river at the foot of the slid
slope was about 7° and it became milder downstream.

All the conspicuous behaviors of the event should have been affected by the characteristics such as the merging angle, the channel slope and the opening angle of the
channel cross-section (Benda and Cundy 1990; Ishikawa 1999). Therefore, we constructed an experimental apparatus that could change these factors independently as
shown in Figure 3.39. The resemblance between the actual situation and the experimental apparatus would be evident by comparing Photo 3.2 with Figure 3.39. The
slope part of the apparatus had a gradient of 30° and it was 500 cm long, 20 cm
wide and 50 cm deep with one side as a transparent glass wall and the bed was made
of rough wooden board. The river channel part was constructed by combining two
wooden boards in a V-shape. The surface of the board was covered by a plastic net to
make the surface hydraulically very rough.

The experimental material was the same as the one used in the previously mentioned experiments for a one-dimensional motion whose size distribution was shown
in Figure 3.36. Before the experiment the water saturated and unsaturated materials
were prepared. The moisture ratio (weight of water/weight of solids) of the saturated
material was 27.1% and that of the unsaturated material was 12.5%. These respective
values correspond to the water contents of 21.3% and 11.1% by the definition in section 3.2.4. The natural soil sampled on the slope adjacent to the landslide scar in the
Harihara River basin contained more fine fractions (<0.075 mm) than the experimental material and it had a moisture ratio as much as 45% when saturated (Yamamoto *et
al.* 1999). This value corresponds to the porosity of 54%. In the experimental material,

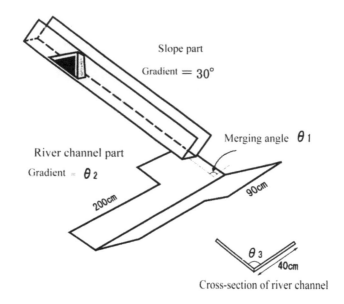

Figure 3.39 Experimental apparatus.

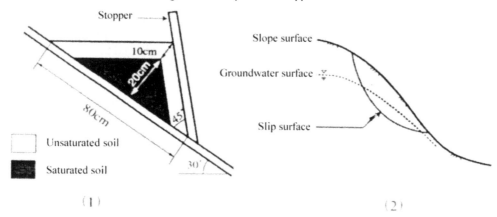

Figure 3.40 Earth mass that starts moving. (1) Experiment, (2) Actual situation.

however, it was difficult to make such a void-rich state and the saturated material in the experiments had, at most, a porosity of 43 to 44%.

It must be noted that the experiment did not intend to do the model test of Harihara Debris Flow but to grasp the characteristics in the processes of a transformation into debris flow and to develop a method to numerically simulate the phenomena. To simulate a specific phenomenon by a physical model test or by a numerical simulation, more strict similarities in the controlling parameters should be pursued.

Prior to the commencement of an experimental run, a triangular earth mass was prepared in the back of a stopper set at 200 cm uphill from the merging point of the slope. The side-view of the mass is shown in Figure 3.40 (1), the core part of the earth

Table 3.4 Kinds of experiments.

Cases	Merging angle θ_1	Channel slope θ_2	Opening angle θ_3	Distal end of deposit from the downstream end of flume
1	45°	5°	100°	185 cm
2	45°	5°	140°	178 cm
3	45°	15°	100°	168 cm
4	45°	15°	140°	174 cm
5	90°	5°	100°	190 cm
6	90°	5°	140°	179 cm
7	90°	15°	100°	120 cm
8	90°	15°	140°	170 cm

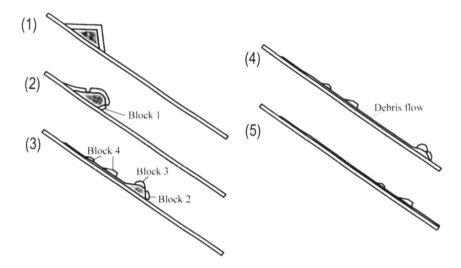

Figure 3.41 Motion and deformation of the earth mass on the slope. Time proceeds from (1) to (5).

mass is composed of the saturated material and it is covered by a superincumbent comprised of the unsaturated material. This soil structure mimics the soil block just before the beginning of slide as shown in Figure 3.40 (2). Immediately after the preparation of the earth mass the stopper was removed and the mass was released downhill. Soon, the mass merged into the river channel. The motion was recorded by video cameras.

The experiments were done under different combinations of merging angle, river channel slope and opening angle of cross-section as shown in Table 3.4. The behaviors and depositing situations were found to be slightly different, even under the same setting of experimental conditions, depending on the varying compactness of the unsaturated material, the consumed time in preparing the earth mass, etc. Therefore, several runs were repeated in each case and the representative behaviors and characteristic values were obtained in each case.

Figure 3.41 shows the general tendency of the earth mass behaviors on the slope. As soon as the stopper was removed, the earth mass began to move and deform, and

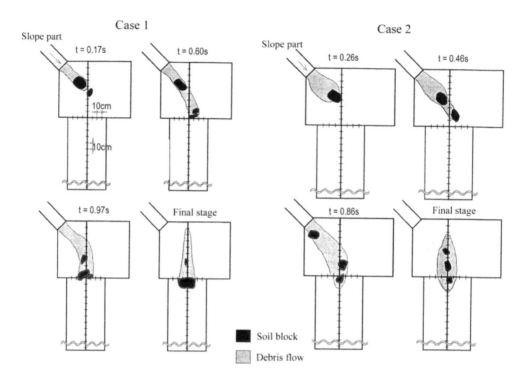

Figure 3.42 Behaviors of soil blocks and debris flow in the river channel. Case 1 and Case 2.

the unsaturated soil crust was divided into several portions. At the head, a lower one third part was pushed out as a block and soon it split into two; blocks 1 and 2. Block 2 proceeded at the head with a velocity of about 250 cm/s, but block 1 was swallowed into the liquefied debris flow part that followed block 2. The other unsaturated parts were also split into several blocks and among them the one that took the lead (block 3) was transported as if floating on the debris flow. The other blocks (blocks 4) were left behind and slowly went downhill. The saturated core part was liquefied making a layer a few centimeters thick. This layer always went after the leading soil blocks, and sometimes, the soil block and the liquefied layer were separated, but soon the liquefied layer formed a bore like shape and continued going down. The saturated soil was almost thoroughly liquefied before merging into the river channel.

Figure 3.42 shows the time sequential behaviors of soil blocks and the liquefied layer (debris flows) for Cases 1 and 2. The origin of time in Figure 3.42 is the instant when the forefront reaches the merging point with the river channel.

In Case 1, blocks 2 and 3 always existed at the leading head, and they climbed up the opposite bank a little. Then, these blocks slid down onto the valley bottom and continued to slide some distance along the channel before stopping. These blocks were laid in contact with each other forming a kind of natural dam. The following debris flow was checked by this dam and formed a muddy pond. Some parts of blocks 4 that came later plunged into this pond and sunk.

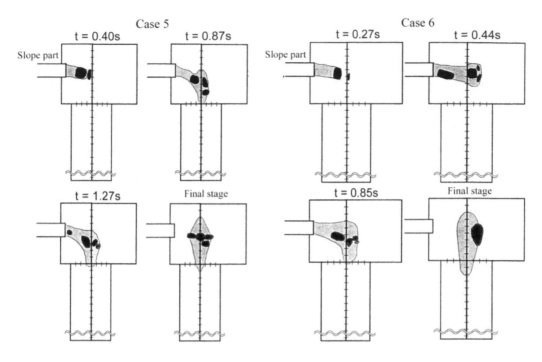

Figure 3.43 Behaviors of soil blocks and debris flow in the river channel. Case 5 and Case 6.

In Case 2, blocks 2 and 3 behaved in much the same way as in Case 1, but they stopped separately on the valley bottom. These blocks checked some parts of debris flow, but the rest could continue to flow by threading their way through and detouring round the stopped blocks helped by the large opening angle of cross-section (θ_3).

The behaviors in Cases 3 and 4 are not shown here, but the behaviors in Case 3 were similar to those in Case 1 except that the natural dam was formed a little further downstream. Debris flow was checked by the dam and its highest stage reached almost to the crest of the dam. The behaviors in Case 4 were similar to those in Case 2, but the debris flow reached a little further downstream.

Figure 3.43 shows the situations for the cases where the earth mass merges into the river channel perpendicularly. In Case 5 the soil blocks stopped immediately after merging into the channel. The following debris flow collided with these stationary blocks and split into two parts; one part flowed downstream and the other part was deposited upstream of the stationary blocks. In Case 6, blocks 2 and 3 behaved similarly to those in Case 5, but they climbed up the opposite bank a little and stopped there. Accordingly, the following debris flow was not much hindered by the blocks and could flow downstream.

In Case 7 (not shown here), blocks 2 and 3 reached further downstream than in Case 5 and the following debris flow could detour round the blocks and reached the furthest of all the experimental cases. In Case 8, blocks 2 and 3 climbed up the opposite bank. Meanwhile, the leading part of the following debris flow proceeded downstream, but soon the once stationary blocks on the opposite bank slipped down to the valley bottom and the following part of the debris flow was checked, forming a muddy pond upstream of the blocks.

The experimental results reveal that the behaviors of soil blocks and debris flow intricately vary depending on the combinations of merging angle, channel gradient and the opening angle of cross-section. Generally, the smaller the merging angle, the smoother the mass flows into the river and further the mass will reach. But, smoother merging will leave the larger soil block to become an obstacle for the following debris flow and consequently the debris flow cannot reach far. On the other hand, if the merging angle is perpendicular to the river channel, the blocks cannot reach far downstream, but by the crushing of blocks due to the collision with the opposite bank (steep bank slope case) or by the higher and further climbing up onto the opposite bank (flatter bank slope case) the blocks will not be a great hindrance for the following debris flow and the debris flow can reach further. The steeper channel gradient generally means that the mass can reach further downstream, but the larger circular motion of a large block (flatter bank slope case) may cause it to hinder the debris flow. The strength, the water content and the volume of the slid earth mass must also be considered as controlling factors.

Numerical simulation of the phenomena

As mentioned earlier, we imagine that the earth mass that starts to move is saturated by water at least in the lower part and this part will be liquefied while in motion. The liquefied part will be left behind the moving but not liquefied earth block as a following debris flow. The majority of debris flow simulation models use Eulerian continuous fluid equations, whereas the motion of a solid earth block will be most easily analyzed as that of a rigid body in the Lagrangian treatment. The earth mass we are considering consists of a rigid body and a fluid in which the rigid body is moving on the surface of fluid flow and the rigid body and fluid interact with one another. Therefore, to analyze the behaviors of such a compound body, a specially devised method is necessary. The objective here is the analysis of such a compound body's motion across a three-dimensional terrain, and the flow as well as the rigid body is treated as a shallow body.

The continuity equation of the liquefied layer in the lower part is:

$$\frac{\partial h}{\partial t} + \frac{\partial (uh)}{\partial x} + \frac{\partial (vh)}{\partial y} - i_c = 0 \tag{3.73}$$

and the equations of motion for the liquefied layer in the x and y directions are, respectively:

$$\frac{\partial u}{\partial t} + u\frac{\partial u}{\partial x} + v\frac{\partial u}{\partial y} = g\sin\theta_{wx} - \frac{\tau_{sx}}{\rho_T h} - \frac{\tau_{bx}}{\rho_T h} \tag{3.74}$$

$$\frac{\partial v}{\partial t} + u\frac{\partial v}{\partial x} + v\frac{\partial v}{\partial y} = g\sin\theta_{wy} - \frac{\tau_{sy}}{\rho_T h} - \frac{\tau_{by}}{\rho_T h} \tag{3.75}$$

where u and v are the velocities in the x and y directions, respectively, i_c is the changing rate into the liquefied layer from the solid earth block, τ_{sx} and τ_{sy} are the x and y-wise shear stresses working on the boundary between the earth block and the liquefied layer (upper boundary), τ_{bx} and τ_{by} are the x and y-wise shear stresses working on the boundary between the liquefied layer and the bottom of flow (lower boundary), θ_{wx} and θ_{wy} are the x and y-wise gradients of the earth block's surface, ρ_T is the apparent density of the liquefied layer. The shear stresses in the horizontal direction are neglected because the thickness of the moving body is shallow.

The boundary shear stresses in Equations (3.74) and (375) are given as (Takahashi et al. 2000):

$$\frac{\tau_{sx}}{\rho_T h} = \frac{9\mu_a^2 (u - u_b)\sqrt{(u - u_b)^2 + (v - v_b)^2}}{\rho_T^2 g h^4 |\sin\theta_w|} \tag{3.76}$$

$$\frac{\tau_{sy}}{\rho_T h} = \frac{9\mu_a^2 (v - v_b)\sqrt{(u - u_b)^2 + (v - v_b)^2}}{\rho_T^2 g h^4 |\sin\theta_w|} \tag{3.77}$$

$$\frac{\tau_{bx}}{\rho_T h} = \frac{9\mu_a^2 u\sqrt{u^2 + v^2}}{\rho_T^2 g h^4 |\sin\theta_w|} \tag{3.78}$$

$$\frac{\tau_{by}}{\rho_T h} = \frac{9\mu_a^2 v\sqrt{u^2 + v^2}}{\rho_T^2 g h^4 |\sin\theta_w|} \tag{3.79}$$

where μ_a is the apparent viscosity of the liquid layer, u_b and v_b are x and y-wise velocities of the earth block, and θ_w is the surface gradient of the earth block that is given by:

$$|\sin\theta_w| = \sqrt{\sin^2\theta_{wx} + \sin^2\theta_{wy}} \tag{3.80}$$

The moving earth block may be considered as an assembly of circular cylinders standing perpendicular to the bed, arrayed in contact in tetrahedral-rectangular rows (the densest packing state). Under such an arrangement, there exist voids among cylinders, and therefore, this arrangement cannot represent the substantial volume of the earth block. To compensate for the false void space among cylinders, the diameter of the cylinders is enhanced so that the cross-sectional area becomes equal to that of the hexagons arranged without voids as shown in the left-hand side of Figure 3.44. Thus, the diameter of the cylinder is about $1.05D_1$ $(=2r_c)$, where D_1 is the original diameter of rigid cylinders closely packed with voids. The cylinders are more or less soft and can be packed without voids when they are assembled as a block, but once a cylinder is separated from the assemblage it becomes a rigid cylindrical block. One may think that if the earth block is assumed as the assemblage of quadratic prisms or hexagonal pillars no such complicated assumption is necessary. But, the assumption adopted here, alternatively excludes the complicated calculation of rotating motions of dispersed earth blocks.

The cylinder supported over the liquefied layer is comprised of lower saturated and upper unsaturated (degree of saturation $= s_b$) layers, and the thicknesses of these

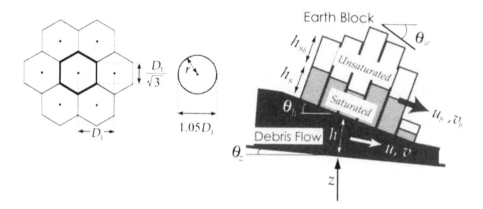

Figure 3.44 Plan and side views of the earth block as an assembly of cylinders and the liquefied layer.

two layers are h_s and h_{sb}, respectively, as illustrated in the right-hand side of Figure 3.44. Therefore, the mass of a cylinder M_b is given by:

$$M_b = \rho_T S_0 (\alpha_1 h_{sb} + h_s) \tag{3.81}$$

where $\alpha_1 = \{C\sigma + (1 - C)\rho s_b\}/\{C\sigma + (1 - C)\rho\}$, S_0 is the bottom area of the cylinder, and the solids concentration in the saturated layer, in the unsaturated layer and in the liquefied layer are all considered to have the same value.

The x and y-wise equations of motion for a cylinder are as follows:

$$\frac{\partial u_b}{\partial t} = g \sin \theta_x + \frac{\tau_{sx} S_0}{M_b} + \frac{1}{M_b} \left\{ \sum f_x + \sum f_{sx} \right\} \tag{3.82}$$

$$\frac{\partial v_b}{\partial t} = g \sin \theta_y + \frac{\tau_{sy} S_0}{M_b} + \frac{1}{M_b} \left\{ \sum f_y + \sum f_{sy} \right\} \tag{3.83}$$

where θ_x and θ_y are the x and y-wise gradients of the liquefied layer's surface (if no liquefied layer exists they are the gradients of bottom), f_x and f_y are the x and y-wise attractive or repulsive forces working between the cylinders, f_{sx} and f_{sy} are the x and y-wise shear stress fractions between the cylinders, and $\sum$ means the summation of these forces. The attractive force is introduced to take account of the strength of the earth block against segmentation and the repulsive force is introduced to make the calculation including collision between the earth blocks possible.

If one considers that the repulsive force acts when the distance between the centers of two cylinders D_b becomes shorter than D_1 and the attractive force acts when D_b

becomes larger than D_1, the absolute values of attractive or repulsive forces between cylinders would be described by the following equations:

$$f_{in} = \begin{cases} \dfrac{f_{inm}}{D_2 - D_1}(D_b - D_1) \; ; & D_b < D_2 \\[2mm] \dfrac{-f_{inm}}{D_3 - D_2}(D_b - D_1) \; ; & D_2 \le D_b \le D_3 \\[2mm] 0 & ; \quad D_3 < D_b \end{cases} \tag{3.84}$$

where D_2 is the distance between cylinders in which the attractive force becomes the maximum, and D_3 is the largest distance between cylinders within which the attractive force operates. Repulsive and attractive forces have positive and negative values, respectively, and the maximum attractive force f_{inm} is assumed to satisfy the following equation:

$$f_{inm} = -2r_c h_{s\,min} c_b \tag{3.85}$$

where $h_{s\,min}$ is the smaller h_s value of the two contacting cylinders and c_b is the adhesive force between the cylinders.

When cylinder A is at the position (x_{ka}, y_{ka}) and cylinder B is at the position (x_{kb}, y_{kb}), the force operating on cylinder A from cylinder B is given by:

$$f_x = f_{in}\frac{(x_{ka} - x_{kb})}{D_b}, \qquad f_y = f_{in}\frac{(y_{ka} - y_{kb})}{D_b} \tag{3.86}$$

where $D_b = \{(x_{ka} - x_{kb})^2 + (y_{ka} - y_{kb})^2\}^{1/2}$.

The shear stress between the two cylinders is given by:

$$f_s = \begin{cases} f_{in}\tan\varphi + 2r_c h_{s\,min} c_b \; ; & f_{in} > 0 \\[1mm] 2r_c h_{s\,min} c_b & ; \quad f_{in} \le 0 \end{cases} \tag{3.87}$$

The position and velocity of cylinder B relative to those of cylinder A are described as $x' = x_{kb} - x_{ka}$, $y' = y_{kb} - y_{ka}$, $u'_b = u_{bb} - u_{ba}$, $v'_b = v_{bb} - v_{ba}$, respectively. Therefore, the x and y-wise components of friction between the two cylinders are expressed as follows:

$$f_{sx} = \frac{-(v'x' - u'y')}{|v'x' - u'y'|}\frac{y'}{\sqrt{x'^2 + y'^2}}f_s, \qquad f_{sy} = \frac{(v'x' - u'y')}{|v'x' - u'y'|}\frac{x'}{\sqrt{x'^2 + y'^2}}f_s \tag{3.88}$$

If $v'x' - u'y' = 0$, $f_{sx} = f_{sy} = 0$ is satisfied.

Referring to the description in section 3.2.4 the rate of liquefaction at the bottom of the cylinder would be given by:

$$i_c = \beta\sqrt{(u - u_b)^2 + (v - v_b)^2} \tag{3.89}$$

Liquefaction will proceed as long as the saturated layer exists in the lower part of the cylinder, but it will not occur in the unsaturated layer. If the liquefied layer no

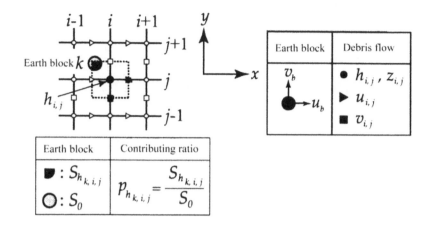

Figure 3.45 The contributing ratio of a cylinder to the boundary shear stresses.

longer exists, the following kinetic friction forces are assumed to operate between the rigid bed and the bottom of the cylinders:

$$\tau_{sx} = -\mu_k \frac{M_b}{S_0} g \cos\theta_{zx} \frac{u_b}{|u_b|} \tag{3.90}$$

$$\tau_{sy} = -\mu_k \frac{M_b}{S_0} g \cos\theta_{zy} \frac{v_b}{|v_b|} \tag{3.91}$$

where μ_k is the kinetic friction coefficient, and θ_{zx} and θ_{zy} are the bed gradient in the x and y directions, respectively.

At an arbitrary time t, some cylinders move as an assembly and others are scattered. If, as an example, a part of one cylinder is within a grid mesh as shown in Figure 3.45, the boundary shear stresses τ_{sx} and τ_{sy} between the earth block and the liquefied layer are calculated by considering the area ratio that is effective to the motion of the liquid in the mesh as explained in Figure 3.45.

The finite difference presentation of the fundamental equations mentioned above makes it possible to numerically simulate the motion of the earth mass. Here, the results of calculations corresponding to Case 1 in the experiments are given. The grid mesh for the calculation of debris flow was a square of 5 cm and the time step was 0.001 second. The radius of a cylinder r_c was set at 2.425 cm, and $D_1 = 4.619$ cm, $D_2 = 4.850$ cm and $D_3 = 5.081$ cm. The sediment concentrations in the liquefied layer and in the earth block were considered equal and set at $C = 0.5$. The degree of saturation in the unsaturated part of the earth block was $s_b = 0.3$, and the density of solids was 2.65 g/cm^3. The other parameters were $\mu_k = 0.5$, $\mu_a = 6.2$ Pas, $c_b = 179$ Pa and $\beta = 0.2$.

Figure 3.46 shows time sequentially the positions of cylinders and thicknesses of the liquefied layer h (debris flow) and cylinders ($h_{sb} + h_s$) after the instant that the earth block was released. Within 2 seconds the earth block was separated into a few blocks. The thinner blocks went downhill at the head. The thicker blocks followed the head and the debris flow ran after these blocks. At 3 seconds the blocks at the head climbed up

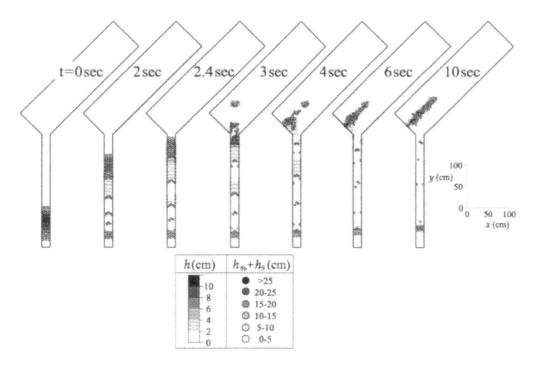

Figure 3.46 Results of calculation for Case 1.

the opposite bank of the river channel but soon they fell onto the valley bottom, and at 4 seconds they moved downstream along the riverbed. The earth blocks and the debris flow left on the slope continued their downhill motion, and by 10 seconds almost all the mass except for a few blocks ran out into the river channel. The time $t = 0$ in Figure 3.42 corresponds to $t = 2.4$ seconds in Figure 3.46, so that by comparing these two figures one will see that the calculation reflects the experiment rather well. However, in the calculation the blocks and debris flow were deposited from the upstream end of the channel, whereas in the experiment deposition began from a little downstream. The reason for this discrepancy is not clear yet. This problem together with the problems how the appropriate size of the cylinders is determined and how large the cohesion and repulsion forces operating between cylinders must be given need to be made clear.

3.3 DEBRIS FLOW AND FLOOD FLOW INDUCED BY THE COLLAPSE OF A NATURAL DAM

A landslide block that is only slightly liquefied falls by chance into a river channel and chokes the river. This is called a natural dam or landslide dam. When a natural dam is formed in a steep slope channel, the collapse of it will cause a debris flow. A large-scale landslide can choke even a big valley and the majority of thus produced natural dams fail catastrophically causing major flooding downstream. For example, in the case of Totsugawa disaster in 1889 in Nara Prefecture, Japan, 53 natural dams

Photo 3.3 Akadani landslide dam.

Photo 3.4 Before and after the emergency work for the Akadani landslide dam.

were formed by the landslides induced by a severe rainstorm and about 70% of these dams failed within a day causing a very severe disaster forcing the residents to move from their home village to Hokkaido (Ashida 1987). In the Aritagawa disaster in 1983, 16 deep-seated landslides formed natural dams and the majority of them failed during the flood, but two of them at Kongoji and Futagawa were left after the flood and they failed during the flood that occurred two months later. The one at Kongoji had a maximum water depth of 65 m and 1.7×10^7 m^3 of stored water. The collapse of this dam induced bores of 5 to 7 m high in the downstream river (Takei 1987). The Totsugawa area was again hit by severe rainfall due to Typhoon No. 12, in September 2011 that caused about 3,000 landslides whose total volume amounted to about a hundred thousand m^3. Many natural dams were formed and among them 5 remained long after the flood and the emergency works to prevent from collapsing were conducted. Photos 3.3 and 3.4 show the situation at the Akadani landslide in that case (Sakurai 2012).

According to the worldwide statistics (Costa and Schuster 1988), 27% of natural dams fail within a day, 56% in a month and 85% in a year. The newly corrected statistics (Peng and Zhang 2012) put these values at 34%, 71% and 87%, respectively. The majority of failures are caused by the overtopping of water and the associated erosion of the dam body. A few examples are due to piping or sliding collapse of the dam body.

For the mitigation of disasters due to natural dam failure, the scales and onset times prediction of debris flow or flood flow caused by the failure is indispensable. The heterogeneity of the inside structure and the diversity in the form and scale of the dam, however, seem to have discouraged the theoretical treatment of the phenomena and the existing studies mainly treat individual failure events or induce some empirical predicting formulae based on the previous events. But, as mentioned earlier, to approach a general and quantitative prediction, the understanding of essential mechanism that controls the phenomena is most important.

3.3.1 Formative conditions and shapes of a natural dam

As discussed in the last section, a landslide earth block is not necessarily deposited immediately at the foot of slope, it depends on its water content and other factors. Therefore, in general, it is necessary to examine first whether the earth block will change into a debris flow or how and where the earth mass will be deposited. Herein, the discussion is focused on the cases where the slid earth mass contains little water, the merging angle to the river channel is nearly perpendicular and the opening angle of the cross-section of the river channel is small. These conditions help with the easy formation of a natural dam. Thereby, the first discussion is on the shape of natural dam (Takahashi and Kuang 1988).

For the sake of simplicity, the sideways supplied soil due to landslide is assumed to be deposited in the river channel in a laterally uniform shape. Then, the dam will be formed following the process from (a) to (b) as depicted in Figure 3.47. If the soil block of W in width and V in volume having a rectangular cross-section as shown by the broken line in Figure 3.47 (a) slides off the slope into the channel, it will at first deform into a quadrangle as shown by the full line, and then, the upstream and downstream sides steeper than the angle of repose will fail to form a trapezoidal or triangular dam as shown in Figure 3.47 (b). When the volume of soil is large it will form a triangular dam, otherwise it will become trapezoidal.

The base line length L_B, the crest length L_T and the maximum height D_{max} of the trapezoidal dam are given by the following formulae:

$$L_B = \frac{W}{\cos\theta} + \frac{V\cos\theta}{2BW}K \tag{3.92}$$

$$L_T = \frac{W}{\cos\theta} - \frac{V\cos\theta}{2BW}K \tag{3.93}$$

$$K = \frac{\cos\theta}{\tan(\varphi+\theta)} + \sin\theta + \frac{\sin(90°+\varphi)}{\sin(\varphi-\theta)} \tag{3.94}$$

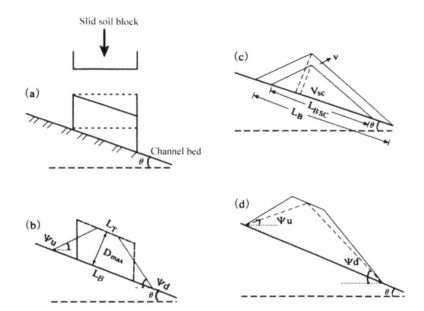

Figure 3.47 Process of formation and the shape of natural dams.

$$D_{\max} = \frac{2V}{B(L_B + L_T)} \tag{3.95}$$

where B is the channel width and φ is the angle of repose of slid soil.

Because the formative condition of a triangular dam is $L_T = 0$, the critical soil volume V_{sc} larger than which a triangular dam will be formed is given from Equation (3.93) as:

$$V_{sc} = 2(W/\cos\theta)^2 B/K \tag{3.96}$$

$$L_{Bsc} = \frac{W}{\cos\theta} + \frac{V_{sc}\sin\theta}{2BW}K \tag{3.97}$$

where L_{Bsc} is the base length of the triangular dam for the critical soil volume.

If $V \geq V_{sc}$, a triangular dam of which both side slope angles are almost equal to the angle of repose as shown in Figure 3.47 (c) is formed. In this case the base length is given by:

$$L_B = \left(\frac{V}{V_{sc}}\right)^{1/2} L_{Bsc} \tag{3.98}$$

If $V < V_{sc}$, a trapezoidal dam shown in Figure 3.47 (d), will be formed, but in this case the positions of the shoulder are not necessarily clear, so that, for the sake of simplicity, the dam is regarded as a triangular one depicted by the broken line in

Figure 3.47 (d). Under such an assumption, the upstream and downstream slope angles of the dam body, respectively, are given by:

$$\psi_u = \arctan\left(\frac{4\alpha \cos^2 \theta}{K + 2\alpha M}\right) - \theta \tag{3.99}$$

$$\psi_d = \arctan\left(\frac{4\alpha \cos^2 \theta}{K + 2\alpha N}\right) + \theta \tag{3.100}$$

where $M = \cos\theta/\tan(\varphi + \theta) - \sin\theta$, $N = \sin(90° + \varphi)/\sin(\varphi - \theta) + 2\sin\theta$, $\alpha = V/V_{sc}$.

Verification by experiments

To verify these predictive formulae and to examine the condition for the formation of a natural dam, laboratory flume experiments were carried out. The transparent acrylic flume modeling a river channel was 4 m long, 10 cm wide and its bottom was roughened by pasting sand of 0.8 mm in diameter. A 30 cm wide rectangular flume, modeling a slope was set perpendicular to the channel, from which the pre-determined volume of soil was supplied into the channel. The characteristic phenomena during the experiments were recorded by video cameras.

The experiments were done under the combinations of landslide widths $W = 10$ cm and 30 cm, landslide volumes $V = 2,700$, $5,400$ and $8,100$ cm³, water discharges in the channel $Q = 0$, 50, 100, 200, 300 and 400 cm³/s, and channel slopes $\theta = 10°$, 12°, 14°, 16°, 18° and 20°. The experimental material was nearly uniform sand with a mean diameter of 0.8 mm.

In the experiments, depending on the combinations of landslide volume and the hydraulic conditions of the river channel, there were cases in which natural dams were formed and not formed by the immediate washing out of supplied soil, where the judgment of dam formation was based on whether the lifetime was longer than at least a few seconds. By analogy with the threshold for the initiation of particle motion due to water flow which was given by the curves A or B in Figure 3.4, the threshold condition for the formation of natural dam was obtained on a plane where the two orthogonal axes were $\sin\theta$ and $X_Q \equiv Q/\{Bg^{1/2}(V/BL_B)^{3/2}\}$, respectively, as shown in Figure 3.48.

The threshold line in Figure 3.48 is given by:

$$\frac{Q/B}{g^{1/2}\{V/(BL_B)\}^{3/2}} = 45 \times 10^{-1.8 \sin\theta} \tag{3.101}$$

This equation has a similar form to the formula for the critical condition for the initiation of a particle motion if the average dam height, V/BL_B, is replaced with the mean diameter. The critical X_Q value from Equation (3.101), however, has a value about 10–20 times larger than that of critical non-dimensional discharge q_* for individual particles. A simple trial calculation of this formula also suggests that the dropping of about a few thousand cubic meters of semi-dry soil into an ordinal mountain ravine almost always forms a natural dam. However, in actual observations, there are few cases that show evidences for the formation of a natural dam even of a very short lifetime. The evidence is that the majority of surficial landslides occur

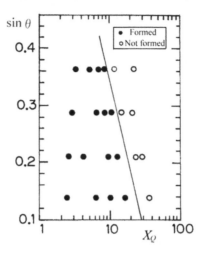

Figure 3.48 The threshold condition for the formation of a natural dam.

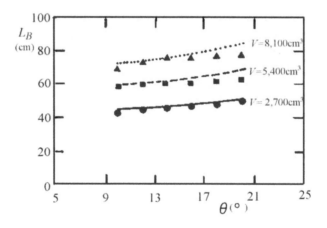

Figure 3.49 Comparison of base length between experiments and theory.

saturated or nearly saturated by water and they easily transform into debris flow without forming natural dams. In addition, experiments revealed that the discharge in the channel does not much affect the shape of a natural dam.

Figures 3.49, 3.50 and 3.51 show the comparison of dam shapes between the experiments and the theory (the plotted points are the results of experiments and the lines are the theory). The agreement between the experiments and theory is good for the base line length (Figure 3.49). Although the agreement for the upstream and downstream slope gradient of the dam body is not so good, at least the tendency of their change depending on the channel slope as well as the ratio V/V_{sc} seems rather well reproduced.

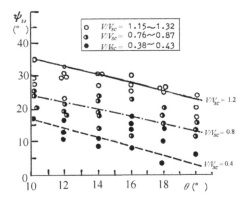

Figure 3.50 Upstream slope gradient of dam.

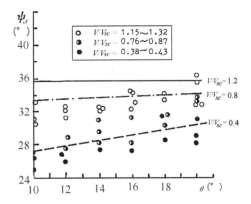

Figure 3.51 Downstream slope gradient of dam.

3.3.2 Life span of a natural dam

As described earlier the major cause of the failure of a natural dam is the overflow of impounded water, so that the time to overflow should be an important factor controlling the life span. Namely, the ratio V_2/Q, where V_2 is the water storage capacity of the reservoir produced by the natural dam and Q is the river discharge, is a controlling factor. The other factor should be concerned with the strength and the volume of the dam body. Here, this factor is assumed to simply be represented by V_1/Q, where V_1 is the volume of the dam body. The river discharge should be the average flow rate within the period of life span. However, this is usually an unknown value, so that the known mean specific discharge of the concerned watershed or that of the neighboring watershed is collected. Then, the river discharge is obtained by multiplying the catchment area of the natural dam to the mean specific discharge. Strictly speaking the river discharge for the calculation of V_2/Q and V_1/Q should be the one subtracted the seepage rate into the dam body but to know seepage rate is another very difficult problem so this is simply neglected.

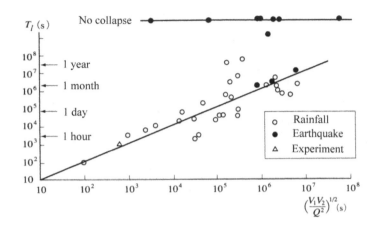

Figure 3.52 Life time of natural dams.

Figure 3.52 shows the life span of the natural dams in Japan as a function of the product of the two factors. The linear line in the figure is given by:

$$T_l = \left(\frac{V_1 V_2}{Q^2}\right)^{1/2}$$

(3.102)

Although the scattering of data is large, the life span seems to be roughly estimated by Equation (3.102) that is given by the inclined solid line. The point plotted around 10^2 in the abscissa is the case of the Kitadera landslide dam produced at the time of the Aritagawa disasters in 1983 and it collapsed within a few minutes after the formation of the dam, then, as the discharge to substitute into Equation (3.102) the peak flood flow instead of the mean flow was used.

The cases plotted far apart from the line indicating Equation (3.102) and the no collapse cases plotted outside may be due to very large seepage rate, very strong or very flat dam body, implementation of some countermeasure works, etc. Examples are Ikari ($V_1 = 3.8E + 06\,m^3$, $V_2 = 6.4E + 07\,m^3$, $T = 1.3E + 09\,s$), Taisho pond ($V_1 = 2.0E + 06$, $V_2 = 1.2E + 06$, no collapse), Akimoto Lake (V_1?, $V_2 = 4.4E + 07$, no collapse), Hibara Lake (V_1 ?, $V_2 = 1.5E + 08$, no collapse), Ontake ($V_1 = 2.6E + 07$, $V_2 = 3.7E + 06$, no collapse, channel works), etc (Mizuyama *et al.* 2011).

3.3.3 Failure in entire channel width and the resulting debris flow

Classifications of the dam failure processes and debris flow formations

The same flume used in the natural dam formation was used for the experiments on the processes of failure and the resulting debris flow formations. A triangular sediment dam was prepared on the flume bed and water flow was given with the pre-determined discharge from upstream. Accordingly, the processes of natural dam failure could be roughly classified into the three types as illustrated in Figure 3.53 (Takahashi and

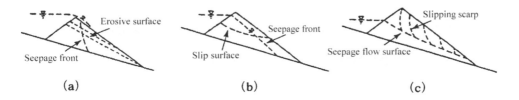

Figure 3.53 Three types of collapse of a natural dam.

Kuang 1988): the erosive destruction due to overtopping, the sudden sliding collapse of the dam body, and the retrogressive failure.

(a) Erosive destruction due to overtopping (Type 1)
     This case takes place when the dam body has a small permeability, large strength and the water supply from upstream is large. As time goes by after the formation of such a characteristic dam, the water stage upstream of the dam rises faster than the seepage water front proceeds downstream in the dam body. The dam itself is so strong that it does not collapse before overtopping occurs. The erosive action of the overtopped flow removes material violently from the top part as well as the downstream slope of the dam body (Figure 3.53 (a)).
(b) Sudden sliding collapse (Type 2)
     When the dam body has a larger permeability and a weaker strength than in Type 1, although the choked water stage behind the dam rises rapidly, the seepage water proceeds both downstream and upward keeping pace with the rise of the backwater stage, and on attaining a critical water stage the dam suddenly collapses due to sliding on a slip surface induced inside the body. The lowering of the dam crest is so sudden that the overtopped flow discharge is very large, resulting in very rapid erosion of the body (Figure 3.53(b)).
(c) Retrogressive failure (Type 3)
     If the permeability of the dam body is very high, seepage appears at a lower point of the downstream face of the body before an evident rise of the blocked water stage. If the dam body is weak, collapse at the point of oozing out of seepage water happens. Such a partial collapse proceeds upward until it ends as a large failure at the moment the partial collapse reaches the location of the blocked water behind the dam (Figure 3.53 (c)).

Depending on which failure type occurs, the debris flow induction processes and the resulting debris flow magnitudes are different.

In the type 1 failure, if the downstream slope of the dam body is steep so as to satisfy the critical condition for debris flow generation and its length is long, the overtopped flow discharge increases by entraining the eroded dam material and finally it develops into a debris flow. The speed of reduction of the dam body is at first rather slow because the dam body is still unsaturated, but the seepage of water from upstream and that infiltrated from the overtopping flow gradually saturate the dam body and once it is saturated erosion proceeds rapidly. If the length of the downstream slope of the dam body is short, the overtopped flow cannot grow completely so as to become a debris flow. Even in such a case, if the downstream river channel has a steep gradient and a thick sediment deposit layer, the flow can develop to a debris flow on the riverbed downstream of the dam. In this case retrogressive erosion proceeds upstream from the riverbed to the dam body and finally all the materials are washed out.

In the type 2 failure, because the slipped block is not thoroughly saturated at the moment of initiation of slippage, it cannot instantaneously be transformed into debris flow but stops around the toe of the dam body. Immediately after that, a massive amount of released water from behind the dam rushes over the now lowered dam and erodes the dam body as well as the momentarily halted dam portion around the toe that was produced by the antecedent slippage, and forms a huge debris flow.

In the type 3 failure, because the discharge of seepage flow that caused the partial failure is usually not enough to mobilize the slipped mass as a debris flow, the slipped block is deposited immediately downstream. Thus, the retrogressive failure merely acts to widen the base length of the dam body. The final collapse that occurs when the force of the blocked water becomes sufficient to push over the small portion of the dam remaining on the slip surface produces an instantaneous release of water and forms a debris flow if the flattened dam slope is still steep enough to generate debris flow. The process after this stage is the same as the Type 1 failure.

Debris flow prediction for each type failure

(a) Overtopping type failure

The process of erosion of a dam body and the development of a debris flow hydrograph can be analyzed using a similar system of equations to that on a varying slope bed mentioned in section 3.1.2 with some slight modifications. Here the modifications must take into account the facts that the flow rate supplied onto the down-slope of the dam is a sum of the flood discharge in the river itself and that produced by the change in the upstream water storage brought about by the lowering of the dam height due to erosion, and that the location of the dam crest, which is the starting point of erosion, moves upstream as the dam body erosion proceeds.

The calculation including the dam formation process proceeds as follows:

1   A constant discharge of water is given from upstream of the river channel,
2   Suddenly the slid earth block is given to the channel and a natural dam is formed. The shape and height of the dam is calculated and then the storage water level change is calculated using the unsteady flood flow equation,
3   Overtopped water proceeds downstream eroding the crest as well as the downstream slope of the dam body and develops to a debris flow (Calculations of the dam body erosion, debris flow discharge, depth and solids concentration, and the change in overtopping flow rate).

The calculation model was applied to an experimental case as follows: the river flow rate per unit width $= 20 \, \text{cm}^2/\text{s}$, channel gradient $\theta = 18°$, the shape parameter of the dam $\psi_u = 30°$, $\psi_d = 36°$, $D_{\max} = 15 \, \text{cm}$, the material of the dam body $d_m = 0.8 \, \text{cm}$. The other parameters that appear in the fundamental equations were: $C_* = 0.65$; $C_{*L} = 0.65$; $C_{*F} = 0$; $C_{*DL} = 0.5$, $K = 0.05$; $\delta_e = 0.0007$; $\delta_d = 0.05$; and $\sigma = 2.65 \, \text{g/cm}^3$. The calculation was carried out by discretizing the fundamental equations by $\Delta x = 2 \, \text{cm}$ and $\Delta t = 0.01$ sec.

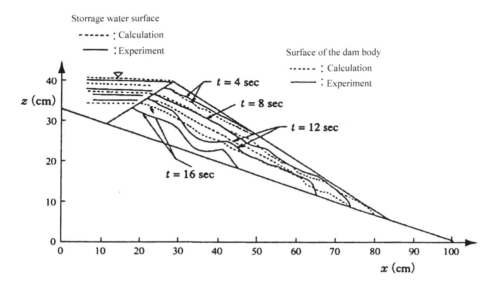

Figure 3.54 Transformation of the dam body with time.

Figure 3.54 compares the process of reduction of the dam body by calculation with the experimental result. Here, t means time in seconds elapsed from the beginning of overtopping. As is clear in Figure 3.54, in the early stage of erosion, the calculation coincides well with the experiment, but later the discrepancy becomes larger. This inconsistency is presumably brought about by the neglect of seepage flow in the dam body. In the course of time, the seepage water stage rises and when it reaches the down-slope surface of dam body the erosion rate must increase. Recently, Awal *et al.* (2008) improved the numerical model to incorporate the seepage flow analysis using Richard's water seepage equation. This model succeeded in wiping out the above mentioned inconsistency.

Figures 3.55 and 3.56, respectively, show the flow depth versus time relationships and the discharge hydrographs at a position 1.4 m downstream of the dam. Here, T_s means the time in seconds elapsed after the sudden formation of the natural dam; therefore, $(T_s - t)$ is the time necessary for the stored water to overspill. Although the measurement in the experiment is not accurate and there are variations in the calculation by some unknown causes, generally the calculation and the experiment fit rather well, and so it can be concluded that debris flow thus generated can be predicted by this method.

(b) Instantaneous slip failure type

Because, in this case, the variation of seepage water stage in the dam body is crucial to the stability of the soil mass, the seepage flow should be known as a function of the blocked water stage behind the dam, and simultaneously, the stability of the dam body should be analyzed to obtain the first slip surface.

The seepage water flow is analyzed by using Darcy's law:

$$v_s = k_s \sin \theta \tag{3.103}$$

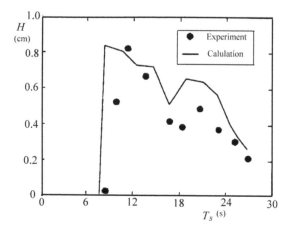

Figure 3.55 Stage-time relationship.

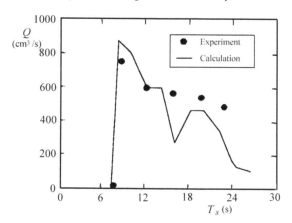

Figure 3.56 Discharge-time relationship.

and the continuity equation:

$$\lambda_0(1 - s_0)\frac{\partial H_{gw}}{\partial T_s} + \frac{\partial v_s H_{gw}}{\partial x} = 0 \tag{3.104}$$

where v_s is the velocity of seepage flow, λ_0 is the porosity of the dam body, s_0 is the initial degree of saturation in the dam body, k_s is the coefficient of permeability, and H_{gw} is the seepage flow depth measured vertically. Substitution of Equations (3.103) into (3.104) gives:

$$\lambda_0(1 - s_0)\frac{\partial H_{gw}}{\partial T_s} + k_s \sin\theta\frac{\partial H_{gw}}{\partial x} = 0 \tag{3.105}$$

in which $\sin\theta$ is assumed to be far larger than $\partial H_{gw}/\partial x$.

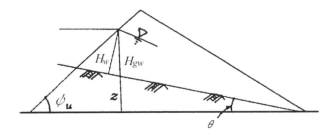

Figure 3.57 Seepage flow in the dam body.

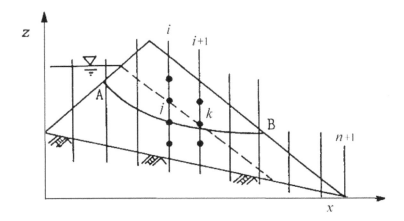

Figure 3.58 Phreatic line and the initial slip surface in the dam body.

The boundary condition in solving Equation (3.105) is the water stage behind the dam H_w in Figure 3.57, and it is obtained by considering the relationship between the stage and the water storage volume under a constant inflow rate per unit width q_w as follows:

$$\frac{dH_w}{dT_s} = \frac{\sin \psi_u \sin \theta}{\sin (\psi_u + \theta)} \frac{q_w - k_s H_w \sin\theta}{H_w} \tag{3.106}$$

The initial slip surface can be obtained by the following procedure. The safety factor F_s for the slippage over the assumed slip surface AB in Figure 3.58 is defined by using the simplified Janbu Method as follows:

$$F_s = \frac{\sum R_i}{\sum T_i} \quad (i = 1, 2, 3, \ldots . n) \tag{3.107}$$

$$R_i = \frac{c L_i \cos \alpha_i + (W_i - P_i L_i \cos \alpha_i) \tan \varphi}{\cos^2 \alpha_i (1 + \tan \alpha_i \tan \varphi / F_s)} \tag{3.108}$$

$$T_i = W_i \tan \alpha_i \tag{3.109}$$

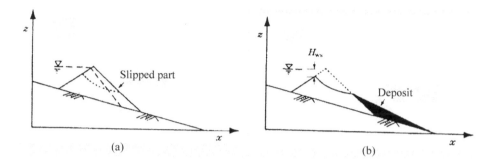

Figure 3.59 The initial transformation of the dam after the instantaneous slip failure.

where L_i is the bottom length of each slice, W_i is the weight of each slice, P_i is the pore pressure acting on the bottom of each slice, α_i is the slope of the bottom of each slice, c is the cohesion of the material of the dam body.

Minimization of F_s is equivalent to minimizing the following subsidiary function G:

$$G = \sum (R_i - F_s T_i) \tag{3.110}$$

If the assumed slip surface AB satisfies:

$$G_m \equiv \min G = \min \left\{ \sum (R_i - F_s T_i) \right\} \tag{3.111}$$

and at that time F_s is less than 1.0, then the assumed slip surface AB can be the real slip surface. To obtain the initial slip surface by this dynamic programming procedure, the dam body is divided into n slices as shown in Figure 3.58 and each slice boundary is divided into m_i segments. An arbitrary line jk which connects points (i, j) and $(i + 1, k)$ is considered as a part of assumed slip surface. R_i and T_i on the surface jk are obtained from Equations (3.108) and (3.109) and the 'return function', $DG_i\,(j, k) = (R_i - F_s T_i)$, is calculated. If $A_i(j)$ is the minimum value of G from the point A to the point (i, j), the minimum G value from A to the point $(i + 1, k)$ is:

$$A_{i+1}(k) = \min \{A_i(j) + DG_i\}, \quad \begin{array}{l}(i = 1, 2, \ \ldots, n, j = 1, 2, \ldots, m_i, \\ k = 1, 2, \ \ldots, m_{i+1})\end{array} \tag{3.112}$$

The boundary conditions of this equation are:

$$A_1(j) = 0, \quad j = 1, 2, \ \ldots, m_1 \tag{3.113}$$

$$G_m = \min \{A_{n+1}(j)\}, \quad j = 1, 2, \ \ldots, m_{n+1} \tag{3.114}$$

If the F_s value calculated by Equation (3.107) after this procedure is less than 1.0, the surface AB, which includes jk, is considered as the initial slip surface. However, Equation (3.108) also contains F_s in the right-hand side. Therefore, iteration is necessary to obtain the exact F_s value. In addition, the phreatic surface level in the dam body changes with time, this procedure should be done for every time step, then the time of initial slip and the shape and position of the initial slip surface can be obtained.

The practical process of calculation starting from the dam formation stage is as follows:

1   A constant discharge of water is given from upstream of the river channel,

2   Suddenly the slid earth block is given to the channel and a natural dam is formed. The shape and height of the dam is calculated and then the storage water level change is calculated using the unsteady flood flow equation,

3   The seepage water flow inside the dam body is calculated using Equation (3.106),

4   Associated with the progress of the phreatic line in the dam body, the stability change and the initial slip surface are calculated by the dynamic programming procedure,

5   Determine the dam shape immediately after the slippage. The slipped soil mass existing above the initial slip surface is not fully saturated, as indicated in Figure 3.59 (a), and it is deposited immediately downstream as shown in Figure 3.59 (b). The surface slope of the deposit may be determined by equating the volume of the deposit with that of the slipped mass,

6   The dam body beneath the slip surface is considered to be fully saturated and the erosion process by the over-spilled water is similarly analyzed as in the case of the overtopping type failure. The overspill depth on the slip surface just after the initial slip is assumed to be equal to H_{ws} in Figure 3.59 (b), which is the difference between the stage of the stored water and the height of the origin of the slip surface at the upstream slope of the dam body.

This model was applied to a case in which the unit width discharge of river flow was $40\,\mathrm{cm^2/s}$, channel slope $\theta = 16°$, the parameters of the dam $\psi_u = 34°$, $\psi_d = 33°$, $D_{\max} = 16.5\,\mathrm{cm}$, and the material $d_m = 0.8\,\mathrm{cm}$. The parameters in the fundamental equations for the process of dam formation were $C_* = 0.65$, $C_{*L} = 0.65$, $C_{*F} = 0$, $C_{*DL} = 0.50$, $k_s = 1.5\,\mathrm{cm/s}$, $\delta_e = 0.0007$, $\delta_d = 0.05$, $\lambda_0 = 0.30$, and $\sigma = 2.65\,\mathrm{g/cm^3}$. The grid sizes in the calculation were $\Delta x = 2.0\,\mathrm{cm}$ and $\Delta t = 0.01$ sec.

Figure 3.60 compares the calculation to the experiment, in which t is the time in seconds after the occurrence of the first slip. Both the shape of the slip surface and the process of erosion after the slippage are successfully predicted.

Figures 3.61 and 3.62 show the depth-time relationships and the discharge hydrographs at various points along the channel obtained by calculation and experiment. The experimental quantities were measured only at $x = 300\,\mathrm{cm}$ downstream of the dam body. The time T_s in Figures 3.61 and 3.62 means the time elapsed after the formation of the dam. The arrival time of the debris flow front and the peak occurrence time at $x = 300\,\mathrm{cm}$ are a little earlier in the experiment than in the calculation. However, the whole shape and the peak discharge of the hydrograph are well predicted. Moreover, the calculations show that the peak discharge attenuates downstream very rapidly.

Awal et al. (2009) developed a three-dimensional transient seepage and slope stability analysis using essentially the same method as the previously mentioned model introducing a slightly modified Richards' 3D seepage flow equation and adopting the factor of safety model for 3D landslide. The comparison of the results by numerical

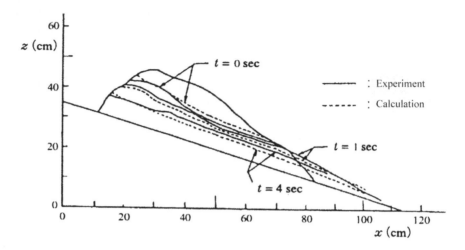

Figure 3.60 Transformation of the dam after the instantaneous slip.

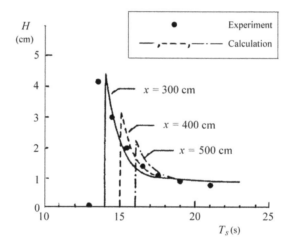

Figure 3.61 Depth change downstream of the dam.

analysis with experiments revealed that the movement of moisture in the dam body and the critical slip surface can be well reproduced.

(c) Retrogressive failure type

According to the observations in the flume experiment, retrogressive failure seems to proceed upstream, maintaining its base level of slippage approximately coincident with the seepage water surface. For the sake of simplicity, the processes are modeled as indicated in Figure 3.63. Namely, the first slip occurs in the neighborhood of the toe, where the seepage water surface intersects the slope surface of the dam body for the first time (Figure 3.63 (a)), and then slips proceed rapidly up to the position of crest of the dam (Figure 3.63 (b)). The failed soil is assumed to be deposited downstream as the part marked out by black. The time necessary to reach this stage is, at the moment, not

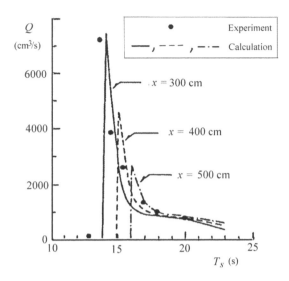

Figure 3.62 Discharge change downstream.

predictable. Herein, this time is neglected. To compensate for this time discrepancy, at least to some extent, the time necessary for the seepage water, which starts at the intersection point of the blocked water surface and the upstream side slope of the dam (point I in Figure 3.63 (b)), to arrive at the scarp right under the crest of the dam (point D in Figure 3.63 (b)) is added before the whole collapse occurs.

The instantaneous overflow depth H'_{ws} just after the collapse is equal to the rise of the blocked water surface during the hypothetical time needed for the progressive failure as stated above (Figure 3.63 (c)).

Figure 3.64 shows the comparison of the hydrograph thus obtained by calculation under the same condition in the experiment with the experimental result; $\theta = 14°$, $\psi_u = 34°$, $\psi_d = 34°$, $D_{max} = 17$ cm. The process and the hydrograph are reasonably well predicted by this simple assumption. However, under an actual scale condition, the estimation of time necessary for retrogressive failure would play a very important role.

3.3.4 Prediction of debris flow/flood flow induced by the overflow in partial width

The crest of a natural dam would be by no means horizontal in the lateral direction but would have irregular undulations or tapering toward a side. If the dam buries a wide valley, the over-spilling water will make a stream channel that occupies only a narrow width within the entire crest width. Therefore, the majority of a natural dam failure will be triggered by the erosion of the incised channel on the dam body thus formed on the crest. In this section, the processes of dam body erosion in such a condition and the prediction of the debris flow or flood flow induced are discussed.

The erosion within the incised channel proceeds vertically as well as laterally, accordingly the cross-sectional area of the channel increases, enhancing the capacity to

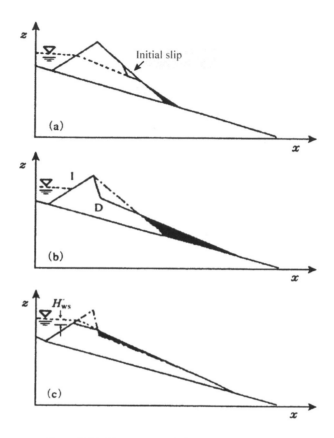

Figure 3.63 Illustrations of the progressive failure.

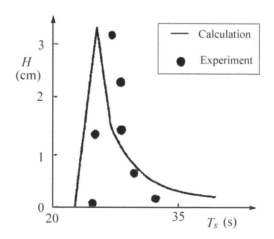

Figure 3.64 Debris flow depth downstream of the dam.

release stored water upstream. Then, the increased flow rate (river discharge + released water from the reservoir) accelerates the erosion and in turn the enhanced cross-section increases the released discharge. Thus, the overflow accompanies a catastrophic failure. The procedure to quantitatively predict the phenomena needs to analyze the flow in the reservoir that concentrates to a lower part on the dam crest, the enlargement of the incised channel and the deposition of sediment around the toe of the dam. Therefore, the theoretical treatment of the phenomena must be two-dimensional. Because the slope of the incised channel is diverse from as steep as the angle of repose of the material to a few degrees, the flow in the channel can be a debris flow on one occasion and a flood flow, highly sediment loaded, on another. Moreover, because the rapid channel slope change may occur in a short time, the equations in the analysis should be applicable to any type of flows.

The horizontal two-dimensional equations of flow were introduced in section 3.2.5 for the analysis of landslide-induced debris flow, but in that case the debris flow was considered as the viscous type. Here, the stony-type debris flow is considered (Takahashi and Nakagawa 1992, 1993).

The x- (down valley) and y-wise (lateral) momentum conservation equations of flow corresponding to Equations (3.74) and (3.75) are respectively given as follows:

$$\frac{\partial M}{\partial t} + \beta' \frac{\partial (uM)}{\partial x} + \beta' \frac{\partial (vM)}{\partial y} = gh \sin \theta_{bx0} - gh \cos \theta_{bx0} \frac{\partial (z_b + h)}{\partial x} - \frac{\tau_{bx}}{\rho_T} \tag{3.115}$$

$$\frac{\partial N}{\partial t} + \beta' \frac{\partial (uN)}{\partial x} + \beta' \frac{\partial (vN)}{\partial y} = gh \sin \theta_{by0} - gh \cos \theta_{by0} \frac{\partial (z_b + h)}{\partial y} - \frac{\tau_{by}}{\rho_T} \tag{3.116}$$

The continuity of the total volume , corresponding to Equations (3.24) and (3.73), is:

$$\frac{\partial h}{\partial t} + \frac{\partial M}{\partial x} + \frac{\partial N}{\partial y} = i\{C_* + (1 - C_*)s_b\} \tag{3.117}$$

The continuity of solid fraction under the assumption that no fine particles are contained is:

$$\frac{\partial (ch)}{\partial t} + \frac{\partial (cM)}{\partial x} + \frac{\partial (cN)}{\partial y} = iC_* \tag{3.118}$$

The equation for the change of bed surface elevation is:

$$\frac{\partial z_b}{\partial t} + i = i_{sml} + i_{smr} \tag{3.119}$$

In these equations, $M(=uh)$ and $N(=vh)$ are the x- and y-wise components of unit width discharge (flux), respectively, z_b is the erosion depth or deposit thickness, θ_{bx0} and θ_{by0} are the x- and y-wise inclination of the original slope surface of the dam body, respectively, β' is the momentum correction coefficient, i_{sml} and i_{smr} are the mean recessing velocity of the left and right-hand side banks of the incised channel, respectively.

The resistance at the bottom of flow τ_{bx} and τ_{by} in Equations (3.115) and (3.116) are described as follows:

For stony-type debris flow:

$$\tau_{bx} = \frac{\sigma}{8}\left(\frac{d_p}{h}\right)^2 \frac{1}{\{(C_*/C)^{1/3} - 1\}^2} u\sqrt{u^2 + v^2} \tag{3.120}$$

$$\tau_{by} = \frac{\sigma}{8}\left(\frac{d_p}{h}\right)^2 \frac{1}{\{(C_*/C)^{1/3} - 1\}^2} v\sqrt{u^2 + v^2} \tag{3.121}$$

For immature debris flow:

$$\tau_{bx} = \frac{\rho_T}{0.49}\left(\frac{d_p}{h}\right)^2 u\sqrt{u^2 + v^2} \tag{3.122}$$

$$\tau_{by} = \frac{\rho_T}{0.49}\left(\frac{d_p}{h}\right)^2 v\sqrt{u^2 + v^2} \tag{3.123}$$

For bed load transportation:

$$\tau_{bx} = \frac{\rho g n_m^2}{h^{1/3}} u\sqrt{u^2 + v^2} \tag{3.124}$$

$$\tau_{by} = \frac{\rho g n_m^2}{h^{1/3}} v\sqrt{u^2 + v^2} \tag{3.125}$$

The erosion velocity for stony debris flow and immature debris flow are given by Equations (3.37) and (3.39), respectively. When the equilibrium solids concentration, obtained by applying Equation (2.144), becomes less than 0.01, the equilibrium concentration in Equation (3.39) should be obtained by using some appropriate bed load formula. The Takahashi's bed load formula that is applicable to steep slope channel is given as follows (Takahashi 1987):

$$\frac{q_B}{\{(\sigma/\rho - 1)g d_p^3\}^{1/2}} = \frac{1 + 5\tan\theta}{\cos\theta}\sqrt{\frac{8}{f}}\tau_*^{3/2}\left(1 - \alpha_c^2\frac{\tau_{*c}}{\tau_*}\right)\left(1 - \alpha_c\sqrt{\frac{\tau_{*c}}{\tau_*}}\right) \tag{3.126}$$

where q_B is the bed load discharge per unit width, τ_* is the non-dimensional shear stress, and τ_{*c} is the non-dimensional critical shear stress, α_c is a coefficient and f is the resistance coefficient. The critical shear stress is given by:

$$\tau_{*c} = 0.04 \times 10^{1.72\tan\theta} \tag{3.127}$$

The coefficient α_c is given by:

$$\alpha_c^2 = \frac{2\{0.425 - \sigma\tan\theta/(\sigma - \rho)\}}{1 - \sigma\tan\theta/(\sigma - \rho)} \tag{3.128}$$

and f is given as follows:

$$\sqrt{\frac{8}{f}} = \begin{cases} A + 5.75 \log\left[\dfrac{(\sigma/\rho - 1)\tau_*}{\tan\theta(1 + 2\tau_*)}\right]; & \tau_* \geq 0.2 \\[3mm] A + 5.75 \log\left[\dfrac{0.2(\sigma/\rho - 1)}{1.4\tan\theta}\right]; & \tau_* < 0.2 \end{cases} \tag{3.129}$$

where $A = 0.04\tan^{-2}\theta$ (when $\tan\theta \geq 0.08$) or $A = 6.0$ (when $\tan\theta < 0.08$). The relationship between f and Manning's roughness coefficient n_m is:

$$n_m = \left(\frac{f}{8g}\right)^{1/2} h^{1/6} \tag{3.130}$$

From Equations (3.126) to (3.127) and Manning's resistance law, the equilibrium solids concentration in the case of bed load transport is:

$$C_{B\infty} = q_B/q_T \tag{3.131}$$

The value of θ in Equations (3.126) to (3.129) is considered as the gradient of flow surface to the direction of velocity vector and is given by:

$$\tan\theta = \frac{u\sin\theta'_{bx} + v\sin\theta'_{by}}{\sqrt{u^2\cos^2\theta'_{bx} + v^2\cos^2\theta'_{by}}} \tag{3.132}$$

where $\tan\theta'_{bx} = \tan(\theta_{bx0} + \theta_{bzhx})$, $\tan\theta'_{by} = \tan(\theta_{by0} + \theta_{bzhy})$, $\tan\theta_{bzhx} = -\partial(z_b + h)/\partial x$, $\tan\theta_{bzhy} = -\partial(z_b + h)/\partial y$.

If the flow on the dam body develops into a stony-type debris flow and it reaches the flat area downstream of the dam, it decelerates and starts to be deposited downstream from the point where the velocity becomes smaller than $p_i U_c$. The deposition velocity is, from Equation (3.42):

$$i = \delta_d \left(1 - \frac{\sqrt{u^2 + v^2}}{p_i U_c}\right) \frac{C_\infty - C}{C_*} \sqrt{u^2 + v^2} \tag{3.133}$$

If $(u^2 + v^2)^{1/2} > p_i U_c, i = 0$.

If the flow on the slope of the dam is an immature debris flow or an ordinary sediment loading water flow, the inertial motion may be neglected and the velocity of deposition will be described as:

$$i = \delta'_d \frac{C_\infty - C}{C_*} \sqrt{u^2 + v^2} \tag{3.134}$$

$$i = \delta''_d \frac{C_{B\infty} - C}{C_*} \sqrt{u^2 + v^2} \tag{3.135}$$

where δ'_d and δ''_d are the coefficients.

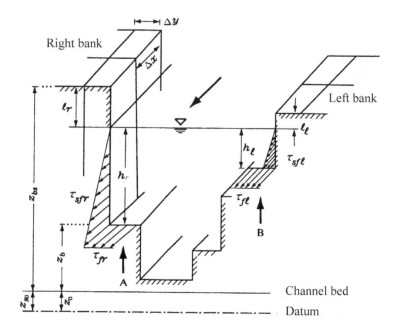

Figure 3.65 Cross-section of the incised channel on the dam.

The side bank erosion velocity is given by Equation (3.43), in which the equilibrium solids concentrations should be the one suitable to the flow characteristics, i.e. debris flow, immature debris flow and bed load transportation.

The right-hand side and the left-hand side bank recessing velocities in the case where the side depths are different as shown in Figure 3.65 are given respectively as:

$$i_{sml} = \frac{h_l}{l_l + h_l} i_{sl}, \quad i_{smr} = \frac{h_r}{l_r + h_r} i_{sr} \tag{3.136}$$

The finite difference calculations of the system of equations shown above can predict the phenomena of dam failure by overtopping. However, because the grid system is spatially fixed, some contrivances are necessary to deal with the continuous widening of incised channel. One may assume that the sediment produced by the recession of bank, e.g. $i_{smr}(z_{bs} - z_b)\Delta t$, is supplied only to the cell adjacent to the bank, e.g. cell A in Figure 3.65, and it instantaneously raises the elevation of that cell to be eroded by the flow in the channel in the next time step. The channel width is considered not to be changed until the total volume of the sediment supplied to the cell becomes equal to the volume of sediment in the side bank cell at the beginning of side bank erosion; $\Delta y \, (h_r + l_r)$. At the moment when the two volumes of sediment become equal, the channel is considered to be widened by one cell width and the bottom elevation of a new cell adjacent to the new bank is set to be equal to that of

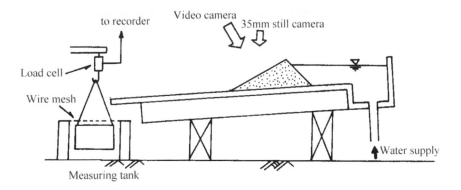

Figure 3.66 Experimental set up.

the former adjacent cell at the beginning of bank erosion, $t = t_0$. The formulation of the procedure is as follows:

$$\int_{t_0}^{t} i_{sr} h_r \Delta x dt = (z_{bs} + z_{s0} - z_b - z_0)\big|_{t=t_0} \Delta x \, \Delta y \, ; \quad then \; z_b = z_b\big|_{t=t_0} \tag{3.137}$$

Experiment

As illustrated in Figure 3.66, a steel flume of 4.97 m long, 40 cm wide and 20 cm deep was set at a gradient of 3°, and in it a triangular natural dam model was installed. The material of the dam was a heterogeneous one with a mean diameter of 2.15 mm. The height of the dam was 16 cm and the longitudinal base length was 140 cm, which corresponds to about 1/250 scale model of the average natural dam in Japan. Water at the flow rate of 100 cm³/s was supplied from the upstream end of the flume and in time it overflowed from the notch, with width 5 cm and depth 1.2 cm, incised at the crest of the dam. The overtopping flow incised a channel on the slope of the dam and that channel increased its cross-sectional area progressively. The change in width and bed elevation of the channel was measured by analyzing the stereo pair of photographs taken every 5 seconds. Run out water and sediment were separated at the downstream end of the flume and measured.

Photo 3.5 shows the time sequential change of the incised channel on the dam slope. Immediately after the over-spilling from the notch the flow elongated downstream with an almost constant width. On arriving at the toe of the dam slope the flow gradually enlarged its width and by that time the failure of side banks of the channel had started in the neighborhood of the crest. By the widening of the channel at the crest the overflow discharge increased and it rapidly eroded the channel. The side bank erosion proceeded downstream keeping the width of the channel almost uniform in an early stage, but later the width downstream became wider. As time went by the upstream width widened and finally the channel had nearly parallel side banks throughout the length on the slope of the dam.

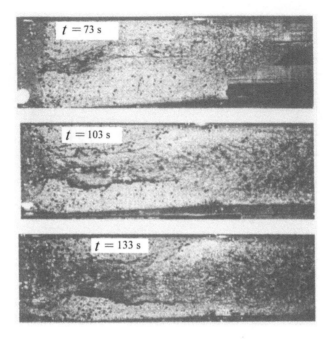

Photo 3.5   Progress of overtopping failure.

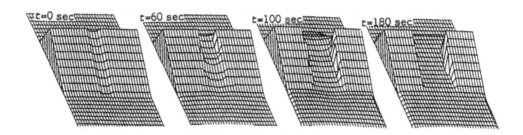

Figure 3.67   Process of enlargement of the incised channel.

Comparison between the experiments and numerical simulation

The numerical simulation was accomplished using the system of Equations (3.115) to (3.136) above, in which the conditions correspond to the experiment (Case A-1 in Table 3.5 that is shown later). The parameters and the grid system used in the calculation were $K = 0.06$, $K_s = 1.0$, $\delta_d = 1.0$, $C_* = 0.655$, $\tan \varphi = 0.75$, $\sigma = 2.65$ g/cm^3, $d_p = 2.15$ mm, $\Delta x = 10$ cm, $\Delta y = 1.25$ cm and $\Delta t = 0.002$ sec.

Figure 3.67 is the bird'-eye view of the process of erosion of the incised channel in the simulation. In the calculation an incised channel of 1 cm in depth and 5 cm in width was set on the entire reach of the downstream side of the dam surface in

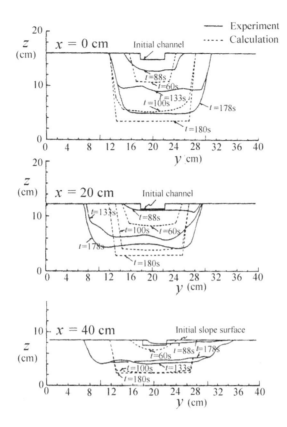

Figure 3.68 Cross-sectional shapes of the incised channel at three points.

advance, whereas only a notch at the crest existed in the experiment. The results of the calculation clearly show the widening and deepening of the channel and a slight deposition around the foot of the dam.

Figure 3.68 shows the changes of the cross-section of the channel at three points on the slope. The instant when the overflow from the crest appeared is defined $t = 0$, the right-hand wall of the flume is $y = 0$, and the bottom of the flume is $z = 0$. In the early stage the vertical erosion is faster and the side bank recession is slower in the calculation than in the experiment, but in general the tendency of the channel enlargement is well explained.

The discharge hydrographs obtained in the experiment and by the calculation are compared in Figure 3.69. The peak discharge as well as the duration of flood is rather well reproduced by the calculation. Note that the supplied water discharge into the flume is only 100 cm^3/s, whereas the peak of outflow discharge is as much as 1,200 cm^3/s. This fact clearly shows how a large flood flow is produced by the natural dam failure.

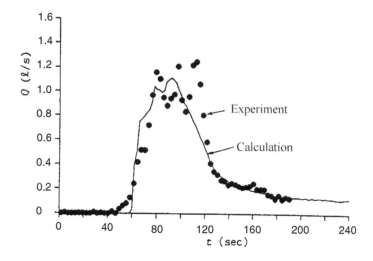

Figure 3.69 Discharge hydrographs due to overtopping failure of a natural dam.

Table 3.5 Effects of the incised channel dimensions on the peak discharges.

Case	Width (cm)	Depth (cm)	Peak discharge (l/s)	Remarks
A-1	5.0	1.0	1.12	
A-2	7.5	1.0	1.13	
A-3	10.0	1.0	1.19	
A-4	5.0	2.0	1.14	
A-5	7.5	2.0	1.20	
A-6	10.0	2.0	1.21	
A-7	5.0	3.0	1.20	
A-8	7.5	3.0	1.20	
A-9	10.0	3.0	1.35	

$\theta = 3°$, $\theta_1 = 16°$, $\theta_2 = 10.76°$
$l = 140$ cm, $H = 16$ cm
Initial storage = 79.8 l

Factors determining the peak discharge

Some systematic numerical simulations were conducted to find out the effective factors to determine the peak discharge of the dam break flood, and calculated peak discharges were compared with the previous empirical data.

The first numerical experiments were concerned with the effects of width and depth of the initially incised channel. The cases examined are listed in Table 3.5. As far as this examination is concerned, the initial width only slightly affects the peak discharge. The deeper the initial depth the earlier and the larger the peak discharge appears.

Table 3.6 Effect of the valley slope on the peak discharge.

Case	Valley lope (°)	H (cm)	I (cm)	θ_1 (°)	θ_2 (°)	Initial storage (l)	Peak discharge (l/s)
B-I	8	16.0	100	16	10.76	18.6	0.52
B-2	13	16.0	140	16	10.76	4.3	0.26
C-I	1	16.3	137	14	12.76	283.1	2.97
C-2	8	13.4	167	21	5.76	16.2	0.44
C-3	13	5.4	416	26	0.76	1.3	0.12

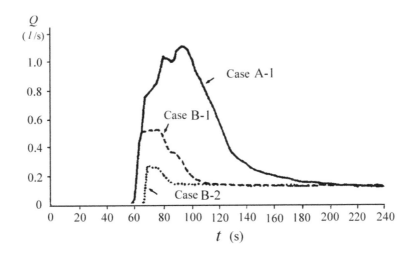

Figure 3.70 Discharge hydrographs on various valley slopes.

In the second numerical experiments, the dams congruous with that in Case A-1 were set on the differently sloped channel as shown Case B-1 and Case B-2 in Table 3.6. The downstream surface slopes of the dam body in these cases satisfy the condition of debris flow initiation. The discharge hydrographs obtained in each case is compared with that of Case A-1 in Figure 3.70. Both the peak discharge and the duration of flood/debris flow become larger the flatter the channel slope.

In the third numerical experiments, the dams were made by the same volumes of sediment on different valley slopes as shown Cases C-1, C-2 and C-3 in Table 3.6. The downstream and upstream surface slopes of these dams were the same (the downstream slope gradient in every case was 13.76°), so that the dam height and the stored water volume behind the dams for the respective dams were different. The resulting peak discharges were very different as shown in Table 3.6. The biggest discharge appeared when the dam was formed on the flattest valley.

The above numerical experiments suggest that the most influential factor in determining the peak discharge is the water storage capacity. Under the same water storage capacity, the peak discharge should be larger the higher the dam. Therefore, as the previous investigations correlate, the peak discharge may change depending on the dam

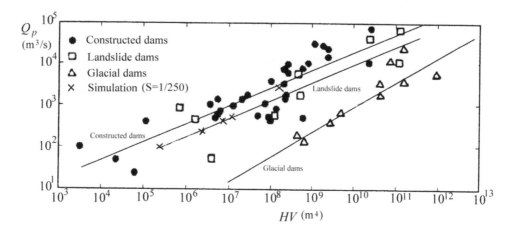

Figure 3.71 Dam factor versus peak discharge.

factor (= dam height × water storage capacity). Assuming that the numerical experiments correspond to the 1/250 scale model experiments and applying the Froude's similitude, the peak discharges are plotted on the figure given by Costa and Schuster (1988) as in Figure 3.71. The peak discharges obtained by numerical simulations are well correlated to the dam factor and they have similar tendency to the previous data.

Chapter 4

Characteristics of fully-developed debris flow

The photograph shows the situation of a debris flow that advanced overflowing and depositing its boulder-rich front part onto the terrace at the left-hand side of the Atsumari River, in the Hogawati area of Minamata City, Kumamoto Prefecture, Japan. The debris flow occurred in July, 2003. Details of that debris flow will be discussed in Chapter 7.

INTRODUCTION

It is not always possible to distinguish between the initiation/development zone and the developed flow zone of debris flow along the river course, as, debris flows can behave in strange ways while advancing in the developing and the flowing zones depending on their types, e.g. inertial or viscous. The accumulation of large stones at the head and the ability to transport huge boulders of stony-type debris flow and the motion and stoppage of the intermittent surges of viscous-type debris flow are examples of particularly interesting characteristics.

In this chapter, the characters of debris flow on its flowage and the mechanics of these characters associated with the numerical simulation models are explained.

4.1 TRANSLATION OF DEBRIS FLOW SURGE AND THE SHAPE OF THE SNOUT

4.1.1 The case of stony-type debris flow

The supply of steady water flow onto a steep uniformly deposited sediment bed produces a steadily progressing debris flow, the mechanism for this was discussed in section 3.1.2 and the conceptual sketch of flow was shown in Figure 3.6. The front part is drawn by a bulbous shape. Photo 4.1 (a) shows the front of a stony debris flow that is produced in an incised channel on the sediment bed. Photo 4.1 (b) shows the front of a stony debris flow that is produced on a laterally horizontal sediment bed, in which the slope gradients are between $\tilde{\theta}_1$ and θ_2. Each has a snout shape and approaches a constant depth rearward. These experiments use nearly uniform materials so that there is no accumulation of larger particles towards the front and even at the forefront the void between particles is saturated by water. The longitudinal profile of the snout of such a quasi-steady debris flow is obtained by the following discussion.

Consider that in the neighborhood of the front of a fully-developed debris flow neither erosion nor deposition takes place ($i = 0$) and there is no lateral input of water ($r = 0$), the one-dimensional mass and momentum conservation Equations (3.24) and (3.25) respectively are rewritten as:

$$\frac{\partial h}{\partial t} + \frac{\partial (Uh)}{\partial x} = 0 \tag{4.1}$$

$$\frac{\partial U}{\partial t} + U\frac{\partial U}{\partial x} = g\sin\theta - g\cos\theta\frac{\partial h}{\partial x} - k'\frac{U^2}{h} \tag{4.2}$$

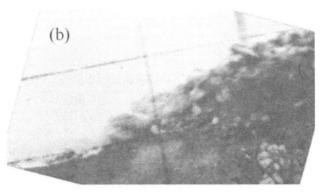

Photo 4.1 Debris flow fronts in the experiments.

where U is the cross-sectional mean velocity and k' is a coefficient to represent the resistance law that is given for stony debris flow from Equation (2.29) as:

$$k' = \frac{25a_i \sin \alpha_i}{4\{C + (1 - C)\rho/\sigma\}\{(C_*/C)^{1/3} - 1\}^2} \left(\frac{d_p}{h}\right)^2 \tag{4.3}$$

Because the purpose of discussion here is to obtain the frontal shape of a uniformly traveling flow (monoclinal wave), let us observe the front from a coordinate system moving at velocity V_f which is equal to the traveling velocity of the front. Namely, let us transform the coordinate system of Equations (4.1) and (4.2) by $X_f = x - V_f t$. This gives:

$$\frac{d}{dX_f}\{(U - V_f)h\} = 0 \tag{4.4}$$

$$(U - V_f)\frac{dU}{dX_f} = g \sin \theta - g \cos \theta \frac{dh}{dX_f} - k'\frac{U^2}{h} \tag{4.5}$$

Since no bed erosion occurs underneath the snout, from Equation (4.4) we obtain $U = V_f$. Substituting this into Equation (4.5), we obtain:

$$\frac{dh}{dX_f} = \tan \theta - \frac{k'}{\cos \theta}\frac{U^2}{gh} \tag{4.6}$$

The solution of Equation (4.6) under the boundary condition:

$$\text{at } X_f = 0; \ h = 0, \text{ and at } X_f = -\infty; \ h = h_\infty \tag{4.7}$$

and under the assumption that k' has a constant value that is obtained by substituting h_∞ into h in Equation (4.3) is:

$$h + h_\infty \ln \frac{|h - h_\infty|}{h_\infty} = X_f \tan \theta \tag{4.8}$$

Figure 4.1 shows the result of the application of this theory for the experiment Run 1 in Figure 2.9. The theoretical longitudinal profile of the snout compares well with the experimental one.

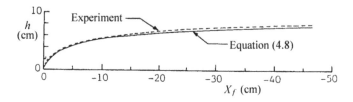

Figure 4.1   Longitudinal profile of the front of debris flow.

The forefront of the debris flow travels down the valley with a constant translation velocity, whereas in the rear part where the flow depth is nearly h_∞, the velocity near the surface is larger than the cross-sectional mean velocity U and the velocity near the bottom is smaller than U as shown in Figure 2.9. Because U is equal to V_f at the forefront, the particles transported near the surface gradually move to the forefront and finally on arriving at the forefront they fall down onto the bed. If the fallen-down particles keep the lead sliding on the bed pushed by the following flow, particles arriving one after another accumulate at the forefront and the constant shape and velocity of snout cannot be maintained. To keep the steady motion, the previously fallen-down particles should be buried under the newly fallen-down particles and these particles should be swallowed into the following flow. Therefore, at the front, the individual particles must move in a rotating motion, as if they were attached to the spokes of a big wheel running down the slope.

If the particle transported to the forefront is far larger than the neighboring particles and if it has a size comparable to the flow depth, it cannot be swallowed into the following flow. If such large particles come up one after another (the reason for this is explained later), large boulders accumulate at the forefront and they are pushed by the following flow as if it were a soil block pushed and moved by the blade of a bulldozer. Because the void space between the particles in such an accumulation of big boulders is large and the water permeability is very large, different from the succeeding debris flow part, the void cannot be saturated with water. Thus, the majority of stony debris flows, comprised of widely size-distributed material, proceed downstream as if the very front part is a granular flow with little water content (see Photo 1.2).

Let us consider the condition for a stony debris flow with a vanguard of big boulders to keep motion ahead. The conceptual sketch of the flow is shown in Figure 4.2. In this case, the motion of the frontal boulder accumulation part can be analyzed by the equation for the sliding motion of the earth block, Equation (3.65), in which the liquefied layer beneath the block does not exist, the depth of the following debris flow is equal to the thickness of the block, and $\alpha_e = 0$. Therefore, the equation of motion is given by:

$$\frac{d}{dt}(\rho_t LHU_e) = \rho_t gLH \sin\theta - \rho_t gLH \cos\theta \mu_k + (1/2)\rho_a gH^2 \cos\theta \\ + \rho_a(u_a - U_e)^2 H \tag{4.9}$$

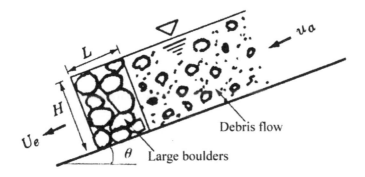

Figure 4.2 Debris flow proceeding downstream with a vanguard of big boulders.

In the steady state where $U_e = u_a$, the left-hand side term and the fourth term in the right-hand side are zero. Then, Equation (4.9) becomes:

$$\frac{H}{L} = 2\frac{\rho_t}{\rho_a}(\mu_k - \tan\theta) \tag{4.10}$$

Considering that ρ_t is the apparent density of the assembly of boulders without void water, we assume $\rho_t/\rho_a \approx 1/2$. Then, from Equation (4.10), if:

$$H/L \geq (\mu_k - \tan\theta) \tag{4.11}$$

is satisfied, the downstream movement can continue. Because the big boulders are continually supplied from the following flow, the head of debris flow would have the surface slope nearly equal to the angle of repose. Therefore, the maximum value of H/L would be the order of $\tan(\phi - \theta)$. The value of μ_k in the right-hand side of Equation (4.11) is nearly equal to $\tan\phi$. Consequently, Equation (4.11) cannot be satisfied. So, as long as boulders are continually supplied to the forefront, the motion will decelerate.

The front of debris flow will be decelerated as it continues down the ravine by the effects of the accumulation of big boulders towards the head. Then, the succeeding faster flow will be hindered on its way by the slower head motion and stored behind the head. Thus, the front part of the stony debris flow will bulge, and by the effects of the third and fourth terms on the right-hand side of Equation (4.9), the front will be accelerated to continue the motion. Namely, the stony debris flow continues its downstream motion concentrating big boulders ahead and increasing the thickness at the frontal part.

If the channel in which debris flow goes downstream has a shallow trench shape, the front of the debris flow will spill over the banks and form natural levees. The front part of debris flow whose bulk is partly broken by overflowing will restore its ability to go downstream. Thus, the stony debris flow can go downstream on a rather flat fan area constructing its way by forming natural levees. The photograph on the front page of this chapter and Photo 1.1 show evidence of such behavior.

4.1.2 The case of viscous-type debris flow

Mechanical characteristics of viscous debris flow could be explained by the Newtonian fluid model and they were verified by the experiments as described in section 2.8. Here, the results of the experiments conducted to examine the processes of flowage, deposition and remobilization of a once deposited bed are described. Details of the experimental set up and procedure are described in section 2.8.

Setting the hopper to supply the experimental material at 7.5 m upstream from the outlet of the flume, 2,500 cm^3 of the material was prepared in the hopper and it was poured onto the dry rigid bed of the flume to generate the first surge. The first surge left a certain thickness of deposit almost parallel to the flume bed. Then, the second surge whose material had the same quality and quantity to the first surge was poured onto now erodible bed, and it eroded and entrained the deposit. As shown in Figure 4.3, if the channel slope was steeper than 11°, the thickness of deposit after the

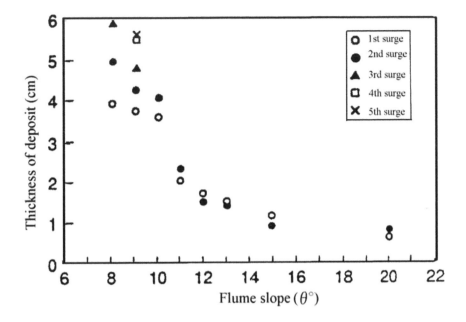

Figure 4.3 Thicknesses of deposit left after the passage of surges.

passage of the first and the second surges did not change much. It was almost the same even after the third and fourth surges. But, if the channel slope was flatter than 9°, the thicknesses of deposit increased surge by surge at least in the early stages of surge recurrence.

In the case when the channel slope was 7°, the first surge stopped in the flume. In the actual case at Jingjia Gully, as mentioned in section 1.3.3, the first arriving surge stops on the rugged stream bed and smoothens the bed upstream by covering the bed by smoother deposit. The next surge goes a little further downstream and elongates the smooth bed. During such phenomena the thickness of deposit increases (see Figure 1.10), but after the passage of many surges the thickness of deposit approaches an equilibrium state. After that, the surges acquire large mobility, and they pass through the channel reach having a gradient of 3°. In the experiments, however, the surges stopped when the channel slope was 7°. More experiments may be necessary to explain the cause of this discrepancy as the material used in the experiments and that sampled from the Jiangjia Gully debris flows resemble each other in composition. In the 9° channel gradient experiment, the deposit thickened surge by surge up to the fourth surge but it did not change after the passage of fifth surge. Although it is not clear that the thickness of deposit arrived at equilibrium by the fifth surge because no more surges were introduced, if one refers to the examples in the Jiangjia Gully, one will understand an equilibrium thickness to be asymptotically attained is sure to exist.

Figure 4.4 is the sketch of the frontal shapes of viscous debris flow in the experiments. When a surge moves on a rigid bed, the front has a blunt nose, where the particles transported in the upper part roll down onto the bed and then they are swallowed in the flow in the lower part. When a surge travels on a soft newly-deposited

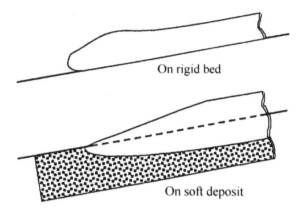

Figure 4.4 Difference in the frontal shapes on rigid bed and on soft deposit layer.

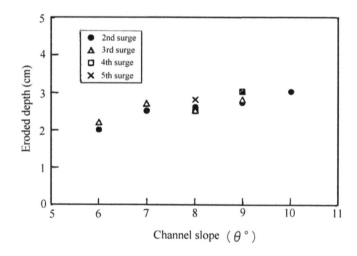

Figure 4.5 Erosion depths at the surge fronts.

layer produced by the preceding surges, the nose has an acute shape, and it erodes and entrains the deposit layer. The depth of the bed erosion depends on the channel slope as shown in Figure 4.5. Comparing these thicknesses with the deposit thicknesses given in Figure 4.3, one will understand that the entire depth of deposit layer is eroded and entrained if the channel slope is steeper than 11°. In such a case the nose begins to have a blunt angle as if it is going down on a rigid bed.

Figure 4.6 shows the examples of temporal surface level changes measured at 3 m and 6 m downstream from the material supply tank, respectively. When the channel slope was 8°, the forefront of the first surge stopped at 6 m, so that the depth measured at 6 m point shown in Figure 4.6 abruptly became almost 4 cm and thereafter it did not change because the flow stopped. The second and third surges passed through the downstream (6 m) point, but because some parts of them were deposited in the

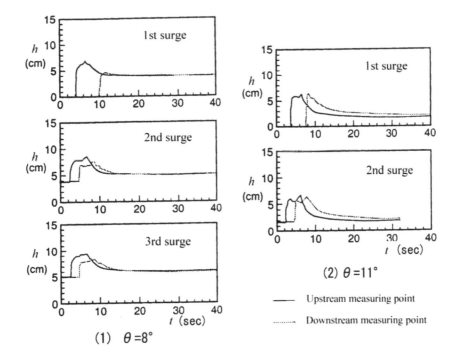

Figure 4.6 Surface level versus time curves at the upstream and downstream measuring points.

reach between 3 m and 6 m points, the amplitude of surface stage variations at the downstream point became smaller than at the upstream point. The thicknesses of deposit left after became larger and larger with the passage of recurrent surges and the peak surface levels became higher and higher. When the channel slope was 11°, the thicknesses of deposit left after the second surge was not much different from that after the first surge, and the amplitudes of stage variations at the upstream and downstream points were almost the same. These characteristics would be associated with the fact that almost the entire deposit layer was eroded and entrained by the front part of the second surge.

The conspicuous characteristic shown in Figure 4.6 other than those pointed out above is the evident difference in the propagation velocities of the surge fronts on the rigid bed and on the soft deposit layer. Namely, the propagation velocities of the second and third surges were far larger than that of the first surge. Figure 4.7 compares the propagation velocities of the forefront on the rigid bed (the first surge) and on the deposit layer (the second surge). The velocity of the second surge is about twice as fast as that of the first surge. Note that the discharge hydrographs supplied at the upstream end for the first and the second surges are almost identical. If one compares the propagation velocity of the second surge with the mean velocity of flow measured at a little behind the front (see Figure 2.53), one will be aware that the second surge proceeds about 1.5 to 2 times faster than the mean velocity. Because the forefront of the first surge propagates by mass transportation, the propagation velocity must be equal

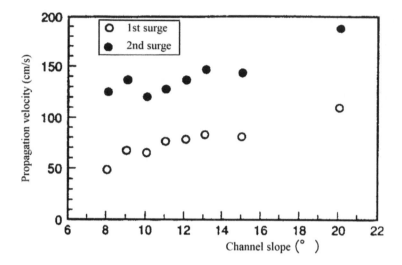

Figure 4.7 Propagation velocities of the forefront of 1st and 2nd surges.

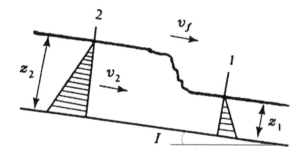

Figure 4.8 A propagating surge.

to the mean velocity of the flow, whereas the forefront of the second surge has more or less the characteristics of a wave that travels faster than the mass transportation.

The propagation of the forefront on the deposit layer can be explained by the application of bore propagation theory. Figure 4.8 illustrates a propagating surge on the deposit layer. Even though the deposit layer before the arrival of a surge is still; its properties are almost identical to the moving status. Therefore, the pressure distribution in the deposit layer would be hydrostatic as shown in Figure 4.8. In such a situation the propagating velocity of the surge front; v_f, becomes equal to that propagating on static water, and in a rectangular channel, it is obtained by applying the momentum conservation equation as follows (Chow 1959):

$$\frac{v_f}{u_*} = \left\{ \frac{(z_1 + z_2)}{2z_1 I} \right\}^{1/2} \tag{4.12}$$

where $u_* = (gz_2 I)^{1/2}$ and I is the energy gradient approximated to be equal to the channel gradient.

If the mean velocity behind the surge front v_2 is approximated by Equation (2.187), it is given by:

$$\frac{v_2}{u_*} = \frac{\rho(1 + \varepsilon \overline{C})z_2}{3\mu_a}\sqrt{gz_2 I} \tag{4.13}$$

Then, one obtains:

$$\frac{v_f}{v_2} = \frac{3\mu_a}{\sqrt{g}I\rho(1 + \varepsilon \overline{C})}\left(\frac{z_1 + z_2}{2z_1 z_2^3}\right)^{1/2} \tag{4.14}$$

If the experimentally obtained values: $\rho(1 + \varepsilon \overline{C}) = 1.92$ g/cm^3, $I = 0.19$ ($= \sin 11°$), $\mu_a = 6.2$ Pas, $z_1 = 2$ cm, $z_2 = 6$ cm, are substituted into Equation (4.14), then $v_f = 108$ cm/s, $v_2 = 70$ cm/s and $v_f/v_2 = 1.54$ are obtained. These values coincide rather well with the experimental values shown in Figure 4.7. Referring to the discussion in section 4.1.1, one will understand that the erosion of soft deposit beneath the forefront causes the wave characters at the surge front.

4.2 BOULDER ACCUMULATION AT THE FOREFRONT OF STONY DEBRIS FLOW

4.2.1 Various concepts for the mechanism

As stated in the previous section, boulders accumulate and tumble down at the forefront of stony debris flows, behind this follows the finer-grained more fluid debris. This phenomenon has attracted the interest of many researchers and quite a few mechanisms have been proposed. The majority of them are associated with the inverse grading structure of debris flow deposits, in which the grain sizes start off fine in the bottom layer and become coarser in the upper layers. So, because the velocity varies from zero at the bottom to a fast value in the upper layer and the mean velocity is somewhere in between the bottom and the surface, if grain-size sorting arises in the flowing layer and coarser particles are transferred to the upper part, they are transported forward faster than the mean propagating velocity of the debris flow front where they finally arrive and tumble down to the bed.

There are different ideas as to the cause of inverse grading. Among them Bagnold (1968) referred to his own concept of dispersive pressure due to grain collision, in which the dispersive pressure is proportional to the square of particle diameter (see Equation (2.18)), and inferred that within the shear flow of heterogeneous sediment mixture the larger particles would be transferred towards the surface of flow where the shearing velocity gradient is smallest. Middleton (1970) took a stand against Bagnold and simply considered that the finer particles fell into the void between coarser particles resulting in relative ascent of coarser particles; the concept of dynamic sieving.

Sallenger (1979) examined which of these two concepts were valid by the following considerations: If Equation (2.18) is valid, then two particles, one with a diameter d_H

and density σ_H, and the other one with diameter d_L and density σ_L, coexist in a layer at a certain height in the flow only when the following relationship is satisfied, in which other parameters in Equation (2.18) are common to the two particles:

$$d_H = d_L(\sigma_L/\sigma_H)^{1/2} \tag{4.15}$$

Namely, the diameter of a dense particle in a certain height measured from the bottom should be smaller than the lighter particle in the same height. Middleton did not consider the effects of particle density, but if the two particles have the same diameter, there is no reason for the heavier particle to move higher than the lighter particle under the effects of dynamic sieving, more likely, the heavier particle would move underneath the lighter particle. Therefore, under the action of dynamic sieving, the lighter particle that can exist at the same height with the heavier particle should have the same or smaller diameter than the heavier particle. This conclusion is opposite to Equation (4.15). Sallenger did a flume experiment of dry granular flow using a heterogeneous grain mixture comprised of two different density materials. The uniform flow in the flume was suddenly stopped by leveling out the flume and the vertical particle size distributions for the respective density materials were measured. The inverse grading was found in each density material, and at the same height from the bottom, the particle diameter of heavier material was evidently smaller than that of lighter material. This result contradicts to the inference based on Middleton's dynamic sieving concept. Moreover, Sallenger converted the diameter of heavier particle to that of the lighter particle by using Equation (4.15) in each layer and compared the size thus obtained with the lighter particle to find them almost in the same size. Thus, he proved that the inverse grading in a dry granular flow was caused by the dispersive pressure proposed by Bagnold.

The discussion above was still in the qualitative state and it was not directly connected to the prediction of boulder accumulation at the forefront in the process of flowage down a valley. Takahashi (1980) developed a theory applicable to the quantitative prediction of such a process based on Bagnold's dispersive pressure concept. Hashimoto and Tsubaki (1983) also proposed a theory based on their own particle collision theory.

Suwa (1988) proposed a theory in which the boulder accumulation at the forefront was not accompanied with the inverse grading but it was caused by the faster longitudinal velocity of big boulders than the mean flow velocity. This theory is based on the equation for the balance of force operating longitudinally on a particle at the bottom of a water or mud flow; the balance of the driving force is due to gravity, the friction force at the base of the particle, and the drag force due to the relative motion between the particle and the surrounding fluid. This theory has been used in the discussion of the initiation of motion of a particle in water flow or in the discussion of the bed-load particle motion, the application of this theory to stony debris flow is, as mentioned in Chapter 2, questionable because the inter-particle collision should be essential. Moreover, in many cases of turbulent muddy-type debris flows and viscous-type debris flows in which the situation is more likely to fit the theory of force balance mentioned above, the accumulation of boulders has not been recognized.

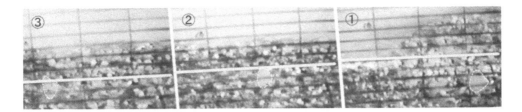

Photo 4.2 A big gravel going up in a stony debris flow. The particle surrounded by a white dotted line underneath the front of debris flow in Photo (1) gradually went up by the entrainment into the flow as shown in Photos (2) and (3); time proceeds from (1) to (3). The white straight lines show the original bed surface.

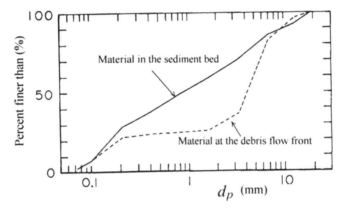

Figure 4.9 Particle segregation in stony debris flow.

4.2.2 The theory of Takahashi (1980)

If water flow is suddenly supplied from upstream on a sediment bed comprised of widely distributed heterogeneous particles and the bed has a gradient steeper than 15°, a bore of debris flow is formed. In the flowing layer of such a debris flow, the larger particles move upwards as shown in Photo 4.2. As the velocity in the upper layer is faster than in the lower layer, the larger particles that were moved upward are also transported ahead, faster than the progressive velocity of the forefront as if they were soil blocks stuck to the caterpillar tread of a tractor moving faster than the speed of the tractor itself. On reaching the forefront these particles tumble down to the bed and are buried in the following flow. But, soon after that, if the buried particle is larger than the surrounding particles, it appears again on the top of flow and moves ahead. The biggest boulders will accumulate at the forefront by the repetition of such processes along the distance of travel. Figure 4.9 compares the particle size distribution in the bed prepared before the experimental run with the particle size distribution in the forefront of debris flow. Fine particles that are less than 0.2 mm move in one body with the interstitial water and are not segregated, but the forefront lacks intermediate size particles between 0.2 mm and 2 mm. This phenomenon is caused by the downward and slower downstream motions of smaller particles (0.2 mm–2 mm) to compensate for the

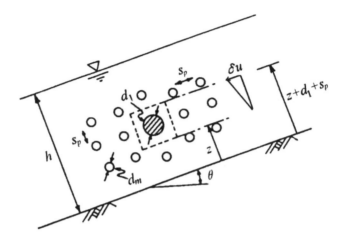

Figure 4.10 Characteristic particle arrangement in debris flow.

upward and faster downstream motions of larger particles (larger than 2 mm). One of the most conspicuous characteristics of stony debris flow is, as shown in Photo 1.2, the accumulation of big boulders at the front and the rear part being comprised of small particles. This is the result of particle segregation along the distance of travel.

Consider a spherical grain whose diameter is d_1 surrounded by uniform particles of diameter d_m in a steady uniform stony debris flow as depicted in Figure 4.10. The mean free distance between particles in the x and z directions is assumed to be a constant s_p (uniform concentration). The equation of motion of the d_1 particle to the direction perpendicular to the bottom is given by:

$$\frac{\pi}{6}d_1^3\left[\sigma + \frac{1}{2}\{(\sigma - \rho)C + \rho\}\right]\frac{dv_{gz}}{dt} = -P_1 - P_2 + P_3 - P_4 \tag{4.16}$$

where v_{gz} is the velocity of the d_1 particle in the z direction, P_1, P_2, P_3 and P_4 are the z components of the submerged weight of the d_1 particle, the dispersive force produced by the collision of the d_1 particle with the d_m particles existing in the layer above, the dispersive force acting at the lower surface of the d_1 particle due to the collision with the d_m particles in the next layer below, and the drag force produced by the motion of the d_1 particle in debris flow, respectively. These forces would be given as follows:

$$P_1 = \frac{\pi}{6}(\sigma - \rho)gd_1^3\cos\theta \tag{4.17}$$

The particle d_1 experiences a momentum change of $\{2r_d/(1 + r_d)\}(\pi/6)\sigma d_m^3 \delta u \cos\alpha_i$ per one collision, in which $r_d = (d_1/d_m)^3$ and δu is the relative velocity between

d_1 and d_m particles. The number of collision per unit time may be written as $f(\lambda)\delta u\{(d_1+s_p)/(d_m+s_p)\}^2/s_p$. Then, referring to Bagnold's notation (Bagnold 1954):

$$P_2 = f(\lambda)\frac{2r_d}{1+r_d}\frac{\pi}{6}d_m^3\sigma\,(\delta u)^2\left(\frac{d_1+s_p}{d_m+s_p}\right)^2\frac{1}{s_p}\cos\alpha_i = \frac{2r_d}{1+r_d}\lambda^2\sigma\,d_m^4$$

$$\times\left\{\frac{1}{2}\left(\frac{d_1}{d_m}\right)+\left(\frac{1}{2}+\frac{1}{\lambda}\right)\right\}^2\frac{(d_1/d_m+1/\lambda)^2}{(1+1/\lambda)^2}\left(\frac{du}{dz}\bigg|_{z+d1+sp}\right)^2 a_i\cos\alpha_i$$

(4.18)

where $(du/dz|_{z+d1+sp})$ is the velocity gradient at the upper surface of the d_1 particle. Similarly, P_3 may be written as:

$$P_3 = \frac{2r_d}{1+r_d}\lambda^2\sigma\,d_m^4\left\{\frac{1}{2}\left(\frac{d_1}{d_m}\right)+\left(\frac{1}{2}+\frac{1}{\lambda}\right)\right\}^2\frac{(d_1/d_m+1/\lambda)^2}{(1+1/\lambda)^2}\left(\frac{du}{dz}\bigg|_z\right)^2 a_i\cos\alpha_i$$

(4.19)

In the case that $d_1 = d_m$, the d_1 particle does not change the layer it moves in so that steady flow state is maintained and Equation (2.20) is satisfied. Therefore, from Equation (4.19), one obtains:

$$\lambda^2\sigma\,d_m^2\left(\frac{du}{dz}\bigg|_z\right)^2 a_i\cos\alpha_i = C(\sigma-\rho)(h-z)g\cos\theta$$

(4.20)

Hence, the resultant dispersive force operating on the d_1 particle is:

$$P_3 - P_2 = \frac{2r_d}{1+r_d}d_m^3\frac{\{d_1/(2d_m)+0.5+\lambda^{-1}\}^2(d_1/d_m+\lambda^{-1})}{(1+\lambda^{-1})^5}\frac{\pi}{6}(\sigma-\rho)g\cos\theta$$

(4.21)

The drag force operating on the d_1 particle is:

$$P_4 = \frac{1}{2}C_D\{(\sigma-\rho)C+\rho\}\frac{1}{4}d_1^2 v_{gz}|v_{gz}|$$

(4.22)

where C_D is the apparent drag coefficient and is assumed to be a constant.

Substituting Equations (4.17), (4.21) and (4.22) into Equation (4.16) one obtains the following equation of motion of the d_1 particle:

$$\frac{d^2z}{dt^2} = K_1g - \frac{K_2}{d_m}\frac{dz}{dt}\left|\frac{dz}{dt}\right|$$

(4.23)

where

$$K_1 = \frac{(\sigma-\rho)\cos\theta}{[\sigma+\{(\sigma-\rho)C+\rho\}/2]}\left\{\frac{2}{1+r_d}\frac{(0.5r_d^{1/3}+0.5+\lambda^{-1})(r_d^{1/3}+\lambda^{-1})}{(1+\lambda^{-1})^5}-1\right\}$$

(4.24)

$$K_2 = \frac{3}{4} C_D \frac{(\sigma - \rho)C + \rho}{[\sigma + \{(\sigma - \rho)C + \rho\}/2]} r_d^{-1/3} \tag{4.25}$$

Solutions of Equation (4.23) under the conditions where at $z = 0$ $v_{gz} = 0$, which is applicable when $v_{gz} \geq 0$, and at $z = h$ $v_{gz} = 0$, which is applicable when $v_{gz} < 0$, are, respectively:

$$\frac{v_{gz}^2}{gd_m} = \frac{K_1}{K_2} \left\{ 1 - \exp\left(-K_2 \frac{h}{d_m} \frac{z}{h} \right) \right\} \tag{4.26}$$

$$\frac{v_{gz}^2}{gd_m} = -\frac{K_1}{K_2} \left[1 - \exp\left\{ -K_2 \frac{h}{d_m} \left(1 - \frac{z}{h} \right) \right\} \right] \tag{4.27}$$

Equations (4.26) and (4.27) imply that if $K_1 > 0$, $v_{gz} > 0$ and if $K_1 < 0$, $v_{gz} < 0$. The cases where $K_1 > 0$ corresponds to $r_d > 1$ and $K_1 < 0$ corresponds to $r_d < 1$, mean that when $d_1 > d_m$, the d_1 particle moves upwards and, when $d_1 < d_m$ it moves downwards.

The qualitative characteristics of the following asymptotic velocities:

$$\frac{v_{gz}^2}{gd_m} = \left| \frac{K_1}{K_2} \right| = \frac{4}{3} \frac{1}{C_D} \frac{(\sigma - \rho) \cos\theta}{\{(\sigma - \rho)C + \rho\}} |V_d| \tag{4.28}$$

in which:

$$|V_d| = r_d^{1/3} \left| \frac{2}{1 + r_d} \frac{(0.5 r_d^{1/3} + 0.5 + \lambda^{-1})^2 (r_d^{1/3} + \lambda^{-1})^3}{(1 + \lambda^{-1})^5} - 1 \right| \tag{4.29}$$

are shown in Figure 4.11, where the calculation is accomplished assuming that the debris flow is in the equilibrium stage, the particle concentration is given by Equation (2.25) and the drag coefficient is equal to 2,000 as described later. Note that the larger the d_1 particle is, in comparison with d_m, the faster the upward velocity becomes. However, in the case when d_1 is smaller than d_m, its downward velocity is small. In the experiment, small particles were found to drop into the void between coarser particles, so that for particles smaller than the surrounding particles, the dynamic sieving effects may surpass the dispersive effects due to particle collision.

To prove the theory, laboratory flume experiments were carried out. Two kinds of materials were used. One had $d_m = 5$ mm and $d_1 = 22$ mm, and the mixing ratio of d_m to d_1 was 3:1 by weight, and the other had $d_m = 5$ mm and $d_1 = 15$ mm, and the mixing ratio was 2:1 by weight. Traces of d_1 particles in the flow observed through the transparent flume wall were recorded by a 16 mm high-speed camera. Experimental and calculated positions of d_1 particles versus time are compared in Figure 4.12. The different symbols in Figure 4.12 indicate that different particles started from different heights in the viewing field of the photograph. The appropriate C_D value was found by trial and error and a very large C_D value of 2,000 was adopted.

In the prototype scale of phenomena, if $d_m = 5$ cm, the ascending velocity of a particle having the diameter of 50 cm will be given by Figure 4.11 as about 35 cm/s. If the depth of flow is 1 m, it will appear on the surface of flow within about 3 seconds. The ratio of the surface velocity to the mean translating velocity in stony debris flow

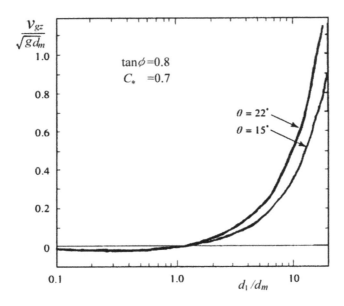

Figure 4.11 The non-dimensional upward/downward velocities versus the ratio of diameters.

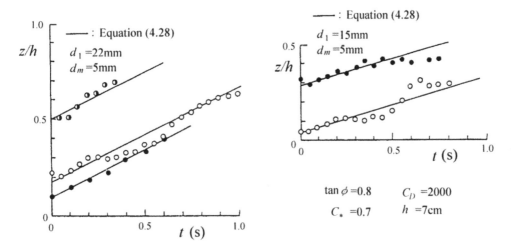

Figure 4.12 Ascent of d_1 particle ($\theta = 22°$).

is 5/3, which is obtained by referring to Equations (2.22) and (2.149). Hence, if the translation velocity of this debris flow is 15 m/s (Equation (2.149) predicts this value when $\theta = 18°$, $\tan \phi = 0.8$ and $C_* = 0.7$), the particle that appeared on the surface of flow will proceed forwards faster than the front by 10 m/s. Thence, if a debris flow surge is from 100 to 200 m long, that particle will arrive at the forefront within 10 to 20 seconds and the distance traveled within this period is from 150 to 300 m. This consideration clearly shows how easily large boulders accumulate at the forefront of stony debris flow. The numerical simulation of the accumulation of large boulders at the front is discussed in section 4.6.

This theory is based on Bagnold's semi-empirical dispersive force equation, but similar results will be obtained based on Takahashi and Tsujimoto's theoretical Equation (2.157). In this case, however, the solids concentration distribution exists in the flowing layer and, to make the discussion more complicated, the velocity distribution at the lower part is convex upwards making the dispersive pressure due to particle collision small, thereby the accumulation of large particles at the forefront will be a little delayed.

4.3 ABILITY TO TRANSPORT LARGE BOULDERS

As stated previously, large boulders accumulate at the front part of stony debris flow. Then the maximum size of boulder that can be transported may be an interesting problem. Photo 4.3 shows the accumulation of large boulders on the deposit surface produced by the debris flow in the case of the Venezuelan disasters in 1999. Many boulders, from a few meters to more than ten meters in diameter were transported and plunged into a building destroying a part of it. Large boulders move down the gorge forming a vanguard, and on arriving at the fan-top they spill-over the channel making natural levees and finally deposit as swollen-type or flat-type lobes. However, many witnesses of the debris flow say that huge stones seemed to be floating on the surface of the flow. In fact, in Photo 1.2 at 5 seconds and 6 seconds, one can see that a large boulder of 3 m in diameter is being transported with almost its entire body protruding from the surface of flow, so it looks like it is floating.

Large boulders are also transported in turbulent or viscous-type debris flows, however in this case, they do not accumulate at the front but they are transported sporadically and are deposited scattered in the deposit area. Photo 4.4 shows the deposit of volcanic mud flow that devastated Armero City at the foot of the Nevado del Ruiz Volcano, Colombia in 1985. The big boulders are scattered in a wide area where the city existed before the disasters.

Photo 4.3 Large boulders runoff from a ravine in Venezuela.

Early consideration of the mechanism of boulder transportation was based on the Bingham fluid model and it was attributed to the strength of the plug part near the surface of flow (Johnson 1970). However, an experiment and the related theoretical consideration revealed that the surface of a clay-slurry flow could only support a particle whose diameter was at most 3 mm (Hampton 1975). This means that to sustain the huge stones usually found in stony debris flow, a much stronger yield strength is needed, and this in turn makes it difficult for flow to occur in thin thickness of a few meters or less, as in the case of usual debris flows. To overcome this difficulty, Rodine and Johnson (1976) suggested that it is not the strength of the plug but the buoyancy that would be large in poorly sorted debris. They suggested that the unit weight of debris flow material effective to support a huge stone of a particular size would be equal to the apparent unit weight of the material from which all the particles equal to and larger than the size of that huge stone are extracted and such a heavy mixture of fluid and solids would behave as a fluid to cause large buoyancy. If this conjecture is true, the fluid pressure in the debris flow should be very much higher than that in plain water flow. According to my laboratory flume experiments, however, the fluid pressure in the stony debris flow that is composed of particles larger than a few millimeters is not different from that in plain water flow; meaning that almost all the particle loads are transmitted not by the enhanced fluid pressure but directly onto the bed via inter-particle encounters. This pressure deficit was also confirmed in the field as stated in section 1.3.1.

In contrast to the above Bingham fluid or the visco-plastic rheological model, the discussion on the upward motion of a large particle in the preceding section (dilatant fluid model) may add a probable reasoning for the boulder transportation. Apart from this kind of particle transportation problem, the problems on the threshold of particle motion in a viscous fluid and the particle motion due to surface wave pressure that is generated by the protrusion of the particle from the surface of shallow rapid flow were discussed by Daido (1971). According to him, if the channel slope is steeper than $10°$ and the depth of flow is deeper than $0.8D_d$, any boulder of diameter D_d will be transported in contact with the bed by the effect of the fluid dynamic force.

Imagine a large boulder moving downstream at the surface of a debris flow sticking a part of its body out of the flow as shown in Figure 4.13. In this case the forces acting on the boulder perpendicular to the main flow direction are its own weight, the

Photo 4.4 Debris flow deposit at Armero City, Colombia.

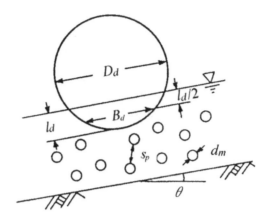

Figure 4.13 A big boulder on the surface of debris flow.

buoyancy and the dispersive upward pressure due to collisions with many other small particles beneath. If the effective bottom area for collision is assumed to be a square of $(B_d + s_p)^2$, which is equal to $[D_d\{1 - (1 - l_d/D_d)^2\}^{1/2} + s_p]^2$, and the velocity of the boulder is assumed to be equal to that of the debris flow at the middle height of the draft l_d, the condition for the equilibrium state of the operating forces to the stone having the diameter D_d is written as:

$$
\frac{2r_d}{1+r_d}C(\sigma - \rho)d_m^3\left(\frac{l_d}{2d_m} + \frac{1}{\lambda} + \frac{1}{2}\right)^2 \frac{\left\{r_d^{1/3}\sqrt{1 - (1 - l_d/D_d)^2} + 1/\lambda\right\}^2}{1 + 1/\lambda}g\cos\theta
$$
$$
= \frac{\pi}{6}\sigma D_d^3 g\cos\theta - \frac{\pi}{8}\rho D_d^3 g\cos\theta\left[\frac{2l_d}{D_d} - \frac{1}{3}\left\{1 - \left(1 - \frac{2l_d}{D_d}\right)^3\right\}\right]
$$

(4.30)

where $r_d = (D_d/d_m)^3$. The left-hand side of Equation (4.30) is the upward force due to the dispersive pressure produced by the particle collisions and the right-hand side is the net weight of the boulder. Substitution of d_m in D_d in Equation (4.30) corresponds to a flow consists of uniform grains and the particle in this case should not move up or down. To satisfy this condition in Equation (4.30) the following relationship must be fulfilled:

$$
\frac{1}{\lambda} = \left(\frac{\pi/6}{C}\right)^{1/3} - 1
$$

(4.31)

If particles distribute homogeneously, referring to Figure 2.6, one particle is involved in the volume of $b^3 d_m^3$. In this case, from Equation (2.14), $C_* = \pi/6$. Therefore, Equation (4.31) is nothing but the definition formula of Bagnold's linear concentration.

The curves in Figure 4.14 are calculated under the condition $\sigma = 2.6\,\text{g/cm}^3$, $\rho = 1.0\,\text{g/cm}^3$, $C_* = \pi/6$, and the plotted points are the results of laboratory experiments, in which quartz sand $d_m = 0.38\,\text{cm}$, $\sigma = 2.61\,\text{g/cm}^3$ was used for the debris

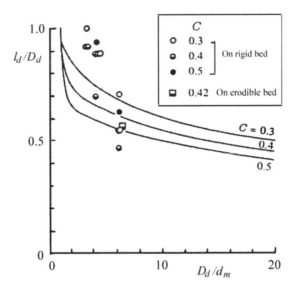

Figure 4.14 Relative draft versus relative diameter of the boulder.

flow material and glass beads $D_d = 2.45$, 1.67, and 1.31 cm, $\sigma = 2.60$ g/cm^3 were used for the big boulders. Although there is considerable scatter in the experimental values, the tendency is rather well explained. The reason why the experimental data demonstrates a larger draft than that predicted by the theory in smaller range of D_d/d_m may be attributable to lower solids concentrations near the surface in the experiments than the theoretical ones in which uniform concentration distribution is assumed.

Figure 4.14 demonstrates that:

1   The denser the solids concentration in a debris flow, the greater the ability of the flow to transport big stones becomes.
2   The larger the diameter of the big boulder the smaller the relative draft l_d/D_d becomes. The tendency of the theoretical curves, however, demonstrates that the draft itself becomes larger, the larger the diameter of the boulder becomes, and consequently at most a boulder whose diameter is up to about twice the depth of flow may be able to be transported by the effect of dispersive pressure.

4.4   THE CAUSES OF INTERMITTENCY

As mentioned in Chapter 1, the conspicuous characteristics of the viscous debris flows at the Jiangjia Ravine, China are their intermittency; the intervals between two successive surges vary irregularly from more than 300 seconds to 10 seconds. But, intermittency in debris flows is not limited to viscous debris flow but is common to all types, as it is recognized in the stony debris flow in Kamikamihorizawa. Takahashi (1983a) focused his attention on the instability inherent in the flow itself and examined it using flume experiments. A rigid bed flume 20 cm in width and 18° in slope

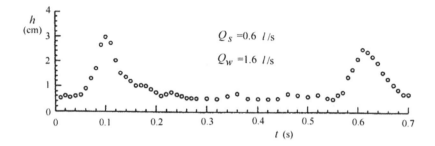

Figure 4.15 Roll waves in a debris flow.

was used. Two kinds of heterogeneous materials, having median diameters of 1.2 mm and 4 mm, were mixed with water at concentrations from 0.18 to 0.45. The respective experimental materials were then poured into the flume.

Figure 4.15 shows the time variation in the flow depth for an experimental debris flow, in which, notwithstanding that a constant rate of debris flow (2,200 cm^3/s) is supplied into the flume, intermittent roll waves are formed. The non-dimensional celerity, cycle, maximum wave height and the ratio of maximum wave height to normal flow depth are plotted in Figure 4.16 versus Froude number calculated at the normal flow part, where h_b is the maximum wave height, T_p is the cycle, h_n is the normal flow depth, C_r is the celerity of roll wave, q_t is the unit width discharge and Fr is the Froude number. The curves in Figure 4.16 are theoretical ones which have been obtained for water flow (Iwagaki 1955). Two theoretical non-dimensional celerity lines for turbulent flow are obtained, giving $\beta' = 1.0$ and $\beta' = 1.25$, in which β' is the momentum correction coefficient introduced to take account of the effect of velocity distribution. $\beta' = 1.0$ corresponds to plain water flow and $\beta' = 1.25$ corresponds to the dilatant fluid flow whose velocity distribution is given by Equation (2.22). The relative depth of flow in the experiments was small even in the case using smaller particle material so that the dilatant flow model was applicable. As is clear from Figure 4.16, the characteristics of roll waves are similar to those of water flow, only if the coefficient β' is appropriately changed to be suitable for a dilatant fluid. For low solids concentrations, however, the β' value would be nearly equal to that of water flow.

For the case of the Jiangjia Ravine debris flow, as a rough standard, a flow depth 1 m, a celerity of 7 m/s and a flow velocity of 5 m/s may be adopted in the observation reach whose channel slope is about 3°, then if the cycle of surges is of the order of 20 seconds, the intermittency might be understood as roll waves in the extent plotted within Figure 4.16. However, the cycle of more than a few hundreds seconds cannot be explained as roll waves, moreover, as mentioned earlier, the flow completely stops between the successive surges and therefore the discussion that attributes the cause of intermittency to the instability of flow does not make sense.

As discussed above, the intermittency of debris flow may sometimes be attributable to the instability of flow, but more simply, it would be due to many spatially and temporally distributed sources in the basin that independently produce surges. For example, in the Jiangjia Ravine there are numerous unstable bare slopes that easily slide in a volume comparable to the volume of a debris flow surge.

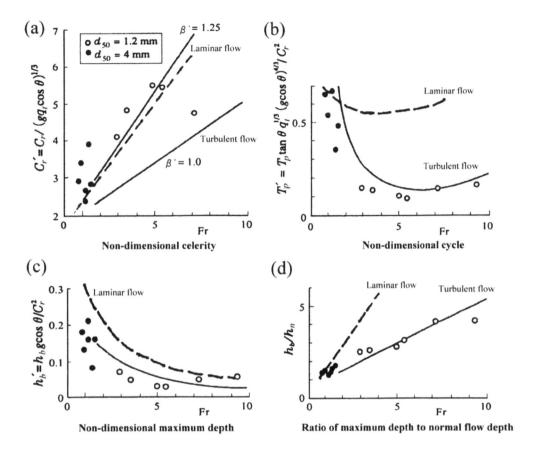

Figure 4.16 Characteristics of roll waves as a function of Froude number.

As pointed out by Okuda *et al.* (1978); referred in section 1.3.1 (page 12), spatially distributed knickpoints where channel gradient become shallower downstream will temporarily store sediment and suddenly release it to produce intermittent surges.

In the case of stony debris flow a cause other than the instability of flow or the distributed sources can be found. That is the sporadic gathering of large boulders along the distance of flow. Shortly after the formation of a debris flow, large boulders accumulate at the front of the bore and thence the traveling velocity slows down and the head swells up. If some of these boulders are deposited along the side banks, the fluidity of flow will recover and the flow will be accelerated to bring about the intermittency. In the rear part of the flow, associated with the reduction of solids concentration, the ability to erode the bed recovers and the highly heterogeneous bed material will be entrained into flow. This process newly gives rise to the particle segregations and large boulders are transported ahead faster than the mean flow velocity, but before arriving at the bore front now traveling far downstream, the large boulders will accumulate in some upstream portions hindered by the channel constrictions, the slope changes or the scattered huge stones that are larger than twice the depth of flow and are slowly moving in contact with the bed or standstill. Each scattered boulder accumulation along the length of flow will swell the flow around it and produces intermittency.

4.5   DEBRIS FLOW AROUND A BEND

A series of experiments was performed to examine the super-elevation along the outer bank of a channel bend (Ashida *et al.* 1981). A straight channel 20 cm in width and 17 m in length was connected to a 45° bending channel as illustrated in Figure 4.17. The bending channels had the radii of curvature at the center line of the channel of 40, 60 and 100 cm. These curved parts were attached to 40 cm straight channels upstream and downstream. The upstream channel slope for various runs was changed from 17° to 20° and the downstream bend channel was fixed at 10° in every run. The channel bed was horizontal in the radius direction and gravels 5 mm in diameter were used as roughness elements. The experimental material was a mixture of 3 mm marble grains and 0.15 mm quartz sand with the ratio of 1:1 by weight. The material was supplied from a sand feeder and water was supplied separately to produce debris flows whose solids concentrations were 0, 0.2 and 0.4 and the discharges were 1, 2 and 3 *l*/s.

The shock waves produced in a bend channel under supercritical flow condition, follow Knapp's famous formula (Knapp 1951):

$$h_g = \frac{U^2}{g} \sin^2\left(\beta_1 + \frac{\theta'_c}{2}\right) \tag{4.32}$$

where h_g is the depth variation along the outer bank of the bend, θ'_c is the bend angle along the outer bank, β_1 is the angle of the crest of the shockwave originating from the outer bank of the entrance of the bend ($\theta'_c = 0$) and this is given by $\sin \beta_1 = 1/F_1$ in

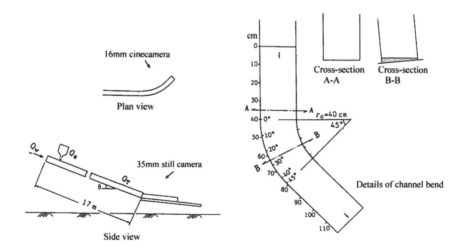

Figure 4.17 Experimental flume.

which F_1 is the Froude number at the entrance of the bend, and U is the velocity at $\theta'_c = 0$. This formula is applicable only in the range:

$$\theta'_c < \theta_{co} = \tan^{-1} \frac{B}{(r_{co} + B/2)\tan\beta_1} \tag{4.33}$$

where B is the channel width and r_{co} is the radius of curvature at the center of the channel.

Lenau (1979) proposed a formula predicting the depth variation along a bend having trapezoidal cross-section, and the formula can be transformed for a rectangular channel as follows:

$$h = h_{no} + E \tag{4.34}$$

$$E = \frac{Uh_{no}^2}{2\sqrt{gh_{no}}r_{co}} \sum_{k=0}^{\infty} (-1)^k \left[\begin{array}{l} \delta\left\{ x' - \sqrt{F_1^2 - 1}\left(kB' + \frac{1}{2}B' - y' \right) \right\} \\ -\delta\left\{ x' - \sqrt{F_1^2 - 1}\left(kB' + \frac{1}{2}B' + y' \right) \right\} \end{array} \right] \tag{4.35}$$

where h_{no} is the normal flow depth at the entrance of the bend, $x' = r_{cc}\theta'_c/h_{no}$, $y' = (r_{cc} - r_{co})/h_{no}$, $B' = B/h_{no}$, and $\delta(x')$ is a function such that when $x' < 0$, $\delta(x') = 0$ and when $x' \geq 0$, $\delta(x') = x'$.

Because $y' = B'/2$ along the outer bank, the maximum of E occurs at $x' = (F_1^2 - 1)^{1/2}B'$, $(F_1^2 - 1)^{1/3}B'$, $(F_1^2 - 1)^{1/5}B', \ldots$, and that the value under the condition $F_1 >> 1$ is given by:

$$E_{\max} = \frac{U^2 B}{2gr_{co}} \tag{4.36}$$

This is equivalent to the one deduced from the ideal balance condition between the centrifugal force and the body force produced by the lateral water surface slope. If the position where $E_{\max}$ arises, $\theta_c = B(F_1^2 - 1)^{1/2}/r_{co}$, is calculated to occur downstream from the outlet of curved channel, the maximum water stage along the outer bank cannot develop as high as $E_{\max}$.

Figure 4.18 compares the steady flow depth (depth subtracted the wave height from the maximum stage) along the outer bank with the calculated results using Knapp's formula and Lenau's formula. Knapp's formula has a tendency to predict too large depths, especially at larger θ'_c. Lenau's formula predicts rather well for all reaches of the bend. The effects of the solids concentration seem not large and the experiments suggested that Lenau's formula could be applicable up to $C = 0.4$.

In the case of supercritical flow roll waves are generated as shown in Figure 4.16, when the Froude number is 5 the maximum wave height becomes 3 times that of normal flow stage. What happens to these roll waves in a channel bend is important in the design of channel works. Figure 4.19 is an example of measurements of water stage variation along the outer bank. The amplitude of the variation increases with increasing deflection angle.

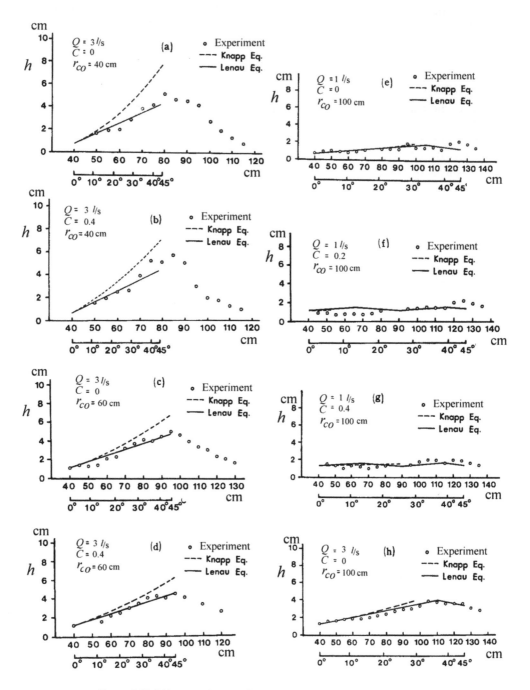

Figure 4.18 Water surface profiles along the outer wall of the bend.

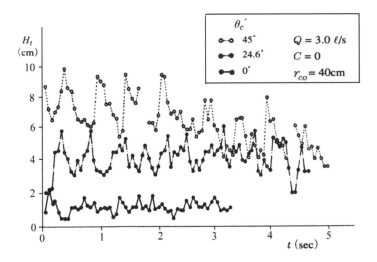

Figure 4.19 Water stage oscillation at three points along the outer bank.

As the centrifugal force plays an important role in the amplification of steady flow depth, it is considered to amplify the roll wave heights as well, and the total depth at the outer bank H_t is written as the sum of the steady flow depth h and the roll wave height h_w. Lenau's formula is also applied to calculate h_w, where the velocity, U, of the wave part at the entrance of the bend is, for the sake of simplicity, given as the same value as that for steady flow part. As for the boundary condition, the wave height at the entrance of bend is assumed to be equal to the significant wave height, defined as the mean of the largest 1/3 wave heights, that is obtained by measurement at the entrance.

Figure 4.20 compares these calculated significant wave heights plus the depth of the steady flow part with the experimentally obtained ones. Although there are some expedient assumptions such as giving the normal flow velocity as the boundary condition for the roll waves, these calculations give relatively good results and this method would be applicable for the design of channel works. Note that the amplification of waves is very large and so the channel must be deep enough to avoid over-spilling.

In Photo 1.2 the flow surface on the left-hand side (right-hand side bank) was raised very much. This is due to the bending of the channel just upstream of this position. This phenomenon is called super-elevation and it is often useful to know the velocity and the discharge passing through that position by measuring the difference in the heights of debris flow marks at the left- and right-hand side banks. It must be noted that if the debris flow accompanied roll waves, the traces might suggest an erroneously large-scale debris flow.

4.6 ROUTING OF DEBRIS FLOW IN THE TRANSFERRING REACH

4.6.1 Kinematic wave method

Debris flow that formed and developed in the occurrence/developing reach comes down to the transferring reach and there it is deformed, and then it runs off to a depositing

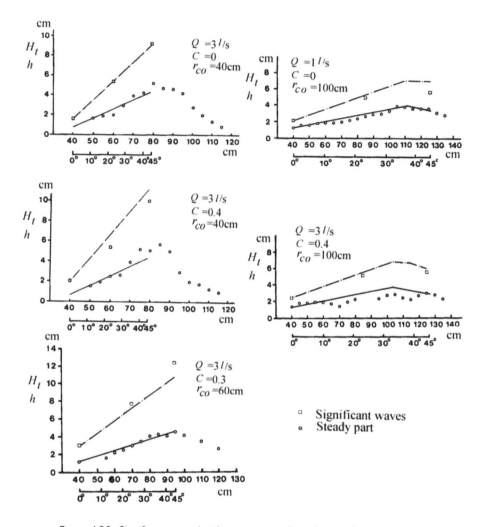

Figure 4.20 Significant wave heights superposed on the steady water stages.

area such as a fan. Debris flow in a narrow gorge moves unidirectionally and behaves like a monoclinal wave. Such a debris flow can be routed by the kinematic wave theory introduced in section 3.1.2.

A simple wide rectangular cross-section is assumed, then, the resistance law of stony-type debris flow is given by Equation (2.149). If neither erosion nor deposition occurs in the process of transfer, the continuity equation can be described from Equation (3.24) as follows:

$$(B_m + B_{dead}) \frac{\partial h}{\partial t} + \frac{\partial}{\partial x}(B_m U h) = 0 \tag{4.37}$$

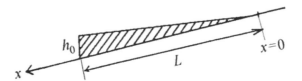

Figure 4.21 The input debris flow to the transferring reach.

Here, the total width of the channel is assumed to consist of the main flow section of width B_m and the stagnation section of width B_{dead}. The storage in the stagnation section works to attenuate the peak discharge of the debris flow. The mean B_m, B_{dead}, θ, and C values do not change within the considering reach.

Let us consider a debris flow having a longitudinally triangular shape as shown in Figure 4.21. Namely, the boundary and initial conditions in solving Equations (2.149) and (4.37) simultaneously are as follows:

$$\left.\begin{array}{ll} h(x,0) = \dfrac{h_0 x}{L}; & 0 \le x < L \\[2ex] h(x,0) = 0; & -\infty < x \le 0,\ L < x < \infty \\[2ex] \dfrac{dx_s(t)}{dt} = U(x_s(t),t); & 0 \le t < \infty \\[2ex] & x_s(0) = L \end{array}\right\} \tag{4.38}$$

where $x_s(t)$ is the position of the debris flow front on the longitudinal axis.

The solution is:

$$x^* = h^* + \frac{5}{2} \frac{B_m}{B_m + B_{dead}} h^{*3/2} t^* \tag{4.39}$$

where the variables with asterisk are the non-dimensional presentations as follows:

$$(h^*,\ x^*,\ t^*) = (h/h_0,\ x/L,\ Ut/L) \tag{4.40}$$

Two successive surges in a debris flow event were measured in the Shiramizudani experimental basin of DPRI at two positions. The Higashitani point at about 1 km and the Shiramizudani point at about 2 km downstream of the debris flow generating area (Ashida *et al.* 1986). The circles in Figure 4.22 show the hydrographs measured at the two measuring points, where the discharge was obtained as the product of the velocity and the cross-sectional area, each of which was measured using the TV video records. The real distance between the two markers recognized in a video frame was measured in the field and the time necessary for the suspended matters or the solid particles on the surface of the flow to pass through that distance was measured in the video records, and then the velocity of debris flow was obtained from these values. The cross-sectional area was obtained by specifying the flow surface stage in the video. Because the riverbed in this reach is rigid lava rock, no erosion takes place and the

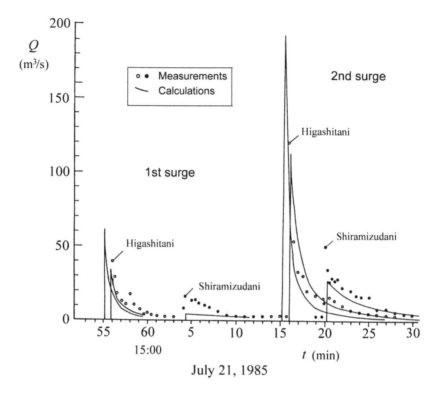

Figure 4.22 Debris flow hydrographs at Higashitani and Shiramizudani points.

stage-area relationship is already prepared by the field measurement. The estimation of discharge by this method may contain about 10% error, but the general characteristic variation of hydrographs should be correctly presented. For the first surge, a total of 3,120 m³ of materials passed through the upstream point and 2,880 m³ of materials passed through the downstream point within about five minutes at each point. For the second surge, 9,630 m³ passed through the upstream point and 8,660 m³ passed through the downstream point. The duration of the debris flow at both measuring points was about ten minutes. The volumes at the upstream point were about 10% larger than those at the downstream point for both two surges, but taking the accuracy of measurement and the longer recessing phase especially at the downstream point into account, we can conclude that almost the same volume of debris flow passed through the two measuring points. In fact, the field survey after the debris flow revealed that little sediment was deposited.

The conspicuous characteristic changes of the hydrographs are the becoming blunt around the peaks and the reduction of the peak discharges by about a half within only 1 km. The riverbed has a U-shape in the cross-section and longitudinally it has many falls, among them the highest one has 60 m drop. Hence, although the general topographic inclination of the riverbed in this reach is about 16°, the inclination between falls is about 6° on the average.

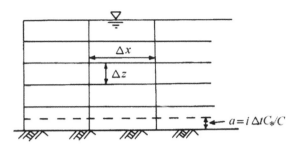

Figure 4.23 Computational grids for debris flow routing.

The application of Equation (4.39) for the respective surges gives the solid line hydrographs in Figure 4.22, where the upstream boundary hydrographs as shown in Figure 4.21 are given at 300 m and 500 m upstream of Higashitani point respectively for the first and second surges. The other conditions for the first surge are: $h_0 = 2$ m, $B_m/(B_m + B_{dead}) = 1$, $C_* = 0.65$, $C = 0.4$, $\rho = 1.3$ g/cm^3, $\sigma = 2.65$ g/cm^3, $\sin\theta = 0.1$, $a_i \sin\alpha_i = 0.02$ and $d_p = 0.1$ m; and for the second surge: $h_0 = 3$ m, $B_m/(B_m + B_{dead}) = 0.7$, and other parameters are the same as for the first surge. For the second surge $B_m/(B_m + B_{dead}) = 0.7$ was adopted to explain the large attenuation, it is reasonable to assume this because the second surge has a deeper depth than the first surge and it flows in a wider section which may have a stagnation area attached to the main flow section. But, referring to the fact that almost no sediment was left after the debris flow, the storage in the stagnation section was only temporal and in the recessing phase it was dried up by the returning flow.

Routing considering the convergence of larger particles to the front

The flow routing described above considers that, as a matter of practice, the debris flow is comprised of uniform particles, but in fact, stony debris flow is comprised of heterogeneous materials and due to the particle segregation while in motion larger particles concentrate at the front. Hence, the accumulation of larger particles in the front affects the hydrograph and the translation velocity of debris flow. Herein, the numerical simulation method that takes the larger particle's convergence to the front into account is explained (Takahashi et al. 1992).

The finite difference calculation of the governing equations in the kinematic wave method described in section 3.1.2 is carried out, by separating a time step into two stages. In the first stage, debris flow is considered to be comprised of uniform material whose particle diameter is equal to the mean diameter of the particles contained in the pillar-shaped space in the flow (which has a height h and bottom area of $\Delta x \times 1$) at a longitudinal position x. In this stage the particle concentration is uniform throughout the depth. A one-dimensional finite difference calculation of the governing equations thus determines the location of the forefront, the mean flow depth along the channel behind the forefront, and the mean velocity and solids concentration at time step t. At the present time step (the second stage) the depth is sliced into layers of thickness Δz (i.e. variable, dependent on the depth) as depicted in Figure 4.23. Because the part beneath the broken line in the lowest layer with thickness a ($= i\Delta t C_*/C$) consists of newly entrained bed materials, the particle-size distribution in this layer is the same as that in the bed. One may assume that particle compositions in the rest of the thickness ($\Delta z - a$) of the lowest layer and in the other layers are the same as calculated for each layer one step earlier. The materials in the two sub-layers of the lowest layer become thoroughly mixed within a negligible time; therefore, we can obtain the sediment composition in each layer at time t.

Particles in a layer are transported downstream with the mean translation velocity at that height, while individual particles move up and down, characterizing the particle interactions. The translation velocity u at height z is approximated by Equation (2.22), in which the mean diameter d_m within the layer at height z is substituted into d_p, and C is assumed to be distributed uniformly over the entire depth. This approximation implies that the particle segregation begins just after the debris flow has become mature. It is assumed that particles larger than the mean size in the grid are moved upward by the action of dispersive pressure. The upward velocity v of the particle having diameter d_k is obtained by Equation (4.28), where $v_{gz} = v$ and $r_d = (d_k/d_m)^3$.

Particles having sizes smaller than the mean are assumed to move downward with a velocity (irrespective of difference in diameter) that compensates for the net upward volume transport of the larger particles.

Thus, the concentration of each size fraction in the grid (i, j); $C_k(i, j)$, is obtained one time step later $(t + \Delta t)$ from the continuity equation:

$$\frac{\partial C_k}{\partial t} + \frac{\partial (C_k u)}{\partial x} + \frac{\partial (C_k v)}{\partial z} = 0 \tag{4.41}$$

The mean diameter d_m is given by:

$$d_m = \frac{\sum C_k d_k}{\sum C_k} \tag{4.42}$$

where Σ means the summation of all particles divided into ranks.

Calculations at the front require special consideration. Because the translation velocity in a higher layer is faster than that in a lower layer, the higher layer at time step $(t + \Delta t)$ protrudes longer than the lower layer from the supposedly same position M of the forefront at time step t, as shown schematically in Figure 4.24. Actually, these protruding portions fall to the bed to form a vertically even forefront that is at a position downstream by $U \Delta t$ from the previous position M, where U is the mean translation velocity of the front obtained by substituting the average particle size over the entire depth into Equation (2.149). At that time, the protruding portions are assumed to fall in sequence from the bottom layer to the top layer, as labeled steps [1] through [4] in Figure 4.24. The particle-size distribution in each layer of length $U \Delta t$ is calculated as follows:

First, a fall is assumed to occur from the second lowest layer to the lowest layer and the leading edges of the two adjacent layers become even (the shaded parts in the two adjacent layers have the same area). Because the fallen part has the sediment composition in the second layer, it is not the same as that in the lowest layer. So in the lowest layer the particles in the shaded part are mixed with those in the rear part

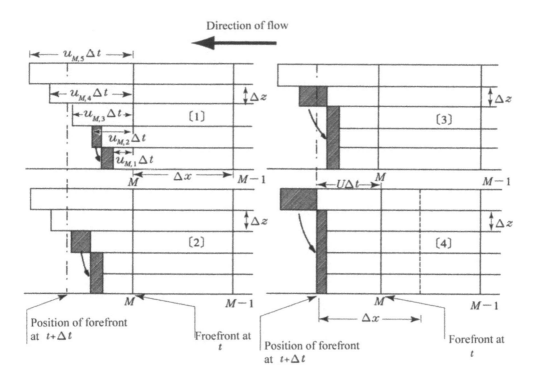

Figure 4.24 Stepwise computational schemes at forefront of debris flow.

with length $U_{M,1} \Delta t$ in step [1]. Thus, the sediment composition at this time step in the lowest layer can be calculated. The sediment composition in the remaining part of the second lowest layer is the same as that in the shaded fall part in the figure in step [1] in Figure 4.24.

Second, a fall from the third lowest layer supplies particles to the second and the first lowest layers as depicted in step [2], and so forth. The sediment composition in each layer is calculated by the same procedure as before. These procedures are repeated in sequence until a fall from the uppermost layer is completed and particles are mingled with the rear parts of the lower layers. This process should take place in a negligible time in Δt.

To return to the one-dimensional calculation of translational debris flow movement, one again divides the flow into the same size grids as before (Δx in length), starting at the new forefront position (i.e. the position of the forefront at $t + \Delta t$ in Figure 4.24), as depicted in step [4]. Then, the particles in each layer with length $U \Delta t$ are mixed with those in the rear part with length ($\Delta x - U \Delta t$). The last calculating operations are carried out throughout the newly divided grids upstream. After that, the whole flow is conceptually moved upstream by $U \Delta t$ so as to coincide with the forefront at time step t if the continuity of the one-dimensional calculation is maintained. During this time the mean solids concentration and the mean diameter throughout the entire depth in each Δx are calculated and stored for subsequent calculations at the next time step. Therefore, the one-dimensional routing of a debris flow with particle segregation is accomplished in Δt, and at the beginning of the next Δt newly calculated particle compositions are used in the computations. Repeating this procedure, one can obtain the size distribution as well as the other flow quantities.

Two series of experiments were carried out. Series A was done to determine the process of particle segregation and series B to compare the predicted discharge hydrograph and other quantities with the experimental results.

In series A an experimental flume 10 cm wide was used. It was equipped with a sediment sampler at the downstream end, which could collect samples by dividing the entire flow depth into four layers. The experiments were conducted for various lengths of erodible bed layer, $L = 1$, 2, 3 and 4 m, maintaining a constant slope of 18°, bed-layer thickness of 10 cm, and water supply discharge of 2,000 cm³/s. The bed material was a mixture of five materials of nearly uniform sizes, and the characteristic values of the mixture were $d_m = 1.66$ mm, $C_* = 0.655$, and $(d_{84}/d_{16})^{1/2} = 3.87$. Prior to an experimental run, the sediment bed was saturated with water. Subsequently, a constant discharge of clear water was introduced suddenly from the upstream end to produce a debris flow. Sediment samples were collected when the forefront, middle, and rear parts reached the position of the sampler. Ideally, the samples were to be collected at longitudinally different positions on a long erodible bed, but such a sampling was difficult and even if it were possible it would affect the later translation of debris flow downstream. It was assumed that the samples collected at the downstream end for different lengths of sediment bed were equivalent to those collected at different longitudinal positions on a long sediment bed.

In series B the same flume was used as in series A, but with the sediment sampler removed. In this series a sediment bed 3.0 m long and 10 cm deep was positioned 5.5 m upstream from the outlet of the channel by installing a partition of 10 cm in height to retain the sediment. Highly heterogeneous material with $d_m = 3.08$ mm, $C_* = 0.65$, and $(d_{84}/d_{16})^{1/2} = 6.12$ was used. The same material was glued on the surface of the rigid bed downstream of the sediment bed. The slope of the channel was 18°. Debris flows were produced as in series A using a steady water supply, with a discharge of 600 cm³/s for 20 seconds. Because the sediment composition and the degree of saturation might not be uniform throughout the sediment layer, experimental runs were repeated six times. In each run, bulk samples for obtaining the temporal variations in discharge, the solids concentration, and the particle-size distribution were collected at the outlet.

Measured solids concentrations and vertical mean particle-size distributions at the forefronts in series A experiments are shown in Figure 4.25. The upper four graphs show the concentration distributions of each size fraction and the lower four the mean particle sizes at various heights. The four paired graphs of each upper and lower graph in Figure 4.25, from left to right, correspond to the processes of changes in the solids concentration and mean size distribution in the front part of a debris flow as the forefront advances 1, 2, 3 and 4 m, respectively. The solids concentration as well as the content of the large fraction at the front increases as the debris flow continues to move downstream. At the beginning (in a short travel distance) the mean particle size in the upper layer of the flow is larger than in the lower layer. Despite being in the front, after moving some distance, it becomes almost uniform as well as larger. This proves that the convergence of large particles towards the front is attributable to size segregation within the flow depth; i.e. inverse grading.

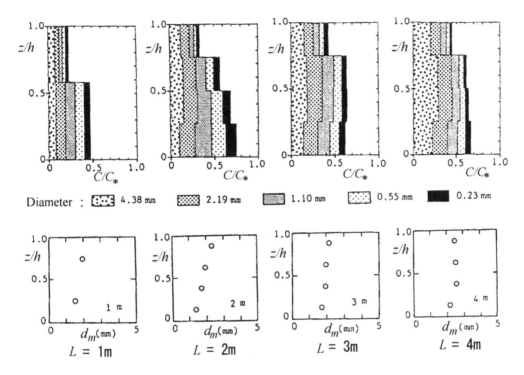

Figure 4.25 Measured solids concentration distribution and vertical mean particle-size distributions at forefront as debris flow moves downstream.

The developing process of a debris flow on a 4 m long sediment layer corresponding to series A experiments was calculated by the method introduced earlier. In the calculation, $\rho = 1.0\,\text{g/cm}^3$, $\sigma = 2.65\,\text{g/cm}^3$, $\tan\phi = 0.75$, $s_b = 1.0$, and $\delta_e = 0.0007$ were used. The appropriate δ_e value was determined after several trial calculations. The calculated results are compared with the experimental results as shown in Figure 4.26. The left three figures correspond to the front part, the middle three to the middle part, and the right three to the rear part of the flow. The major difference between the experimental and calculated solids-concentration distributions is apparently caused by ignoring the solids-concentration distribution across the flow depth in the calculation; a uniform distribution is assumed despite the gradual decrease in the concentration towards the upper layer in the experiment. Nevertheless, the trend in the relative composition of each size fraction is fairly well reflected in the calculation; the predominance of large particles and nearly uniform size distribution throughout the depth in the front part of the flow in contrast to the predominance of small particles having inversely grading sediment size distributions across the depth in the rear part of the flow.

The comparison of the calculated debris flow discharge hydrograph at the outlet with the experimentally obtained discharge hydrograph in series B experiments is shown in Figure 4.27. Measured points are very much scattered, perhaps resulting from fluctuations in the degree of saturation, degree of compaction, and non-uniformity of the sediment composition in the sediment bed. In general, however, the calculated

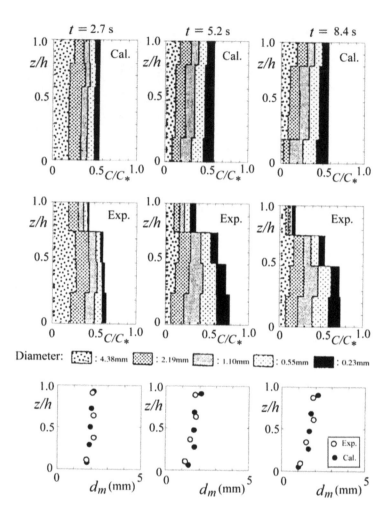

Figure 4.26 Comparison of calculated and measured changes in solids concentration and mean particle-size distributions with time at the downstream end of flume.

results show a trend similar to the measured ones. The input values in the calculations were the same as for the case of series A, in addition to setting $\Delta x = 10$ cm, and $\Delta t = 0.002$ s. A transition in the reach from the erodible sediment bed to the rigid bed was considered to be gradual even though there was actually a sudden drop of 10 cm. Moreover, the particle segregation process was calculated only when the thickness of the flow exceeded five times the value of the mean diameter. The broken line in Figure 4.27 is calculated on the assumption of uniform-size sediment, in which the particle diameter is the same as the mean diameter of the heterogeneous material ($= 3.08$ mm) and the values of other variables are set equal to those for the heterogeneous bed. The peak discharge of the debris flow with non-uniform size sediment is larger than that for uniform size sediment because the convergence of large particles in the front part can increase the maximum discharge of flow that happens immediately after the

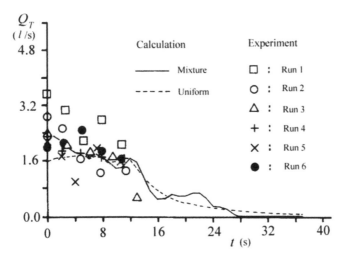

Figure 4.27 Calculated and measured hydrographs at the downstream end of flume.

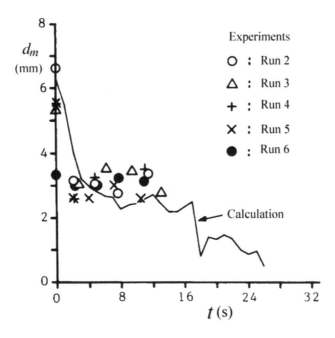

Figure 4.28 Temporal changes of mean particle sizes in debris flow at the downstream end of flume.

arrival of its forefront. The discharge hydrographs plotted in Figure 4.27 also show that the peak discharge of debris flow is far larger than the supplied water discharge at the upstream end.

Temporal change in the mean diameter of the run-out sediment is shown in Figure 4.28, from which the following characteristics are evident: The large particles gather at

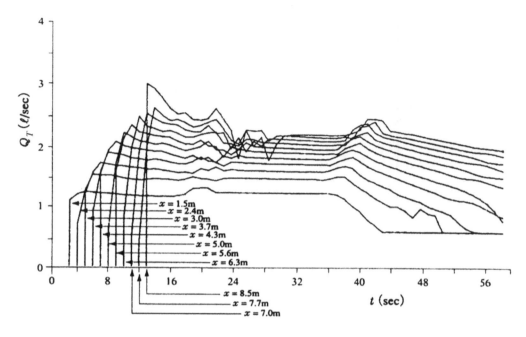

Figure 4.29 Calculated hydrographs at various points along the channel.

the front part of the flow. After the passage of the front, a rather rapid size reduction with time occurs to the extent that the size becomes a little smaller than the mean particle size in the original bed. Soon, the size returns to a little larger one that is approximately the same as the mean diameter in the original bed (in the calculation this characteristic is not evident as in the experiment) and this state continues for a while. The experiments ceased at this stage, but in the calculation particles become smaller again corresponding to the washout of the bed material.

For the sake of a deeper understanding of debris flow processes with particle segregation, a numerical experiment of debris flow formation and flowing down was performed, in which water was supplied from the upstream end of flume at the rate of $600 \, \mathrm{cm^3/s}$ for 60 seconds on a sediment bed of 10 cm thick and 8.5 m long; the channel condition and sediment material were the same as the series B experiments except for the length of bed and the duration of the water supply (Takahashi *et al.* 1991). Figure 4.29 shows the calculated results of hydrographs at various points along the channel. The debris flow develops downstream to the extent that the discharge becomes much larger than the supplied water discharge, but with the exposure of the rigid bed in order from upstream ($t = 43$, 51 and 55 seconds at 1.5 m, 2.4 m and 3.0 m, respectively) the discharge becomes equal to the supplied water discharge. The peak discharge at a point along the channel arises at the instant the debris flow front passes through that point and the peak discharge increases downstream. The discharge observed at a point begins to recess just after the peak and the recessing rate becomes larger if the observation point is located further downstream. The discharge recession ends by approaching to a steady flow stage at each point. The flow rates at the steady flow stages are larger

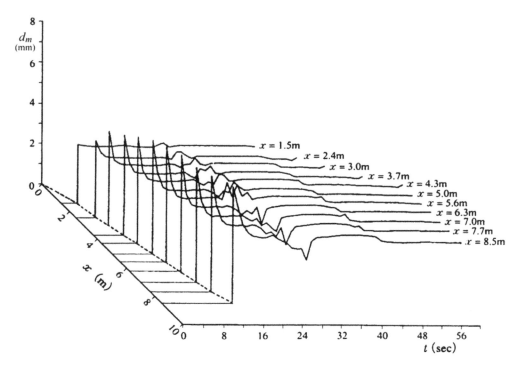

Figure 4.30 Temporal changes of the mean diameters along the channel.

downstream (the cause of irregular variation at the end of recession is not clear but is perhaps induced by a defect in the computation technique). The steady flow stage continues for a while and then the flow rate begins to decrease again. The duration of the first recessing phases almost correspond to the time necessary for the debris flow to travel the whole channel length of 8.5 m. Although it is still in the developing process, the translation velocity of the debris flow front is almost constant; it was mentioned in section 3.1.2 that such a characteristic had been confirmed by experiments. According to Equation (2.149), the increase in solids concentration and/or in particle diameter slows the velocity, but, the increase in flow depth accelerates the velocity. Therefore, it may be concluded that in the developing process debris flow propagates with nearly constant velocity under the balance of effects between deceleration and acceleration.

Figure 4.30 shows the temporal changes in the mean particle diameters in the flow at various points along the channel. At the 1.5 m point, from the beginning to the end of debris flow, the mean particle diameter is still almost the same as that of the bed material, but it becomes larger in the front part moving downstream. The mean particle diameter in the front at the 5.6 m point becomes as large as about 6 mm (85% diameter of the bed material), but downstream of that point the tendency of particle size increase becomes less evident. Soon after the passage of front, the particle diameter begins to decrease and approaches to a minimum corresponding to the time of the end of the first recession in the discharge depicted in Figure 4.29. The extent of the mean diameter change from the maximum at the front to the minimum at an end of discharge decrement becomes larger and larger with the distance of travel. The minimum of the mean diameter in the flow becomes much smaller than that in the original bed at the observation point located far downstream; at the 8.5 m point it is as small as 2.1 mm whereas the mean diameter in the original bed is 3.08 mm. This phenomenon is associated with the particle segregation that induces the large particle's convergence to the forefront, and in compensation, small particle's accumulation in the rear part. As mentioned earlier, at $x = 8.5$ m the debris flow is still in the developing stage, but as is understood from Figure 3.6, the erosion rate in the neighborhood of the front is so small that the supply of bed material is minimal and it results in the near completion of particle segregation in this range. Thus, a surge which has a coarse head and a fluid tail is formed as long as the front is within the channel. Therefore, the length of a surge would be at most equal to the length of the channel; 8.5 m in this case, and the duration of a surge at a point would be nearly equal to the time necessary for it to pass through the channel reach. After the passage of a surge, the flow becomes more or less steady (because a steady water flow is continually supplied), and even though the particle segregation takes place vertically in the flow, the composition of the flow in the entire depth becomes almost the same as that of the original bed material due to rather high and steady erosion rate of the bed.

4.6.2 Dynamic wave method

The case of turbulent-type debris flow

The kinematic wave method can be applied to the turbulent-type debris flow as well, but here, considering the case where the time variation of the hydrograph is rapid, the application of the dynamic wave method is explained, where the acceleration terms are taken into account.

The governing equations; the momentum conservation equation and the mass conservation equation, are already appeared in section 3.3.4 for the case of a two-dimensional flow. The one-dimensional versions of these equations are as follows:

$$\frac{\partial M}{\partial t} + \beta' \frac{\partial (UM)}{\partial x} = gh \sin \theta_{bx0} - gh \cos \theta_{bx0} \frac{\partial H_s}{\partial x} - \frac{\tau_{bx}}{\rho_T} \tag{4.43}$$

$$\frac{\partial h}{\partial t} + \frac{\partial M}{\partial x} = i \tag{4.44}$$

where $H_s = z_b + h$, and other notations are the same as those that appeared in Equation (3.115), etc.

The third term in the right-hand side of Equation (4.43) represents the friction resistance, and if the coefficient of resistance is written as f, it is given by:

$$\frac{\tau_{bx}}{\rho_T} = \frac{f}{8} U|U| \tag{4.45}$$

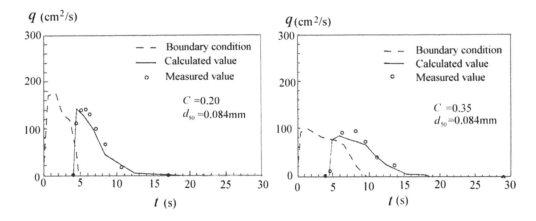

Figure 4.31 Deformations of turbulent type debris flow along the channel.

and if the resistance law is given by Equation (2.148), f is given by:

$$f = \frac{8}{\kappa} \left(\ln \frac{1 + \sqrt{1 + \Phi_1^2}}{Z_0 + \sqrt{Z_0^2 + \Phi_1^2}} - \sqrt{1 + \Phi_1^2} + \Phi_1 \right)^{-2} \tag{4.46}$$

Arai (1991) conducted the experiments using a smooth bed flume 8 m long, 15 cm wide and 18° in gradient. Quartz sand whose 50% diameter was 0.088 mm, was mixed with water and installed in an inverted cone-shaped container hung over the upstream end of flume. That material was supplied into the flume by opening the valve at the bottom of the container. The debris flow discharge hydrograph that run out from the flume was calibrated by a measuring box. No deposition took place in the experiments.

Equations (4.43) and (4.44) were solved to obtain the runoff discharge hydrograph from the flume by giving the input hydrograph as the boundary condition at the upstream end, and compared with the experimentally obtained ones in Figure 4.31. One can understand that the calculation and experiment agree well.

The case of viscous-type debris flow

A typical viscous-type debris flow has the following conspicuous characteristics: as the velocity becomes small towards the end of a surge, deposition proceeds from the lower to upper parts and finally flow stops, thereafter the next surge comes down, it entrains the still soft deposit formed by the last surge, but again toward the end of this surge deposition takes place and finally stops. An event of viscous debris flow repeats these processes tens or hundreds times. Therefore, to trace the behaviors of viscous debris flow in a transferring reach, the mechanisms of deposition and stoppage and of entrainment of the newly deposited layer should be built in the system of governing equations. Herein, such an analyzing method is explained (Takahashi et al. 2000).

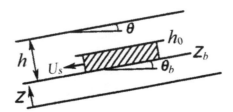

Figure 4.32 Depositing process model for a viscous debris flow.

Although viscous debris flow can transport highly concentrated sediment, the flow that is fully developed on a very steep channel upstream must be overloaded when it comes down to a flatter reach, where the shearing velocity becomes too small to disperse coarse particles and the coarse particles together with the interstitial fluid will stop. The volume change of material before and after the stoppage is negligibly small. Figure 4.32 illustrates such a depositional process in which the hatched part will deposit in the next unit of time.

The equation of motion of this hatched volume is given by:

$$m_s \frac{dU_s}{dt} = F + \frac{dm_s}{dt} u \tag{4.47}$$

where m_s is the mass of the hatched volume, U_s is the velocity of the hatched volume, F is the external force operating on the hatched volume, and u is the relative velocity of the depositing part that is decoupled from the hatched volume observed from the hatched volume.

Writing the equilibrium coarse particle concentration for the depth h in the reach under consideration as C_e, and assuming a uniform solids concentration throughout the entire depth, the excess pressure that is transmitted through the coarse particles and operates on the bottom of the hatched volume is given by $(\sigma - \rho_T)(C - C_e)gh \cos \theta$. This pressure causes the Coulomb friction force. The fluid dynamic force, $(1/2)\rho_T f U_s^2$, also operates on the bottom of the volume, but this is almost balanced by the driving force operating on the same plane, $\rho_T gh \sin \theta$. Therefore, the external force that contributes to decelerate the hatched volume is written as:

$$F = -(\sigma - \rho_T)(C - C_e)gh \cos \theta \tan \varphi \, \Delta x \tag{4.48}$$

where Δx is the length of the hatched volume.

The mass variation of the hatched volume is given by:

$$\frac{dm_s}{dt} = -\rho_T \frac{dz_b}{dt} \Delta x \tag{4.49}$$

If the depositing part that is decoupled from the hatched part is assumed to stop immediately, the relative velocity u is given as:

$$u = -U_s \tag{4.50}$$

Substituting Equations (4.48), (4.49) and (4.50) into (4.47), gives:

$$\frac{dU_s}{dt} = \frac{1}{h_0 - z_b}\left\{-\frac{(\sigma - \rho_T)(C - C_e)gh\cos\theta\tan\varphi}{\rho_T} + \frac{dz_b}{dt}U_s\right\} \tag{4.51}$$

Let us assume that the velocity of the hatched volume becomes zero when z_b coincides with h_0 and during that process the velocity decreases linearly. Then, the depositing velocity $i(=dz_b/dt)$ is, writing the time necessary for this process as T, $i = (h_0 - z_b)/T$. Because $T = -iU_0/(h_0 - z_b)$, the following formula is obtained:

$$\frac{dU_s}{dt} = -\frac{U_s}{h_0 - z_b}i \tag{4.52}$$

The relationship between the mean velocity of the debris flow, U, and U_s depends on the velocity distribution, but, here, the following simple relationship is assumed:

$$U_s = \alpha_p U \quad (\alpha_p < 1) \tag{4.53}$$

Substituting Equations (4.52) and (4.53) into Equation (4.51) gives:

$$i = \frac{gh\sin\theta(\sigma - \rho_T)(C - C_e)\tan\varphi}{2\alpha_p U_s\rho_T\tan\theta} \tag{4.54}$$

Consequently, substituting Equation (2.187) into Equation (4.54), gives:

$$i = \frac{3\mu_a(\sigma - \rho_T)(C - C_e)\tan\varphi}{2\alpha_p h\rho_T^2\tan\theta} \tag{4.55}$$

The deposit thus formed by the stoppage of a surge will again be activated and mixed up with the following surge, and the deposition and erosion processes are repeated. In the flume experiments described in section 4.1.2, if the channel slope is shallow, the entire depth of the deposit is not necessarily reactivated but a certain thickness is left inactive. The activated thickness seems to change with the time elapsed after the deposition, but at the moment it is difficult to determine. If the channel slope is steep the entire deposit is reactivated, and such a thorough removal of the deposit by an arriving surge is confirmed in the Jiangjia gully. Moreover, in the Jiangjia gully, sometimes, even no surge arrives, minimal disturbance leads to a swelling of the deposit layer and it develops to a bore. Thus, the deposit is very unstable, at least during the time of repetition of the surges, and so, in the following discussion, the deposit is assumed to be immediately eroded to the depth of original bed on the arrival of next surge.

One-dimensional routing of the viscous debris flow surges is fulfilled by solving the following equations of continuity and momentum conservation:

$$\frac{\partial h}{\partial t} + \frac{\partial(Uh)}{\partial x} = -i \tag{4.56}$$

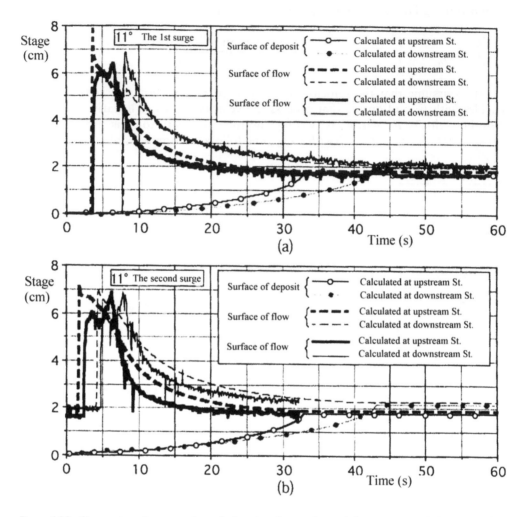

Figure 4.33 Comparison between the calculated and experimental flow stages and deposit surface stages for viscous debris flow in the case of 11° channel slope. (a): the first surge on the rigid bed; (b): the second surge that traveled on the deposit made by the first surge.

$$\frac{\partial(Uh)}{\partial t} + 1.2\frac{\partial(Uh)}{\partial x} = gh\sin\theta_b - gh\cos\theta_b\frac{\partial h}{\partial x}$$

$$- \frac{9\mu_a^2 U^2}{\rho_T^2 gh^3\{\sin\theta_b - \cos\theta_b\partial h/\partial x\}} \tag{4.57}$$

where θ_b is the original bed slope, the coefficient 1.2 of the second term on the left-hand side of Equation (4.57) is the momentum correction coefficient for the parabolic velocity distribution form applicable to Newtonian laminar flow, and i is given by Equation (4.55).

Among the experimental data explained in section 4.1.2, the one whose bed slope was 11° are compared with the calculation using Equations (4.56) and (4.57) in Figure 4.33. In the calculation the following values are used: $C_e = 0.41$, $\alpha_p = 0.9$, $\tan \phi = 0.7$, $\rho_T = 1.92\,\mathrm{g/cm^3}$, $\sigma = 2.65\,\mathrm{g/cm^3}$, $\rho = 1.38\,\mathrm{g/cm^3}$ and $\mu_a = 6.2\,\mathrm{Pas}$.

Figure 4.33(a) shows the results for the first surge, which traveled on a rigid bed (no deposit layer exists before arrival of the surge). The translation velocities of the front, both in the calculation and in the experiment, are the same 73.7 cm/s and it takes 4 seconds to travel between the two stations, a distance of 3 m. Apart from the calculated front stage being a little higher than the experimental result, the general tendency of the stage variation is well reflected by the calculation. With time, both calculated and experimental stages approach constant heights at respective stations. This means that the flow has already stopped. The calculated time of stopping and the thickness of deposit coincide well with the experiment.

Figure 4.33(b) shows the propagation of the second surge that traveled on the deposit made by the first surge. The hydrographs as the boundary conditions of the first and second surges in the calculation are identical. There are some differences in the front arrival time between the calculation and the experiment, but it is probably due to difference in setting of the starting time, because the time lags between the two stations are almost the same, about 2.7 seconds. Stage variation in the experiment has dual peaks, whereas in the calculation it has only one peak. This probably occurred due to the difficulty in controlling the supply discharge in the experiment. Except for these discrepancies, the calculation generally explains well the experimental results. The time required for the first surge to travel the 3 m reach between the observing stations is 4 seconds and that for the second surge is only 2.7 seconds. The translation velocity of the first surge is almost equal to the mean cross-sectional velocity at the peak. Therefore, the existence of the deposit contributes, in this case, to making the front velocity faster than the mean velocity of the following part. The cause of this phenomenon has been explained in section 4.1.2.

Note that the final stages of the deposit surface after the first surge and after the second surge are almost the same. This is one of the peculiar behaviors in the Jiangjia gully as well.

Chapter 5

Processes and geomorphology of deposition

The photograph shows the debris flow cone (fan) formed by debris flows produced in the Kamihorizawa and Kamikamihorizawa ravines draining the mountain slope of the Yakedake Volcano (the mountain on the left-side of the picture). Taisho Pond, which is seen at the lower part of the photo, has been narrowed and made shallow by debris flows, and it is now appropriate to call it a river rather than a pond. The Kamikamihorizawa ravine where debris flow observations have long been carried out is the one that originates from the saddle part on the ridge seen in the center and flows to the right-hand side of the photo. The fan having the flattest slope is on the extreme right of the photo. It is the one formed by Kamikamihorizawa and it is still growing.

INTRODUCTION

The debris flow mechanisms enable solids to be carried in dense volumes due to particle dispersion caused by the inter-particle collision, turbulent mixing or laminar dispersion. Because the magnitude of these particle dispersion effects depends on the shearing velocity (vertical velocity gradient), for the flow to continue maintaining its high solids concentration requires sufficient shearing velocity. The shearing velocity is, irrespective to the debris flow type, proportional to the channel gradient and it becomes small with the flattening of the slope. Therefore, if the debris flow comes out to a gradually flattening area, it tends to continue its motion diluting the solids concentration by depositing the excess solids over the equilibrium concentration that is a function of the topographical conditions of the coming out place. However, as often the cases, the slope gradient suddenly flattens from the mountain ravine to the debris flow/alluvial fan, the coming out debris flow cannot continue moving and it stops *en masse*. Even in such a place, if the cross-sectional area of the channel is large enough, the deposition is within the channel, but often the capacity of the channel is so small that over-spilling takes place and alluvial (debris) cone develops. It is this process that causes debris flow disasters.

In this chapter, first, the simple method to estimate the debris flow arrival distance and the shape of the deposition is introduced, where the channel suddenly flattens. Then, the numerical simulation method of the debris flow fan formation processes is explained.

5.1   ONE-DIMENSIONAL STOPPAGE/DEPOSITING PROCESSES OF STONY DEBRIS FLOW

5.1.1   The arrival distance at the sudden change in channel slope

When a stony debris flow enters a place where the slope of the channel suddenly levels out, it decelerates, thickens and finally stops. The profile of such a debris flow at time t and $(t + \Delta t)$ may be modeled by the trapezoidal shape in Figure 5.1 (Takahashi and Yoshida 1979). It is assumed that all the parts of the flow proceed at the same velocity u and have the same concentration equal to C_u, which is the concentration in the upstream channel. It is also assumed that the flow continues to be inertial until just before it stops. Then, the momentum conservation between Section I (at the forefront) and Section II (at the change in slope) is written as:

$$\frac{d}{dt}\left\{\frac{1}{2}(h + h_{fr})x\rho_T u B_{do}\right\} = \frac{1}{2}(h + h_{fr})x\rho_T g \sin\theta_d B_{do} + \rho_T q_T u u_u \cos(\theta_u - \theta_d)B_{up}$$
$$+ \frac{1}{2}gh_u^2 \cos\theta_u \cos(\theta_u - \theta_d)\{(\sigma - \rho_m)C_u\kappa_a + \rho\}B_{up} - F_b$$

$$(5.1)$$

where the left-hand side expresses the temporal momentum change, the first term of the right-hand side is the driving force due to gravity, the second is the input of momentum from Section II, the third is the sum of the hydrostatic and earth pressures operating on Section II, and the fourth is the friction at the bottom, κ_a is the coefficient similar to the active earth pressure coefficient, B_{up} and B_{do} are the upstream and downstream channel widths, respectively, and the subscript u and d indicate the value upstream and downstream from the change in slope, respectively.

The dilatant fluid model for stony debris flow considers, as stated in sections 2.2 and 2.3, the almost all of the shearing stresses at an arbitrary height in the flow is assigned by inter-particle collision stress and it is described by Equation (2.19) under the condition of Equation (2.18). Therefore, considering Equation (2.20), the shearing stress at the bed surface, τ_0, for a uniform flow with depth h, particle concentration C, bed slope θ and the density of the interstitial fluid ρ_m is described as:

$$\tau_0 \approx (\sigma - \rho_m)Cgh\cos\theta\tan\alpha_i \qquad (5.2)$$

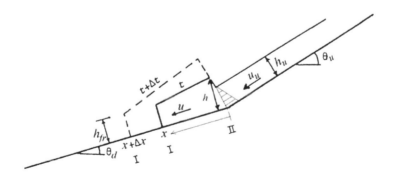

Figure 5.1   Process of the stoppage of a debris flow front.

This is a similar expression to the Coulomb friction stress if $\tan \varphi$ is substituted for $\tan \alpha_i$. But, it must be noted that, under the expression of Equation (5.2), the total particle load affected by the buoyancy of the interstitial fluid is sustained by the repulsive force produced on the collision of particles disengaged from each other, whereas the Coulomb stress arises only when particles are under quasi-static motion where they are always in contact. The two concepts are completely different.

Thus, F_b in Equation (5.1) is written as:

$$F_b = \frac{1}{2}(\sigma - \rho_m)gC_u(h_{fr} + h)x \cos \theta_d \tan \alpha_i B_{do} \tag{5.3}$$

The equation of continuity is written by neglecting small terms as follows:

$$\frac{1}{2}(h + h_{fr})xB_d = q_T t B_u \tag{5.4}$$

Substituting Equations (5.3) and (5.4) into Equation (5.1), gives:

$$\frac{du}{dt} + \frac{1}{t}u = \frac{V_e}{t} - G \tag{5.5}$$

where:

$$V_e = u_u \cos(\theta_u - \theta_d)\left[1 + \frac{\{(\sigma - \rho_m)C_u\kappa_a + \rho_m\}\cos \theta_u}{2\{(\sigma - \rho_m)C_u + \rho_m\}}\frac{gh_u}{u_u^2}\right] \tag{5.6}$$

$$G = \frac{(\sigma - \rho_m)gC_u \cos \theta_d \tan \alpha_i}{(\sigma - \rho_m)C_u + \rho_m} - g \sin \theta_d \tag{5.7}$$

The solution of Equation (5.5) under the initial condition:

$$t = 0; \quad u = V_e \tag{5.8}$$

is:

$$u = -\frac{1}{2}Gt + V_e \tag{5.9}$$

Then, the travel distance x is given as follows:

$$x = -\frac{1}{4}Gt^2 + V_e t \tag{5.10}$$

Therefore, the distance x_L within which the debris flow stops is given by:

$$x_L = \frac{V_e^2}{G} \tag{5.11}$$

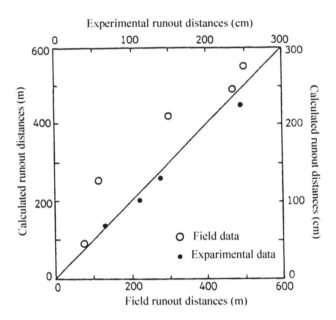

Figure 5.2 Comparison of run out distances calculated and observed.

Equation (5.11) means that if $G \leq 0$, the debris flow does not stop. This condition is given from Equation (5.7) as:

$$\tan \theta_d \geq \frac{(\sigma - \rho_m)C_u}{(\sigma - \rho_m)C_u + \rho_m} \tan \alpha_i \tag{5.12}$$

Equation (5.12) means that the smaller the particle concentration and the larger the density of the interstitial fluid, the flatter the area on which a debris flow can continue its motion.

Figure 5.2 compares the distances calculated by Equation (5.11) and the actual ones observed in the experiments (Takahashi 1980) as well as in the fields (Hungr *et al.* 1984).

5.1.2 Topography of deposit formed at a sudden slope change

When debris flow stops the particle concentration must change from C_u to C_*, so that the excess pore water will be squeezed out of the deposit and flow on its surface. If the debris flow is a viscous one, the volume change before and after deposition is minimal, therefore the surface water flow produced by volume change is negligibly small. But, if it is a stony debris flow the surface water flow thus produced may be able to flatten the surface slope of the deposit formed just before by a fully-developed debris flow until it becomes approximately equal to the critical slope for the occurrence of an immature debris flow. The reason for this surface-slope formation is considered as follows: Although the bed load transportation may be able to flatten the once deposited surface-slope, it is too gradual to be finished within the short time of the debris flow depositing process, and therefore, the virtual flattening process ends when the slope becomes approximately equal to the critical slope for an immature debris flow on which slope flattening occurs drastically. Contrary, if the surface slope of the deposit is flatter than that critical slope, deposition continues on the pile of newly deposited sediment as long as the debris flow continues from upstream. In fact, the stoppage and the remobilization of the deposit surface would occur simultaneously, and the deposition would proceed, from the beginning, so as to form the surface slope approximately equal to the critical slope for the immature debris flow.

Hence, the surface slope of deposit γ_d is given by substituting d_m into a_c in Equation (3.9) as Equation (5.13), in which the surface flow depth h_o can be described as $(f/8 \sin \gamma_d)^{1/3}(q_o^2/g)^{1/3}$, where f is the resistance coefficient and q_o is the unit width discharge of surface water flow:

$$\tan \gamma_d = \frac{C_*(\sigma - \rho_m) \tan \varphi}{C_*(\sigma - \rho_m) + \rho_m \left\{ 1 + \left(\dfrac{f}{8 \sin \gamma_d} \right)^{1/3} \left(\dfrac{q_o^2}{g d_m^3} \right)^{1/3} \right\}} \tag{5.13}$$

This equation contains γ_d in terms on both the left and right-hand side and f is unknown. Therefore, γ_d cannot directly be obtained. As stated in section 3.1.1 the experiments using a nearly uniform material of $d_m = 5$ mm revealed that the relationship between the surface water flow depth and the water discharge per unit width can be described by Equation (3.13). This means that Equation (5.13) can be approximately described as follows:

$$\tan \gamma_d = \frac{C_*(\sigma - \rho_m) \tan \varphi}{C_*(\sigma - \rho_m) + \rho_m \left\{ 1 + 0.52 \left(\dfrac{q_o^2}{g d_m^3} \right)^{1/3} \right\}} \tag{5.14}$$

Figure 5.3 is the comparison of the semi-theoretical surface slopes of deposit with the experimental ones, in which the experimental flume has the same widths in the upstream flowing reach and in the downstream depositing reach and the experimental material has an approximately uniform size of $d_m = 5$ mm. The experimentally obtained surface slopes for the debris flow in which the interstitial fluid is plain water agree well with Equation (5.14). Although no data for smaller specific particles (denser interstitial fluid cases) are available, the tendency of solid lines in Figure 5.3 shows that the larger the surface water flow discharge (the larger the discharge of debris flow) and the denser the interstitial fluid then the flatter the surface slope of the deposit becomes.

Describing the flattest slope that satisfies Equation (5.12) as θ_c, the depositing and flowing processes downstream from the change in slope are divided into the following three cases:

1   $\theta_d < \gamma_d$: The debris flow deposits with a surface slope γ_d. A very fluid and short duration debris flow can deposit even with a surface slope less than γ_d.
2   $\gamma_d < \theta_d < \theta_c$: Some parts of a rundown debris flow deposit and other parts continue to flow. The slope of the deposit is between θ_c and γ_d.
3   $\theta_d > \theta_c$: The debris flow continues to rundown.

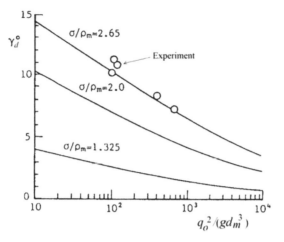

Figure 5.3 Slopes of deposit surface.

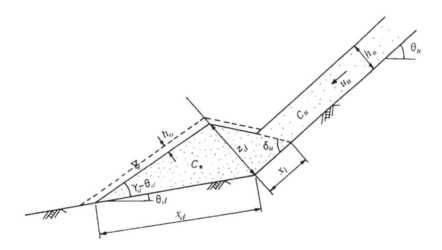

Figure 5.4 Early stage of depositing process at the channel slope change.

The process of deposition when $\theta_d < \gamma_d$

If the arrival distance x_L of a debris flow front is small, the short decelerating and stopping process of the front as described in section 5.1.1 may be neglected in comparison with the subsequent comparatively long depositional process. Therefore, here we consider the depositing process to begin as soon as a debris flow appears at the change in slope. As the debris flow continues to run down from upstream, it flows over the already halted portion, and the deposit grows in both length and height. An early stage of this process is modeled as in Figure 5.4, in which the upstream confronting slope δ_u is unknown but is assumed to be a constant.

Referring to Figure 5.4, the continuity equation of solids phase is given by:

$$\frac{1}{2}x_d^2 A = q_t \frac{C_u}{C_*} t \tag{5.15}$$

$$A = \frac{\tan(\gamma_d - \theta_d)}{\tan(\gamma_d - \theta_d)\tan(\theta_u - \theta_d) + 1}$$
$$+ \frac{1}{\tan \delta_u \cos^2(\theta_u - \theta_d)} \left\{ \frac{\tan(\gamma_d - \theta_d)}{\tan(\gamma_d - \theta_d)\tan(\theta_u - \theta_d) + 1} \right\}^2 \tag{5.16}$$

From Equation (5.15), the variables that determine the shape of deposit are given as follows:

$$x_d = \sqrt{\frac{2q_t C_u}{C_* A} t} \tag{5.17}$$

$$z_d = \frac{1}{\cos(\theta_u - \theta_d)} \frac{\tan(\gamma_d - \theta_d)}{\tan(\gamma_d - \theta_d)\tan(\theta_u - \theta_d) + 1} x_d \tag{5.18}$$

$$x_l = \frac{z_d}{\tan \delta_u} \tag{5.19}$$

In this model, as long as the debris flow from upstream continues downward, the height of the deposit at the change in slope monotonously increases. However, the energy the debris flow has is finite, and so after it exceeds a certain height, the flow as a whole can no longer override it and some of the debris begins to be deposited upstream. Thus, in time the deposition moves upstream. This stage of the depositing process may be modeled as shown in Figure 5.5. Some experiments have shown that

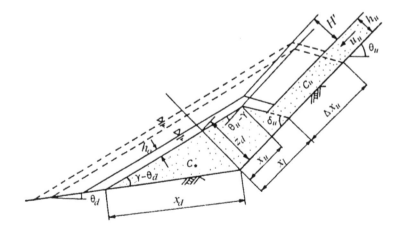

Figure 5.5 Rebounding deposition at the channel slope change.

there is negligible energy loss in the process of overriding. Therefore, the height of the rebound H' can be given by the following equation:

$$H' = \alpha' \frac{u_u^2}{2g} \cos\theta_u - \frac{u_o^2}{2g} \cos(\theta_u - \gamma_d) \tag{5.20}$$

where α' is the energy coefficient equal to 1.7 for a dilatant stony debris flow, u_o is the velocity of the surface water flow over the deposit.

Referring to Figure 5.5, the equation of continuity is obtained by neglecting the small terms as follows:

$$\frac{dx_d}{dt} = \frac{C'}{A'x_u + H'B'} \tag{5.21}$$

where:

$$A' = \frac{\cos(\theta_u - \theta_d)}{\sin(\gamma_d - \theta_u)} \tan(\theta_u - \gamma_d) + \frac{1}{\cos(\theta_u - \gamma_d)} \tag{5.22}$$

$$B' = \frac{\cos(\theta_u - \theta_d)}{\sin(\gamma_d - \theta_d)} + \frac{1}{\sin(\theta_u - \gamma_d)} \tag{5.23}$$

$$C' = q_t \frac{C_u}{C_*} \frac{1}{\sin(\theta_u - \gamma_d)} \tag{5.24}$$

Equation (5.21) indicates that the rebounding velocity gradually decreases with the progress of deposition. Integration of Equation (5.21) under the initial condition; $t = t_0$, $x_u = 0$ gives:

$$x_u^2 + H' \frac{2B'}{A'} x_u - \frac{2C'}{A'}(t - t_0) = 0 \tag{5.25}$$

where t_0 is the time of the initiation of the rebounding deposition and it is the time when z_d becomes $(h_u + H' - h_o)$. The location of the distal front of deposit is given by:

$$x_d = \{H' + x_u \tan(\theta_u - \gamma_d)\} \frac{\cos(\theta_u - \gamma_d)}{\sin(\gamma_d - \theta_d)} \tag{5.26}$$

Figure 5.6 compares the above mentioned deposition model to the experimental results, in which the experimental and calculating conditions are the same; $d_m = 5$ mm, $\sigma/\rho_m = 2.65$, $C_* = 0.7$, $\varphi = 36°$, $\theta_u = 17°$, $\gamma_d = 10.8°$, $q_t = 418\ \text{cm}^2/\text{s}$, $u_u = 110\ \text{cm/s}$, $h_u = 3.8$ cm, $H' = 10$ cm, $\delta_u = 25°$ and θ_d's are three kinds of 7°, 5°, and 3°. In the early stage of deposition the difference between the calculations and experiments is considerably large, possibly due to the effect of neglecting the stopping process in the calculating model, but the rebounding depositions are satisfactory calculated.

The process of deposition when $\gamma_d < \theta_d < \theta_c$

In this slope condition, the decelerated flow will continue to run down, depositing some parts of the solid fraction. If such a partial deposition occurs, squeezed water

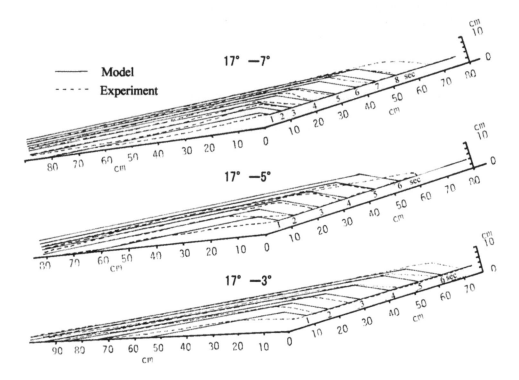

Figure 5.6 Temporal changes of depositional profiles in the model and in the experiments.

from the deposit will dilute the still flowing part to increase its fluidity. The continuity equations for water and solid fraction in the debris flow are, respectively:

$$(1 - C)\frac{\partial h}{\partial t} + (1 - C_*)\frac{\partial z_b}{\partial t} + \frac{\partial q_{wt}}{\partial x} = 0 \tag{5.27}$$

$$C\frac{\partial h}{\partial t} + C_*\frac{\partial z_b}{\partial t} + \frac{\partial q_s}{\partial x} = 0 \tag{5.28}$$

where z_b is the thickness of deposit, q_{wt} and q_s are the water and sediment discharges per unit width in debris flow, respectively, and they satisfy the following relationship:

$$q_s = \frac{C}{1 - C}q_{wt} \tag{5.29}$$

From Equations (5.27), (5.28) and (5.29), one obtains:

$$(C_* - C)\frac{\partial z_b}{\partial t} + \frac{q_{wt}}{1 - C}\frac{\partial C}{\partial x} = 0 \tag{5.30}$$

For the solids concentration C in the flowing part, it is assumed that the equilibrium concentration is attained at an arbitrary position and time, and moreover, the concentration equation for a mature debris flow, Equation (2.25), is applicable to an immature debris flow as well, where θ is not the channel slope but the surface slope of the deposit. Then:

$$C = \frac{\rho_m(\tan\theta_d - \partial z_b/\partial x)}{(\sigma - \rho_m)(\tan\varphi - \tan\theta_d + \partial z_b/\partial x)} \qquad (5.31)$$

Substituting this in Equation (5.30), and considering that $\partial z_b/\partial x$ is small in comparison to $(\tan\phi - \tan\theta_d)$, one obtains:

$$\frac{\partial z_b}{\partial t} = \frac{q_{wt}}{(C_* - C)(1 - C)} \frac{\rho_m}{\sigma - \rho_m} \frac{\tan\varphi}{\tan\varphi - \tan\theta_d} \frac{\partial^2 z_b}{\partial x^2} \qquad (5.32)$$

Provided that the difference between the slopes of the upstream and downstream channel is small, as is the case we are now considering, the change in C and q_{wt} are small. Therefore, one can substitute the constant C_m and q_{wm} for C and q_{wt}, respectively in Equation (5.32). Then, Equation (5.32) becomes an equation of diffusion with a constant diffusion coefficient D_f as follows:

$$\frac{\partial z_b}{\partial t} = D_f \frac{\partial^2 z_b}{\partial x^2} \qquad (5.33)$$

where:

$$D_f = \frac{q_{wm}}{(C_* - C_m)(1 - C_m)} \frac{\rho_m}{\sigma - \rho_m} \frac{\tan\varphi}{\tan\varphi - \tan\theta_d} \qquad (5.34)$$

$$C_m = (C_u + C_d)/2, \quad q_{wm} = (q_{wu} + q_{wd})/2 \qquad (5.35)$$

The values of C_u and q_{wu} are given as the debris flow properties in the upstream channel, but C_d and q_{wd} must be obtained to satisfy the conditions in the downstream channel. Herein, the following relationships are assumed:

$$C_d = \frac{\rho_m \tan\theta_d}{(\sigma - \rho_m)(\tan\varphi - \tan\theta_d)} \qquad (5.36)$$

$$q_{wd} = \frac{C_*(1 - C_d)}{C_* - C_d} q_0 \qquad (5.37)$$

where q_0 is water discharge per unit width supplied at the upstream end of the channel.

If the surface gradient of the deposit at the change in slope ($x = 0$) is assumed to be equal to the slope of the upstream channel, the boundary condition to solve Equation (5.33) is given as:

$$x = 0; \quad \partial z_b/\partial x = \tan\theta_d - \tan\theta_u \qquad (5.38)$$

The initial condition is:

$$t = 0; \ z_b = 0 \quad \text{or} \quad \partial z_b / \partial x = 0 \tag{5.39}$$

The solution of Equation (5.33) under these boundary and initial conditions is:

$$\varsigma = 2\,\text{ierfc}(\chi) \tag{5.40}$$

where:

$$\varsigma = \frac{z_b}{M\sqrt{D_f t}}, \chi = \frac{x}{2\sqrt{D_f t}}, M = \tan\theta_u - \tan\theta_d \tag{5.41}$$

and $\text{ierfc}(\chi)$ is the inverse of the error function.

Figure 5.7 is an example of a comparison of theory with experiment, the given debris flow in the upstream channel has the same characteristics as the one shown in Figure 5.6 but the downstream channel gradient is 11°. In the early stage the theory does not fit to the experiment due to neglecting the stoppage process, but with the elapse of time the theory approximately fits to the experiment. But, in general, the thickness of the deposit at the change in slope is shallower and thicker at the middle reach than the theory and the broken line in Figure 5.7 fits to the experiment better. This broken line is:

$$\varsigma = -\chi + 1 \tag{5.42}$$

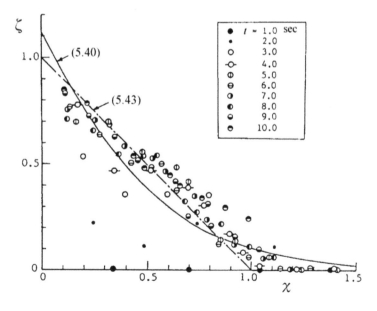

Figure 5.7 Profiles of deposition on a channel steeper than the stable slope of deposition.

which is equivalent to:

$$z_b = -\frac{M}{2}x + M\sqrt{D_f t} \tag{5.43}$$

This means that the surface slope of the deposit is nearly equal to the average of the upstream and downstream channel slopes.

5.1.3 Numerical simulation of depositing process

A one-dimensional analysis of debris flow using the dynamic wave theory is accomplished by simultaneously solving the momentum conservation equation, Equation (4.43) and the continuity equation, Equation (4.44), in which the shearing stress on the bed is given by Equation (5.2) for the case of mature stony debris flow. In the case of immature debris flow the bed shear stress is given from Equation (2.143) by:

$$\tau_{bx} = \frac{\rho_T}{0.49}\left(\frac{d_L}{h}\right)^2 u|u|. \tag{5.44}$$

The analysis including the erosion or deposition processes requires, in addition to the continuity equation for the total mass of water plus sediment, the continuity equations for coarse and fine particle fractions given by Equations (3.30) and (3.31), respectively.

Here, we are discussing the deposition process, and it must be noted on the expression of depositing velocity. As stated in section 5.1.1, when a debris flow reaches a sudden flattening in a channel slope, it does not stop immediately but continues decelerating awhile and then stops. During the deceleration the velocity gradient in the flow is larger than that which is sufficient to generate the dispersive pressure to sustain all the coarse particles in the flow. Because the depth-averaged velocity gradient is represented by the ratio of mean velocity U and depth h, a debris flow having the depth h will continue to flow without deposition as long as U is larger than a critical value U_c.

Consider a debris flow, whose coarse particle concentration is C_L, coming out of a steep upstream channel. Referring to the discussion in section 5.1.1, if the downstream channel slope is flatter than θ_c, the debris flow can be deposited on the downstream channel, where θ_c is given by the following equation:

$$\tan\theta_c = \frac{C_L(\sigma - \rho_m)\tan\alpha_i}{C_L(\sigma - \rho_m) + \rho_m} \tag{5.45}$$

Therefore, the critical velocity U_c for a debris flow, with depth h and concentration of coarse particle fraction C_L, to commence deposition would be given by the steady flow velocity of that characteristic flow on the critical slope channel, namely:

$$U_c = \frac{2}{5d_L}\left[\frac{g\sin\theta_c}{a_i\sin\alpha_i}\left\{C_L + (1 - C_L)\frac{\rho_m}{\sigma}\right\}\right]^{1/2}\left\{\left(\frac{C_{*DL}}{C_L}\right)^{1/3} - 1\right\}h^{3/2} \tag{5.46}$$

where C_{*DL} is the coarse particle fraction in the total volume of deposit. In fact, the debris flow will have some more inertial motion after it approaches the critical velocity and the deposition will begin after the velocity becomes $p_i U_c$ in which p_i is

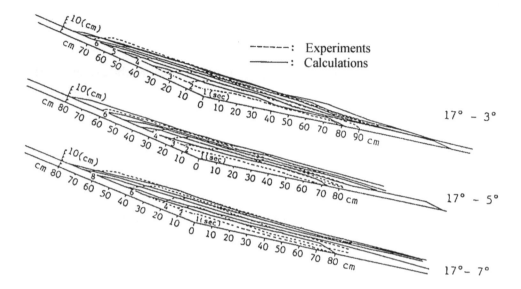

Figure 5.8 Comparison of the results of experiments with the calculations.

a coefficient less than 1.0. This is the reason for introducing Equation (3.41) as the depositing velocity of stony debris flow. Of course, if $(1 - u/p_i U_c)$ is negative, $i = 0$. The appropriate value of p_i obtained by experiments is about 1/3.

Figure 5.8 compares thus calculated profiles of deposit with the experimental results shown in Figure 5.6. The boundary conditions necessary for the calculation were obtained by the measurement in the experiments. Although the calculated results are vague at the beginning of the rebounding deposition and in showing the sudden step at the upstream boundary of the rebound, the general profile and the thickness of deposit are rather well simulated.

5.2   ONE-DIMENSIONAL DEPOSITING PROCESS OF TURBULENT MUDDY DEBRIS FLOW

The relationship between the channel slope and the solids transport concentration for a turbulent muddy-type debris flow is given in Figure 2.41 or Figure 2.43 similarly as in the case of stony-type debris flow. Therefore, the depositing velocity of a turbulent debris flow at a position would also be described similar to Equation (3.41), i.e. the depositing velocity is proportional to the difference in concentrations between the equilibrium one at that position and the one in the coming out debris flow. But, herein, we consider the case where the debris flow comes out to a nearly horizontal place where almost all the particles contained in the debris flow become the excess constituents (Arai and Takahashi 1988).

In such a case, the particle deposition would be due to the gravitational particle settling, and if the dynamic effect of flow is ignored, the sediment volume that settles down onto the bed in a unit time would be described as Cw_s, in which w_s is the settling velocity of particles in a static debris flow material. Therefore, the parameter that determines the depositing velocity would be Cw_s/C_*. As long as the coming out debris flow still has sufficient velocity to be able to carry sediment with the concentration C, deposition will not occur. Hence, if we write the critical velocity of debris flow to commence deposition as U_c, by analogy to the stony debris flow case earlier mentioned, the depositing velocity would be described as follows:

$$
i = \begin{cases} -\left\{1 - \left(\dfrac{u}{U_c}\right)^m\right\}\dfrac{C}{C_*}w_s; & u_* < u_{*c} \\ 0 & ; \quad u_* \geq u_{*c} \end{cases}
\tag{5.47}
$$

where u_{*c} is the shear velocity corresponding to U_c.

Then, u_{*c} and U_c must be given. To obtain these values the following energy balance equation is assumed:

$$
\rho_T u_*^2 \frac{du}{dz}\left(1 - \frac{z}{h}\right) = (1 - C)\frac{\rho u_{*o}^3}{\kappa_o}\left(1 - \frac{z}{h}\right)\frac{1}{z} + \rho\left(\frac{\sigma}{\rho} - 1\right)w_s g C
\tag{5.48}
$$

The left-hand side term is the energy production, the first term of the right-hand side is the turbulent energy dissipation within the interstitial fluid and the second term is the energy dissipated to suspend particles. The smaller terms such as the energy dissipation due to the relative motion between particles and fluid are neglected. The suffix o means the values for plain water and κ_o is the Kármán constant for plain water. The sediment concentration is assumed to be uniformly distributed throughout the entire depth.

When $d_p << 1$, the velocity distribution formula, Equation (2.147) becomes:

$$
\frac{du}{dz} = \frac{u_*}{\kappa}\frac{1}{z}
\tag{5.49}
$$

Substituting this into the left-hand side of Equation (5.48) and integrating from $z = \delta$ to $z = h$, one obtains the following shear velocity that corresponds to u_{*c}:

$$
u_*^3 = \frac{(\rho/\rho_T)\kappa_o(\sigma/\rho - 1)gw_s C(h - \delta)}{(\kappa/\kappa_o)\{\ln(h/\delta) - 1\}\{1 - (1 - C)f_n(C)\}}
\tag{5.50}
$$

where δ/h is neglected in comparison to 1, and:

$$
\frac{\kappa}{\kappa_o}\frac{\rho}{\rho_T}\left(\frac{u_{*o}}{u_*}\right)^3 = f_n(C)
\tag{5.51}
$$

Arai and Takahashi (1986) gave the following relationships:

$$
\frac{\kappa_o}{\kappa} = \frac{1 + 2C - 4C^2}{2}\left[1 + \{1 + 52\kappa_o(1 + 2C - 4C^2)s_1\}^{1/2}\right]
\tag{5.52}
$$

$$
s_1 = \frac{g(\sigma/\rho - 1)w_s C(h - \delta)}{u_*^3 \ln(h/\delta)\{1 + (\sigma/\rho - 1)C\}}
\tag{5.53}
$$

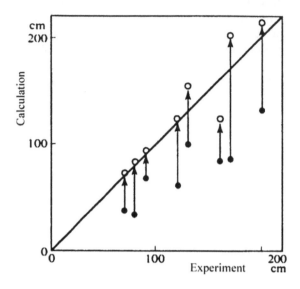

Figure 5.9 Comparison of the calculated distances to the deposit initiation with the experiments.

On the other hand, experiments revealed that if $f_n(C)$ was described as:

$$f_n(C) = 1 - 1.47C + 7.87C^2 \tag{5.54}$$

the position to initiate deposition was well explained, where that position was calculated by substituting:

$$G = g(\sin \theta_c - \sin \theta_d) \tag{5.55}$$

into Equation (5.11), in which the critical energy slope, $\sin \theta_c$, to suspend all the particles in the debris flow supplied from the upstream channel was assumed equal to the gradient of friction loss corresponding to the friction velocity, u_*, given by Equation (5.50).

Figure 5.9 compares the calculated positions of deposit initiation with those in the experiments, where the calculations are accomplished based on two different concepts in applying Equation (5.11). The black circles indicate the cases in which the distance between the change in slope and the deposit initiation is calculated by substituting Equation (5.7) into G assuming that the formulae for stony debris flow are applicable to turbulent debris flow as well, whereas, the open circles indicate the cases in which the distance is calculated by substituting Equation (5.55) into G in Equation (5.11). As is clear in Figure 5.9, the open circles are plotted closer to the line of perfect fitting to the experimental results than the black circles. Thus, the deposition model for turbulent debris flow introduced herein is valid, and, at the same time, the inertial motion before the initiation of deposition for turbulent debris flow is more important than in the case of stony debris flow in estimating the deposition process and the resulting profile of deposition. However, the discussion here is based on the very simplified energy

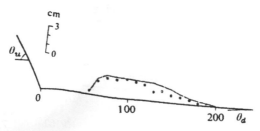

Figure 5.10 Longitudinal profiles of deposition by calculation (line) and experiment (circle).

balance equation that should be applicable to a steady uniform flow notwithstanding that the process is just out of an equilibrium state. Therefore, a more strict discussion is required.

The profile of deposition would be obtained by solving the governing equations as in the case of stony debris flow. In the following calculation, however, the momentum coefficient is considered to be equal to 1 and the resistance law is Equation (2.148) that is suitable to turbulent debris flow. The depositing velocity is given by Equation (5.47), after some trial calculations, the exponent m in this equation is set equal to 0.3; the value of m is not so sensitive to the calculated results. Figure 5.10 compares the calculated profile at 10 seconds later than the coming out instant of the debris flow forefront at the change in slope with the experimental one, in which the upstream and downstream channel slopes are 18° and 0.573°, respectively, $d_{50} = 0.016$ cm, unit width discharge is 53.3 cm^2/s, and solids concentration is 0.3. One can see the general depositing profile is well reproduced by the calculation, and at the same time, one will notice that the deposition begins considerably downstream from the change in slope.

5.3 FORMATION OF A DEBRIS FLOW FAN

5.3.1 Description of the experimental results for stony debris flow and empirical presentations of the feature of a debris flow fan

A series of experiments to examine the formation processes of a debris flow cone/fan at the mouth of a mountain stream was carried out (Takahashi 1980a). In the experiments a steel flume, 20 cm wide, 40 cm deep, and 10 m long was connected to a board, 2 m wide and 6 m long. The slope of the flume was variable from 0° to 30° and that of the board from 0° to 10°. Gravels 5 mm in diameter were glued to the bottom of the flume, as a roughness element, whereas the board was covered by a smooth Styrofoam plate. After the flume and the board were set to the prescribed slopes, a constant rate of water was suddenly supplied using an electromagnetic valve at the upstream end of the flume. When its bore front reached the position of sediment supply, a constant rate of sediment was poured into the flow via the sediment feeder to make a debris flow with the prescribed sediment concentration and discharge. Thus, it debouched onto the board and formed a debris flow fan. The processes of fan formation were recorded by a 35 mm motor-driven camera and a 16 mm high-speed cine-camera. The sediment

Table 5.1 Experimental conditions and experimental and calculated results.

Run No.	Water discharge Q_o (l/s)	Sediment discharge Q_s (l/s)	Channel slope (°) Upstream/ Downstream	Duration (s)	Arrival distance (cm) Experiment/ Calculation	Surface slope by experiment γ_1/γ_2	Surface slope by calculation γ_1/γ_2
1	1.4	0.8	18/0	35	72/73	7.2/14.7	6.0/12.8
2	1.6	0.6	18/0	50		5.3/12.8	5.3/11.9
3	1.4	0.8	18/4	35	110/102	7.6/13.1	6.0/12.8
4	1.6	0.6	18/4	50		6.0/15.0	5.3/11.9
5	1.4	0.8	18/6	35	130/126	7.1/16.0	6.0/12.8
6	1.6	0.6	18/6	50		5.7/16.0	5.3/11.9
7	1.4	0.8	18/10	35	240/227	7.6/13.8	6.0/12.8

material used had the following characteristics: $d_{50} = 1.2$ mm, $(d_{84}/d_{16})^{1/2} = 2.58$, $\sigma = 2.65$ g/cm^3, $\tan \varphi = 0.73$. Provided that the experiments are considered as the physical model tests of 1/100 scale, the debris flow corresponds to a large-scale flow having a discharge of 220 m^3/s, a duration of six to eight minutes, and a total runoff sediment volume of 46,000 m^3. The debris flow is so large that it forms almost a complete fan shape, whereas the actual debris flow in Kamikamihorizawa formed only several lobes, as explained in Chapter 1.

Table 5.1 demonstrates the experimental conditions and the experimental and calculated results for the each experimental run.

Figure 5.11 shows the experimental results under the combination of an 18° upstream channel slope and a 4° board slope, Figure 5.11(a) is for run 3 and Figure 5.11(b) is for run 4. The curves in Figure 5.11 show the temporal variations in the shapes of the debris flow fringes, and the number on each curve indicates the time in seconds elapsed from the moment that the forefront of the debris flow arrived at the mouth.

In the early stage, the path of the flow downstream from the mouth is straight, and the width of it is at most about twice the width of the mouth, but it quickly reaches its maximum length (4 seconds in the case of Figure 5.11(a)). Then, it begins to deposit the debris, accompanied by a hydraulic jump at the boundary of the deposit and the flow and the deposit goes upstream. After a short time, when the hydraulic jump arrives at the mouth (approximately 5 seconds in the case of Figure 5.11(a)), the flow changes its direction to the right, then to the left, and so on. Consequently, at the end of rather long sporadic channel shifting process as in the case of a large-scale debris flow, the deposit is circular, with a diameter approximately equaling the distance from the outlet to the distal end of the first flow. In this stage, the thickness of the deposit at the mouth becomes too high for the flow to climb over, and then the deposition rebounds within the upstream channel. The experiment ended on reaching this stage, but if debris flow continued, soon the deposit would arrive back at the debris supply position and the debris flow would start to overflow the deposit and flow downstream increasing the size of the debris flow fan.

The arrival distances of the debris flow front were calculated using Equation (5.11). Comparing these with the values in Table 5.1, confirmed that the calculated values

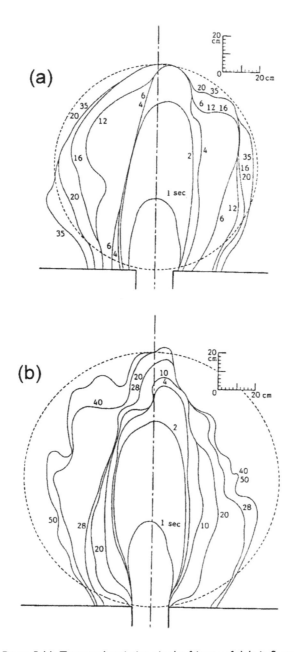

Figure 5.11 Temporal variations in the fringes of debris flows.

well coincided with the experiments, where $\kappa_a = 0.5$ was assumed and $h_u = 0.76$ cm was obtained from the experiment.

Figure 5.12 shows the contour maps of the debris flow fans formed in the experiments corresponding to runs 1 and 3, respectively, where the datum point of altitude

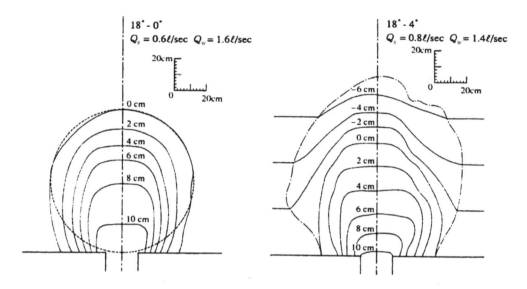

Figure 5.12 The contour maps of debris flow fans.

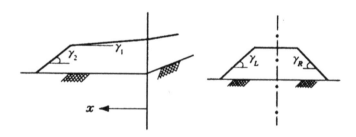

Figure 5.13 Schematic longitudinal and cross-sectional shapes of a debris flow fan.

is at the bed of the mouth. The deposit shows a tapering plane along the center line, which has nearly uniform width, and both the outside edges of this plane have steep slopes.

The schematic longitudinal profile and the cross-sectional shape at the midpoint between the mouth and the distal end of the fan are shown in Figure 5.13. The formation of steep slopes outside the central plane is related to the particle segregation while the debris flow is flowing in the upstream channel by which coarse particles gather at the front part. The prediction of spatial particle size distributions within the debris flow fan is discussed in section 5.4, but herein, only the longitudinal surface slope along the central axis of the fan is discussed using the experimentally obtained particle size distributions along the surface of the deposit.

The comparatively stable slope of the fan under the effects of squeezed surface water flow is given by Equation (5.14). During the process of rebounding deposition within the upstream channel, the water discharge coming out to the fan, Q, becomes smaller than the supplied water discharge at upstream end, Q_o, due to the trapping of water within the rebounding deposit, as follows:

$$Q = Q_o - \frac{1 - C_*}{C_*}Q_s \tag{5.56}$$

Therefore, neglecting the seepage flow into the deposit, the unit width discharge on the surface of the planer portion of the fan is given by:

$$q_o = Q/B_f \tag{5.57}$$

where B_f is the width of the planer portion. Substituting $C_* = 0.7$, $B_f = 40\,\text{cm}$, $d_m = 0.08\,\text{cm}$, $Q_o = 1400\,\text{cm}^3/\text{s}$ and $Q_s = 800\,\text{cm}^3/\text{s}$ into Equation (5.14), one obtains $\gamma_1 = 6.0°$. Similarly, for the combination of $Q_o = 1600\,\text{cm}^3/\text{s}$ and $Q_s = 600\,\text{cm}^3/\text{s}$, one obtains $\gamma_1 = 5.3°$. If experimentally obtained particle size at the front part of the deposit $d_m = 0.3\,\text{cm}$ is used, the combinations of $Q_o = 1400\,\text{cm}^3/\text{s}$ and $Q_s = 800\,\text{cm}^3/\text{s}$, and $Q_o = 1600\,\text{cm}^3/\text{s}$ and $Q_s = 600\,\text{cm}^3/\text{s}$ give $\gamma_2 = 12.8°$ and $\gamma_2 = 11.9°$, respectively. These surface slopes approximately agree with the experimental values listed in Table 5.1.

From the point of view of hazardous zone prediction, it is very important to predict the deposition area for the case of a large-scale debris flow that cannot be stored in the circular area whose diameter is equal to the distance between the mouth of the gorge and the distal end of the bore front stoppage. Let us assume that the deposition area for such a large-scale debris flow is also stored in a circular area, but in this case, the diameter of the circle is equal to the distance between the mouth and the maximum arrival distance of the debris flow deposition. Then, one must know the maximum reach of the debris flow, which should be related to the total volume of runoff sediment. For the calculation of the total runoff sediment volume, the deposition area is simplified as a square that envelops the circular depositing area, as shown in Figure 5.14. The cross-section of the deposit perpendicular to the axis of the channel is trapezoidal as shown in Figure 5.14. The width of the plane in the central part of the fan that inclines with the gradient γ_d is assumed to be given by the following regime theory:

$$B_f = 3Q^{1/2} \quad (\text{m-sec unit}) \tag{5.58}$$

We further assume that the surface gradient of the deposit within the upstream channel is also γ_d notwithstanding the channel width is B_u. The total volume of the deposit V_d is, then, given as:

$$V_d = \frac{1}{2}x_L^2 \tan(\gamma_d - \theta_d)\left\{\frac{1}{2}(x_L + B_f) + B_u\frac{\tan(\gamma_d - \theta_d)\cos\theta_u\cos\gamma_d}{\sin(\theta_u - \gamma_d)}\right\} \tag{5.59}$$

From Equation (5.59), provided the total volume of the debris flow is known, x_L can be obtained.

If the second term in the curly brace of the right-hand side of Equation (5.59), that is the deposit volume in the upstream channel, is neglected and $B_f = (2/3)x_L$ is assumed, Equation (5.59) becomes:

$$V_d = \frac{5}{12}x_L^3 \tan(\gamma_d - \theta_d) = \frac{5}{12}A^{3/2}\tan(\gamma_d - \theta_d) \tag{5.60}$$

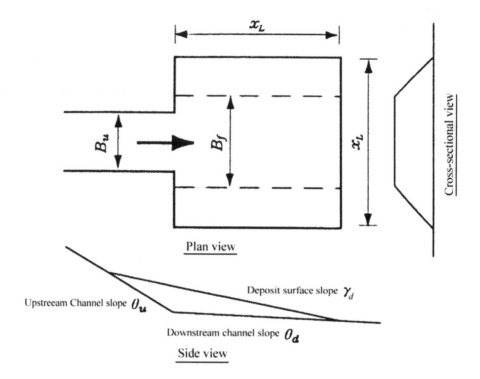

Figure 5.14 Conceptual diagram of simplified debris flow fan shape.

where A is the area of deposition. Then:

$$A = \left\{ \frac{12}{5} \frac{1}{\tan(\gamma_d - \theta_d)} \right\}^{2/3} V_d^{2/3} \tag{5.61}$$

Crosta *et al.* (2003) described the relationship between A and V_d as:

$$A = k V_d^{2/3} \tag{5.62}$$

and demonstrated that the previous field data were well explained if k was appropriately given.

The relationship between A and V_d that is described as Equation (5.62) is reasonable because the area is proportional to the square of the length scale and the volume is proportional to the cube of the length scale. Then, the problem is what elements determine the k value. According to the previous data (Crosta *et al.* 2003), k for a volcanic mud flow (lahar) is the largest, as large as 200; the muddy debris flow at Sarno, Italy had a value of 32.5 and a stony debris flow in the Alps had a value of 6.2. Such a wide variety of k values should be caused by the degree of fluidity of the debris flows. Equation (5.61) has a similar expression to Equation (5.62), but the former gives how k values change depending on the fluidity of flow. If a generally recognized γ_d value

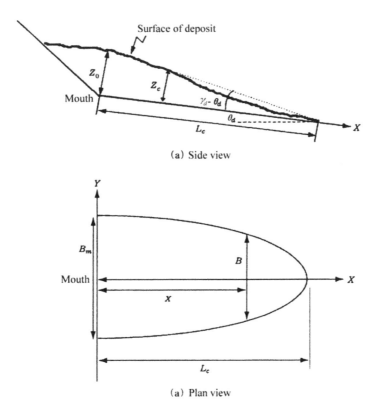

(a) Side view

(a) Plan view

Figure 5.15 Simplified debris flow fan shape.

between $5°$ and $7°$ for stony debris flows (Shih *et al.* 1997) and $\theta_d = 2°$ are substituted into Equation (5.61), one obtains $k = 9 - 13$. If a debris flow has high fluidity and the difference between γ_d and θ_d becomes small, for example $\gamma_d - \theta_d = 0.05°$, k becomes 196. These k values well agree with the data for stony debris flows and lahars, respectively as mentioned above.

Shieh and Tsai (1997) carried out many experiments similar to Takahashi's and they concluded that the fan shape was better represented by a semi-ellipse shown in Figure 5.15 rather than a circle.

The shape is described mathematically as:

$$\left(\frac{B}{B_m}\right)^2 + \left(\frac{X}{L_c}\right)^2 = 1 \tag{5.63}$$

The notations are given in Figure 5.15. They gave the thickness of deposit along the channel axis; Z_c, as follows:

$$\frac{Z_c}{Z_0} = \exp\left\{-\frac{1}{2C_1^2}\left(\frac{X}{L_c}\right)^2\right\} \tag{5.64}$$

and they also gave the thickness Z at a distance Y from the channel axis where the width of deposit was B as follows:

$$\frac{Z}{Z_c} = \exp\left\{-\frac{1}{2C_2^2}\left(\frac{Y}{B}\right)^2\right\}$$

(5.65)

Furthermore, they assumed the volume of deposit as:

$$V_d = \alpha_v L_c B_m Z_0$$

(5.66)

where C_1, C_2 and α_v are the coefficients given as:

$$C_1 = 0.393 + 0.011 \log(d_{84}/d_{16})$$

(5.67)

$$C_2 = 0.212 + 0.024 \tan^{-1}\{1.2(d_{84}/d_{16} - 1)\}$$

(5.68)

$$\alpha_v = 0.234 + 0.033 \tan^{-1}\{0.5(d_{84}/d_{16} - 1)\}$$

(5.69)

Using these relationships one can obtain the shape and dimension of a debris flow fan if L_c, B_m and Z_0 are given. Shih *et al.* (1997) proposed to use Takahashi's Equation (5.11) for obtaining L_c, and the empirically obtained 5 to 7 degrees for γ_d. Then, Z_0 is given as:

$$Z_0 = L_c \times \tan(\gamma_d - \theta_d)$$

(5.70)

The value of B_m is obtained by getting the total volume of debris flow V_d, using Equation (5.66).

5.3.2 Numerical simulation of fan formation process and its verification

The depth-wise averaged two-dimensional momentum conservation equations are given by Equations (3.115) and (3.116), respectively, and as we are dealing with the depositing process, the total mass conservation equation, Equation (3.117), can be modified as follows:

$$\frac{\partial h}{\partial t} + \frac{\partial M}{\partial x} + \frac{\partial M}{\partial y} = i$$

(5.71)

where i is the volume of water plus sediment subtracted in unit time from the flow by deposition.

The conservation equations for the coarse and fine solid fractions, respectively, are similar to Equations (3.30) and (3.31):

$$\frac{\partial V_L}{\partial t} + \frac{\partial(C_l M)}{\partial x} + \frac{\partial(C_L N)}{\partial y} = iC_{*DL}$$

(5.72)

$$\frac{\partial V_F}{\partial t} + \frac{\partial\{C_F(1 - C_l)M\}}{\partial x} + \frac{\partial\{C_F(1 - C_L)N\}}{\partial y} = i(1 - C_{*DL})C_F$$

(5.73)

As for the shear stress on the bed, Equations (3.120) and (3.121) were given previously, but here to make the relationship with Equation (5.2) clearer, the following expressions are used:

$$\tau_{bx} = \frac{u}{\sqrt{u^2 + v^2}}(\sigma - \rho_m)ghC_L \cos\theta_b \tan\alpha_i \qquad (5.74)$$

$$\tau_{by} = \frac{v}{\sqrt{u^2 + v^2}}(\sigma - \rho_m)ghC_L \cos\theta_b \tan\alpha_i \qquad (5.75)$$

However, for immature debris flow the expressions in Equations (3.122) and (3.123) are used without any modification. The equilibrium concentrations of coarse particle fractions, $C_{L\infty}$ or $C_{sL\infty}$, are obtained from Equation (3.36) or by substituting $C_{L\infty}$ into C_∞ in Equation (2.144), at that time, $\tan\theta$ is given by Equation (3.132), that considers the inclination of the flow surface to the direction of velocity vector. Then, corresponding to Equation (3.41), the depositing velocities of stony and immature debris flows, respectively are given by:

$$i = \delta_d \left(1 - \frac{\sqrt{u^2 + v^2}}{p_i U_c}\right) \frac{C_{L\infty} - C_L}{C_{*DL}} \frac{\sqrt{M^2 + N^2}}{d_L} \qquad (5.76)$$

$$i = \delta_d \left(1 - \frac{\sqrt{u^2 + v^2}}{p_i U_c}\right) \frac{C_{sL\infty} - C_L}{C_{*DL}} \frac{\sqrt{M^2 + N^2}}{d_L} \qquad (5.77)$$

The numerical integration of the above system of equations which is composed of Equations (3.115), (3.116) and from (5.71) to (5.77) renders possible to predict the processes of debris flow fan formation in which the hydrograph and sediment concentrations of respective coarse and fine fractions at the outlet of the gorge are given as the boundary conditions under the given initial topographical conditions of the depositing area.

For the verification of the mathematical model mentioned above, laboratory flume experiments were conducted. A steel flume connected to a board as shown in Figure 5.16 was used. A material, with median diameter $d_{50} = 1.28\,mm$ and density $\sigma = 2.65\,g/cm^3$, was laid along a distance of 3 m from 5.5 m to 8.5 m measured from the debouching point with a thickness of 10 cm and soaked by the seepage flow. Then, a sudden discharge of plain water, $600\,cm^3/s$, was introduced at the upstream end of the bed to make a bore of a mixture of gravel and water. The velocity and the depth of flow just upstream of the flume outlet were measured by a high-speed TV-video camera via the transparent glass-wall of the flume. The variations of the shape and thickness of the deposit were measured by two other video cameras, in which the thickness of deposit was measured on the video image by reading out the elevations of the deposit surface using the gauging rods set on the board. As the information necessary for the boundary conditions for the calculation, the time variations of the depth and velocity in the channel just upstream of the outlet were directly measured on the visual record of the high-speed video. The time variation of solids concentration in the flow was indirectly given by substituting the measured depth, velocity and

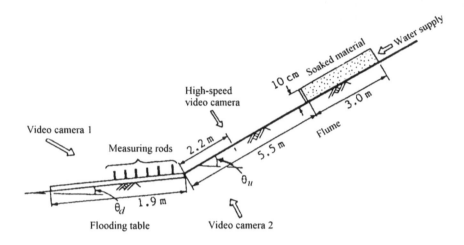

Figure 5.16 Experimental set up.

slope in Equation (2.149), in which $d_L = 1.36$ mm, $C_{*DL} = 0.7$, $\rho_m = 1$ g/cm^3, $\varphi = 36°$ are used.

Figure 5.17 compares the shapes and thicknesses (flow depth + deposit thickness) in the process of a debris flow fan formation in a laboratory experiment with those obtained from the calculation by the mathematical model. Numbers on the contour lines indicate the thickness in centimeters measured from the surface of the flooding board, and t on the respective figures indicate the time elapsed from the moment of the initiation of debouching from the upstream flume. The slopes of the channel and the flooding board in this case were 17° and 7°, respectively. The mesh intervals in the computation were set $\Delta x = \Delta y = 5$ cm, $\Delta t = 1/500$ sec, and the values $\delta_d = 0.05$, $\tan \alpha_i = 0.45$ and $p_i = 1/3$ were used. There is comparatively good agreement between the experiment and the calculation, except for the existence of a hump at a little downstream of the outlet in the experiment, possibly being formed by erosion of the deposit by a dilute later flow which was not taken into account in the calculation.

5.3.3 Numerical simulation of fan formation by turbulent debris flow

The two-dimensional (in horizontal direction) momentum conservation equations for turbulent debris flow are similar to those for stony debris flow, given by Equations (3.115) and (3.116), and the conservation of the total mass of water and sediment is given by Equation (5.71). Assuming a uniform size material, the conservation of mass for solids fraction is given as:

$$\frac{\partial V_s}{\partial t} + \frac{\partial(CM)}{\partial x} + \frac{\partial(CN)}{\partial y} = iC_* \tag{5.78}$$

where V_s is the substantial volume of solids contained in a columnar space in the flow whose height is h and the base has a unit area.

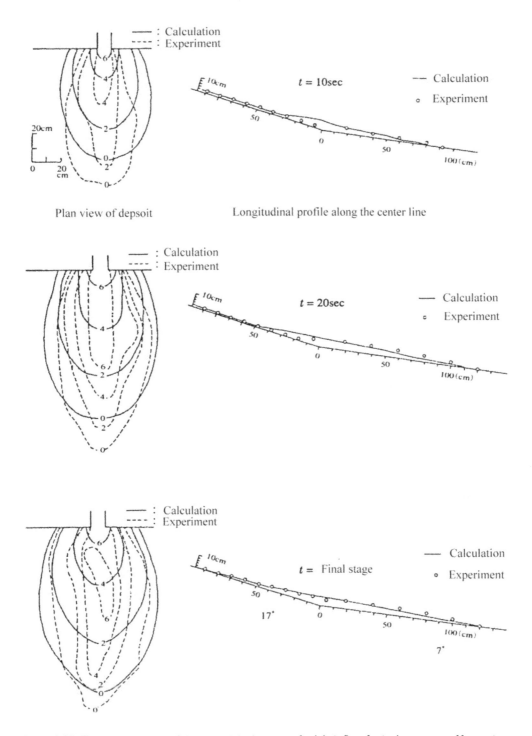

Figure 5.17 Temporal changes of shapes and thicknesses of a debris flow fan in the process of formation.

The depositing velocity is given by Equation (5.47), and the temporal variation in the thickness of deposit is given by Equation (3.32).

If the x and y-wise frictions on the bed are described by:

$$\frac{\tau_{bx}}{\rho_T} = \frac{f}{8} u \sqrt{u^2 + v^2} \tag{5.79}$$

$$\frac{\tau_{by}}{\rho_T} = \frac{f}{8} v \sqrt{u^2 + v^2} \tag{5.80}$$

the friction coefficient f is given, from Equation (2.148), as:

$$f = 8 \left[\frac{1}{\kappa} \left\{ \ln \left| \frac{1 + \sqrt{1 + \Phi_1^2}}{Z_o + \sqrt{Z_o^2 + \Phi_1^2}} \right| - \sqrt{1 + \Phi_1^2} + \Phi_1 \right\} \right]^{-2} \tag{5.81}$$

If the base of flow is smooth, $Z_o = (a_o v_o)/(u_* h)$, but if the base has a rough surface, $Z_o = b(k_r/h)$, where k_r is the equivalent roughness height and $b = 1/30$.

Because Equation (5.81) is applicable only to the flow maintaining the state of turbulent debris flow, one must make clear the applicable condition (Arai and Takahashi 1986). Among the shearing stresses on the bed, the one shared by turbulence can be obtained by subtracting the shearing stress attributable to the inter-particle collisions from the total stress operating on the bed. Thus, substituting Equation (2.19) associated with Equation (2.20) into the second term of the right-hand side of Equation (2.146), one obtains the turbulent shearing stress on the bed as follows:

$$\tau_{0f} = \rho g h \sin \theta \left\{ \left(\frac{\sigma}{\rho} - 1 \right) \left(1 - \frac{\tan \alpha_i}{\tan \theta} \right) C + 1 \right\} \tag{5.82}$$

For the particles to be suspended by turbulence even at the neighborhood of bed, the shear velocity of fluid phase u_{*f} $(= (\tau_{0f}/\rho)^{1/2})$ should be larger than the settling velocity of the particle w_s. If Rubey's particle settling velocity equation, Equation (2.168), is used in the above mentioned condition, Equation (5.82) gives the following formula:

$$\frac{h}{d_p} \geq \frac{(\sigma/\rho - 1)F_R^2}{\sin \theta \{(\sigma/\rho - 1)(1 - \tan \alpha_i/\tan \theta)C + 1\}} \tag{5.83}$$

This is the condition for the flow to be a turbulent muddy-type debris flow. If Equation (5.83) is not satisfied, the flow can no longer suspend particles by turbulence and the following formula obtainable from the logarithmic resistance law for plain water flow is used:

$$f = 8 \left\{ 8.5 + \frac{1}{\kappa} \ln \left(\frac{h}{k_r} \right) \right\}^{-2} \tag{5.84}$$

The above mentioned Equations (3.115), (3.116), (5.47), (5.71) and from (5.78) to (5.84) will predict the debris flow fan formation processes by a turbulent debris flow whose hydrograph and other characteristics are given at the outlet of the gorge.

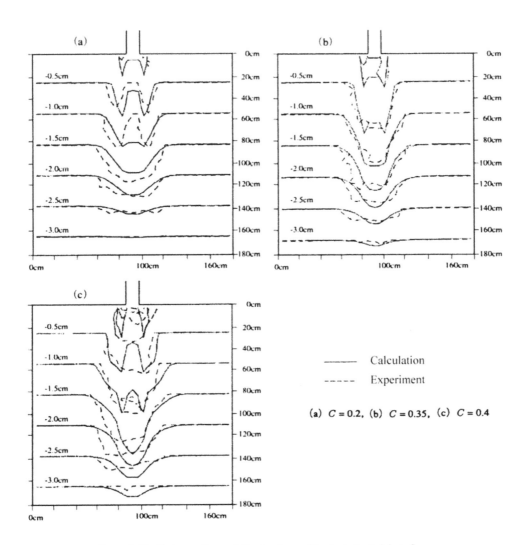

Figure 5.18 Contour lines of the fan formed by turbulent debris flow.

Experiments to verify the simulation method were conducted using the combination of a steep upstream channel and a flat downstream flooding board. Here, the turbulent debris flows in the upstream channel, which was 6 m long, 15 cm wide and 18° in slope, plunged onto the flooding board, which had a slope of 1°. The respective debris flows had prescribed solids concentrations but had a constant discharge of 800 cm³/s for the duration of 5 seconds. The median diameter of the debris flow material was $d_{50} = 0.2$ mm.

Figure 5.18 compares calculations and experimental results for the topographies of the fan at the final stages of deposition by comparing the patterns of contour lines. Figure (a) corresponds to $C = 0.2$; figure (b) corresponds to $C = 0.35$; and figure (c) corresponds to $C = 0.4$. The datum altitude 0 of the contour lines is set at the height of the surface of the flooding board at the outlet of the channel. As shown in Figure 5.18, the debris flow does not immediately deposit at the outlet of the channel but throws off like a jet and starts to deposit at a distance downstream of the outlet. This characteristic behavior is rather well reflected in the calculation, and moreover, the general topography of the formed fan is also rather well explained.

Figure 5.19 shows the cross-sectional shapes of deposit at three representative positions in the calculation as well as in the experiment for the case of $C = 0.35$. Figure 5.20 shows the longitudinal profiles along the centerline by calculation and experiment. Except for some discrepancies in the thickness of the deposit near the outlet, the general characteristic pattern of the fan is rather well reproduced by the calculation. This is notable, especially, as the proof of validity of the deposition velocity formula, Equation (5.47).

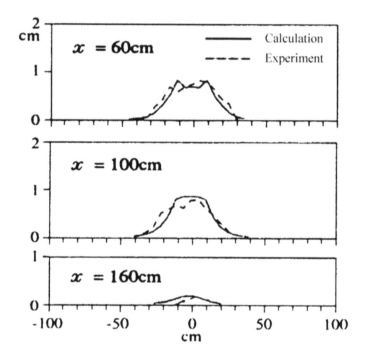

Figure 5.19 Cross-sections of a debris flow fan.

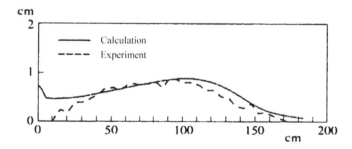

Figure 5.20 Longitudinal profile along the center axis of a debris flow fan.

5.4   PARTICLE SIZE DISTRIBUTION IN THE FAN FORMED BY STONY DEBRIS FLOW

5.4.1   General situations found in the field and experimental data

In an actual stony debris flow flowing in the upstream channel, the mean diameters are the maximum at the forefront and diminish rearward. The spatial distribution of particle sizes in a deposit reflects these characteristics. Results of a survey in one small-scale debris lobe at the Kamikamihorizawa debris flow fan was referred to in Chapter 1. Herein, another example of a field survey on the surface of a debris flow fan which was produced by one episode of debris flow is introduced.

Takei and Mizuhara (1982) carried out a survey on a debris flow fan at the junction of a small tributary of the Yoshino River. The debris flow fan had runoff from a tributary with a basin area of $0.74\,km^2$, mean longitudinal slope of $25°$ and a main channel length of $1.05\,km$. They measured on the surface of deposit the lengths of the three orthogonal axes of all the stones larger than $15\,cm$ in the longest axis, and defined the particle diameter as the third root of the product of the three orthogonal axes lengths.

Figure 5.21 (a) shows the variations in the diameters versus the distance measured downward from the proximal site to the distal end of the deposit at the junction with the Yoshino River, the length of measuring reach is $220\,m$. The diameter obtained by averaging the diameters of particles depositing in the total width of the fan at a cross-section does not change much along the direction of main flow, but the mean diameters in $10\,m$ intervals only along the centerline axis of the fan have a maximum value in the reach $110–130\,m$, and downstream they become smaller than in the upstream reach. As for the biggest stones in the cross-sections and along the centerlines, the largest stones accumulate at a reach of about $120–150\,m$. These facts

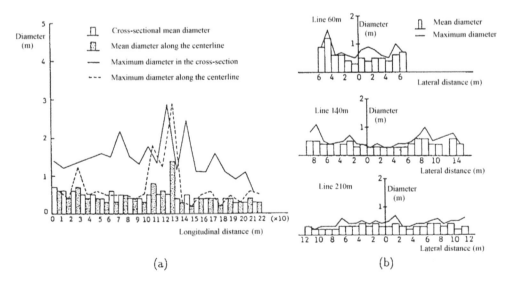

(a)   (b)

Figure 5.21 Particle size distributions on the surface of a debris flow fan.

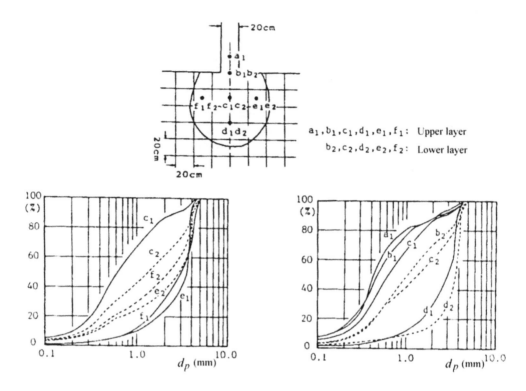

Figure 5.22 Particle size distributions in an experimental debris flow fan.

suggest that the forefront of the debris flow stopped around this reach and some parts of the following flow deposited upstream and the other parts overrode the stopped forefront and deposited downstream. Figure 5.21(b) gives some typical size distributions at representative cross-sections. At the 60 m and 140 m sections both the mean diameter and the maximum diameter become larger with increasing distance from the centerline of the fan, but at the 210 m section this tendency is obscure. These facts suggest that, while the debris flow flows down over the depositing area enlarging its width downstream, the largest stones that move at the front of flow in the upstream channel, shift their positions to the peripheral zones of a tongue shaped lobe and stop in time.

Figure 5.22 shows the particle size distributions in the samples collected from the positions demonstrated in the figure for an experimental case explained in section 5.3.1. By analyzing the samples collected at a position on the fan from the upper layer and the lower layer, the processes of flow and deposition are more clearly understood than by the analyses of samples collected only from the surface. In the peripheral zones (positions d, e and f), the coarse particles larger than d_{70} of the supplied material are deposited, and in the central parts (positions c, b and a) fine particles are deposited. In the central parts, if the particle sizes are compared with those in the upper and lower layers, those in the lower layer are coarser, and if those in the lower layer are compared between upstream and downstream samples, those in the upstream are finer. From these facts one can conclude that the coarsest particles come down at a bore front, form a thick circular levee-like deposit at first, then the following comparatively shallow flow comprised of smaller particles stops inside the levee and deposition proceeds from downstream to upstream gradually increasing the thickness by the overriding of the succeeding flow, and finally, the smallest particles that come down last are deposited inside the upstream channel. At the positions e and f, particles are coarser in the upper layer than in the lower layer; the reason for this inverse grading may be due to wash out of small particles in the upper layer by the effects of a final dilute overland flow.

5.4.2 Mathematical model for the particle size distributions

If, in considering the depositing processes on the fan area, the processes of particle size segregation are neglected, and if we are only concerned with the mean particle diameters at a three-dimensional position, the mean particle size distributions on the surface as well as under the fan can be obtained by adding only an equation for the conservation of particle number to the system of governing equations introduced in section 5.3.2. A change in the particle number in the flow on an infinitesimal area is caused by flow-in, flow-out and deposition, and that is described by the following equation (Takahashi *et al.* 1988):

$$\frac{\partial}{\partial t}\left(\frac{V_L}{k_v d_L^3}\right) + \frac{\partial}{\partial x}\left(\frac{Q_{Lx}}{k_v d_L^3}\right) + \frac{\partial}{\partial y}\left(\frac{Q_{Ly}}{k_v d_L^3}\right) = \frac{iC_{*DL}}{k_v d_L^3} \tag{5.85}$$

where Q_{Lx} and Q_{Ly} are the discharges of the coarse particle fraction in a unit area in the x and y directions, respectively, and $k_v d_L^3$ is the volume of a particle whose diameter is d_L. The temporal change of the mean diameter in the coming out debris flow is given as a boundary condition.

So far no comment has been done on the method of numerical calculation of governing equations. This is essentially similar to the one used in the overland flood flow analysis; the upwind finite difference equations are solved by the leap-frog staggered scheme, the details of the method can be found elsewhere (Takahashi *et al.* 1986). The upstream boundary conditions are given at a position within the upstream channel which the rebounding deposition does not affect. The forefront of flow is a moving boundary, but for the sake of simplicity, a threshold thickness is given, and only when the depth in a mesh is calculated to be deeper than that threshold value, can the flow proceed to the next mesh. Herein, the threshold value is set equal to the mean diameter at that position, but if the particle concentration becomes less than 0.01, it is considered as a water flow and the threshold is set to 1 mm.

Examination of depositing characteristics by numerical simulations

The calculation is carried out under the conditions: the upstream channel width is 10 cm and the slope is 17°; the slope of the downstream flooding board is 7°; $\Delta x = \Delta y = 5$ cm, $\Delta t = 1/500$ sec.

The boundary conditions given at 1 m upstream of the debouchment are shown in Figure 5.23.

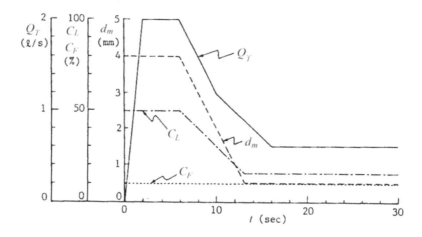

Figure 5.23 Boundary conditions for the calculation.

Figure 5.24 (a) shows the distributions in the thicknesses of flow depths plus the deposit thicknesses at 5, 15 and 30 seconds after the arrival of the debris flow at the boundary 1 m upstream of the debouchment. Figure 5.24 (b) shows the mean particle size distributions in the flowing layer which flows on the newly deposited and developing debris flow fan. The time variations of the mean particle size within the deposit and the thickness of the deposit at some locations are given in Figure 5.25. Figure 5.25 is rearranged in Figure 5.26 to show the relationship between the thickness of deposit and the particle diameter. The representative locations are given in Figure 5.27.

Figures 5.24, 5.25, 5.26 and 5.27 indicate the following characteristics: On the centerline of the fan (a, d, k, r, t), deposition occurs at first at r, then it occurs in the order of k, t and d, and except for the point d, the depositing particles are all 4 mm which is the largest size in the supplied material. The maximum size particles accumulate at the front of the debris flow, and when the forefront arrived at around the point r (about $t = 5$ s) deposition arises at first around this point. Whereas at the point k the front arrives at about $t = 3$ s, the flow passes through without deposition until $t = 6$ s (the broken line of the figure k in Figure 5.25 starts at $t = 3$ s but the solid line starts at $t = 6$ s). The forefront arrives at the point t at $t = 8$ s, here, deposition takes place as soon as the forefront arrives. Thus, along the centerline of the fan, deposition commences at first around the position of the stoppage of forefront and after that deposition proceeds upstream. As shown in Figure 5.26, deposition at the point f begins after the diameter diminished to 3 mm. This means that the forefront of the debris flow passed through the point f without deposition. The situation at the point d is the same with the situation at f and at the point a no deposition occurs at all. At the point k, r and t, deposition takes place only during the flow that contains particles as large as 4 mm, after that, even when the flow contains smaller particles it flows over the deposit without forming a new deposit. On the other hand, at the points d and f, although deposition begins after the particles in flow become a little less than 4 mm, deposition continues even after the particles become as small as 0.5 mm.

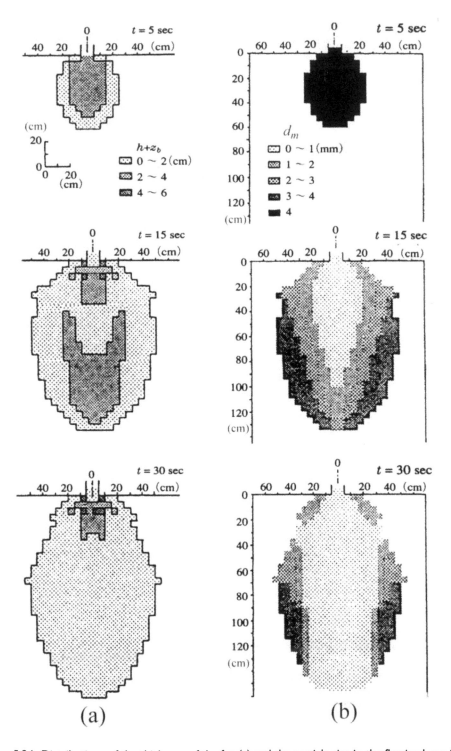

Figure 5.24 Distributions of the thickness of the fan (a) and the particle size in the flowing layer (b).

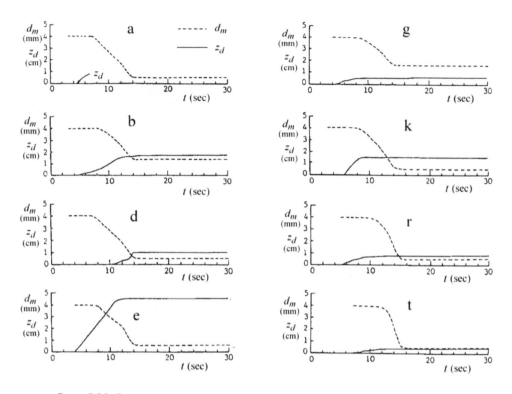

Figure 5.25 Particle size and deposit thickness variation at some reference points.

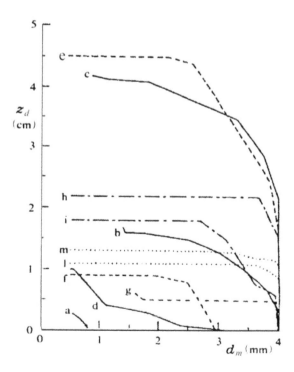

Figure 5.26 Particle size in the deposit versus thickness of deposition.

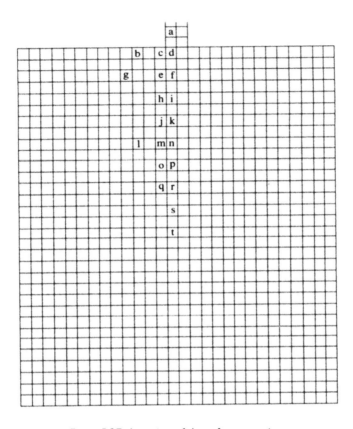

Figure 5.27 Location of the reference points.

At the points b and g, coarse particles enters in the early stage and deposition begins almost simultaneously with the entrance of the flow. At these points deposition continues for long time and the upper layer of the deposit becomes to be comprised of smaller particles. Also at the points d and f, particles become finer upward through the deposit.

At the points c, e and h located at one mesh distance apart from the centerline meshes, the thicknesses of deposit are the largest and deposition starts as soon as the arrival of the front. Because the large particles deposit in a large quantity in the early stage, the lower half of the deposit layer is occupied by coarse particles. By comparing the tendency of the pairs of curves representing adjacent points in Figure 5.26; c and d, e and f, h and i, respectively, it can be seen that the thickness of the deposit at a little away from the centerline is larger than on the centerline, and if particle sizes are compared at the same altitude between the pair points, the particles in the mesh a little away from the centerline are larger. Note that the coarser particles come out in the earlier stage; one will understand that the natural levee-like deposits comprised of coarser particles are formed on both sides of the centerline in the earlier stage of deposition. However, as is evident in Figure 5.24(a), the thickness of the deposit plus the flowing layer is always larger on the centerline than at circumferential positions.

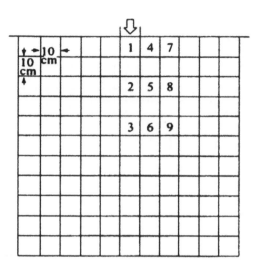

Figure 5.28 Location of the sampling points.

Therefore, in the ongoing process of deposition, the levee-like deposits are formed beneath the flow surface, in other words, the flow along the centerline is deeper than the flow over the levee-like deposit. This would be the reason why a deep incision occurs along the centerline of a debris flow fan, the phenomena is often confirmed in the fields, because smaller particles accumulated on the centerline are easily eroded by normal floods.

Comparison with experiments

A 2 m wide, and 6 m long board inclined at 5° longitudinally was connected to the downstream end of the channel used in the series B experiments described in section 4.6. During the experiments on the deposition process on a debris flow fan, debris flows were generated using the same material and procedures as in the series B experiments. The experimental material was glued on the surface of the board to roughen the surface. After a debris flow came to complete stop, the deposited sediment was sampled at nine locations, as shown in Figure 5.28, at various depths, in 1 cm intervals. The experiments were repeated five times under the hypothetical identical conditions.

A typical temporal change of the deposit shape and thickness (deposit thickness plus flow depth on the deposit) of a debris flow fan is shown in Figure 5.29. The numbers on the contour line indicate the thickness in centimeters measured from the surface of the board, and t is the time measured from the moment the forefront of the debris flow comes out of the channel outlet. The numerical calculations were performed with $\Delta x = \Delta y = 5$ cm; $\Delta t = 0.002$ sec; $\delta_d = 1.0$; and $p_i = 2/3$, where Equation (3.42) was used as the depositing velocity equation. The upstream boundary conditions were the hydrograph and the temporal variation of mean diameters shown in Figures 4.27 and 4.28, and also, although no graphical presentation was given, the temporal variations of coarse particles' concentrations simultaneously obtained in the routing of the debris

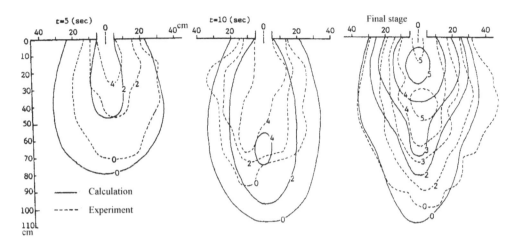

Figure 5.29 Time-varying shapes and thicknesses of debris flow fan surface.

flow in the upstream channel. The agreement between the calculated and experimental results is generally good, although the calculated values predicted the position of a deposited mound slightly closer to the outlet than the measured ones.

The vertical distributions of the mean particle sizes at the nine positions in the debris flow fan are shown in Figure 5.30. If one compares the particle sizes embedded in the lower half of the deposit in the experiments along the centerline of the fan; points 1, 2 and 3, one finds a tendency for the finer particles to deposit close to the outlet (point 1) and for the larger particles to deposit near the distal end (point 3). If one further compares the particle sizes embedded in the lower half of the deposit in experiments laterally; e.g., points 2, 5 and 8, one finds that the size becomes larger toward the edge. Namely, in the early stage of deposition (the lower part of the deposit corresponds to this stage), as referred to in the results of the numerical simulation, the bore front carrying the largest particles continues its motion, bulldozing the coarse particles laterally as well as longitudinally and then stops forming a crescent deposit. The following finer grained more fluid flow is deposited inside the crescent levee-like deposit, thus, as typically seen in Figure 5.30 for point 2, the particle size in the lower half becomes smaller upwards corresponding to the deposition of the belated run-off of finer particles. However, there is a tendency in the experiments that the particle diameters in the upper layers of the deposit become larger upwards to the extent that the diameters become as large as the mean diameter of the debris flow material (3.08 mm) or a little larger than that.

Next, we examine the results of the calculation. If one compares the particle diameters deposited adjacent to the surface of the board along the centerline of the fan; points 1, 2 and 3, the largest particles whose mean diameter is, as shown in Figure 4.28, about 6 mm and accumulated at the bore front are deposited around point 3, and the mean diameter of deposited particles decreases upstream. Presumably because of the poor accuracy in measuring the mean diameters within deposit layers in every 1 cm thick interval, the calculations do not seem to agree well with experiments,

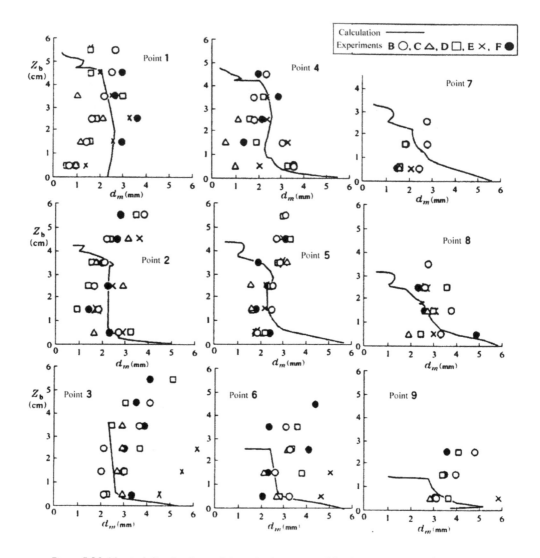

Figure 5.30 Vertical distributions of deposited mean particle sizes at nine sampling stations.

but the experimental tendency is reasonably well explained by the calculation. The experimentally obtained particle diameters at point 1 adjacent to the bed are as small as about 1 mm, whereas, as shown in Figure 4.28, the particle diameter given as the upstream boundary condition at the moment deposition commences at point 1 is larger than about 2 mm. Therefore, the deposition of such a small particle cannot be explained by this numerical simulation. If one compares the particle diameters laterally at points 2, 5 and 8, one finds that the largest size particles are deposited thickly toward the edge. Looking at the size distributions vertically at points 4 and 5, one finds that the particle size at first decreases upwards from the largest at the lowest layer to a smaller size than the mean diameter of the experimental material, and after that, particles as large as the mean diameter of the experimental material are again deposited on the already deposited layer, and finally, the diameter becomes very small. This tendency is in accord with the calculated results of temporal particle size variations shown in Figure 4.28.

However, the evident discrepancies in the thicknesses of the deposit and the particle sizes in the upper layer exist between the calculations and the experiments. Namely, the thicknesses of the deposit especially at the points 2, 3, 5, 6 and 9 in the experiments are far larger than those by the calculations. This fact may suggest that, in the experiments, the given debris flows to the flooding board had longer duration than the one given by Figures 4.27 and 4.28. The tendency of deposition of particles larger than the mean diameter of debris flow material near the surface of the deposit in the experiments, like at points 3 and 6, may be the results of the washing out of the finer particles by the surface flow that appears towards the end of an experimental run, because points 3 and 6 are located at the steep edge of the deposit. For a more appropriate explanation of these discrepancies and for more accurate prediction of the geomorphology of the debris flow fan, the model modification especially to include the particle segregations during the depositing process is necessary.

5.5   EROSION AND DEFORMATION OF A DEBRIS FLOW FAN

5.5.1   Experiments for the process of erosion

A debris flow fan was produced on the flooding board of 2° in slope, connected to the upstream debris flow formation channel with width of 10 cm and slope of 18°. The debris flow was generated by supplying a plain water flow of 1,000 cm³/s on a soaked sediment layer laid on the channel bed, the material was a mixture of particles; $d_{50} = 2.03$ mm and $(d_{84}/d_{16})^{1/2} = 3.8$. Then, after removing the sediment deposited in the upstream channel, a prescribed rate of water flow was supplied via the upstream channel onto the fan to erode it freely. Figure 5.31 shows the processes of formation of the incised channel by erosion and the redeposit of sediment downstream (the topography shown in Figure 5.31 was produced by a water flow of 500 cm³/s), where the

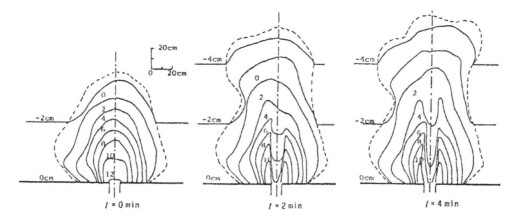

Figure 5.31   Processes of erosion and redeposit on a debris flow fan.

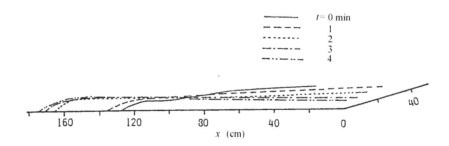

Figure 5.32 Temporal change in the longitudinal profile of the incised channel and redeposit.

broken line indicates the fringe of the deposit and the topography is shown by contour lines whose datum is set at the bed of the debouch. As is evident in Figure 5.31, a straight incised channel without conspicuous widening is formed, and at a nearly fixed position from the channel outlet within the incised channel, the so-called 'intersection point' appears, downstream of which a new alluvial fan is formed.

The temporal change in the longitudinal profile of the channel produced along the centerline of the fan is, as demonstrated in Figure 5.32, almost linear and gradually flattens. The crossing point of the profiles in Figure 5.32 is the intersection point. The distal end of the redeposit part has a similar thickness to that of debris flow fan, whose longitudinal surface slope is nearly 20°. When the longitudinal sediment transportation becomes small due to the flattening of the newly formed deposit, the downstream propagation of the deposit front virtually stops and the flow begins to expand laterally and a new alluvial fan is formed. According to the observation in the experimental runs, the infiltration of water into the newly formed deposit seems to play an important role in determining the thickness at the distal end. But, it should be checked in more detail by further experiments.

5.5.2   Model and its verification for the fan comprised of uniform material

The processes of the channel formation and the redeposit should be analyzed by the two-dimensional mathematical model introduced in section 3.3.4. Herein, however, considering the dilute sediment concentration in the flow during the processes, alternative to the erosion velocity equation the conservation equation of sediment discharge is used, where the river bed variation is brought by the difference between the flow-in and the flow-out sediment discharges. Further, we assume that there is no widening of the incised channel.

The governing equations are: the momentum conservation equations, Equations (3.115) and (3.116), in which as the resistance terms Equations (3.120) and (3.121) are used, the continuity equation of water, Equation (3.117), in which the right-hand side term is set equal to zero, the sediment discharge equation that may be used in the case of immature debris flow as well; Equation (3.126) and the accompanying

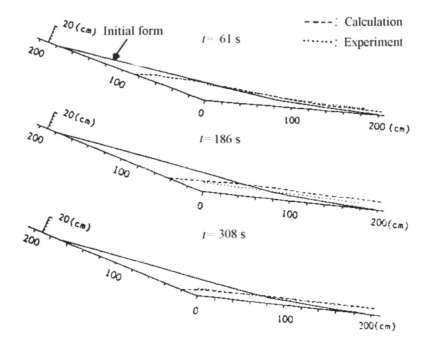

Figure 5.33 Erosion and redeposit in a one-dimensional channel.

equations from Equations (3.127) to (3.132), and the continuity equation of sediment discharge that is given as:

$$\frac{\partial z_b}{\partial t} + \left(\frac{\partial q_{bx}}{\partial x} + \frac{\partial q_{by}}{\partial y}\right) \frac{1}{C_{*DL}} = iC_{*DL} \tag{5.86}$$

where q_{bx} and q_{by} are respectively the x and y component of sediment discharge and they are assumed to be divided by the ratio of velocity components as follows:

$$q_{bx} = \frac{u}{\sqrt{u^2 + v^2}} q_b, q_{by} = \frac{v}{\sqrt{u^2 + v^2}} q_b \tag{5.87}$$

To verify the application of this mathematical model, let us first compare the calculations with the results of a one-dimensional experiment. The experiment was done using a jackknife flume whose upstream channel is 18° and the downstream channel is 2°, and the channel width is 20 cm. The debris flow was deposited by the same procedure as the experiments shown in Figure 5.8, then, the deposit was eroded by a water flow of 1,000 cm³/s. The calculation was performed by setting $\Delta x = 10$ cm, $\Delta t = 1/500$ sec, $n_m = 0.025$, and the initial condition was the topography of the debris flow deposit before erosion took place. Figure 5.33 compares the results of calculation with the experiment. We can see that this mathematical model can rather well reproduce the one-dimensional erosion and redeposit processes. But, concerning

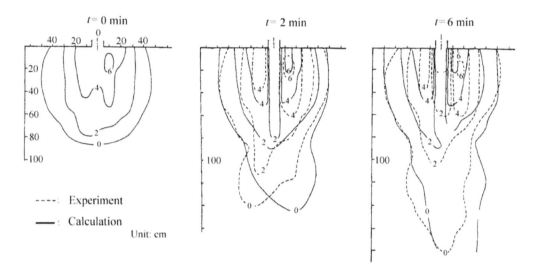

Figure 5.34 Erosion and redeposit of a debris flow fan (uniform material case).

the situations in the upstream channel, we cannot compare the results because no measurement in the upstream channel was carried out.

Next, we compare the calculations with the experiment for the erosion and the redeposit of a debris flow fan. A debris flow fan was produced by the procedure explained in section 5.3.2, where the debris flow was produced by supplying water with a discharge $600 \, cm^3/s$ for 20 seconds under the combination of the upstream channel slope of $17°$ and the flooding board slope of $2°$, and the experimental material was nearly uniform with $d_{50} = 1.28$ mm. The sediment that rebounded to the upstream channel was removed and a slight depression was artificially made along the centerline of the fan to guide the water flow straight. The removal of sediment in the upstream channel was implemented to avoid the formation of a sand bar just downstream of the channel outlet that causes the deviation of stream channel on the fan. After such artificial arrangements, a plain water flow of $300 \, cm^3/s$ was introduced into the upstream channel, and the flow depth in the incised channel and the thicknesses of deposit were measured every two minutes, the continuous change of the fan shape was recorded by video.

Figure 5.34 shows the initial fan morphology after the arrangements mentioned above and compares the results of the calculations with the experiments at two minutes and six minutes later than the introduction of the water flow. After two minutes, the incision proceeds straight along the centerline and an immature debris flow fan is formed downstream; and after six minutes, this newly produced immature debris flow deposit is also eroded. The calculation seems to reproduce the phenomena rather well.

The calculated value of the erosion velocity near the outlet of the upstream channel is, however, a little faster than that in the experiment. The reasons for this are as follows: In the experiment the infiltration of water from the wetted perimeter of the incised channel becomes considerably large, hence the actual discharge in the channel may be less than the assumed $300 \, cm^3/s$; although the experimental material is nearly uniform ($d_{50} = 1.28$ mm), it contains particles as large as about 3 mm and an armor coated bed may appear in the channel; although the channel expansion is neglected in the calculation, as is seen in Figure 5.34 at $t = 6$ minutes, actually the side walls fall on the bed resulting in some channel enlargements and thus the supplied material on the bed retards the erosion of the bed. The last possibility is investigated by the analysis introduced in section 3.3.4.

5.5.3   Model and its verification for the fan comprised of heterogeneous material

The small assemblages of particles with different mean diameters distribute on and underneath the incised bed along the centerline of the fan reflecting the mechanism during the fan producing process. With the progress of erosion, the particles buried underneath will appear whose mean diameter is different from that of original bed surface material. Furthermore, at a location in the channel, particles having a different mean diameter to that on the bed will be supplied from upstream and if erosion takes place particles on the bed will be entrained into flow changing the mean diameter of particles in the flow. Hence, by the result of erosion, particle diameters on the bed and that in the flow will be spatially as well as temporally different. If the mere exchange of particles between the coming-in particle and the going-out particle without any bed variation is neglected, the process of particle diameter change can be formulated, for a one-dimensional case, as follows:

In the case of erosion:

$$f_i q_i = f_{i-1} q_{i-1} + d_s (q_i - q_{i-1}) \tag{5.88}$$

In the case of deposition:

$$f_i = f_{i-1} \tag{5.89}$$

where f_i is the particle diameter going out from the mesh i, f_{i-1} is the particle diameter coming into the mesh i, q_i is the sediment discharge going out from the mesh i, q_{i-1} is the sediment discharge coming into the mesh i, and d_s is the particle diameter on the bed of the mesh i. For the two-dimensional formulation, considerations by classifying the cases depending on the directions of sediment transportation are necessary. Herein the descriptions of the formulae are omitted, but in the following calculations the two-dimensional version are used.

The addition of the particle size variation formulae to the system of equations used in the uniform material case enables the analysis of erosion and redeposit of the debris flow fan produced by a heterogeneous material.

An experiment was conducted to verify the mathematical model. The debris flow fan was produced under the same conditions as described in section 5.4.2, and it was eroded by a water flow of 300 cm³/s for ten minutes. Different from the previous case for uniform material, neither the removal of sediment in the upstream channel nor the modification of the fan shape was done. During the initial ten minutes or so, sediment that accumulated in the upstream channel ran out and it covered the fan, then, an incised channel was formed. By the effects of the deposit formed just downstream of

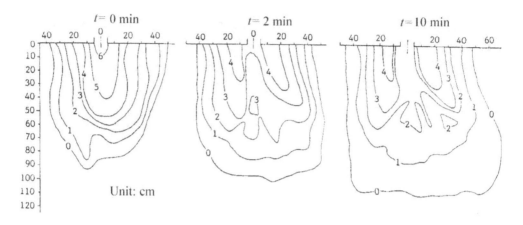

Figure 5.35 Processes of erosion and redeposit on a debris flow fan (heterogeneous material case).

the debouchment, the stream channel downstream deviated to the right or to the left depending on the runs. Several runs under the same conditions were repeated, and among them the case of a nearly straight channel was formed is discussed thereinafter.

Figure 5.35 shows the time sequential variations in the morphology of the debris flow fan. The channel formed in the central part was divided by a mound comprised of coarser particles into two branches and the mound became an island. The main flow appeared in the left-hand side branch. By about six minutes, the island became obscure, and after that gradual erosion proceeded generally in the entire central portion. The clear side banks were formed close to the debouchment, and after about six minutes no conspicuous morphological change took place.

Before the numerical calculation of morphological changes on a debris flow fan, a fan was produced experimentally again under the same condition as the one that produced the fan depicted in Figure 5.35. The whole fan area was divided into 10 cm × 10 cm meshes and at all the grid points the particle size distributions at a 1 cm depth intervals were measured. Thus, the three-dimensional distributions of mean particle sizes in the debris flow fan were obtained. This information together with the topographical information of the fan, consisted of the initial conditions of the numerical calculation of morphological changes of debris flow fan due to water flood. Because the experimental run to obtain Figure 5.35 and the run to obtain the initial conditions of calculation are different, the morphology and the particles size distributions in the calculation are not exactly the same as those of the fan shown in Figure 5.35. In the calculation, to ensure the initial formation of incised channel, a channel of 10 cm wide with a bed thickness of 4 cm parallel to the surface of flooding board was preset connecting with the upstream flume.

Figure 5.36 shows the temporal variations of the fan morphology by calculation, where the supplied water discharge is 300 cm^3/s. Because of the reasons mentioned above, we cannot compare directly the results of calculation with the results of experiment shown in Figure 5.35. But, the general tendency of the channel morphology on the fan seems to be reproduced. Notwithstanding that the side bank erosion was

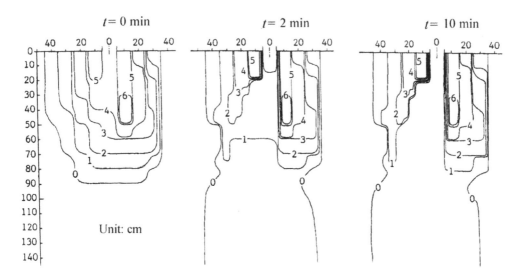

Figure 5.36 Variations of the fan morphology by calculation.

not taken into account in the calculation, the downstream part of the initial channel was enlarged due to flooding, deposition and erosion around the downstream end. The most conspicuous difference between the calculation and the experiment was that in the calculation the vertical erosion velocity of the incised channel was very large and the surface of the flooding board was exposed by ten minutes, whereas in the experiment a bed layer about 2 cm thick was left at ten minutes. This is perhaps due to the use of a mean diameter as a representative diameter to calculate the bed erosion rate irrespective of the existence of larger particles. At the central part of the fan fine particles are deposited whose mean size particles can be transported on a mild slope as flat as the flooding board; 2°. In the actual case, however, as mentioned in the uniform material case, some particles larger than the mean size may prevent erosion. The method to take the size distributions into account in the calculation of riverbed variation is explained in Chapter 7.

Chapter 6

Sediment runoff models that include debris flow processes

The photographs above show the same ravine at two locations. The left shows the situation where a vast amount of sediment is flowing into a reservoir. The photograph on the right shows a newly produced talus deposit at the foot of a landslide upstream from the location shown on the left. Such erosion of talus and riverbed is the source of the sediment runoff into the reservoir. In the steep upper reaches of the ravine, the main mode of sediment transportation is debris flow, and in the downstream reaches just up from the reservoir this mode shifts to individual particle transportation. Therefore, the sediment runoff model in such a ravine must take into account the mode change in sediment transportation during the runoff processes.

INTRODUCTION

In this chapter a distributed sediment runoff model that is suitable for a devastated steep and high mountain watershed is presented. This physics–based model, named 'SERMOW' (SEdiment Runoff from MOuntain Watershed), includes the processes of sediment production (loosening and detachment of particles on a slope), transport, and deposition. The sediment production on a mountain slope in an extreme event such as a landslide, and its transformation into a debris flow, are processes and mechanisms that are discussed in Chapter 3. Although such an extreme event can be taken into account as one of the phenomena that supply sediment from the mountain slope to the stream channel, SERMOW focuses on the rather normal events such as freeze-thaw cycles, raindrop splash, and rill and inter-rill flows that cause the sediment production and motion on the slopes. Such phenomena are more common on bare-land slopes than on vegetated slopes. Because the majority of the sediment produced and moved by these phenomena was once stored on the talus existing on the foot of the bare-land slope, SERMOW focuses on the erosions of talus and riverbed, massive and individual sediment transportation, and deposition. Sediment supply to the talus controls the advance and retreat of the talus, and so it indirectly controls the amount of sediment transportation in the river channel. In this context, the estimation of the erosion rate on the bare-land slope is an important matter and an estimation method is necessary.

SERMOW aims at predicting the sediment runoff from a watershed in terms of the sediment volume and also the changes in the particle size distributions contained both in the flow as well as in the bed depositions both over a long time period and for a single event. The conditions to be given in the simulation are the hyetograph and the initial watershed conditions such as the topography, the distributions of bare-land slope, sediment particle size distributions in the bed and the talus. Although the fresh occurrence of a landslide is not considered, debris flows must be the common phenomena generated in the river channel due to very steep channel gradient. SERMOW includes the phenomena of debris flow, immature debris flow, and individual particle motion, and it is capable of handling the mode changes of sediment transportation from debris flow to individual particle motion during the process of run down.

The original SERMOW assumes the material to be transported is coarse and one of the equations of motion for the stony debris flow, immature debris flow and the bed load is chosen to be used depending on the channel slope and the sediment concentration in flow. To extend the applicability of SERMOW to both coarse and fine material, SERMOW ver.2 was developed, where either the generalized equation for the inertial debris flow or the total bed-load equation is used depending on the channel slope and the sediment concentration in flow.

The applicability of SERMOW and SERMOW ver.2 is examined by the application to an actual watershed.

6.1   THE VIEWPOINTS FOR THE PROCESS-BASED MODELING OF SEDIMENT RUNOFF

6.1.1   What kinds of phenomena are considered?

The identification of the dominant phenomena in a particular watershed under an inducing force causing sediment runoff is crucially important for the construction of a reliable quantitative prediction model of sediment runoff. Identifying the phenomena relating to the sediment runoff that takes place in a watershed is problematic, because the phenomena are diverse, e.g. surface soil erosion, stream channel erosion, underground erosion, debris flow, surface landslide, deep-seated landslide, debris avalanche, creep, detachment, dissolution, and so forth. When extremely large events such as deep-seated landslides and debris avalanches are excluded, the sediment runoff processes due to a rainfall in an ordinary watershed can be modeled as shown in Figure 6.1.

When an objective watershed is small, flat and no channel system develops within it, neither the gully erosion nor the shallow landslide depicted in Figure 6.1 occur and only the phenomenon of surface soil erosion arises. Many of the existing sediment runoff models focused only on the surface soil erosion since they were concerned with the on-site hazards due to the erosion of farm lands and pasturelands. The Universal Soil Loss Equation (USLE), for example, relates the rainfall to the surface soil erosion without considering the hydrological transformation processes to the surface flow, subsurface flow, etc., and as the model is not concerned with the off-site hazards that occur at a remote downstream area, it does not consider the soil erosion as the sediment supplying source to the river channel (Wischmeier and Smith 1965). The other models

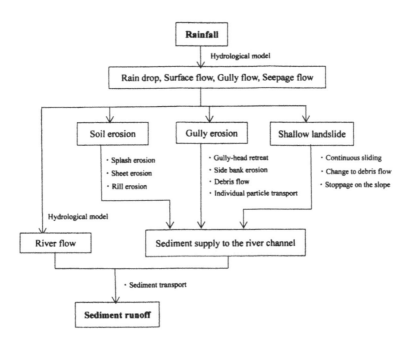

Figure 6.1 Sediment runoff processes due to rainfall in a watershed.

such as EUROSEM (European Soil Erosion Model) (Morgan *et al.* 1998) that were developed to be more faithful to the physical mechanism, consider the rills and the planes between the rills as the places to be eroded by the rain drop splash and/or the particle pick-up and transportation by surface sheet water flow. These models can treat the sediment runoff phenomena from a small and relatively smooth watershed, but the gully erosion and the surface shallow landslide are not taken into account.

Here, the sediment runoff phenomena both within and from a very steep and devastated watershed are the main point of interest. Therefore, the phenomena of gully (ephemeral channel) erosion and in some cases surface shallow landslides must be incorporated in the model.

6.1.2 Which of the two is to be predicted: average runoff in a period, or runoff at every moment?

The prediction of sediment runoff volume, over a year or for the duration of a flood, is the basis with which to grasp the sedimentation characteristics of a river basin and its tendencies over a long time span. This information is useful for discussing and managing the problems concerning the reservoir sedimentation, riverbed variation and coastal erosion. The existing models aiming to grasp the sedimentation characteristics of a river basin were developed under the governing principles of simple and easy-to–apply. For example, for a rather large watershed, the simplified distributed runoff models were developed, and the entire watershed area was divided into several unit slopes. The surface erosion volume in each unit slope was estimated applying USLE. Then the runoff sediment volumes from every unit slopes were added to obtain the total runoff volume from the watershed (e.g. Pilotti and Bacchi 1997; Van Romaey *et al.* 2001). The simple addition of the sediment runoff volumes from the unit slopes, however, over-estimates the total sediment runoff volume from a watershed because USLE does not take the deposition into account. Therefore, techniques for depositing sediment that is in excess of the ability of the river flow to transport sediment must be created.

The other physics-based models such as CREAMS (Chemicals, Runoff and Erosion from Agricultural Management Systems), SEM/SHE (Soil Erosion Model attached to Système Hydrologique Européen) and WEPP (Water Erosion Prediction Project) simulate a temporally continuous surface soil erosion event using the total rainfall or average flow discharge and the total sediment runoff during a rainfall event. These models assume the phenomena take place on a small slope surface, and hence they are usually not applicable to a large watershed.

It is desirable to obtain simultaneously the flood hydrograph and the sediment graph (sedigraph) that runoff from a watershed during a flood event. Models belonging to this category are, for instance, ANSWERS (Aerial Nonpoint Source Watershed Environment Response Simulation), KINEROS2 (Kinematic Erosion Simulation), GUEST (Griffith University Erosion System Template), Erosion 2D/3D, LISEM (Limburg Soil Erosion Model), EUROSEM (European Soil Erosion Model) and SHETRAN (Système Hydrologique Européen TRANsport). These models dynamically simulate the physical phenomena in a short period, but they have difficulties in setting the initial condition of watershed and require a long calculation time. The majority of them do not consider the runoff of infiltrated rain water and so they are not applicable to a large watershed.

SHETRAN (Bathurst 2002) is a distributed model concerning the water flow, sediment transport and pollutant transport in a vegetated watershed. It includes almost all factors depicted in Figure 6.1. Therefore, it is a comprehensive model suitable to a long-term simulation as well as a simulation for a rainfall event. It uses the SHE model as the hydrological model that converts rain to overland and subsurface water flows. Thereby the phenomena of evapo-transpiration, interception by trees, surface sheet water flow, channelized water-flow, ground-water flow, thawing, and the exchange of the surface flow with the groundwater are included. In this model, overland water flow appears as the rainfall in excess of the infiltration capacity of the ground and/or the rise of the groundwater level within the surface soil layer. Thus, the model is suitable to apply to a watershed as large as several hundred square kilometers.

The surface soil erosion in SHETRAN is caused by rain drop splash and sheet flow. The erosion rate by rain drop splash is given as a function of the momentum of rain drop, and the erosion rate by surface sheet flow is given as a function of shear stress due to surface flow. The coefficient of these functions must be given empirically. The bed and side-wall erosion rate and sediment transportation and deposition rate in the channel system are calculated by the bed-load equations classifying the particles into several size-classes.

The position and the occurrence time of surface landslide in SHETRAN are predicted both by the combination of the analyses of variations in water content within the surface soil layer and by the safety factors against landslide for an infinitely long slope. The calculation of the hydrological model is carried out discretizing the area by a grid system of $1–2\,km^2$, whereas the stability of slope is calculated in $10–100\,m$ grids. The soil block incorporated in the unstable grid cell is assumed to be immediately transformed into debris flow and it moves downslope until the slope flattens to less than $4°$. If the debris flow reaches to the river channel before the slope becomes flatter than $4°$, it moves obeying the model for sediment transport in the river channel.

Gully erosion in SHETRAN is calculated under the grid-size that is comparable to a typical gully size (about $100\,m$). The runoff sediment volume from a gully is given by a function of the water flow that is produced by the oozing out from the gully wall and/or the concentrations of surface flow into the gully. However, at the moment, this function is not determined yet.

SHETRAN aims to predict sediment runoff from an ordinary mountain watershed, in addition to considering landslide and gully erosion as well. This would seem to be a promising model to apply to our aim of predicting sediment runoff phenomena within and from a very steep and devastated watershed. The difficulty in determining the many coefficients and parameter values, however, is one of the many problems to be reconsidered. Besides this problem, there are more crucial points as the assumptions including the immediate transformation of slid earth block into debris flow (see section 3.2.1), the occurrence of debris flow as only by landslide, landslide volume depends on the grid-size, the sediment transportation in river channel is controlled only by the bed-load equation, and the mechanism of gully erosion that is often the source of debris flow as in the Kamikamihorizawa mentioned in Chapter 1 is left uncertain. This is a comprehensive model including surface soil erosion and landslide (debris flow), but landslides occur only under very severe rainfall, so, except for the case of severe rainfall, the analysis for the possibility of landslide often results in the unnecessary consumption of calculating time and computer capacity. Therefore, the severe rainfall case that may cause a landslide would be best analyzed separately from the other medium-scale rainfall cases. The occurrence of landslides and the debris flow that is induced by a landslide are discussed in Chapters 7 and 8 (the fundamental mechanism of the transformation of landslide to debris flow is discussed in Chapter 3), and in this Chapter 6 the sediment runoff model for the case where no new landslide occurs is introduced.

6.2   A DISTRIBUTED SEDIMENT RUNOFF MODEL FOR A DEVASTATED MOUNTAIN WATERSHED: SERMOW

A distributed sediment runoff model: SERMOW (Takahashi *et al.* 2000b) was developed by the author and colleagues. This is essentially a river system sediment routing model and does not consider the detailed processes on the surface of mountain slopes. The sediment produced on a slope by landslide, debris flow or individual particle motions such as rainfall splash or rill erosion is considered to be carried onto the talus at the foot of the slope or directly to the channel bottom and there it is temporarily stored. After a while, it is eroded as debris flow or bed material load to yield from the watershed. The major source of sediment supply to a talus under normal climate conditions is the surface erosion of the bare slope that is left after a landslide. Therefore, the estimation of how much sediment is eroded and supplied to the talus under a given climate condition is crucial. A new landslide might supply a vast amount of sediment directly to the river channel and could be a major source of sediment. Such a sediment supply model including the phenomena of a natural dam formation and its destruction can be incorporated into the model if the position, volume and time of the landslide are known, but herein, due to the difficulty in the precise prediction, and since we are interested in the ordinal climate conditions, the occurrence of landslide is ignored.

SERMOW applies the kinematic wave method in the analysis of water discharge at an optional point in the river system. If the sediment concentration in the flow at that point is below the capacity of the flow to carry sediment, the erosion of the riverbed and the side banks including the talus takes place. If, on the contrary, the sediment concentration in the flow is denser than the capacity, deposition onto the bed takes place. As a consequence of numerous repetitions of erosion and deposition cycle down the river system, the sediment-laden flood flow appears in the outlet of the watershed. Thus, the model predicts not only the flood hydrograph, but also the sediment graph with the temporal change in the particle size distributions. The conditions to be given beforehand are the hyetograph and the watershed conditions such as the topography, the distributions of the existing bare slopes that are the landslide scars, and the quantities and the particle size distributions of the sediment that is initially accumulating on the riverbed and as talus deposit.

6.2.1   The constitution of SERMOW

Flood runoff analysis by the kinematic wave method

The kinematic wave method considers a watershed that is composed of a group of the unit slopes, and the flow on a unit slope is described by Manning's uniform flow equation and the continuity equation. Hence, the fundamental equations of flow on a unit slope are as follows:

$$q_{so} = \frac{1}{n_e} h^{5/3} \sin^{1/2} \theta_s \qquad (6.1)$$

$$\frac{\partial h}{\partial t} + \frac{\partial q_{so}}{\partial x} = r_e \qquad (6.2)$$

where q_{so} is the unit width discharge of the surface flow, θ_s is the slope gradient, n_e is the equivalent roughness coefficient, and r_e is the effective rainfall intensity.

The unit width discharge on a slope obtained by Equations (6.1) and (6.2) is supplied as the input to the river channel discharge per unit length; q_{in}, that is converted by the following formulae:

For the river-side slope:

$$q_{in} = q_{so} \frac{l'}{l}, \; l' = \frac{S_s}{L_s} \qquad (6.3)$$

For the upstream-end slope:

$$q_{in} = q_{so} \frac{l'}{B} \qquad (6.4)$$

where S_s is the area of the unit slope, L_s is the length of the unit slope, and B is the width of the river channel at the upstream end. The definition of l and l' is clear in Figure 6.2.

Neglecting the increase and decrease in water discharge that is accompanied by the erosion and deposition of sediment, the flood runoff in the river channel is given by the following equations:

$$q = \frac{1}{n_m} h^{5/3} \sin^{1/2} \theta \qquad (6.5)$$

$$\frac{\partial h}{\partial t} + \frac{1}{B} \frac{\partial (q_o B)}{\partial x} = \frac{q_{in}}{B} \qquad (6.6)$$

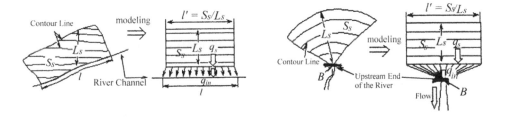

Figure 6.2 Modeling of slopes. Left: river-side slope, Right: upstream-end slope.

where q_o is the unit width discharge in the river channel, n_m is the Manning's riverbed roughness coefficient, B is the channel width, and θ is the water surface gradient.

Resistance law for the flow in the river channel

Generally, the riverbed is comprised of erodible material and the river flow entrains that material to change into sediment loading flow. When it meets with a talus at the foot of a bare-land slope, it erodes the talus, increasing the sediment concentration. The maximum possible sediment concentration in the flow is mainly determined by the gradient of the channel, and the flow can be a debris flow, immature debris flow, bed load and suspended-load flow depending on the concentration. The resistance law for the flow with bed-load and suspended-load is the same for that of the plain water flow; i.e. Equation (6.5). If the value of n_m is appropriately estimated, Equation (6.5) is also applicable to the turbulent muddy-type debris flow. But, for a stony debris flow and for an immature debris flow, the other resistance laws that appeared in section 3.1.2 must be used. These resistance laws are given as follows:

For stony debris flow (when $h/d_{mL} < 20$, $C_L \geq 0.4C_{*L}$ or $C_L > 0.2$):

$$q = \frac{2}{5d_{mL}} \left\{ \frac{g}{0.02} \frac{\sigma C_L + (1 - C_L)\rho_m}{\sigma} \right\}^{1/2} \left\{ \left(\frac{C_{*L}}{C_L} \right)^{1/3} - 1 \right\} h^{5/2} \sin^{1/2} \theta_w \qquad (6.7)$$

For immature debris flow (when $h/d_{mL} < 20$, $0.01 < C_L < 0.4C_{*L}$ or $0.01 < C_L < 0.2$):

$$q = \frac{0.7\sqrt{g}}{d_{mL}} h^{5/2} \sin^{1/2} \theta_w \qquad (6.8)$$

For bed-load / suspended-load or turbulent muddy-type debris flow (when $C_L \leq 0.01$ or $h/d_{ml} \geq 30$):

$$q = \frac{1}{n_m} h^{5/3} \sin^{1/2} \theta_w \qquad (6.9)$$

where q is the total flow rate of water plus sediment in unit width, d_{mL} is the mean diameter of the coarse particles in the flow, and C_{*L} is the volume concentration of coarse particles in the sediment deposit.

Preparation for the calculation of particle diameter change

To consider the variations in the particle-size distributions in the flow and on the bed, grain size is divided into k_e groups, and the diameter of the kth group grain is written d_k. Particles from group $k = 1$ to $k = k_1$ are defined fine and are considered to constitute a fluid phase if contained in the flow. Particles from groups $k = k_1 + 1$ to $k = k_e$ are defined as coarse particles.

The volumetric concentration of the coarse and fine fractions, the density of the interstitial muddy fluid, and the mean diameter of the coarse particles in the flow are expressed by:

$$C_L = \sum_{k=k_1+1}^{k_e} C_k \tag{6.10}$$

$$C_F = \left(\sum_{k=1}^{k_1} C_k \right) / (1 - C_L) \tag{6.11}$$

$$\rho_m = \rho + \frac{\sigma - \rho}{1 - C_L} \sum_{k=1}^{k_1} C_k = \rho + (\sigma - \rho) C_F \tag{6.12}$$

$$d_{mL} = \left(\sum_{k=k_1+1}^{k_e} d_k C_k \right) / C_L \tag{6.13}$$

where C_k is the volumetric concentration of group k particles in the total volume of water and sediment.

Because the particle size composition of the riverbed material is not necessarily the same as the particle size composition in the flow, the composition in the flow will change through the entrainment of the bed material. To determine the particle size composition of runoff sediment, the particle composition of the riverbed with which the flow exchanges particles must be known.

Assuming that the total volume of the group k particles on the bed is V_k, the existing ratio of this group's particles to the total particles (coarse plus fine ones) is:

$$f_{bk} = \frac{V_k}{V_L + V_F} \tag{6.14}$$

where V_L and V_F are the total volumes of coarse and fine particles, respectively, and they are given by:

$$V_L = \sum_{k=k_1+1}^{k_e} V_k, \quad V_F = \sum_{k=1}^{k_1} V_k \tag{6.15}$$

The existing ratio of group k particles (group k signifies coarse material) to the total coarse particles, f_{bLk}, is:

$$f_{bLk} = \frac{f_{bk}(V_L + V_F)}{V_L} = \frac{f_{bk}}{\sum\limits_{k=k_1+1}^{k_e} f_{bk}} \equiv \frac{f_{bk}}{F_p} \tag{6.16}$$

The following points must be noted about the structure of the bed. If the total volume of fine particles is small, the coarse particles form a skeletal structure; and if the fine particles are stored only in the void space of the coarse particle's framework, the volume ratio of the fine particles to the volume of the void space, C_{*k} $(k = 1 \sim k_1)$, is obtained via the following considerations: Provided the total volume of the bed is V and the volume of void space is V_v, then, $V_L + V_v = V$. Writing $V_L/V = C_{*L}$, the relation $V_v = V - V_L = (1 - C_{*L})V_L/C_{*L}$ is obtained, where C_{*L} is equal to the volume concentration of all the coarse particles when the bed is composed of only coarse particles. Consequently,

$$C_{*k} = \frac{C_{*L}}{1 - C_{*L}} \frac{V_k}{V_L} = \frac{C_{*L}}{1 - C_{*L}} \frac{f_{bk}}{\sum\limits_{k=k_1+1}^{k_e} f_{bk}}, \quad k = 1 \sim k_1 \tag{6.17}$$

Because this formula is deduced under the assumption that fine particles are stored only in the void space of the framework formed by coarse particles, for the achievement of Equation (6.17) the following formula should be satisfied:

$$\sum_{k=1}^{k_1} C_{*k} = \frac{C_{*L}}{1 - C_{*L}} \frac{\sum\limits_{k=1}^{k_1} f_{bk}}{\sum\limits_{k=k_1+1}^{k_e} f_{bk}} \leq C_{*F} \tag{6.18}$$

where C_{*F} is the volume concentration of all the fine particles when it is composed only of fine particles. By introducing the definition of F_p that appears in Equation (6.16) into Equation (6.18), Equation (6.18) is rewritten as:

$$F_p \geq \frac{C_{*L}}{C_{*F} + C_{*L} - C_{*L}C_{*F}} \tag{6.19}$$

For simplicity, both C_{*F} and C_{*L} are assumed to be 0.65, then, the smallest F_p that satisfies Equation (6.19) is 0.74. Namely, when the ratio of the fine particles is greater than 26%, coarse particles can no longer form a skeleton and will be scattered among the accumulated fine particles. These fine particles form a skeletal structure with the volume concentration C_{*F}, but its void space is too small to store coarse particles. In this case, because $(V - V_L)$ is the bulk volume of fine particles plus the void space between fine particles, the substantial volume of the fine particles, V_F, is given by $(V - V_L)C_{*F}$. The definition of C_{*k} is the volume ratio of group k particles (group k signifying fine material) to the volume $(V - V_L)$. Therefore, in this case:

$$C_{*k} = \frac{V_k C_{*F}}{V_F} = \frac{V_k C_{*F}}{(V_L + V_F)\sum\limits_{k=1}^{k_1} f_{bk}} = \frac{f_{bk}C_{*F}}{1 - F_p} \tag{6.20}$$

Then, the volume concentration of coarse particle fraction on the bed, C_{*L}, is, from the relationships $V_F = (V_L + V_F)(1 - F_p)$ and $V_L = (V_L + V_F)F_p$:

$$C_{*L} = \frac{F_p C_{*F}}{C_{*F}F_p + 1 - F_p} \tag{6.21}$$

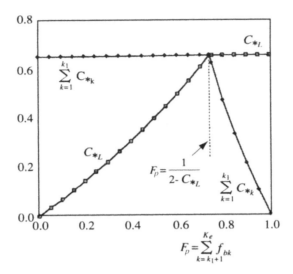

Figure 6.3 Volume concentrations of coarse and fine particles on the bed as functions of F_p.

Summarizing the above discussion, the following relationships are obtained:

When $F_p \geq 0.74$:
   Coarse particle concentration; $C_{*L} = 0.65$
   Fine particle concentration; $C_{*k} = C_{*L}f_{bk}/\{(1 - C_{*L})F_p\}$
When $F_p < 0.74$:
   Coarse particle concentration; $C_{*L} = F_pC_{*F}/(C_{*F}F_p + 1 - F_p)$
   Fine particle concentration; $C_{*k} = C_{*F}f_{bk}/(1 - F_p)$

Figure 6.3 shows how coarse and fine sediment concentrations change with the variation of F_p.

Continuity equations

The continuity equation for the total volume of water plus sediment is:

$$\frac{\partial h}{\partial t} + \frac{1}{B}\frac{\partial(qB)}{\partial x} = \frac{q_{in}}{B} + i_{sb}K_1 + i_gK_2\frac{h}{B} \tag{6.22}$$

where q_{in} is the inflow discharge of water plus sediment per unit length of channel except for the reach where a talus exists, B is the stream channel width within the valley (see Figure 6.4), i_{sb} is the erosion ($i_{sb} > 0$) or the deposition ($i_{sb} < 0$) velocity of the stream channel bed, i_g is the lateral erosion velocity of the bank that is formed by the talus deposit, and K_1 and K_2 are the coefficient given as follows:

$$\left.\begin{array}{ll} K_1 = C_{*L} + (1 - C_{*L})\{C_{*F} + (1 - C_{*F})s_b\} ; & i_{sb} > 0 \\ K_1 = 1 & ; \quad i_{sb} \leq 0 \end{array}\right\} \tag{6.23}$$

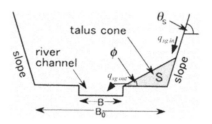

Figure 6.4 Sediment balance of a talus.

$$K_2 = C_{*gL} + (1 - C_{*gL})\{C_{*gF} + (1 - C_{*gF})s_g\} \tag{6.24}$$

where C_{*gL} is the volume concentration of coarse particles in the talus deposit, C_{*gF} is the volume concentration of the fine particles in the void space formed by the skeletal structure of coarse particles, s_b and s_g are the degree of saturation of the stream channel bed and that of the talus, respectively.

The lateral input of sediment in Equation (6.22) is divided into the one that is directly supplied to the stream channel and the one that is supplied by the erosion of the talus. If the stream channel width is almost equal to the valley width, the produced sediment on the slope is directly given to the stream channel as q_{in}, and if the produced sediment is added to the talus as shown in Figure 6.4 the sediment yielded from the slope is indirectly supplied to the stream channel as $i_g K_2 h/B$.

The continuity equation for each particle group is:

$$\frac{\partial(C_k h)}{\partial t} + \frac{1}{B}\frac{\partial(q C_k B)}{\partial x} = \frac{C_k q_{in}}{B} + i_{sbk} + i_{gk}\frac{h}{B} \tag{6.25}$$

where i_{sbk} is the bed erosion or deposition rate of kth group particles, and i_{gk} is the erosion rate of kth group particles of the talus. It must be noted that the summation of i_{sbk} from $k=1$ to $k=k_e$ is not equal to i_{sb}, because i_{sb} is the bulk erosion or deposition rate that include voids but i_{sbk} is the substantial erosion or deposition rate without voids. Because of the same reason, the summation of i_{gk} from $k=1$ to $k=k_e$ is not equal to i_g.

The variation of the bed arises in the stream channel, but due to the lateral shifting of the stream channel, it occurs in the whole width of the valley, so the following equation is satisfied:

$$\frac{\partial z_b}{\partial t} + \frac{B}{B_0}i_{sb} = 0 \tag{6.26}$$

where z_b is the elevation of the bed, and B_0 is the valley width. The reason why the right-hand side of Equation (6.26) is zero is that the lateral inflow of sediment is assumed to be thoroughly entrained into the flow.

Erosion and deposition velocity

The equation of erosion velocity that can be used regardless the type of sediment transport; debris flow, immature debris flow, or bed load transport is Equation (3.37). Because this equation is applicable for the case where the bed is composed of uniform material, d_L in this equation is replaced by the mean diameter of coarse particles d_{mL} and i is replaced by i_{sb0} which is the erosion velocity of the bed when the bed is composed of uniform d_{mL} particles.

In debris flow and immature debris flow, as mentioned in section 3.1, the largest particle that can be moved by the effect of surface flow is assumed to have the diameter that is the same as the depth of flow. Under this assumption, if $d_{k2+1} > h \geq d_{k2}$ is satisfied, the ratio of erodible coarse sediment to all coarse particles is $\sum_{k=k_1}^{k_2} f_{bLk}$, and the bulk erosion velocity is $i_{sb0}\sum_{k=k_1}^{k_2} f_{bLk}$. The substantial volume of coarse particles belonging to groups $k_1 < k \leq k_2$ is $C_{*L} f_{bLk}$. Therefore, the erosion velocity for each group particle when $d_{k2+1} > h \geq d_{k2}$ is:

$$
\left.
\begin{aligned}
k_1 < k \leq k_2 \; ; \quad i_{sbk} &= i_{sb0}\, f_{bLk}\, C_{*L} \sum_{k=k_1+1}^{k_2} f_{bLk} \\
k > k_2 \qquad ; \qquad\qquad i_{sbk} &= 0
\end{aligned}
\right\}
\tag{6.27}
$$

In bed load transport case ($\tan\theta < 0.03$), the critical tractive force of flow determines the size of the erodible particles on the bed and the erosion velocity of particle larger than that size is zero. Whether the particle of diameter d_k is movable or not is determined by the following modified Egiazaroff's equation:

When $d_k/d_{mL} \geq 0.4$:

$$
\frac{u_{*ck}^2}{u_{*cmL}^2} = \left\{ \frac{\log 19}{\log(19 d_k/d_{mL})} \right\}^2 \frac{d_k}{d_{mL}}
\tag{6.28}
$$

When $d_k/d_{mL} < 0.4$:

$$
\frac{u_{*ck}^2}{u_{*cmL}^2} = 0.85
\tag{6.29}
$$

where u_{*ck} is the critical friction velocity of the kth group particle, u_{*cmL} is the critical friction velocity of the particle whose diameter is d_{mL} ($u_{*cmL} = \{\tau_{*c}(\sigma/\rho_m - 1)g d_{mL}\}^{1/2}$).

Because the substantial volume ratio of fine particles is $C_{*k}(1 - C_{*L})$, the erosion velocity of fine particles ($k \leq k_1$) is given as:

$$
i_{sbk} = i_{sb0}(1 - C_{*L})C_{*k} \sum_{k=k_1+1}^{k_2} f_{bLk}
\tag{6.30}
$$

When the volumetric concentration of coarse particles in flow C_L at a certain position is larger than the equilibrium concentration $C_{L\infty}$ at that position, the coarse particles will be deposited on the riverbed. The bulk deposition velocity i_{sb0} will be given, corresponding to Equation (3.42) and neglecting the effects of inertial motion, as:

$$i_{sb0} = \delta_d \frac{C_{L\infty} - C_L}{C_{*L}} \frac{q_t}{h} \tag{6.31}$$

The deposition velocity of each coarse particle group is then given by:

$$i_{sbk} = i_{sb0} \frac{C_k}{C_L} C_{*L\,max}, \quad (k > k_1) \tag{6.32}$$

where C_{*Lmax} is the volume concentration of the coarse particles in the maximum compacted state.

If settling due to its own density is neglected, the fine particle fraction mixed with water in the flow is considered to constitute a fluid phase and it is trapped within the voids of the coarse particle's skeleton produced by the deposition of coarse particles. Then, the deposition velocity for fine particles is:

$$i_{sbk} = i_{sb0}(1 - C_{*L\,max}) \frac{C_k}{1 - C_L}, \quad (k \leq k_1) \tag{6.33}$$

But, such as at an estuary or immediately upstream of a check dam, if the shear velocity at a position is less than the settling velocity w_{sk} of kth group particle, deposition due to particle settling will also arise. In such a case the deposition velocity for fine particles is:

$$i_{sbk} = -w_{sk}C_k + \frac{i_{sb0}(1 - C_{*L\,max})C_k}{1 - C_L} \tag{6.34}$$

The settling velocity can be taken into account for coarse particles as well.

Consequently, erosion velocity in bulk that includes a void space is given by:

$$i_{sb} = \frac{1}{C_{*L}} \sum_{k=k_1+1}^{k_e} i_{sbk} \tag{6.35}$$

and the deposition velocity in bulk is:

$$i_{sb} = \frac{1}{C_{*L\,max}} \sum_{k=k_1+1}^{k_e} i_{sbk} + \frac{1}{C_{*F\,max}} \sum_{k=1}^{k_3} (-w_{sk}C_k) \tag{6.36}$$

where k_3 is the largest particle grade that satisfies $u_* < w_{sk}$, and C_{*Fmax} is the volume concentration of the fine particles in the maximum compacted state.

The value of $C_{L\infty}$ necessary to determine erosion and deposition velocities is different depending on whether the flow is a stony debris flow, an immature debris flow, or a bed load transportation. The types of flow are approximately classified by channel gradient; If $\tan \theta > 0.138$, the flow is stony-type debris flow and $C_{L\infty}$ is given by Equation (3.36), if $0.03 < \tan \theta < 0.138$, the flow is an immature debris flow and

$C_{L\infty}$ is given by Equation (2.144), and if $\tan\theta < 0.03$, the flow is a normal bed load transporting water flow and $C_{L\infty}$ is given, using Ashida, Takahashi and Mizuyama's bed load function (Ashida et al.'s 1978), as:

$$C_{L\infty} = \frac{(1 + 5\tan\theta)\tan\theta}{\sigma/\rho_m - 1}\left(1 - \alpha_0^2\frac{\tau_{*c}}{\tau_*}\right)\left(1 - \alpha_0\sqrt{\frac{\tau_{*c}}{\tau_*}}\right) \qquad (6.37)$$

where:

$$\tau_{*c} = 0.04 \times 10^{1.72\tan\theta} \qquad (6.38)$$

$$\alpha_0^2 = \frac{2\{0.425 - (\sigma/\rho_m)\tan\theta_w/(\sigma/\rho_m - 1)\}}{1 - (\sigma/\rho_m)\tan\theta_w/(\sigma/\rho_m - 1)} \qquad (6.39)$$

$$\tau_* = \frac{h\tan\theta_w}{(\sigma/\rho_m - 1)d'_{mL}} \qquad (6.40)$$

in which d'_{mL} is the mean diameter of movable particles on the bed. This is obtained, selecting only the particles whose u_{*ck} (given by Equation (6.28) or (6.29)) is smaller than the u_* of the flow, as:

$$d'_{mL} = \frac{\sum\limits_{k=k_1+1}^{k_4} d_k f_{bk}}{\sum\limits_{k=k_1+1}^{k_4} f_{bk}} \qquad (6.41)$$

where k_4 is the grade of the largest movable particle. The mean movable particle's size for stony debris flows and immature debris flows is also given by Equation (6.41), in that case $k_4 = k_2$.

Talus is formed by the falling particles from the upper bare slope, so the slope gradient of the talus is close to the angle of repose. Therefore, if the foot of the talus is eroded and taken away, that volume of removed sediment will be automatically replenished to the talus. The recession velocity of the foot of the talus lateral to the channel bed is assumed to be half of the bed erosion velocity as:

$$i_g = \frac{1}{2}i_{sb} \qquad (6.42)$$

The erosion velocity of the kth group particles are given as follows:

For the coarse particles:

$$i_{gk} = i_g\frac{f_{gk}}{\sum\limits_{k=k_1+1}^{k_e} f_{gk}}C_{*gL} \qquad (6.43)$$

For the fine particles:

$$i_{gk} = i_g(1 - C_{*gL})C_{*gk} \tag{6.44}$$

where f_{gk} is the existing ratio of kth group particles in the talus, C_{*gL} is the volume concentration of coarse particles in the talus deposit, and C_{*gk} is the volume concentration of fine particles ($k = 1 \sim k_1$) in the voids in the talus deposit and this is given as:

$$C_{*gk} = \left(\frac{C_{*gL}}{1 - C_{*gL}}\right)\frac{f_{gk}}{\displaystyle\sum_{k=k_1+1}^{k_e} f_{gk}} \tag{6.45}$$

No sediment supply from the stream channel flow to the talus is considered, so i_g and i_{gk} have always positive values.

Variation of riverbed grain size

The volume of solids in the surface layer of the riverbed V_s is related to the total volume of that layer V_t as:

$$V_s = V_t J, \quad J = C_{*L} + (1 - C_{*L})C_{*F} \tag{6.46}$$

The substantial volume of grains that is eroded from the surface area of S in a time Δt is $i_{sb}JS\Delta t$. When that volume is removed, the layer that existed below that layer becomes a new surface layer whose substantial volume of solids is $i_{sb}J_0 S\Delta t$, where J_0 is the J value of the lower layer. It is written as:

$$J_0 = C_{*L0} + (1 - C_{*L0})C_{*F0} \tag{6.47}$$

Writing V_s and the total volume of kth group particles V_k after the time Δt as V_s' and V_k', respectively, and if the existing ratio of kth group particles to the total particles in the lower layer is f_{0k} and the substantial erosion velocity of kth group particles is i_{sbk}, the existing ratio of the kth group particles f_{bk}' after Δt is given as:

$$f_{bk}' = \frac{V_k'}{V_s'} = \frac{V_k + (i_{sb}J_0 f_{0k} - i_{sbk})S\Delta t}{V_s + (i_{sb}J_0 - i_{sb}J)S\Delta t} \tag{6.48}$$

When the valley bottom width, B_0, is different from the stream channel width, B, the erosion of the entire valley bottom proceeds by the shifting of the stream channel and the erosion velocity becomes (B/B_0) times that in the case $B_0 = B$. Then, the differential form of Equation (6.48) gives the variation in the ratio of group k particles on the surface layer of the bed as follows:

$$\frac{\partial f_{bk}}{\partial t} = \frac{B}{B_0}\frac{i_{sb}J_0 f_{0k} - i_{sbk} - i_{sb}(J_0 - J)f_{bk}}{\delta_m J} \tag{6.49}$$

where $\delta_m = V/S$ is the thickness of particle exchange layer.

When deposition takes place, the group k particles that settle down to the bed of the unit area in unit time is $-i_{sb}Jf_{bk}$, where $i_{sb} < 0$. Noticing that the total settling volume is $-i_{sb}C_*$, where $C_* = C_{*Lmax} = C_{*Fmax}$, we obtain the existing ratio of kth group particles f'_{bk} after Δt as:

$$f'_{bk} = \frac{V'_k}{V'_s} = \frac{V_k + (i_{sb}Jf_{bk} - i_{sbk})S\Delta t}{V_s + (i_{sb}J - i_{sb}C_*)S\Delta t} \tag{6.50}$$

Considering the difference in B_0 and B, and adopting the differential form of Equation (6.50), we obtain the following equation for the variation in the ratio of group k particles on the surface layer of the bed:

$$\frac{\partial f_{bk}}{\partial t} = \frac{B}{B_0} \frac{-i_{sbk} + i_{sb}C_*f_{bk}}{\delta_m J} \tag{6.51}$$

6.2.2 Variation of grain size beneath the surface of the riverbed

Equations (6.49) and (6.51) renew the particle size distributions of the riverbed surface for each time step of the calculation, so the information on the old bed's characteristics is lost. For a rather long period of time, erosion and deposition repeat and as a consequence of those phenomena, a certain thickness of the sediment bed is left, whose particle size composition must be different from that of the original bed. The calculation of the riverbed variation by the method described above, however, needs only the particle size distributions of the bed surface that is renewed in every time step. Therefore, it does not memorize the particle size composition within the bed layer beneath the bed surface whose particle composition should be changed from that of the original bed. Thus, one cannot predict the sediment runoff with its particle size distributions from a newly formed sediment bed layer that is caused by the erosion and deposition cycle. To eliminate this shortcoming a method to memorize the grain size distributions within the bed layer that is affected by the erosion and deposition cycle is introduced.

Divide the riverbed into several layers of equal thickness of δ_s as shown in Figure 6.5. If the riverbed surface at a time exists in the layer m, the thickness of the surface layer, δ_a, is given by:

$$\delta_a = (z - z_s) - (m - 1)\delta_s \tag{6.52}$$

where z_s is the height of the rigid bed measured from the datum plane. If it becomes $\delta_a = 0$, the bed surface is considered to exist in the one layer above (add 1 to m) and it is set $\delta_a = \delta_s$.

When deposition takes place, the total volume of sediment in the mth layer increases from $(\delta_a \Delta x B_0 J_0)$ by $(-i_{sb}\Delta t \Delta x BJ)$ in the time Δt. Thus, the volume of kth group particles in the mth layer increases from $(\delta_a \Delta x B_0 J_0 f_{0k})$ by $(-i_{sb}\Delta t \Delta x BJf_{bk})$, thereby, the new existence ratio of the kth group particles in the new mth layer, $f_{0k\,new}$, is given by:

$$f_{0knew} = \frac{\delta_a \Delta x B_0 J_0 f_{0k} - i_{sb}\Delta t \Delta x BJf_{bk}}{\delta_a \Delta x B_0 J_0 - i_{sb}\Delta t \Delta x BJ} = \frac{\delta_a J_0 f_{0k} - i_{sb}\Delta t (B/B_0)Jf_{bk}}{\delta_a J_0 - i_{sb}\Delta t (B/B_0)J} \tag{6.53}$$

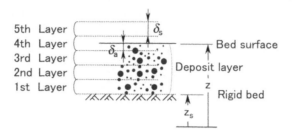

Figure 6.5 A thick erodible bed model.

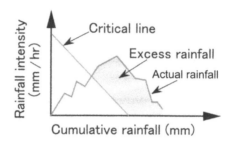

Figure 6.6 The definition of the excess rainfall.

When erosion takes place, if $\delta_a > i_{sb}\Delta tB/B_0$, $f_{0k\,new} = f_{0k}$, and if $\delta_a \leq i_{sb}\Delta tB/B_0$, f_{0k} in the mth layer is set to zero, since all the sediment in the mth layer is eroded.

6.2.3 Growth and recession of the talus

The growth and recession of the talus shown in Figure 6.4 are described as:

$$\frac{\partial S}{\partial t} = q_{sg\,in} - q_{sg\,out} \tag{6.54}$$

where S is the cross-sectional area of the talus, $q_{sg\,in}$ is the supply rate of sediment from the upper bare slope per unit width, and $q_{sg\,out}$ is the erosion rate of the talus by the flow in the stream channel per unit width. Here, the value of $q_{sg\,out}$ is assumed to be equal to $i_g h$ and it is given from Equation (6.42) as $0.5i_{sb}h$. Then, the problem is how to estimate $q_{sg\,in}$.

Generally, under a given rainfall intensity, the sediment production on a bare slope caused by the detachment due to rain drop splash and the surface flow erosion becomes greater with increasing cumulative rainfall. In Figure 6.6, a zigzag line of the actual rainfall record is drawn on a graph whose abscissa is the cumulative rainfall amount

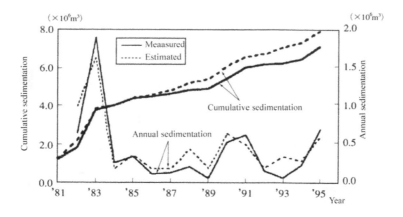

Figure 6.7 Comparison between the measured and estimated sedimentation volumes.

from the beginning of rainfall and the ordinate is the rainfall intensity in every one hour. A critical line descending towards the right may be drawn on this graph. We assume that only if the zigzag line crosses the critical line, will sediment be produced and supplied to the talus. The intensity of sediment supply is assumed to be proportional to the excess rainfall intensity that is defined as the surplus to the critical value. A discretionary critical line may be selected by giving the various combinations of the gradient and the intercept on the rainfall intensity axis. But, we set a valuation standard that the critical line must be the one that gives the highest correlation between the annual mean erosion depth of the bare-land slopes and the annual mean summation of the excess rainfalls. Using the sedimentation data of the Takase reservoir from 1982 to 1995 the annual mean erosion depth of the bare-land slopes was obtained by dividing the annual sedimentation volume by the total bare land area in the watershed. Thereby, we get an appropriate critical line that has the gradient of -0.2 (1/hr) and the intercept of 30 (mm/hr).

When the excess rainfall intensity at a time is r_0, the sediment supply rate per unit width onto the talus, $q_{sg\,in}$, is given as:

$$q_{sg\,in} = k_g r_0 \frac{A_g}{l_g} \tag{6.55}$$

where k_g (m/mm) is a coefficient, A_g (m^2) is the area of the bare-land that is connected to the concerned talus, and l_g is the length of the concerned talus along the stream channel. Equation (6.55) gives $q_{sg\,in}$ in one hour, so in the calculation it must be converted into that in a second. Figure 6.7 compares the calculated annual sediment supply from the Nigorisawa and Fudosawa basins by Equation (6.55) setting $k_g = 9.89 \times 10^{-4}$ (m/mm) with the actual annual sedimentation in the Takase reservoir. One can see that not only the annual and cumulative sedimentation volumes but also the yearly variations of sedimentation are well reproduced by this simple method.

6.3   INVESTIGATION OF ACTUAL SITUATIONS OF SEDIMENT RUNOFF PHENOMENA IN THE TAKASE DAM BASIN

The investigations based on the field survey on the characteristics of sediment production, transportation and deposition processes in a mountain river basin which has steep topography and severe erosion problems are comparatively few. Here, an investigation is introduced, carried out in the Takase dam basin, which is one of the severest sediment yielding watersheds in Japan (Inoue *et al.* 1999; Takahashi 2002).

The Takase dam and the Nanakura dam consist of a pumped-storage power generation system that was completed in 1978 (Figure 6.8). From 1978 to 1996, a total of about 12,900,000 m³ of sediment was deposited in the Takase dam reservoir (The total catchment area is 131 km²). Of the 12,900,000 m³ sediment; 4,000,000 m³ came from the upper Takase River (including the Yumata River) that is upstream of the reservoir (90 km² in the catchment area) and 7,400,000 m³ came from the Nigorisawa and Fudosawa ravines (12.8 km² in the total area of the two ravines). The specific sediment yield of the total Takase dam basin amounts to 5,470 m³/km²/year. This value is far greater than the average value for all mountain areas in Japan, i.e. 300 m³/km²/year. The specific sedimentation rate for the Nigorisawa-Fudousawa basin is as large as 32,043 m³/km²/year.

6.3.1   Characteristics of the Takase dam basin

The Takase dam basin is surrounded by mountains higher than 2,800 m in elevation. The boundary of the basin lies above the forest line. The area consists primarily of bare slopes, grass lands, and alpine shrubs. In winter, the entire basin is covered with snow, and snow remains even in summer in some sections of the basin. Average annual precipitation and temperatures at the dam site are around 2,100 mm and from −16 to 30°C, respectively.

Most of the Takase river basin areas form slopes steeper than 30°; the average mountainside slope gradient of the Nigorisawa basin is 36.32°, that of the Fudousawa basin is 33.33°, and that of the Yumata river basin is 28.55° (Nishii 2009). Geologically, more than 90% area of the Takase river basin consists of Cretaceous to Paleogene granites. Partially, there are volcanic rocks, alluvial sand and gravel deposits. Landslide sites, which are the major source of sediment, are highly concentrated in the Yumata River, the Nigorisawa Ravine and the Fudousawa Ravine, whose landslide area ratios are 15%, 25% and 20%, respectively. In these basins, there are hydrothermally altered and fragile granites whose steep slopes are easy to slide.

Figure 6.9 shows the temporal changes in the area of landslides in the Yumata, Fudosawa and Nigorisawa basins. As is clear in Figure 6.9 the landslide area increased by about a factor of 1.5 in 1969. This was due to the very heavy rain in August 1969. Except for this event, although some conspicuous rainfalls occurred, the landslide area has repeated small ups and downs in area. Therefore, the vast amount of sediment runoff into the Takase reservoir after 1978 must be mainly caused by the very active erosion of these bare land surfaces. However, no matter what things looked like at the small-scale, landslides either occurred or their areas were enlarged during that period. This fact may be used to determine the threshold for the occurrence of landslides in this watershed. In this context, in Figure 6.10, the conspicuous rainfall records were plotted

Figure 6.8 Takase dam basin.

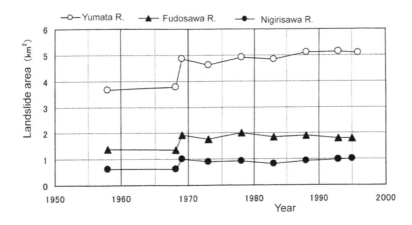

Figure 6.9 Changes in the area of landslides in the major sediment yielding basins in the watershed of the Takase reservoir.

in the same graph as that referred to in Figure 6.6. Because some landslides occurred in 1969, 1977, 1988, 2003, 2004 and 2005, a curve that intersects with the zigzag lines of these conspicuous rainfalls just at the acute peaks–except for the case of 1969–was drawn and defined as the threshold for the occurrence of landslide. One can see how severe the rainfall in 1969 was and only in such a case does the occurrence of new landslides increase in frequency. Furthermore, by drawing in Figure 6.10, the threshold

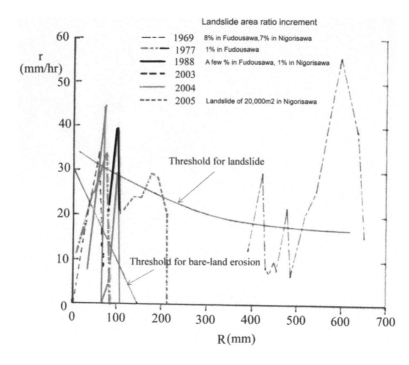

Figure 6.10 Conspicuous rainfalls and the threshold lines for the occurrence of bare-land erosion and landslides.

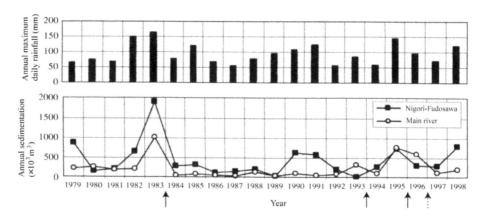

Figure 6.11 Annual maximum daily rainfalls and the annual sedimentation volumes. Arrows show the year in which detailed photogrammetry was carried out.

line for the occurrence of bare-land erosion that was obtained in section 6.2.1, one can understand that the surface erosion on the previously existing bare-lands would have been very active during this period of conspicuous rainfall.

Figure 6.11 shows the annual maximum daily rainfalls at the Eboshidake rainfall measuring station and the annual sedimentation volumes caused by the sediment runoffs from the Nigorisawa-FudosawaRavine and the main Takse River, respectively (see Figure 6.8). The Nigorisawa Ravine and the Fudosawa Ravine were originally two independent ravines, but the severe sediment runoffs from these two ravines made a compound fan at the entrance to the Takase reservoir, so we often consider them as a single ravine. The delta deposits from the Nigorisawa-Fudosawa Ravine and that from the main Takase River are still the independent deposits as can be seen in Figure 6.8. Because the Eboshidake rainfall station is located near the Nigorisawa-Fudosawa basin, the correlation between the annual maximum daily rainfall and the sedimentation volume from the Nigorisawa-Fudosawa Ravine seems better than for the main river. However, the degree of correlation is not high. This means, as a matter of course, we cannot explain the annual sedimentation volume by relying on the annual maximum daily rainfall alone, but need some more reliable models such as SERMOW.

6.3.2 Erosion of bare-land slope and riverbed variation in a medium time span analyzed by digital photogrammetry

Changes in the surface elevations of the bare-lands left after landslides and the riverbeds during some medium time spans were analyzed in the Nigorisawa-Fudosawa and the main river basins using aerial photographs. The 1:15,000-scale aerial photographs were taken in October 1983, 1993 and 1995 (that of the Yumata river basin was in 1996). See the arrows in Figure 6.11. The photographs in 1983 were taken immediately after a largest amount of sediment inflow to the reservoir since the completion of the reservoir. The 1993 photographs were taken after the two relatively dry and small sediment inflow years. The 1995 (1996 in the Yumata river) photographs were taken immediately after the intense rainfall and large amount of sedimentations both from the Nigorisawa-Fudosawa basin and the Yumata river basin.

Table 6.1 shows the balance between the sediment movement within the watersheds and the sediment deposition in the reservoir. The movement of sediment within the watershed was measured by photogrammetry and the sediment deposition was measured by sounding. The negative values in Table 6.1 mean the volumes eroded during the indicated periods of time and the positive values are the deposited volumes during the respective periods. It was assumed that when erosion takes place on the bare-land slope the sediment volume swells 1.3 times in the Nigorisawa-Fudosawa basin and 1.2 times in the Yumata river basin when deposited. The difference in the swelling rates between the two basins is the reflection of the degree of weathering. The surface of the bare-land in Yumata river basin contains much finer particles than in the Nigorisawa-Fudosawa basin.

Table 6.1 shows that almost all the bare-land slopes were eroded throughout the period of 1983 to 1996. Although there were some places of temporary deposition on these slopes, they were limited to the places subjected to the concentration of eroded sediment such as the hollow-shaped areas, knick-points towards gentler slopes, and the talus on the foot of bare slopes. The sediment produced on the bare slopes including that temporarily stopped on the slope was eventually transported via the channel system in the watershed. The river channel reach in Table 6.1 is divided into the reaches that steeper than 15° and that flatter than 15°. The sediment stored in the reach steeper than 15° would easily be flushed away as debris flow on the occasion of a comparatively large-scale flood. The debris flow may stop in the gentler downstream reaches,

Table 6.1 The balance sheets between the sediment movement within the watersheds and the sediment deposition in the reservoir.

$(\times 10^3 m^3 +: deposition, -: erosion)$

Basin Name	Location	1983~1993	1993~1995	1983~1995	Remarks
Nigorisawa	Bare slope	−1,143	−464	−1,561	volume expansion
	Riverbed ($\theta \geq 15°$)	222	−258	−36	factor = 1.3
	Riverbed ($\theta < 15°$)	280	289	568	mostly in the alluvial
	Subtotal	−641	−433	−1,029	fan area of $\theta < 7°$
Fudosawa	Bare slope	−1,563	−853	−2,303	volume expansion factor = 1.3
		(−126)	(82)	(73)	( ): deposition on
	Riverbed ($\theta > 15°$)	64	−104	−33	the slope
	Riverbed ($\theta < 15°$)	147	439	586	mostly in the alluvial
	Subtotal	−1,352	−518	−1,750	fan area of $\theta < 7°$
Nigirisawa-Fudosawa total		−1,993	−951	−2,779	corresponding to the to sediment inflow the reservoir
Measured by sounding		2,176	651	2,827	

Basin Name	Location	1983~1993	1993~1996	1983~1996	Remarks
Yumata R.	Bare slope	−1,081	−1,328	−2,340	volume expansion factor = 1.2
		(342)	(431)	(394)	( ): deposition on the
	Riverbed ($\theta \geq 15°$)	146	61	202	slope
	Riverbed ($\theta < 15°$)	64	70	135	
	Subtotal	−871	−1,197	−2,003	
Other main river basins	Bare slope	−223	−320	−530	volume expansion factor = 1.2
		(37)	(25)	(17)	( ): deposition on the
	Riverbed ($\theta \geq 15°$)	38	32	69	slope
	Riverbed ($\theta < 15°$)	149	216	365	
	Subtotal	−36	−72	−96	
Main river basin total		−1,307	−1,269	−2,099	corresponding to the sediment Inflow to the reservoir
Measured by sounding		672	1,296	1,968	

or on occasion may flow directly into the reservoir. The sediment once captured in the gentler slope reach is transported again as the bed load and suspended load and finally it reaches to the reservoir. Consequently, in the twelve years from 1983 to 1995, in the Nigorisawa-Fudosawa basin, the majority of the sediment that was produced on the bare-land slopes was transported into the reservoir (the negative value in the row indicated as 'Nigorisawa-Fudosawa total' in Table 6.1 means the sediment volume that was lost from the watershed upstream of the reservoir so that that volume was considered to flowed into the reservoir at a place lower than the high water level of the reservoir, HWL), with the remainder being deposited in the alluvial fan area upstream of the reservoir where the elevation is higher than HWL of the reservoir and flatter than 7°. The absolute value in the row 'Nigorisawa-Fudosawa total' in Table 6.1 that

Table 6.2 Average annual erosion depths of the bare-land slopes based on photogrammetry.

Basin Name	1983~1993	1993~1995	1983~1995	(cm/year) Remarks
Nigorisawa	8.8	17.8	10.0	
Fudosawa	6.7	18.2	8.2	
Nigorisawa-Fudosawa	7.4	18.1	8.8	
	(6.4)	(13.1)	(7.6)	from reservoir sedimentation
Basin Name	1983~1993	1993~1996	1983~1996	
Yumata R.	1.8	7.2	2.9	
Others	1.0	4.8	1.8	
Whole main river	1.6	6.6	2.6	
	(0.9)	(5.6)	(2.0)	from reservoir sedimentation

is considered to have flowed into the reservoir almost coincides with the value measured in the reservoir by sounding, a value that is also shown in Table 6.1. The similar discussion can be applied to the main Takase river basin including the Yumata River for the thirteen years from 1983 to 1996. However, if the period is divided into the ten years from 1983 to 1993 and the rest two or three years (for main river basin), the situation is different between the watersheds as well as within a particular watershed.

The average annual erosion depths of the bare-land slopes in the respective watersheds and in the respective periods were obtained by photogrammetry as shown in Table 6.2. The annual erosion depths converted from the volumes of annual reservoir sedimentation are also shown in Table 6.2 in the parentheses. The average annual erosion depth during the former 10 year period in the Nigorisawa-Fudosawa basin is 7.4 cm and that during the latter 2 year period is 18.1 cm. Those for the main river basin are 1.6 cm and 6.6 cm, respectively. The annual erosion depths in the Nigorisawa-Fudosawa basin are far larger than that in the main river basin, and in the Nigorisawa-Fudosawa basin the erosion depth during the latter period is 2.4 times that during the former period, however, in the main river basin the erosion depth in the latter period is 4.1 times the former period. That is, whereas the absolute value of the annual erosion depth in the Nigorisawa-Fudosawa basin is far larger than that in the main river basin, the range of fluctuation of the annual erosion depths is larger in the main river basin than in the Nigorisawa-Fudosawa basin. A theory that is able to predict the absolute value of the erosion rate of bare-land slope is needed. That theory, at the same time, must be able to make clear the reason why the erosion rate between the two watersheds and during the two periods differs. One clue to the problem may be as follows: the surface layer of the bare-land slope in the Takase Dam basin is loosened by the action of freezing and thawing during every winter and some of the thus produced particles fall onto the talus or to the riverbed only by the action of gravity during snow melting season. The other particles are transported down slope by the action of rainfall splash and surface flow. As is discussed in the following section 6.3.3, due to the flatter slope gradient in the main river basin than in the Nigorisawa-Fudosawa basin, both the sediment production rate and the transporting rate of the produced particles onto the talus or to the riverbed in the main river basin are less than in the Nigorisawa-Fudosawa basin. This must be the cause of larger erosion rate of the

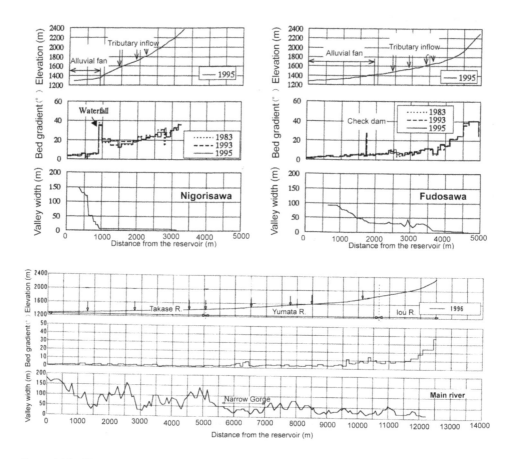

Figure 6.12 Characteristics of the river channels of the Nigorisawa, Fudosawa and the main river.

Nigorisawa-Fudosawa basin than the main river basin. Because of the flatter topography of the main river basin, the relative contribution of the particle transportation due to only the effect of gravity to that due to rainfall may also be less in the main river basin than in the Nigorisawa-Fudosawa basin. As shown in Figure 6.11, the maximum of the annual maximum daily rainfall since 1984 occurred in 1995 and the relative contribution of this rainfall to the total annual erosion depth for the main river basin should be larger than for the Nigorisawa-Fudosawa basin. Thus, an increase of erosion depth during the latter period in the main river basin should have occurred.

In Table 6.2, in the both basins, the average annual erosion depths obtained by photogrammetry are larger than the converted values from the reservoir sedimentation volumes. These differences are close to the volumes of sediment that was deposited on the riverbeds in the alluvial fan and other areas.

Figure 6.12 shows the longitudinal variations in the characteristics of the river channels, and Figure 6.13 shows the distributions of the landslide areas in the Nigorisawa-Fudosawa basin. The riverbed of the Nigorisawa abuts to the large-scale bare-land slopes in the upstream reach of the waterfall that is located at about 1,000 m

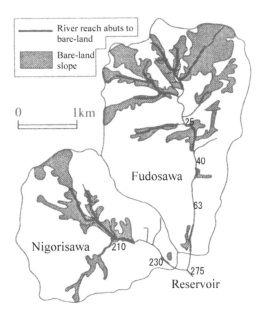

Figure 6.13 Distributions of the bare-slopes left after landslides in the Nigorisawa-Fudosawa basin.

from the reservoir and the riverbed of the Fudosawa also abuts to the large-scale bare-land slopes upstream about 3,800 m from the reservoir. In these river reaches the channel gradient is steeper than 15° meaning that debris flow can arise. Because the catchment areas are small, the water discharges in these reaches are small in ordinary times. In addition, the valley widths are very narrow and occasionally a big boulder that fallen down from a bare-slope beside chokes the valley, so these reaches are apt to store sediment that will be washed out as debris flow in the case of a severe rainfall.

Figure 6.14 shows the riverbed variations during some time spans that were measured from the aerial photographs. During the periods of 1983~1988, 1988~1993 and 1995~1997 when the average annual rainfall are relatively low, sediment storage in the reach steeper than 15° was large, and during the periods of 1993~1995 and 1995~1997 in which plenty sediment flowed into the reservoir, an enormous bed degradation took place in the same reach. The volumes of sediment deposited in the reaches steeper than 15° in the Nigorisawa and in the Fudosawa during the relatively dry years from 1983 to 1993 were about 220,000 m³ and 60,000 m³, respectively. The volumes of sediment eroded during the period from 1993 to 1995 which includes a severe rainfall in 1995 were about 260,000 m³ and 100,000 m³, respectively(see Table 6.1). Thus, the role of the reach that is abutting to the bare-slope and steeper than 15° is clear; it stores sediment gradually in the ordinary years and it flushes the storage all at once as debris flow at a chance of severe rainfall.

On the Fudosawa riverbed from 1,700 m to 3,800 m (bed slope: 7 to 15°), bed aggradation and degradation were repeated particularly at the junction of the tributaries, then, the slope and the valley width changed, but the range of their variation was not large. This reach plays a role as if a gutter that passes through the sporadic debris flow generated in the upstream reach steeper than 15°.

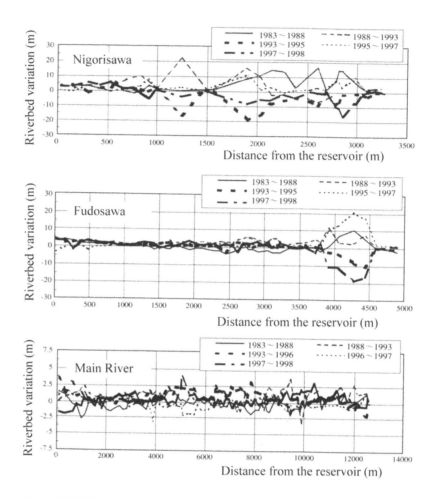

Figure 6.14 Riverbed variations in the Nigorisawa, Fudosawa and the main river.

The Nigorisawa river reach from the reservoir to the waterfall and the Fudosawa river reach downstream of the point 1,700 m to reservoir (bed slope: 2 to 7°) form a compound alluvial fan and sediment deposition predominates throughout the entire periods.

The main Takase River upstream of the reservoir is flatter than the Nigorisawa and Fudosawa. A long reach of about 10,000 m with the large fluctuations of valley widths exists upstream of the reservoir. The bed aggradation and degradation repeat at the junction of tributaries, slope and river width changes. A natural control of sediment runoff operates in this reach.

6.3.3 Measurements of bare-land erosion in the Takase River basin

Iron bars 40 to 60 cm long and 12 to 16 mm in diameter were drove into several bare-land slopes of the Fudosawa and Yumata River basins to measure the erosion

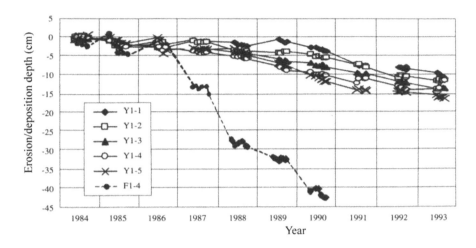

Figure 6.15 Results of measurements at the upper part of the slope.

and deposition thicknesses in a period. The slopes with $35°\sim50°$ gradients were the scars of landslide and they were softened by the effects of hydrothermal alteration and fracturing. The routine measurements were carried out once a month from June (rainy season) to October (before the onset of snow) and the extra measurements were taken immediately after a severe rainfall. The examples of measured results are shown in Figures 6.15 and 6.16 in which measured values are plotted by the different marks for the different measuring sticks and the group of points in one year is connected by a line segment to the points group obtained in the next year, so that no measured values are available on these line segments that correspond to winter season.

In Figure 6.15, the Y1-1 to Y1-5 zigzag lines show the results measured at five points on the slope 1 in the Yumata basin and the F1-4 line shows the results measured at the 4th stick on the slope 1 in the Fudosawa basin. All these points are located close to the ridge or in the relatively smooth surface areas where the once moved particles can hardly be stopped. According to Figure 6.15, although the reason for the small erosion rates from 1984 to 1986 is not clear, the erosion at F1-4 after 1986 is very active, and that the erosion rate in winter to snow melt season (from November to the beginning of June) is higher than in the rainy season. The average erosion rates at F1-4 point in the rainy season and the winter-snowmelt season in 1986 to 1990 are 1.7 cm/year and 8.6 cm/year, respectively. This tendency of variation in the seasonal change in the erosion rate and the value of the annual mean erosion depth are similar to the observations carried out at the Yakedake Volcano, Japan (Sawada and Takahashi 1994), where the measurements were carried out on a slope of unsolidified deposit of volcanic ejecta at about 2,000 m in altitude. The surface layer section (about 2 cm deep) of the slope is very loose in spring because of the heaving of frost columns, but this section is not evident in autumn thereby indicating a one year cycle fluctuation in the surface layer portion.

On the slope of the Yumata river basin, however, the annual mean erosion rate is about 1.8 cm/year, which is far less than that in the Fudousawa basin, and there is no conspicuous difference between those of the rainy season and winter-snow melt

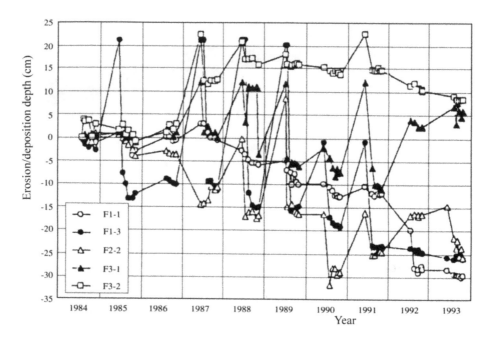

Figure 6.16 Results of measurement at the middle part of the slope.

season. In some years the erosion rate in rainy season is larger than that in the winter-snowmelt season. One reason for the evident difference in the erosion rates between the Y1 slope and the F1 slope would be attributable to the difference in slope gradient. At the Yakedake Volcano, Sawada and Takahashi (1994) discovered the relationship as follow:

$$E = a \sin^3 \theta \quad \text{(cm/year)} \tag{6.56}$$

where E is the erosion rate; θ is the slope gradient, and a is a coefficient that reflects the geological, geotechnical and climatological factors. Although Equation (6.56) was obtained on a slope flatter than the angle of repose, if it is assumed to be applicable to even steeper slopes and a is common to the Fudosawa and Yumata River and if the representative gradient of F1 is 50° and that of Y1 is 35°, the erosion rate of F1 can be about 2.4 times of that of Y1. The other reasons for small erosion rate in the Yumata river basin would be attributable to the geothermal frost melting effects and the higher clay content due to hydrothermal weathering in the Yumata river basin.

Figure 6.16 shows the results measured at the middle part of the slope in the Fudosawa basin where the topography is somewhat hollow-shaped. In such an area the sediment produced and transported from the upper part by the effects of gravity is deposited as shown by the sharp rises of the line segments in Figure 6.16. The deposited sediment is, then, washed out by the concentrated water flow that occurs on the occasion of comparatively heavy rainfall in the rainy season as shown by the sharp falls of the line segments in Figure 6.16. Thus, repeating a rather large fluctuation

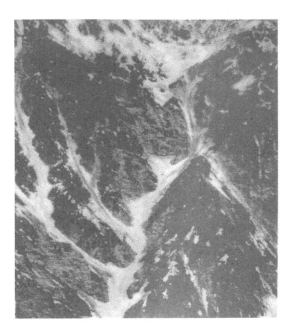

Photo 6.1 Sediment deposits on the snow gorge in the early snow-melting season.

of erosion and deposition, at some locations a tendency of the long-term erosion is observed and at other locations a tendency of the long-term deposition takes place. It is in this middle part of the slope that the sediment supply to the talus is controlled.

On some occasions, the sediment produced in the early snow-melting season is directly supplied by the action of gravity onto the bed of the gorge where plenty of snow is still remaining as shown in Photo 6.1. Such a deposit is easily entrained into debris flow that occurs for the first time in the rainy season.

6.4 APPLICATION OF SERMOW TO THE TAKASE DAM BASIN

6.4.1 Modeling of the watershed

The solid line system drawn on a 1:5,000-scale topographic map is taken to represent the river channel system and the boundaries of the catchment area of these channels demarcated the unit basins. Figure 6.17 shows these divisions for the Nigorisawa-Fudosawa watershed and Figure 6.18 shows the river channel system corresponding to Figure 6.17. In Figure 6.17, the fat solid lines indicate channels, the tetragons are the slopes abutting on the river channels, the triangles are the upstream-end basin and numbers are attached to the individual unit basins. The unit basins have various sizes and the whole Takase Dam basin (131 km²) was divided into 461 unit basins. Each basin area was measured on the topographic map, and the length and the slope gradient of each basin were obtained by drawing a line towards the steepest direction in the basin. The length of the river channel reach that abuts on a unit basin was set

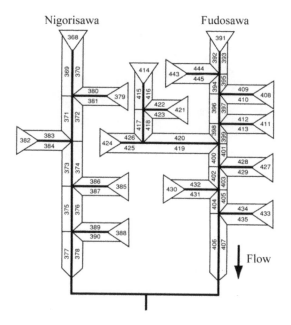

Figure 6.17 Division into unit basins for the Nigorisawa-Fudosawa basin.

equal to the product of the numbers of the calculating grids involved within the reach and the grid interval that was fixed to 50 m. The numbers attached to the river channel in Figure 6.18 indicate the calculating grid number. The point 2983 in Figure 6.18 is at the position 600 m off the entrance to the reservoir, and at this position, the water stage was given as the downstream boundary condition for the calculation.

The elevation of the riverbed of every channel reach was obtained from 1;5,000-scale topographic map and the valley widths were read from the aerial photographs of 1:15,000-scale. However, at some points near the upstream-end and mountain shadows, it was difficult to read the width. At such a reach the following regime concept was used: If the basin area upstream of the objective point is A and the effective rainfall intensity is r_e, the water flow discharge, Q, at that point would be written as kAr_e, and from the regime theory:

$$B_0 = \alpha Q^{1/2} \tag{6.57}$$

where B_0 is the valley width. Thus, the following relationshipis obtained:

$$B_0 = \left(\alpha k^{1/2} r_e^{1/2}\right) A^{1/2} = \beta A^{1/2} \tag{6.58}$$

The relationships between B_0 and A were read and plotted in Figure 6.19. Although the plotted points widely scattered, $\beta = 0.0064$ was obtained.

The particle size distributions of the riverbed material were measured at some representative positions as shown in Figure 6.20, where the abscissa is the particle diameter and the ordinate is the existing ratio of d_k particles by weight to the weight

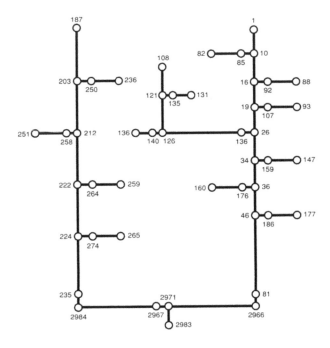

Figure 6.18 River channel networks of the Nigorisawa-Fudosawa basin.

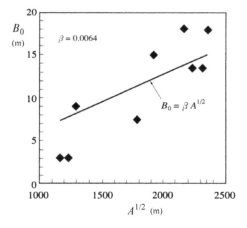

Figure 6.19 Relationship between the basin area and the valley width.

of the entire material. As a matter of fact, there were many stones larger than a few meters in diameter on the riverbed, but the size fraction that is larger than 30 cm was excluded in the measurement. The material that composes the talus was also sampled and size distributions were measured as shown in Figure 6.21.

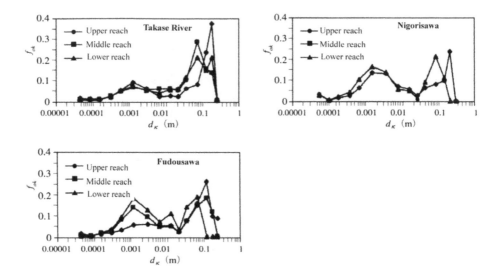

Figure 6.20 Particle size distributions of the riverbed material.

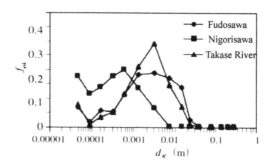

Figure 6.21 Particle size distributions of the talus deposit material.

6.4.2 Implementation of the first calculation

Conditions for the calculation

Calculations were carried out by transforming the governing equations that appeared in section 6.2 into discrete finite forms. The water discharge, sediment discharge, particle size distributions in the flow as well as on the bed, bed variation, etc. at an arbitrary river reach were obtained by giving the runoff discharges from the abutting slopes as the lateral input. At the junction of a tributary the water and sediment discharges of the tributary were forcibly added to those of the main river.

Here, as mentioned earlier, the sediment that was produced by surface erosion was considered to be once stored as a talus deposit and then due to erosion of the talus it was entrained into the river flow. When the sediment supply onto the talus is larger than the consumption due to erosion, the base of the talus lateral to the valley's cross-section develops towards the center of the stream channel. If there are two taluses on both sides of the channel, the maximum base widths to be able to grow are considered as the half width of the valley. If there is only one talus on one side of the channel, the possible maximum base width is set equal to the width of the valley. When the base width attains the maximum the growth in the volume of the talus ceases and the sediment supplied from the slope is directly given to the river flow.

The downstream ends of the Nigorisawa-Fudosawa and the Takase river basins are a reservoir and in the reservoir the water surface gradient is so small that the kinematic wave method that uses the resistance law of a steady normal flow may not be applicable. But, herein, the purpose of calculation is to obtain the inflow discharges of water and sediment from the upstream basins, so, for the sake of simplicity, the kinematic wave method is used by giving a constant water stage at 600 m off the entrance to the reservoir for the Nigorisawa-Fudosawa basin and at 750 m off the entrance to the reservoir for the Takase river basin as the downstream boundary conditions.

The grid size, Δx, was 50 m and the time step, Δt, was set from 2.0 s to 0.1 s depending on the maximum velocity in the last time step.

The total number of the particle-size division, k_e, was 15, in which 0~3 belong to the fine groups and the maximum particle diameter of the fine particles was 0.162 mm. The river channel reaches were divided by the gradient into three. The gradient of the downstream reach is from 0 to 3°; the middle-stream reach is from 3° to 7°; and the upstream reach is steeper than 7°. The particle-size distributions in the respective reaches for the respective rivers are given in Figure 6.20. The Manning's roughness coefficients were also changed for the respective reaches: 0.05 for the upstream reach; 0.04 for the middle-stream reach; and 0.03 for the downstream reach.

The solids concentration within the sediment deposit was set at a constant value of $C_* = 0.65$ irrespective to the size distributions. The density of particles, σ, was 2,650 kg/m³. The coefficient of erosion velocity, K, was 0.5, and the coefficient of deposition velocity, δ_d, was 0.0002. The degree of saturation of the bed, s_b, was set at 0.5 for the upstream and middle stream reaches and at 1.0 for the downstream reach. The initially deposited bed thickness was unknown, but for the time being, it was set at 2 m, and if the channel gradient was steeper than the angle of repose, it was set at 0 m. Here, the angle of repose of the bed was set at 35°.

Prior to the calculation, water flow depths at all the reaches were set at zero. A threshold depth of 1 mm was given to a grid cell to allow the advance of flood flow to the next cell.

Results of calculation and some evaluations of SERMOW

The first trial calculation was carried out for the rainfalls in 1994 and 1995, in which the initial bed data were obtained by the field survey done in October, 1993. Three rain gauge stations exist in the Takase Dam basin; Eboshidake, Takase Dam, and Takago. The rainfall data of the nearest station to the respective basins was used. Namely the data of the Eboshidake station were used for the Nigorisawa-Fudosawa basin, that of the Takago station for the main Takase river basin, and for other basins that of the Takase Dam station were used. Figure 6.22 shows the rainfall records at the Eboshidake station, in which a series of rainfall without a break of ten hours or

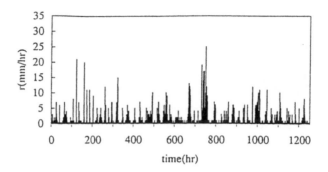

Figure 6.22　Rainfall events in the Nigorisawa-Fudosawa basin in 1994 and 1995. The hyetograph of the rainfall events in order ignores the actual arising dates and time, in which the values from 0 to 421 on the abscissa mean the events occurred in 1994 and others are the events that occurred in 1995. The time interval between the adjacent events was set as 11 hours except for the events that actually occurred within a lesser interval time.

more is considered as one rainfall event and among such rainfall events the one which contains the rainfall intensity heavier than 5 mm/hr were picked up. For the Takago and Takase Dam stations, the rains that fell within the same periods to the one picked up at the Eboshidake station were adopted regardless the breaking time and rainfall intensity.

The effective rainfall intensity; r_e, is assumed to be given by:

$$r_e = ar + b \tag{6.59}$$

where r is the actual rainfall intensity (mm/hr), a and b are coefficients. Numerous combinations of the values of a and b may be possible, but, here, $a = 0.5$ and $b = 0.3$ (mm/hr) were adopted, since by that combination the discharges flowed into the reservoir in July 1992 were most appropriately reproduced.

The stream channel width, B, was assumed, similar to the valley width, to be proportional to the 1/2 power to the basin area. The data read from the 1:5,000-scale topographic map of the main Takase river are given in Figure 6.23. The following relationship:

$$B = \beta A^{1/2} \tag{6.60}$$

seems to fit rather well by putting $\beta = 0.0024$. However, the width of the stream channel given in the topographic map may be that in the low water period. According to the field survey the discharge in a flood time was about nine times the low water discharge, so, here, the stream channel width is assumed to be given by putting $\beta = 0.0072$. If thus obtained stream channel width is larger than the valley width, the stream channel width is set equal to the valley width.

Figure 6.24 compares the calculated inflow discharges of the sum of water plus sediment to the Takase reservoir with the measured ones. Since the time intervals between the two successive rainfall events are different from the actual intervals, the fitness

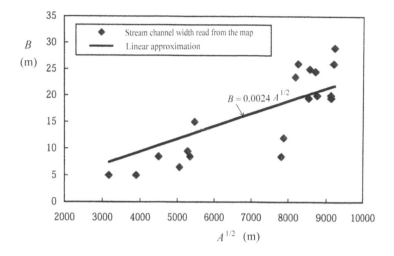

Figure 6.23 Relationship between the basin area and the stream channel width.

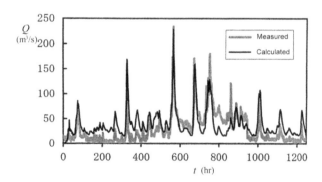

Figure 6.24 Comparison between the measured and calculated inflow discharges to the Takase reservoir.

of calculation to the actual discharge is not so good particularly for low discharges, but the fitness to the flood discharges larger than $150\,\mathrm{m^3/s}$ is good. Fitness to large discharge is important in respect to the prediction of sediment runoff.

Figure 6.25 shows the calculated discharges of the sum of water plus sediment, the total of fine and coarse sediment discharges, concentrations of fine and coarse particles in the flow for the Nigorisawa-Fudousawa and the Takase rivers at the entrances into the reservoir. For the Nigorisawa-Fudousawa basin, a high peak sediment discharge is conspicuous around $t = 700 \sim 800\,\mathrm{hr}$, but no other peaks are evident. The high peak arose corresponding to the occurrence of severe rainfall and the very high coarse sediment concentration suggests that the large quantity of sediment was transported as a debris flow. In the Takase River many peaks in the sediment discharge occurred corresponding to every peak in the water discharge hydrographs. This fact suggests

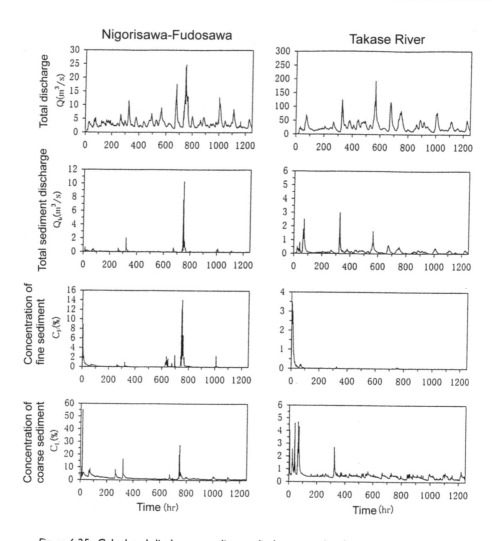

Figure 6.25 Calculated discharges, sediment discharges, and sediment concentrations.

that, in the comparatively large watershed with a long river reach like the Takase River, sediment is temporarily stored on the way of transportation at the junction with the tributaries and in the flatter channel gradient reaches. Then, with the occurrence of the next flood it is moved and transported downstream. Thus, in a long river channel reach, sediment transportation process naturally controls to leveling the sediment runoff.

A very high sediment concentration in the flow appeared at the start of the calculation as shown in Figure 6.25 although the water discharge is small. This would be an erroneous results caused by giving an inappropriate initial condition for the channel bed. This issue and the resolution of it will be discussed in Section 6.4.3.

Figure 6.26 shows the calculated cumulative discharges of water plus sediment, sediment, fine particles and coarse particles at the respective outlets of the watersheds.

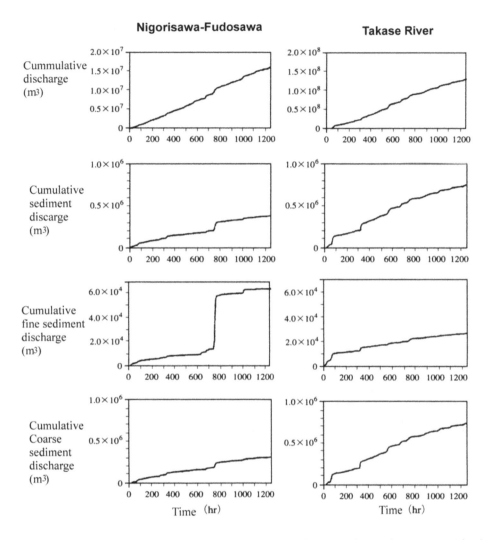

Figure 6.26 Cumulative discharges of water, total sediment, fine particles, and coarse particles by calculation.

In these watersheds the ratio of the fine sediment (smaller than 0.162 mm) discharge to the whole sediment discharge is as small as 1/6 ~ 1/25 of that of the coarse sediment. The cumulative sediment runoff volumes in the two years count up as 400×10^3 m^3 and 75×10^3 m^3 in the Nigrisawa-Fudousawa and in the Takase River, respectively. These values correspond to 610×10^3 m^3 and 115×10^3 m^3 in the bulk volumes and they are close to the measured values of 650×10^3 m^3 and 95×10^3 m^3, respectively. This is not surprising because the k_g values for the respective watersheds (9.89×10^{-4} m/mm and 2.92×10^{-4} m/mm, respectively) were determined so as to explain the mean annual erosion depths of the bare-land slopes in the respective watersheds in 12 years from 1983 to 1995 in which the concerned two years are involved. However, it is noteworthy

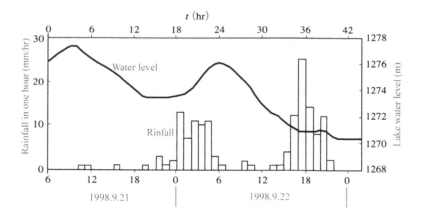

Figure 6.27 Hyetograph and lake water level in the occasion of the 1998 flood.

that, under the wide scattering of annual sediment yields as shown in Figure 6.11, the sediment yield in a specific year can be reasonably well calculated by giving only the hyetographs.

6.4.3   Improvement of the model by the second calculation

Sensitivity analysis concerning the initial talus size and bed thickness

The detailed three-dimensional surveys of the lake-bottom topography that is under the influence of the sediment yield from the Nigorisawa-Fudousawa basin were carried out on 27 May 1998 and 26 November 1998. The difference between the two bottom levels calculates the sediment yield during this period that is $298 \times 10^3 \text{ m}^3$. The conspicuous rainfall in this period occurred on 22 September as shown in Figure 6.27. The application of SERMOW to this rainfall event gives a useful data to improve the model.

The method and conditions for the calculation except for the initial conditions are the same as before. The initial conditions of sediment size distributions of the bed are given as the ones on the end of 1995 that were obtained by the last calculation. But, the initial bed thicknesses and the volumes of the taluses are given variously to examine the sensitivity of these factors to the sediment runoff.

Figure 6.28 shows the hydrographs and the sediment graphs of the Nigorisawa Ravine and the Fudosawa Ravine at 250 m upstream from the entrance to the reservoir under various initial conditions. The total runoff sediment volume described by a numerical value on each right-hand side figure in Figure 6.28 is the sum of those from the Nigorisawa and the Fudousawa. Case 1 uses the calculated results of the bed thicknesses and the volumes of taluses on the end of the previously mentioned calculation; on the end of 1995, as the initial conditions. Namely, the initial bed thickness is almost 0. Case 2 neglects the loss of rainfall but its initial conditions are the same as Case 1. Case 3 assumes the initial volume of taluses is ten times that in Case 1. Case 4 assumes the volumes of the taluses are the maximum whose individual

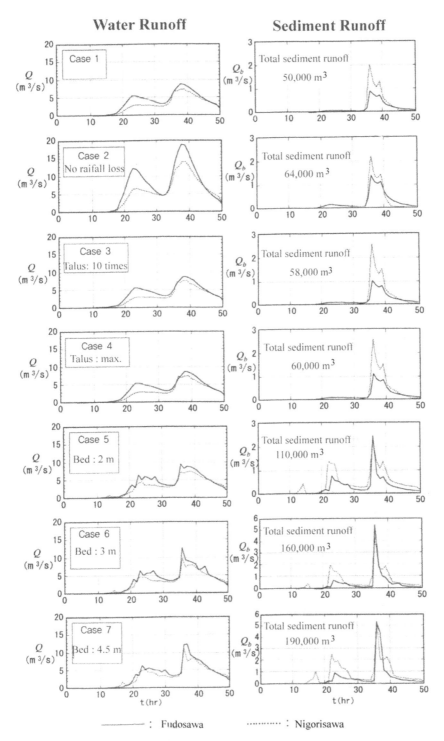

Figure 6.28 Calculations of hydrographs and sediment graphs for the Nigorisawa-Fudosawa watershed under various channel conditions.

quantity is given under the rule explained previously in section 6.4.2. Cases 5 to 7 change the initial bed thicknesses to 2 m, 3 m and 4.5 m, respectively, but the volumes of the taluses are the same as Case 1. From Figure 6.28 one can understand that the volume of taluses does not influence much on the volume of sediment yield but the initial thickness of the bed affects much. Because the bed thickness for the cases $1 \sim 4$ is almost 0 and the sediment produced on the bare-land slopes is stored on the taluses, the calculated results explains that the quantity of sediment produced on the slopes and stored on the taluses is not directly related to the sediment volume that inflows to the reservoir during a flood. In this flood case the sediment flowed into the reservoir which originates from the taluses is about 50 to 60 thousands m^3 in substantial volume.

The total sediment volume that was deposited in the reservoir during the flood is 298×10^3 m^3 in bulk, so this is almost equivalent to 194×10^3 m^3 in the substantial volume. This means Case 7 explains the actual situation best. If this is true, about 140×10^3 m^3 of sediment should have originated from the riverbed. Therefore, before the flood the riverbed should have stored sediment more than 215×10^3 m^3 in bulk volume. In the calculation the riverbed sediment is initially given only in the river reach steeper than 7° and if 4.5 m thick bed is given in this reach as Case 7 the amount of stored sediment is about 330×10^3 m^3. The problem is how such an amount of sediment could have stored before the 1998 flood.

One reason for the sediment storage on the bed is, as mentioned earlier, due to the shortage of vacant room on the talus into which sediment can be put because the talus has already developed up to the maximum capacity, and another reason may be due to the direct supply of sediment onto the riverbed in the snow melting season as shown in Photo 6.1. Small scale floods that erode the talus and deposits sediment on the riverbed nearby the talus may also be a cause. Among such causes the direct supply of sediment in the snow-melting season due to gravity action is independent of the sediment supply due to rainfall.

According to Figures 6.7 and 6.11, in the Nigorisawa-Fudosawa watershed, after the large amount of sediment runoff in 1983, the twelve year period can be divided into the relatively scanty sediment runoff years (1984 ~ '89 and '92 ~ '94) and the plenty sediment runoff years ('90, '91 and '95). The mean annual sediment runoff from 1984 to 1995 is 273×10^3 m^3 and that in the relatively scanty sediment runoff years is 164×10^3 m^3. If one assumes that the difference between the two runoff amounts; i.e. 109×10^3 m^3, is left on the riverbed during a scanty rainfall year and that amount is brought to the bed during the early snow-melt season. The river reach steeper than 7° can store 109×10^3 m^3 with the average thickness of 1.5 m. There was no conspicuous flood during the period of three years from 1995 to 1998, so the riverbed before 1998 flood was likely to have the three years equivalent deposit of 4.5 m. If this conjecture is sound, in the case of SERMOW application, 1.5 m thick bed deposit should be given before every year's flood season.

Effects of minute topography of the bed

The sediment graph of the Nigorisawa Ravine for Case 7 shown in Figure 6.28 has a small but evident peak between 10 hours and 20 hours after commencement of rainfall. As is clear in Figure 6.27 rainfall in this period is still trivial, so this peak could be

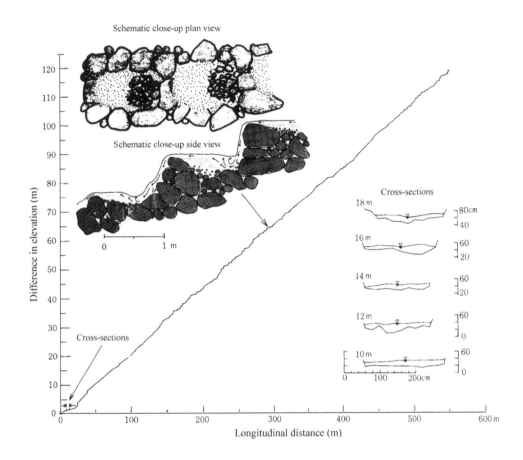

Figure 6.29 Stair-like bed morphology in the Hirudani Ravine.

fictitious. This result may be brought by the neglect of minute bed topography such as the existence of stable boulders and step forms.

Stair-like riverbed morphology is often formed naturally. Figure 6.29 shows the bed profile of the Hirudani experimental watershed of the Hodaka Sedimentation Observatory, DPRI (Ashida *et al.* 1976). The average topographical slope is 12°~14°, but looking in detail, as shown in the detail drawing part in Figure 6.29, a stair-like bed morphology is formed, and in the normal flood flow the local water surface slope on the pool is far less than the average topographical gradient, thence only fine particles accumulated in the pool are transported. This bed morphology has been referred to in the explanation of Figure 3.4.

If similar bed morphology exists on a steeper channel bed where a debris flow can be generated, this morphology would restrain the onset of debris flow in the case of small-scale flood runoff. Hence, the assumptions that the bed profile originally had a stair-like morphology as shown in Figure 6.30 and the water surface slope θ changes from being quite flat to a steeper general topographical slope θ_0 with the building up of depth were made as shown in the right-hand side graph in Figure 6.30.

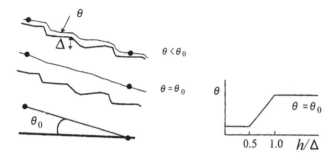

Figure 6.30 Local water surface slope versus relative depth on a stair-like bed profile.

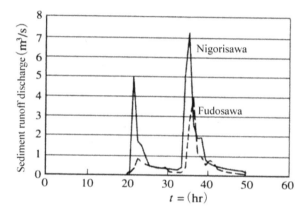

Figure 6.31 The sediment graph for Case 7 after the reworking concerning the water surface gradient.

Accordingly, a functional relationship between the relative depth h/Δ and the water surface gradient as shown in Figure 6.30 was introduced and the recalculation of SERMOW was carried out, where Δ is the height scale of the irregularity; θ_w is the local water surface gradient; and θ_{w0} is the general topographic gradient. The calculated results are shown in Figure 6.31. The unnatural first peak disappeared, but instead, the second and the third peaks increased their sediment discharges to be larger than the case without the stair-like morphology.

6.5 SERMOW VER.2

6.5.1 The constitution of SERMOW ver.2

SERMOW considers the debris flows that possibly arise in the watershed to be typical stony debris flows or typical turbulent muddy debris flows. In the calculation the specific resistance formulae applicable for the respective stony-type mature debris flows or immature debris flows and for turbulent-type debris flows are switched, during the process of calculation, depending on the relative depth (= flow depth/mean particle diameter in the flow) and the coarse particle concentration in the flow. However, the sediment composition in a watershed are widely distributed; from large stones to silt size material, and the flow may gradually change its mode from stony debris flow or turbulent-muddy-type debris flow to individual particle motion via hybrid-type during its motion, by successively depositing particles. To increase versatility and simplicity SERMOW ver.2 changes its fundamental equations for debris flows to that for a generalized inertial debris flow that needs no switching of equations throughout the change in the flowing mode during the course of run down, except for the switching from the debris flow equation to the equation of the individual particle motion. In the SERMOW ver.2 model, for the individual particle motion that occurs after the stopping of the debris flow or the gradual phase change after the deposition of coarse particles, a Brown-type total load formula is used taking the large amount of suspended sediment into account.

In SERMOW ver.2 Equation (2.175) is used as the equation of motion instead of switching use of Equations (6.7), (6.8) and (6.9) in SERMOW. Because the kinematic wave method is used in routing the flow, the total unit width discharge of water plus sediment must be known by giving only the water surface gradient θ_w and the flow depth h. To meet this requirement following three conditions are attached to Equation (2.175):

1   The shear velocity is given as $u_* = \sqrt{gh \sin \theta_w}$.
2   As the transport concentration C_{tr}, the average coarse particles' concentration C_L in the flow is used.
3   As the representative particle diameter d_p, the mean diameter of coarse particles d_{mL} in the flow is used.

Then, Equation (2.175) is written as:

$$q = Uh = \sqrt{gh \sin \theta_w}\, hF_1 \left[A_1 + A_2 \log_{10}\left(\frac{h}{d_{mL}}\right) + A_3\left\{\log_{10}\left(\frac{h}{d_{mL}}\right)\right\}^2 \right] \quad (6.61)$$

where $A_1 = 1.1632 - 3.0374C_L + 1.0589C_L^2$, $A_2 = 2.6898 - 4.7747C_L + 3.9167C_L^2$, $A_3 = 0.8313 - 2.2134C_L + 0.6818C_L^2$, when $d_{mL} > 0.1$ cm, $F_1 = 1$, and when $d_{mL} < 0.1$ cm, $F_1 = -13.72B_1(C_L - 0.270)^2 + B_1 + 1.0$, in which $B_1 = A_4 + A_5\log_{10}(h/d_{mL}) + A_6\{\log_{10}(h/d_{mL})\}^2$, $A_4 = 0.50372 - 0.2773 \log_{10}d_* + 0.036998\{\log_{10}d_*\}^2$, $A_5 = 0.5530 - 0.24564 \log_{10} d_* + 0.030294\{\log_{10} d_*\}^2$, $A_6 = -0.20525 + 0.097021 \log_{10} d_* - 0.012564 \{\log_{10} d_*\}^2$ and $d_* = (\sigma/\rho_m - 1)gd_{mL}^3/\nu^2$.

For the application of Equation (6.61) the sediment concentration should be larger than a threshold value. We set the lowest limit to apply Equation (6.61) as C_{Lmin} and when $C_L < C_{Lmin}$ Manning's Equation (6.9) is used instead of Equation (6.61). Even when $C_L > C_{Lmin}$ is satisfied, if q obtained by Equation (6.61) becomes larger than that obtained by Equation (6.9), Equation (6.9) is used instead of Equation (6.61).

In SERMOW, the equations that give the values of $C_{L\infty}$ were switched between Equation (3.36) , Equation (2.144) and Equation (6.37) depending on the mode of flow. But, in SERMOW ver.2 between the two equilibrium concentration equations for the inertial debris flows; Equation (2.174), and for the individual particle transportation; Brown formula, are switched depending on the water surface gradient, where $C_{tr} = C_{L\infty}$ and $x = \sin\theta_w$ are assumed. Namely, Equation (2.174) is written as:

$$C_{L\infty} = a_1 + a_2 x + a_3 x^2 + a_4 x^3 \qquad (6.62)$$

where the respective coefficients are given by:

$$a_1 = b_1 + b_2 y + b_3 y^2, \quad a_2 = b_4 + b_5 y + b_6 y^2,$$
$$a_3 = b_7 + b_8 y + b_9 y^2, \quad a_4 = b_{10} + b_{11} y + b_{12} y^2$$

in which $y = \log_{10}(h/d'_{mL})$ and d'_{mL} is the mean coarse particle diameter on the surface of the bed that is less than the depth of the flow; i.e. the mean diameter of movable particles on the bed as inertial debris flow. Defining $d_* = (\sigma/\rho_m - 1)g(d'_{mL})^3/\nu^2$ and by expressing $\zeta = \log_{10} d_*$, $b_1 - b_{12}$ for the case of $d'_{mL} \leq 0.1$ cm are given as shown in Table 2.4 and for the case $d'_{mL} > 0.1$ cm,

$$b_1 = -0.000851, \quad b_2 = 0.00023019, \quad b_3 = -0.0023895, \quad b_4 = 0.41059,$$
$$b_5 = -0.74452, \quad b_6 = 0.58275, \quad b_7 = -4.5674, \quad b_8 = 14.767$$
$$b_9 = -6.6375, \quad b_{10} = 25.953, \quad b_{11} = -36.505, \quad b_{12} = 14.236$$

Brown formula is given by:

$$\frac{q_B}{u_* d_{mL}} = 10 \left\{ \frac{u_*^2}{(\sigma/\rho_m - 1)g d_{mL}} \right\}^2 \qquad (6.63)$$

where q_B is the sediment discharge of coarse particles per unit width.

As mentioned earlier the formula of inertial debris flows cannot be applicable for a shallow slope channel and for a thin sediment concentration. If θ_{wmin} is defined as the flattest limit of the gradient steeper than that Equation (6.62) is applicable, then when $\theta_w < \theta_{wmin}$, the equilibrium concentration is obtained from Equation (6.63) instead of Equation (6.62). However, even when $\theta_w > \theta_{wmin}$, if the concentration calculated by Equation (6.62) is less than the concentration obtained by Equation (6.63), the concentration obtained by Equation (6.63) is adopted as the equilibrium concentration.

For the sake of obtaining the appropriate θ_{wmin} and C_{Lmin} values, several SERMOW ver.2 calculations for a simplified river channel model that was hypothetically set in the Fudosawa watershed with various values of θ_{wmin} and C_{Lmin} were carried out. Consequently, the most reasonable results in view of the riverbed variations, water and sediment discharges and sediment graphs were obtained under giving $\theta_{wmin} = 4°$ and $C_{Lmin} = 0.05$. Therefore, in the following discussions these threshold values are used.

6.5.2 Application of SERMOW ver.2 model to the Nigorisawa-Fudosawa watershed

Conditions for the calculation

A rainfall event was defined, different from the previous SERMOW model application, as one having no interruption longer than 15 hours and having a rainfall intensity

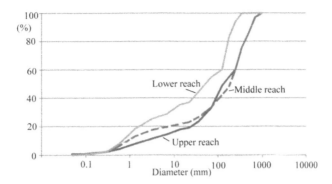

Figure 6.32 Particle size distributions on the bed surface of the Fudosawa Ravine.

greater than 5 mm/hr. The rainfall events thus sampled from 1980 to 2009 were put in an order similar to Figure 6.22 and water runoff analysis was carried out applying the kinematic wave method, in which the coefficients of Equation (6.59) that determines the effective rainfall were set as $a = 0.8$ and $b = 0$ mm/hr.

For the initial condition of the riverbed, 2 m thick bed was laid in the reach flatter than the angle of repose and initially the volume of all the taluses was set zero. Before the commencement of the water and sediment runoff calculations, an amount of sediment equivalent to the erosion thickness of 2 cm on the abutting bare-land slope was given to every talus. This was done in order to take into account the rainfall-independent sediment supply in the early snow-melting season. This thickness is considered to be already swollen due to frost-columns heaving and no volume increase occurs when added onto the talus. Thus, in the actual calculation, at the beginning of every year a $q_{sg\,in}$ that is equal to $0.000833A_g/l_g$ is first given to every talus for 24 hours, in which 0.000833 (m/hr) is the value 2 cm/24 hours. During this period if a base flow is given to the channel, the erosion of the riverbed as well as that of the talus may occur.

The field survey of the riverbed was carried out recently and the particle size distributions of the bed surface as shown in Figures 6.32 and 6.33 were obtained. These size distributions are by no means the same as the conditions in 1980, but were used as the initial bed conditions. The results from field survey on the particle size distributions of the talus concluded that the distribution was too coarse. Since the material that is deposited in the reservoir must be supplied from the taluses, this result cannot explain the characteristics of reservoir sedimentation. Therefore, taking the coring data of the delta deposit in the reservoir and the field survey data of the talus into account, the possible size distribution of the material produced on the bare-slope was made as shown in Figure 6.34. In the Nigorisawa watershed the upper reach where the taluses exist was not accessible so no field survey data is available for this watershed. Therefore, the same size distributions as those in the Fudosawa were assumed.

The value of k_g that is crucially important to determine the rate of erosion on the bare-land slope was determined as 10.8×10^{-4} m/mm after some trial calculations.

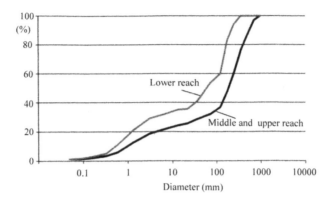

Figure 6.33 Particle size distributions on the bed surface of the Nigorisawa Ravine.

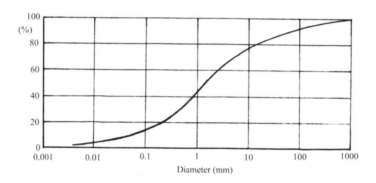

Figure 6.34 Particle size distribution of the talus.

Results of calculation for the long term sediment runoff

Giving the time series of rainfall events sampled from the actual records of rain gauge, the calculations of water and sediment runoffs from the Nigorisawa-Fudosawa watershed in 30 years from 1980 to 2009 were carried out. The results of calculation of annual sediment runoffs are shown in Figure 6.35 as the comparison between the measured and calculated results. The calculated total sediment runoff in 30 years is 106% of the measured one and the differences between the measured and calculated values in the respective years are good, given the uncertainty of the phenomena and the simplifications in the model. Simplifications such as where the rainfall in the watershed is represented by the records at a point, the quantity of bare-land erosion is estimated by giving only one common value of k_g for all the slopes, the sediment size distribution of all the taluses is common, etc. For example, in 2005, a landslide of about 20,000 m^2 occurred near the point I = 210 indicated in Figure 6.13 and it transformed into debris flow reaching to the downstream end of the ravine. Such a phenomenon is not considered in this analysis. If the sediment runoff volume by such a phenomenon is added, the calculated runoff volume in this year approaches to the measured one. Note that the

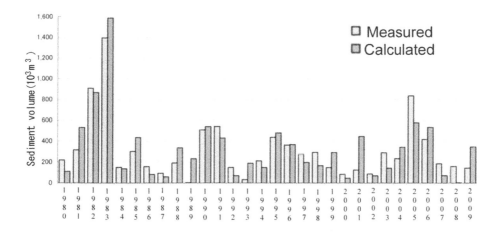

Figure 6.35 Annual sediment runoffs from the Nigorisawa-Fudosawa watershed.

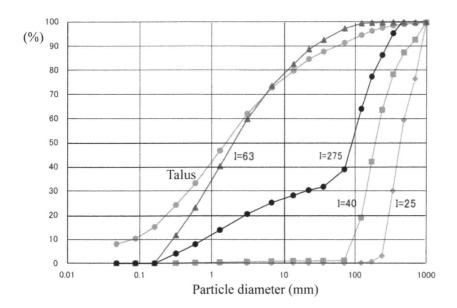

Figure 6.36 Particle size distributions on the riverbed of the Fudosawa Ravine.

rainfall in 2005 exceeded the threshold line for the occurrence of landslides as shown in Figure 6.10.

Figures 6.36 and 6.37 show the particle size distributions of the riverbed surface material in the Fudosawa and Nigorisawa, respectively, at the end of year of the SERMOW ver.2 calculations from 1980 to 2009. The positions of the point numbers indicated on the figures are shown in Figure 6.13. $I = 25$ and $I = 210$ correspond to the upper reaches of the respective ravines, $I = 40$ corresponds to the middle reach of the

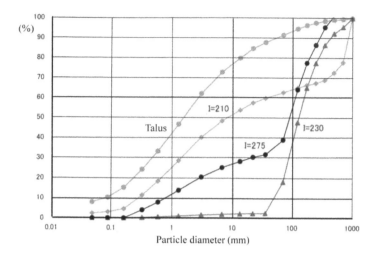

Figure 6.37 Particle size distributions on the riverbed of the Nigorisawa Ravine.

Fudosawa ravine, I = 63 and I = 230 correspond to the lower reaches of the respective ravines, and I = 275 is the junction point of the two ravines.

In the Fudosawa Ravine, the particle sizes are the larger the upper the reach is and at I = 25 and 40 the armor coated bed prevails, but at I = 63 the particle size distribution is similar to that of the taluses; i.e. the particles larger than 10 cm in diameter scarcely come down to this place. This is due to the existence of long flat and wide channel in the lower reach of the Fudosawa Ravine.

In the upper reach of the Nigorisawa Ravine, I = 210, the finer particles supplied from the bare-land slope are dominant and the riverbed has little tendency to be armor coated. At the lower reach of the Nigorisawa the armor coating develops.

At the junction of the Nigorisawa and the Fudosawa ravines, although little amount of coarse material larger than 10 cm is supplied from the Fudosawa, plenty of coarse material is supplied from the Nigorisawa. Therefore, the typical difference in material coming to this place makes up a peculiar size distribution in which particles coarser than 7 cm and finer than 1 cm in diameters prevail but the particles in between the two sizes are absent.

Hydrographs and sediment graphs produced by a conspicuous rainfall

Figure 6.38 shows the hyetograph of the rainfall event from July 30 to August 3 in 1982. This event was the 9th largest cumulative rainfall amount (278 mm) and the 1st strongest in one hour rainfall intensity (38 mm/hr) during the last 30 years. Figure 6.39 shows the calculated hydrograph and sediment graph at just downstream of the waterfall in the Nigorisawa basin; I = 226, and the sediment graph for each size group particle. The nominal diameters shown in Figure 6.39 are as follows: d1 (0.08 mm): <0.323 mm; d2(2 mm): 0.324 ~ 6.72 mm; d3(2 cm): 6.73 ~ 70.6 mm; d4(15 cm): 70.7 ~ 244 mm; d5(34 cm): 245 ~ 692 mm; and d6 (1 m): >693 mm. The

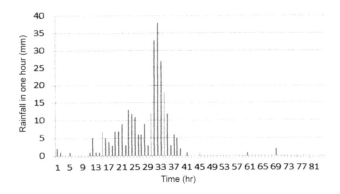

Figure 6.38 Hyetograph of the rainfall event from July 30 to August 3, 1982.

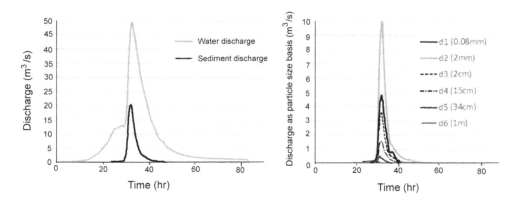

Figure 6.39 Calculated hydrograph and sediment graph (left), and the sediment graph for each particle group (right) in the Nigorisawa basin.

total water plus sediment runoff volume amounts to 2.18 million m³ and the total sediment runoff volume amounts to 0.325 million m³ in the substantial volume (about 0.52 million m³ in bulk volume), so the average sediment concentration in the flow is about 0.15 in volume, whose 24% is d1, 48% is d2, 18% is d3, 7.5% is d4, 2% is d5 and 0.1% is d6 particles. At the peaks of hydrograph and sediment graph the sediment concentration is as large as 0.4, that is the concentration of general debris flow even though the majority of sediment is less than a few centimeters. If the sediment particle size involved in the flow is compared with the bed particle size at I = 230 in Figure 6.37, one understands that the flow passes through this point without deposition, albeit that at the junction of Nigorisawa and Fudosawa Ravines some sediment less than 7 cm in size was deposited.

Figure 6.40 shows the calculated hydrograph and sediment graph at the point I = 50 in the Fudosawa Ravine for the same rainfall event as before. The total water plus sediment runoff volume amounts to 1.15 million m³ and the total sediment runoff volume amounts to 0.247 million m³ in the substantial volume (about 0.38 million m³

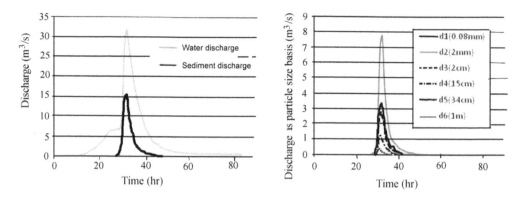

Figure 6.40 Calculated hydrograph and sediment graph (left), and the sediment graph for each particle group (right) in the Fudosawa basin.

in bulk volume), so the average sediment concentration in the flow is about 0.21 in volume, whose 22% is d1, 50% is d2, 18% is d3, 7.6% is d4, 2% is d5 and 0.6% is d6 particles. At the peaks of hydrograph and sediment graph the sediment concentration is as large as 0.45 that is the concentration of general debris flow even though the majority of sediment is less than a few centimeters. If the sediment particle size involved in the flow is compared with the bed particle sizes at I = 40 and at I = 63 in Figure 6.36, one understands that the flow passes through the point I = 40 without deposition but in the lower reach it deposits sediment of all the size groups except for the d1 group.

Chapter 7

Debris flow disasters and their reproduction by computer simulations

The photograph shows the sediment disasters that occurred at a resort area on the Caribbean coast of Venezuela in December 1999. Cities that had grown up on the fans formed by rivers penetrating through the zonal area along the coast for about 20 km suffered catastrophic damages due to debris flows runoff from the rivers. The particular city taken in this photograph is the largest one; Caraballeda, that is on the fan of the San Julian River. The low-rise houses existed among the high-rise buildings and in the midst of the fan were flushed out or buried under the runoff sediment of about 1.8 million cubic meters that covered an area of 1.2 km^2 with a deposit about 5 m thick containing many boulders as large as 5 m in the largest diameter.

INTRODUCTION

I have done many field investigations on the actual conditions of flood and sediment disasters, starting with the occurrence at Okuetsu, Japan, in 1965. I have studied not only the physical phenomena but also on the issues of psychology and social science. Because the disasters are so interlaced with complicated factors, there still remain a number of problems to be solved. However, by piling up and preserving the questions that arose in the surveys, I, helped by the achievements of others, have progressively found a number of approaches to solve the problems.

In this chapter, I will look at some examples of my field surveys and, using the basic investigations discussed in the previous chapters, try to reproduce only the physical aspects of them by the numerical simulations. In fact, the sequence of the development of the investigations is reversed; the fruits of efforts to generalize the particular phenomena discovered in the fields are the basic physics of debris flow described in the previous chapters. This chapter shows how the basic investigations can be used to understand the actual phenomena, but it also introduces new problems that cannot be explained by the mathematical models introduced in this book. This is an indispensable step for the level up of investigations.

It can be seen that a viable simulation model for a debris flow is one that is able to quantitatively analyze the quality change, growth and degeneration during its motion down the river channel like the models introduced in this chapter.

7.1 THE RAINSTORM DISASTERS AT OKUETSU

7.1.1 Outline of the disasters

From 13 to 15 September 1965, a severe rainstorm struck the expanse of the boundary between Fukui and Gifu Prefectures in central Japan, the district is called Okuetsu. The rainfall statistics were as follows:

Over the three days the total rainfall was 1,044 mm;
the daily rainfall from 9 a.m. on the 14th to 9 a.m. on the 15th was 844 mm;
and the maximum hourly rainfall that occurred around 10 p.m. on the 14th was 90 mm/hr.

A village named Nishitani in Fukui Prefecture was the most severely damaged and residents were forced to abandon the village itself.

I joined the disaster survey team for the first time in my career, and I was deeply impressed by seeing the houses on the fan of Kamatani Ravine; Nakajima district, were buried up to the second floor by the sediment runoff from the ravine. The most severely damaged districts in Nishitani Village were Nakajima that was the nerve center of the village and Kamisasamata that was adjacent to Nakajima. Among the 154 houses at Nakajima, 58 were washed away and 86 were buried; and among the 40 houses at Kamisasamata 21 were washed away and 16 were buried. At Nakajima, the community facilities such as the village office and school were also completely destroyed or buried. As the number of households in the whole of the Nishitani Village was only 272, the village was truly catastrophically destroyed. Now, the old Nakajima district is a campground on the upstream end of Managawa Reservoir that was constructed after the disaster.

The disasters were not brought about by a single cause. Nakajima was located on the junction of the Sasou River and the Kumo River, and Kamatani Ravine penetrated the center of the district (Figure 7.1). It was located on the fan formed by the Kamatani Ravine and the foot of the fan had been under the effects of side erosion due to the flows in the Sasou and the Kumo Rivers. According to what we were told on the spot and to literary documents, the rainfall became severe from about 7 p.m. on the 14th and from midnight to about 2 a.m. on the 15th; the greater part of Nakajima was buried by the sediment runoff from the Kamatani Ravine. Meanwhile, the flood flow in the Sasou River washed out the north-eastern part of the fan and there was severe bank erosion, as all the houses on the eastern third of the fan area fell into the river.

On the morning of the 15th (the exact time is unknown) a big landslide arose at a topographical depression called Kowazotani and it rushed down to choke the Mana River. The suddenly blocked flood water reversed its original direction and it engulfed and swept away the houses on the left-bank of the Kumo River. Soon, the reservoir made by the landslide was filled up by water and it overtopped presumably from the left-bank side. The overtopped flow incised the channel into the dam body and enlarged it to release a larger discharge than ever. Thus, the big flood, which had increased its power by the effect of the aggradation of the riverbed due to deposition of sediment eroded from the dam body and transported by the flood, struck Kamisasamata located about 1 km downstream of the dam and swept away 21 houses. The other 16 houses were buried in the runoff sediment from the ravine at the back of the district.

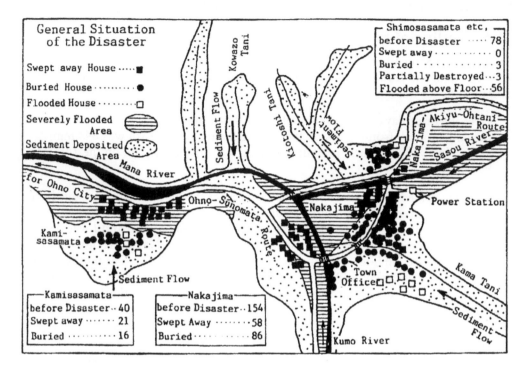

Figure 7.1 Disasters that occurred at Nishitani Village in 1965.

The houses were washed away mainly due to side-bank erosion and the landslide dam-break flood, whereas the houses were buried mainly due to sediment runoff from the back of the districts. Although the houses were buried up to the second floor, and these were not-so-strong wooden houses, the frame work of the majority of houses were safe. Thus, the sediment runoff from Kamatani and the back of Kamisasamata was assessed as being not a high-powered debris flow but an immature debris flow that caused a shallow flow with a rather long duration.

At Nakajima no one was killed by those severe events because people had evacuated their houses to take refuge. It was by virtue of the warning issued by the village authority who recognized that the flood from Kamatani had overflowed a bridge and that the muddy flow smelled like rotten garbage.

At that time, there was no means to reproduce the processes of disasters quantitatively. A table-top computer that could perform only the arithmetic operations of the four rules had just been invented, and therefore, the calculation of sediment discharge was limited to the cases of bed load and suspended load in the one-dimensional shallow slope channels. The calculations of debris flow or immature debris flow and moreover the reproduction of two-dimensional flooding processes of these flows remained little more than fantasy. After several years, motivated by the issues concerning the safety of natural dams produced by the debris avalanche that accompanied the eruption of Mount St. Helens in 1980 and that which accompanied the earthquake at Ontake in 1984, we carried out a series of investigations on the destruction of natural dams as mentioned in section 3.3. By that time, the quantitative estimation of two-dimensional flooding processes had gradually became possible, so that we had done the numerical reproductions of the sediment disasters due to sediment runoff from Kamatani, the washing away of houses upstream within the reservoir of the natural dam, and the flood disasters due to the collapse of the natural dam (Takahashi 1989; Takahashi and Nakagawa 1994). Because the reproductions of sediment flooding will be discussed in the later sections, although for another basin, herein, only the phenomena concerned with the natural dam formation and destruction are explained.

7.1.2 The natural dam formation and the damage done by backwater

On the morning of 15 September (possibly later than 8 o'clock) a topographical depression about 50 m in width and about 600 m in length on a slope of about 30°, suddenly collapsed, and that earth block made a natural dam choking the Mana River. The exact volume of the dam body was unknown but it was estimated to be about 330,000 m³. The aerial photograph taken on 1 October 1965 reveals some small-scale flow-mounds scatter on the opposite side valley bottom, suggesting that the natural dam had a plan form as illustrated in Figure 7.2. The thick black curves in Figure 7.2 show the stream channel pattern as it was at the time of taking the pictures, and the cut end of the dam

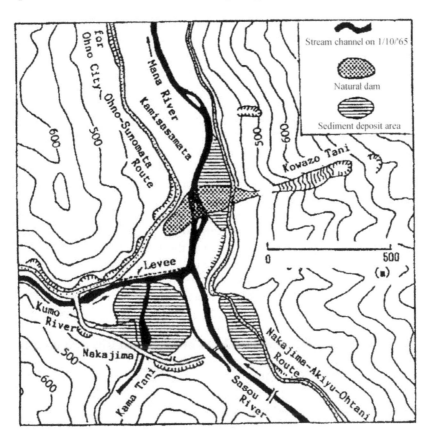

Figure 7.2 The natural dam and its neighborhood.

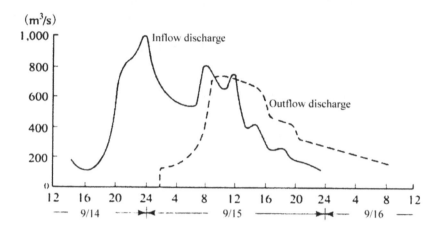

Figure 7.3 Inflow and outflow hydrographs at the Sasou-gawa dam.

on the right-hand side of the stream channel seemed to make a cliff of ten meters or so; to the right of this cliff the deposit made an ascending slope that seemed to be the original dam body. On the left-hand side of the stream channel the thickness of deposit was thin suggesting that the original dam crest was inclined to the left bank and the first overtopping flow appeared adjacent to the left bank slope.

Figure 7.3 shows the inflow and outflow discharges at that time at the Sasou-gawa dam, which is about 6 km upstream of the Sasou River from the junction with the Kumo River. Figure 7.3 shows that the discharge from the Sasou River at the moment of the natural dam formation was about 800 m^3/s. The basin areas of the Sasou and the Kumo Rivers at Nakajima are 96.6 km^2 and 90.7 km^2, respectively, so that the discharge from the Kumo River would also have been about 800 m^3/s. Therefore, the discharge at the point of the natural dam formation would have been 1,600 m^3/s.

The exact cross-sectional shape of the valley before the natural dam formation is unknown, but it is assumed to have had a compound section whose main channel width is about 63 m. The longitudinal gradient of the valley was about 1.5%. If Manning's roughness coefficient is assumed as 0.04 and if the main channel is at its bank-full stage throughout the reach with the discharge of 1,600 m^3/s, the depth of flow in the main channel at the instant of dam formation is 4 m.

Taking these data into account, the geometry of the natural dam just after the formation may be modeled as shown in Figure 7.4. The contour map of the ground level of the valley immediately after the dam formation would have been as shown in Figure 7.5. There was a levee on the left bank of the Kumo River, so no water flow across this levee is assumed in the following calculation.

The phenomena of backwater just after the sudden choking of flow can be reproduced by the depth-averaged two-dimensional flow analysis method (Takahashi *et al.* 1986). In this method the governing equations are the same as for the flooding of debris flow except that the erosion and deposition velocity are set to 0 and Manning's equation is used as the resistance to flow.

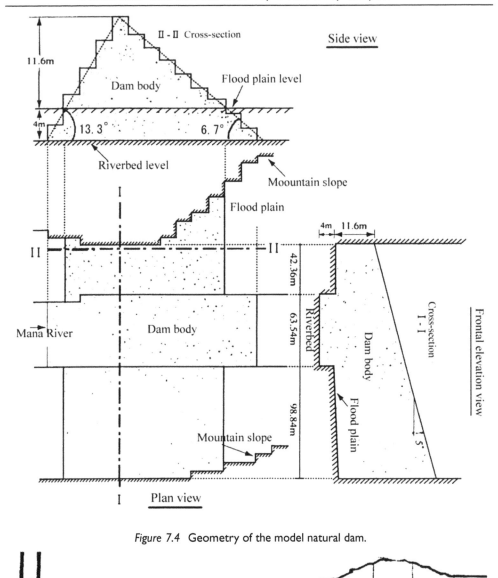

Figure 7.4 Geometry of the model natural dam.

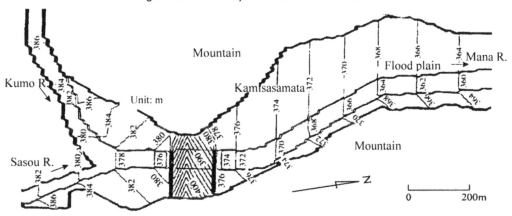

Figure 7.5 Model topography of the Mana River valley just after the dam formation.

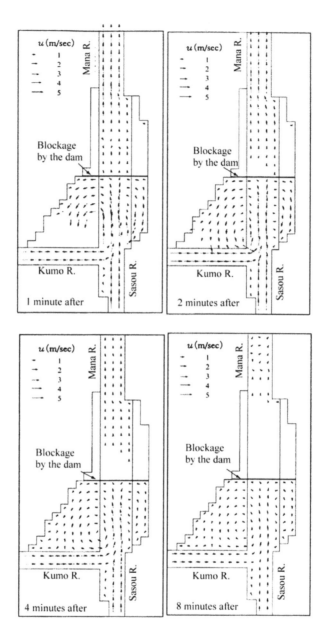

Figure 7.6 Temporal change in the pattern and scale of velocity vectors.

As the initial condition for the calculation, a steady flow of discharge 1,600 m³/s that is flowing in the bank-full stage of the low-water flow channel is given, where as a technique to smoothly transit to the unsteady flow after the dam formation, initially the river flow is set to zero and gradually increased up to a steady 1,600 m³/s. In this case, both in the Kumo and the Sasou Rivers, discharges were initially set to zero and linearly increased to 800 m³/s in a period of four hours and after that, an additional

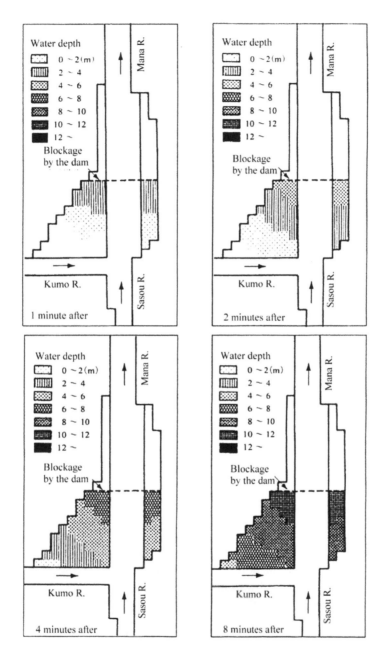

Figure 7.7 Time varying backwater depths after closure.

four hours were given to the steady flows of 800 m³/s in the respective rivers. After this time the flow was suddenly shut down at the location of the dam formation. The blocked off flow changed its way to the lateral direction and then went back upstream on the flood plain. Figure 7.6 shows the flow situation 1, 2, 4 and 8 minutes after

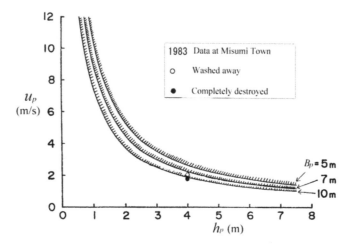

Figure 7.8 Criteria for the washing away of wooden houses.

the blockage. The flow direction and the velocity at one minute after the blockage clearly show the appearance of a high speed flooding flow on the left-hand side of the flood plain. Although the situations after that time may be somewhat different from the actual one because of the assumption that no flow spills over the levee on the left bank of the Kumo River, in the early stage, a strong circulating flow whose maximum velocity attains 3–4 m/s is produced. Downstream of the dam, on the other hand, a dry riverbed elongates its length with time. Figure 7.7 shows the time-varying depths on the flood plain, whose respective figures correspond to the respective ones shown in Figure 7.6.

Around eight minutes after the closure, the depth at the dam site reaches 12 m, and from that time the flow overtops the dam crest and the severe erosion of the dam body begins.

According to observers on the ground, the houses in the triangular area surrounded by the Kumo River, the Mana River and the mountain slope, were swept away upstream. The direction and strength of the circulating flow and the depth around this area suggest that this is true, but, herein, we will try a more quantitative discussion on the risk to the houses. Takahashi *et al.* (1985) obtained the criterion for the destruction of Japanese frame-structured wooden houses based on the theory and the measurements in the experiments as shown in Figure 7.8, in which u_p is the flow velocity, h_p is the inundated depth, and B_p is the width of the house perpendicular to the direction of flow. The formula of the hyperbolic curves in Figure 7.8 is given by:

$$u_p h_p = \sqrt{M_V/(h_c/h_p \cdot C_D/2 \cdot \rho_T)}/\sqrt{B_p} \tag{7.1}$$

where M_V is the critical bearing moment of the house which for a typical Japanese wooden house is about 418 kNm, h_c is the height of the force operating point due to the flow (according to the flume experiment $h_c/h_p = 0.732$), C_D is the drag coefficient

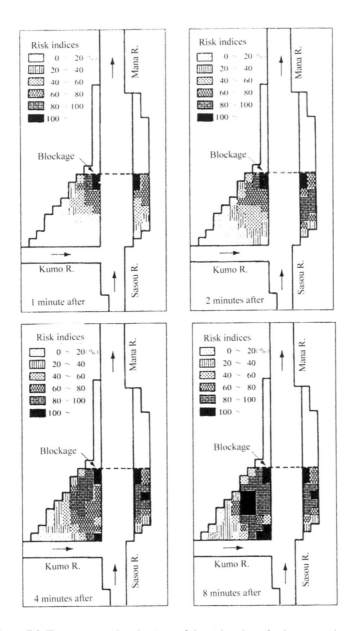

Figure 7.9 Time varying distributions of the risk indices for house wash out.

$(C_D/2 = 1.064)$, and ρ_T is the density of the fluid. If $B_p = 5\,\text{m}$ and $\rho_T = 1,000\,\text{kg/m}^3$ are assumed, the critical discharge of flow per unit width F_c is given as:

$$F_c = (u_p h_p)_c \approx 10\,\text{m}^2/\text{s} \tag{7.2}$$

Therefore, the ratio of $(u_p h_p)$ in the actual flow and F_c gives an index to describe the risk for the washing away of houses. The data for Misumi Town in Shimane

Prefecture given in Figure 7.8 show the results of calculations of velocity and depth of flooding flow at that town. The plotted points correspond to the actual washed out and completely destroyed houses, respectively, and these points are plotted near the critical line.

Figure 7.9 shows the results of calculations concerning with the time varying distributions of the risk indices mentioned above. At one minute after the closure, the zones above a hundred percent of wash away risk are limited to the area nearby the dam, but at eight minutes after the closure the places and the zones having more than 80% of wash away risk prevail in a wide area. The inundation depth is well over the roofs. This risk criterion was obtained for modern wooden houses, anchored on a concrete foundation, but in 1965 the common wooden houses in a mountainous area were merely set on footing stones, thereby they could easily be washed away by a smaller velocity and depth confirming the wash away of houses upstream.

7.1.3 Processes of destruction of the natural dam and the damage downstream

The method to predict the natural dam failure processes has already been introduced in section 3.3.4. Herein, the processes of damage at Kamisasamata will be discussed based on the numerical simulations. The parameter values used are: $\beta' = 1.0$, $s_b = 0.8$, $K = 0.06$, $K_s = 1.0$, $\delta_d = 1.0$, $\tan\phi = 0.75$, $\rho = 1.0\,\text{g/cm}^3$, $\sigma = 2.65\,\text{g/cm}^3$, $d_L = 5\,\text{cm}$, $C_* = 0.655$, $n_m = 0.03$ (main channel) and $n_m = 0.04$ (flood plain). The calculating grids are $\Delta x = 14.12\,\text{m}$, $\Delta y = 7.06\,\text{m}$ and $\Delta t = 0.03\,\text{sec}$.

Figure 7.10 shows the time varying depths upstream and downstream of the dam from the time immediately after the dam formation. As mentioned above, the rebound of water upstream arises immediately and around ten minutes after the dam formation the overflow near the left side bank begins. The maximum flow depth of about 1 m at the Kamisasamata flood plain is attained some twenty minutes later than the dam formation.

The processes of dam failure and the depositions downstream are shown in Figure 7.11. No sediment load is considered to be contained in the coming-in flood flows from both of the tributaries, and therefore no deposition takes place within the naturally produced reservoir. However, the flood flow coming-out from the reservoir contains much sediment from the erosion of the dam body and it deposits sediment from just downstream of the dam and raises the riverbed as well as the flood plain at Kamisasamata. The deepening and the widening of the channel formed on the crest of the dam body on the left-hand side clearly occur. The remaining part of the dam on the right-hand side of the incised channel on the dam body at $t = 90$ minutes makes a cliff facing the stream whose height and location well agree with the ones estimated from the aerial photograph taken after the flood. The thickness of the deposit laid on the Kamisasamata flood plain reached as thick as 0.6–1.6 m. The thickness of the deposit within the main channel in front of the Kamisasamata community increases downstream and suddenly it becomes zero. The maximum thickness is almost equal to the depth of the channel, so that the main channel is nearly buried.

Figure 7.12 shows the velocity vectors of the flow. The riverbed downstream of the dam becomes almost dried for a while before overtopping takes place. About ten to twenty minutes after the beginning of overtopping, the maximum flow velocity of

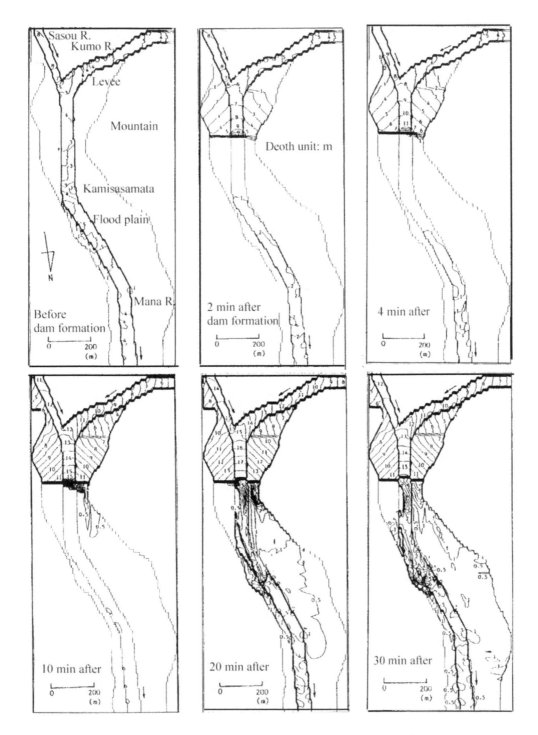

Figure 7.10 Distributions of water depth before and after the dam formation.

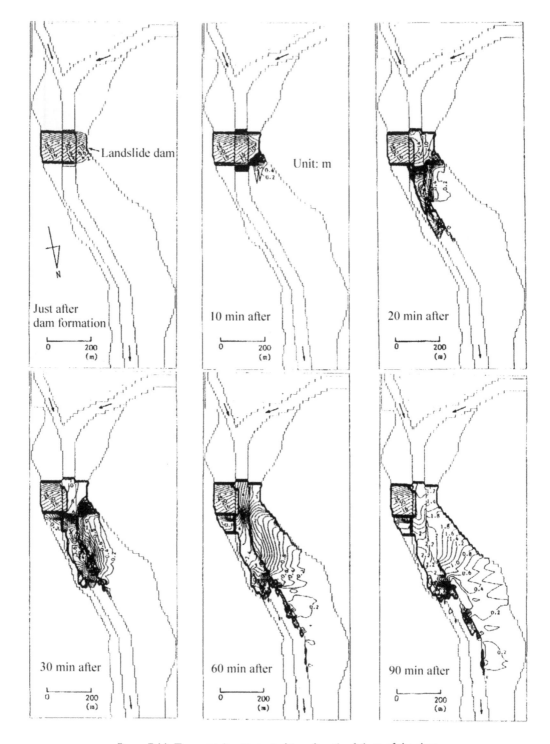

Figure 7.11 Time varying topographies after the failure of the dam.

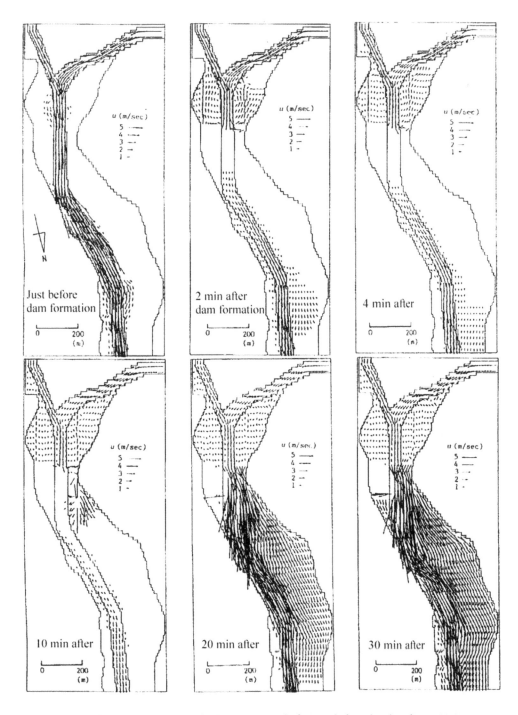

Figure 7.12 Distributions of velocity vectors before and after the dam formation.

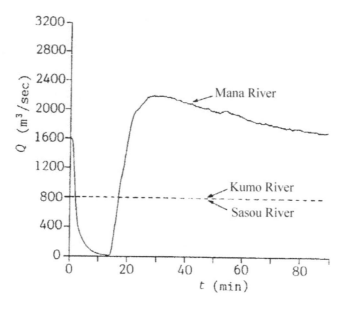

Figure 7.13 Calculated discharge hydrograph downstream of the natural dam.

4–5 m/s occurs on the Kamisasamata flood plain. The maximum flow velocity of 5 m/s and the maximum flow depth of 1.6 m seemingly do not satisfy the threshold for the washing away of wooden houses shown in Figure 7.8, but, as mentioned earlier, the houses at that time would have been weaker than the standard of modern houses, and in addition to that, because the calculated velocities in the main channel well exceed the velocities on the flood plain, the sporadic deviations of flow direction in the main channel during the process of sediment deposition may easily give much higher velocities on the flood plain. Therefore, the results of calculation may not contradict the fact that many houses were washed away.

The discharge hydrograph downstream of the dam is calculated as shown in Figure 7.13. The maximum inflow discharge was about 1,600 m³/s, whereas the maximum outflow discharge was about 2,200 m³/s and it was attained starting from nearly zero in only ten minutes.

7.2 HORADANI DEBRIS FLOW DISASTERS

7.2.1 Outline of the disaster

A debris flow occurred in the basin of a small mountain ravine named Horadani in Tochio, Kamitakara Village, Gifu Prefecture, Japan at 7:45 a.m. on 22 August 1979. Rainfall began at about 12 o'clock 21 August. It amounted to more than 100 mm when the strongest rainfall intensity occurred to trigger the debris flow which struck the Tochio community. This community had grown up on the debris flow fan of

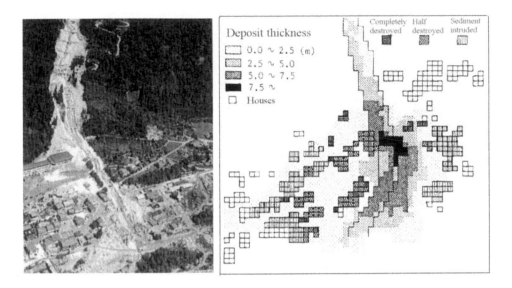

Figure 7.14 The situation just after the disaster and the distributions of damages on the fan.

Horadani. A car that was passing through the area was wrecked and three people on board were killed. The other damage was seven houses completely destroyed, 36 houses half destroyed, and 19 houses inundated. The Horadani ravine had penetrated the central part of Tochio by the channel works of 14 m in width and 4 m in depth. The first overflowing from the channel works had occurred at a channel bend. The total amount of runoff sediment by bulk is estimated at about 66,000 m³. Figure 7.14 demonstrates the distributions of deposit thicknesses and damaged houses (Sabo Work Office, Jinzu River System, Ministry of Construction, and Tiiki Kaihatsu Consultant 1979), in which the squares correspond to the finite meshes in the calculation covered the area and coalesced with the houses, but each square does not necessarily mean a house.

The basin of the Horadani Ravine is, as shown in Figure 7.15, a tributary of the Gamata River in the Jinzu River System. The altitude at the outlet is 800 m and that at the highest point is 2,185 m, the vertical drop of 1,400 m is connected by a very steep stem channel of 2,675 m long. The basin area is 2.3 km². The outlet of the ravine forms a large debris flow fan of 500 m wide and its longitudinal gradient is about 9.5° on which Tochio is located. The upstream end of the channel works that penetrated Tochio was at the fan top, and upstream of the fan top it formed a wide torrent of 6–12° in longitudinal gradient for about 500 m long within which 11 check dams had been constructed. The storage capacities of these dams were unknown but almost all dams were destroyed and washed away by the debris flow.

The debris flow occurred at around 7:50 a.m. According to the witnesses of a person who passed the bridge on Horadani at about 7:30 a.m. and others, the water discharge in the channel was a little larger than normal, when suddenly there was a debris flow accompanied by a ground vibration like an earthquake and a noise like

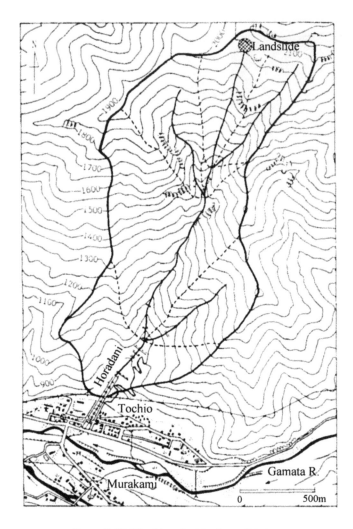

Figure 7.15 The Horadani basin and Tochio.

thunder. Therefore, no one made refuge beforehand; there was a person racing against the running down debris flow on a road safely and many people barely survived by moving quickly to the second floor in their houses. The casualties killed in a car were tourists. There were many hotels in the basin of the Gamata River upstream of Tochio. The only road to these hotels was a dangerous mountain road and there was a rule to block it with a gate when it rained severely. On that day of the disaster, it rained severely from early morning, and an announcement of imminent road blocking was transmitted to the respective hotels urging the tourists to go down the mountain. This situation highlights the difficulties in blocking the road, the methods to announce the warning notice and the implementation of evacuation.

The place of the initial generation of the debris flow in the basin is hard to assess. At an upstream end of the basin, as shown in Figure 7.15, a landslide occurred (according to the report referred to earlier, the volume was 8,740 m^3) and from the scar of this landslide the trace of debris flow continued downstream. Therefore, one possibility of debris flow generation is the induction due to landslide. However, the longitudinal gradient of the channel bed is steeper than 20° in the upstream reach, which guarantees the possibility of the generation of debris flow by the erosion of the bed under a sufficient flood discharge on it. Therefore, the definite cause of the debris flow should be investigated based on many points of view.

The characteristics of the debris flow materials as they were on the slope and on the riverbed before the occurrence and the characteristics of the debris flow deposit around the Tochio debris flow fan are difficult to know precisely, and the quantity of material accumulated on the riverbed before the debris flow is unknown. But, here, the following representative values are assumed: $d_L = 10$ cm, $C_* = 0.65$, $C_{*F} = 0.2$, $C_{*L} = C_{*DL} = 0.56$, $\tan \phi = 0.75$, $\sigma = 2.65$ g/cm^3 and the thickness of deposit on the riverbed $D = 4$ m.

7.2.2 Hydrograph estimation of the debris flow

Evaluation of flood runoff discharge

In the debris flow analysis, the evaluation of runoff water discharge to the river system is indispensable. Because no record of the actual flood discharge in Horadani is

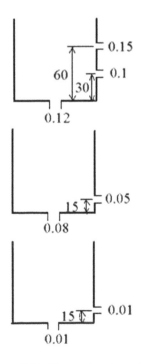

Figure 7.16 The used tank model.

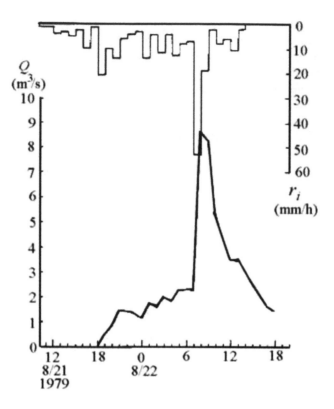

Figure 7.17 Given rainfall and the obtained hydrograph.

available, it must be estimated based on an appropriate runoff analysis. Takahashi and Nakagawa (1991) obtained the flood hydrograph at the outlet of the Horadani basin using the known 'tank model' under the assumption that no debris flow had arisen in the basin. The three-storied tanks used had holes whose positions and discharge coefficients are as shown in Figure 7.16. The actual rainfall, observed at the Hodaka Sedimentation Observatory, DPRI located in the neighborhood depicted in Figure 7.17, was supplied to these tanks to obtain the hydrograph at the outlet of the basin as shown in Figure 7.17.

For the routing of debris flow we must know the flood hydrographs at arbitrary locations along the channel, for that purpose the calculated hydrograph at the outlet was scaled down proportionally to the basin areas at the respective locations neglecting the time to travel along the channel. Figure 7.18 shows the longitudinal discharge variation along the stem channel when the discharge at the outlet is Q_0. Namely, the discharge at the junctions of main tributaries changes discontinuously but in other parts of the channel a uniform lateral inflow along the channel is assumed.

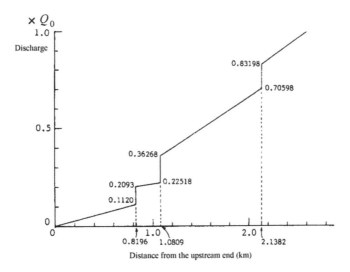

Figure 7.18 Discharge variation along the stem channel.

Calculation of the debris flow hydrograph

The method to calculate the debris flow hydrograph under arbitrary topographical and bed material conditions due to arbitrary lateral water inflow is given in section 3.1.2. Here, the same method is applied under the parameter values, in addition to the previously given bed material's characteristics, as follows: $K = 0.05$, $\delta_e = 0.0007$, $\delta_d = 0.1$, $p_i = 1/3$ and the width of channel is a uniform 10 m. The degree of saturation s_b of the bed material is assumed to be 0.8 in the channel reach steeper than 21° and 1.0 in the reach flatter than 21°. The calculations are carried out by discretizing into the meshes of $\Delta x = 50$ m and $\Delta t = 0.2$ s. In the reach of 500 m immediately upstream of the fan top in which check dams were installed the bed erosion is neglected (Takahashi and Nakagawa 1991; Nakagawa *et al.* 1996).

Figure 7.19 shows the calculated time-varying flow depths along the channel under the assumption of no landslide occurrence at the upstream end. At 7:50 a.m. a small-scale debris flow occurred at 1800 m downstream from the upstream end. This position corresponds to the upstream end of the reach where the check dams are installed. The position of debris flow initiation does not move upstream with time, and the debris flow travels downstream attenuating its maximum depth with distance. As shown here, debris flow can occur even without the landslide, but its magnitude is too small to explain the total volume of runoff sediment, and moreover, the severe erosion to the bedrock that actually occurred in the reach from just downstream of the landslide to the upstream end of check dams installation reach can not be explained. Therefore, the landslide at the upstream end should have played very important role.

The shallow landslide that often occurs synchronizing with the severest rainfall intensity tends to contain sufficient water to be able to immediately transform into debris flow. Figure 7.20 shows the results of the calculation under the conditions that

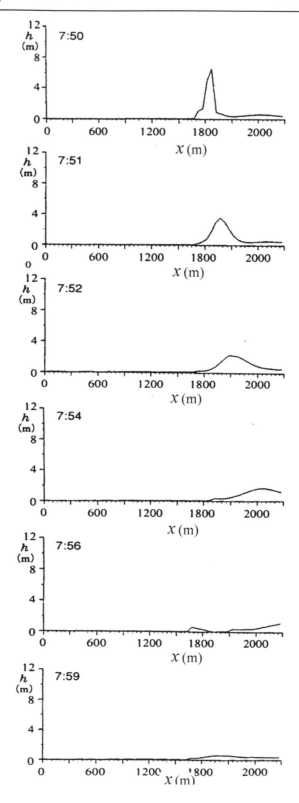

Figure 7.19 Time-varying flow depths along the channel (no landslide case).

at 7:50 a.m. the landslide at the upstream end occurred and it changed to a debris flow before arriving at the channel bed. The debris flow thus supplied onto the channel bed had the characteristics $q_t = 40\,\mathrm{m^2/s}$ for ten seconds, $C_L = 0.5$, and $C_F = 0$. In this case the supplied substantial sediment volume was 2,000 m³ that may be about 3,000 m³ in bulk. Although this is considerably less than the volume described in the survey report, this calculating condition was adopted because some amounts of sediment should have remained on the slope. Initially, the same kind of debris flow as the no landslide case shown in Figure 7.19 occurs in the downstream reach, but, soon, the one generated by the landslide proceeds downstream accompanying the erosion of bed and thus developing to a large-scale.

Figures 7.21 and 7.22 show the temporal changes in flow depths, discharges, concentrations of coarse particle fraction and concentrations of fine particle fraction at the outlet of the basin when no landslide occurs and when a landslide occurs at the upstream end. When no landslide occurs, the maximum depth is a little shallower than 2 m and the mean depth is about 50 cm, the peak discharge is about 50 m³/s and the mean discharge is about 20 m³/s. These values are far smaller than the actual values. However, when the landslide occurs, the maximum depth and discharge are about 4 m and 150 m³/s, respectively, and they have clear peaks. If the total substantial sediment runoff volumes are compared, it is about 8,000 m³ when no landslide occurs but it is about 50,000 m³ when a landslide occurs. The concentrations of coarse particle fraction in both cases suddenly increase, synchronizing with the arrival of the debris flow, and they reach the maximum corresponding to the time of peak discharges, then, they decrease. The coarse particle concentration when no landslide occurs decreases to less than 20% in a short time, whereas, when a landslide occurs, a high concentration continues for about thirty minutes up to well after the decrease of the discharge. A witness record says that in the five to seven minutes after the passage of the front, the debris flow over-spilled from the channel works and flooded onto road and then after thirty to sixty minutes the debris flow ceased. Another witness record says that the debris flow came down as a stone layer as thick as 3 m. These witness records suggest that the case including the landslide gives adequate results.

The calculation further revealed that the variation of the riverbed along the stem channel, in which the thickness of the deposit before the debris flow occurrence from the upstream end to about 1,950 m downstream of it was assumed to be a uniform 4 m, was as follows: When no landslide occurred, the riverbed was eroded to the bedrock only in the downstream reach beginning at a distance 1,500 m from the upstream end. When the landslide occurred, in the reach from the upstream end to a distance of 600 m some thicknesses of deposit were left, but downstream of that reach the riverbed was eroded thoroughly to the bedrock. The latter situation is in agreement with the actual situation. Generally, the erosion of the bed within the reach from the upstream end to a distance of 600 m increases its depth downstream. The sediment concentration in the debris flow just transformed from the landslide is too high to have the ability to erode the bed, but by the addition of water from the side downstream, the concentration is diluted to enhance the ability to erode the bed and the debris flow develops. Thus, the debris flow that is triggered by a shallow landslide can runoff a sediment volume far larger than the landslide volume itself.

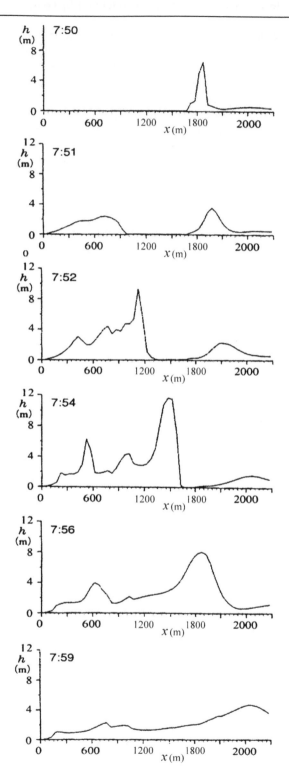

Figure 7.20 Time-varying flow depths along the channel (landslide occurred).

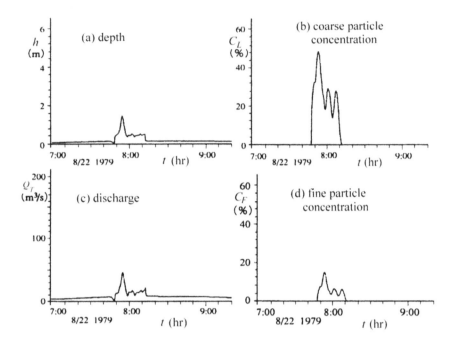

Figure 7.21 Time-varying depth, discharge and fine and coarse particle concentrations at the fan top (no landslide case).

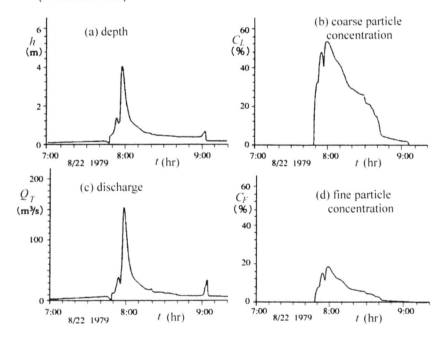

Figure 7.22 Time-varying depth, discharge and fine and coarse particle concentrations at the fan top (landslide occurred case).

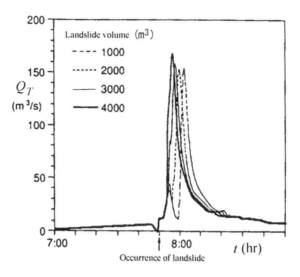

Figure 7.23 Effects of slid volume to the hydrograph downstream.

Examination of the effects of landslide volume

Figure 7.23 shows the results of an examination of how the scale of a landslide at the upstream end of the basin affects the debris flow hydrograph at the fan top. The four different size landslides were assumed to have happened at 7:50 a.m. at the same upstream end of the basin and to have transformed immediately into debris flows; with a duration of ten seconds, $C_L = 0.5$ and $C_F = 0$ as was the case previously calculated, but, the supplied debris flow discharges per unit width were changed to 20, 40, 60 and 80 m²/s. These discharges correspond to the substantial slid sediment volumes of 1,000, 2,000, 3,000 and 4,000 m³. Figure 7.23 shows that the larger the scale of the landslide, the earlier the arrival time at the fan top and the larger the peak discharge become. But, the difference between the cases is not so large. This fact suggests that if the location of landslide is far from the position in question and the channel between these two locations has plenty of deposited materials the scale of the landslide does not have much effect on the hydrograph at that position.

7.2.3 Reproduction of debris flow depositing area on the fan

The hydrograph at the fan top obtained by the routing of debris flow was used as the boundary condition for the calculation of the flooding and deposition in the fan area. The governing equations were the ones described in sections 3.3.4 and 5.3.2, in which flow was assumed to be horizontally two-dimensional. The parameter values and grid mesh sizes used in the calculation were; $\Delta x = \Delta y = 5$m, $\Delta t = 0.2$ s, $\delta_d = \delta'_d = \delta''_d = 0.1$, $d_L = 10$ cm, $C_* = 0.65$, $C_{*F} = 0.2$, $C_{*L} = C_{*DL} = 0.56$, $\tan \phi = 0.75$, $\sigma = 2.65$ g/cm³, and $n_m = 0.04$.

Figure 7.24 shows the time-varying distributions of surface stages (deposit thicknesses plus flow depths) on the fan. At five minutes after the onset of landslide (7:55 a.m.) the debris flow is already in the channel works but it is still before the peak

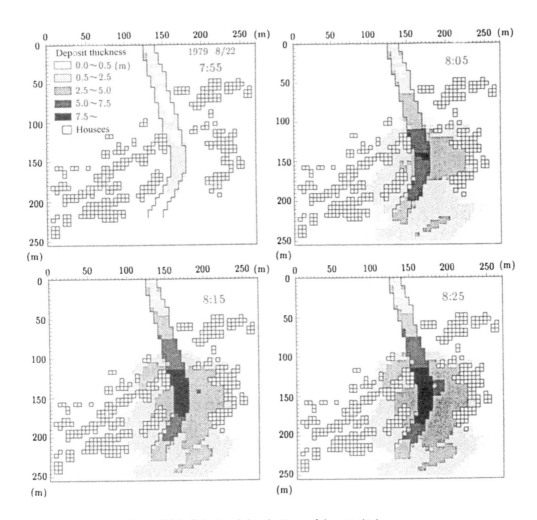

Figure 7.24 Calculated distributions of deposit thicknesses.

and no deposition occurs. At fifteen minutes after the onset of landslide (8:05 a.m.), the peak discharge has just passed through the channel and plenty of sediment has deposited around the channel bend from where sediment flooded mainly towards the left-hand side bank area. At twenty-five minutes after the onset of landslide (8:15 a.m.), the deposition in the channel works develops further and the increase in the deposit thicknesses in the left-hand side bank area is evident. The deposition also proceeds in the right-hand side bank area. By this time the debris flow discharge has decreased considerably and the enlargement of debris flow deposition area has almost ceased. Therefore, the difference between the situations at 8:15 a.m. and 8:25 a.m. is marginal. If the calculated result at 8:25 a.m. shown in Figure 7.24, is compared with the actual situation, shown in Figure 7.14, the calculation seems to predict a little thicker deposit in the channel works. The reasons for this can be attributable to several factors: the differences in the magnitudes and the characteristics of debris flows between the prediction and the actual one; the poor accuracy in the data of ground levels; and the errors in the measurement of the deposit thicknesses in the field. However, in general, the distributions of damaged houses and the calculated deposit thicknesses correspond to each other rather well. Therefore, we can conclude that this reproduction by the calculation is satisfactory.

In this reproduction, the sediment size distributions in the debris flow material was not considered, therefore, the effects of boulder accumulation in the frontal part of the debris flow were neglected. Because the numerical reproduction taking the effects of particle segregation into account is possible by the method introduced in the previous chapter, this may be an interesting point to investigate.

7.3 COLLAPSE OF THE TAILINGS DAMS AT STAVA, NORTHERN ITALY

7.3.1 Outline of the disasters

At 12:23 p.m. on July 19 1985, two coupled tailings dams, used to store calcium fluorite waste, collapsed at the village of Stava in Tesero, Dolomiti district, northern Italy. The stored tailings associated with the materials of the dam bodies formed a gigantic muddy debris flow that swept away 47 buildings, including three hotels, in the Stava valley, and 268 people were killed (Muramoto *et al.* 1986).

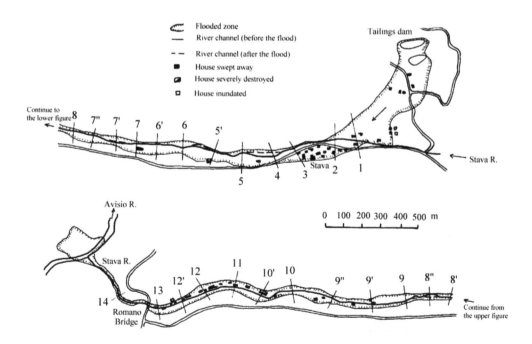

Figure 7.25 Plan view of flood marks and damaged buildings.

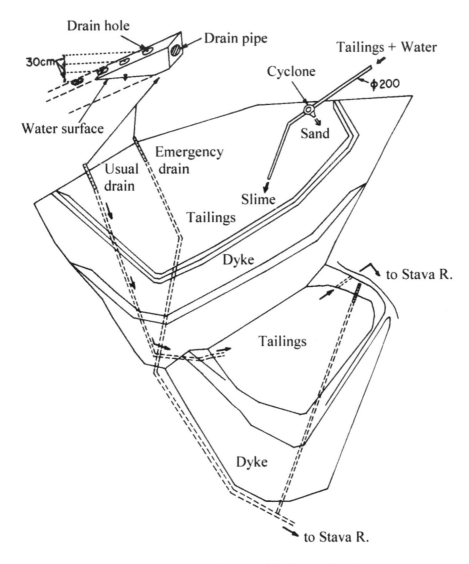

Figure 7.26 Conceptual diagram of the Stava tailings dams.

The muddy debris flow went along the Stava River as shown in Figure 7.25 and deposited at the junction with the Avisio River. Along the valley bottom of the Stava River there had been farm houses for 300–400 years, but the only experience of disasters was a small-scale flooding in 1966. After that flood, the channel works were installed. Because the area abounds in natural beauty, there has been active tourism development since 1970, especially in the area of Stava, where the hotels and villas were located. These were swept away as shown in Figure 7.25.

The tailings dams had a two-deck structure as shown in Figure 7.26. It is speculated that the upper dam collapsed first and the fall of mud debris from the upper dam caused

Photo 7.1 Frontal view of the collapsed tailings dams.

the lower dam to collapse almost instantaneously. Photo 7.1 is the frontal view of these failed dams.

The exact cause of the upper dam's collapse is not known but it might be due to the following causes:

1   overloading due to the additional elevation of the embankment that was under construction;
2   liquefaction of slime due to water seeping from the slope;
3   toe failure of the upper embankment caused by seepage;
4   failure of the upper embankment due to blockage of a drainage pipe in January;
5   insufficient separation between the pooled water and the embankment.

The increase of seepage water might have been affected by the massed rainfall that occurred two days before the collapse. As is usual, the tailings had been transported via the pipe depicted in Figure 7.26 to the upper reservoir. These tailings had a solids concentration of 25–35% by the discharge of 50 m³/hr. The pipe was equipped with a cyclone separator and the coarse fraction (fine sand) was deposited nearby and the fine fraction (silt slime) was deposited at a distance. The supernatant liquid was drained via the drainage pipe laid on the original mountain slope. The pipe had drain holes every 30 cm and during the process of deposition the supernatant water was drained from the nearest hole to the surface of deposit. Hence, if it worked normally, the depth of the supernatant liquid should never exceed 30 cm.

According to the survey done after the collapse, the total substantial volume of sediment flowed out is estimated to be 88,325 m³ (11,996 m³ of fine sand and 76,329 m³ of silt), and by estimating the volumes of the supernatant water and the interstitial water the total volume of 185,220 m³ of mud debris with a solids concentration of 0.476 is considered to have been released as a debris flow. If the peak discharge, Q_p, of this debris flow is estimated by applying the Ritter's dam collapse function:

$$Q_p = \frac{8}{27} c_0 h_0 B, \quad c_0 = \sqrt{g h_0} \tag{7.3}$$

$Q_p = 28,160$ m³/s for $h_0 = 40$ m, $B = 120$ m is obtained, where h_0 is the initial depth and B is the width of collapse. If the hydrograph has a right-angled triangular

Photo 7.2 View of the hazards at Stava area.

shape with the peak discharge arising at the instant of collapse, the total volume of 185,220 m³ gives the duration of debris flow as 13.2 seconds.

The debris flow surged down the 600 m long mountain slope of about 10° which was covered by grass in the upper part and by the forest in lower part and changed the flow direction perpendicularly by clashing with the cliff on the left bank of the Stava River. At that time it destroyed many hotels and other buildings. Then, it flowed along the Stava River channel for about 3.8 km with a surface width of 50–100 m and a depth of 10–20 m. It flowed weaving right and left and arrived at the junction with the Avisio River within about 6 minute, where it was deposited. A witness record says that the bore front was about 20 m high, trees were transported standing, sand-cloud was kicking up at the edge, and it passed through within about 20 seconds.

Photo 7.2 is the situation at the Stava area just after the disaster. The debris flow flowed into the Stava River from the upper left-hand side of the photograph and it

Photo 7.3 Situation at the middle part of the Stava River (Section.6-Section.7).

changed direction towards the lower left-hand side. To the right of a remaining tree at the photograph's center, there were hotels and buildings all of which were swept away. But, the channel consolidation structures kept almost their original shapes and the levee used as the road remained almost safe, suggesting that although the debris flow was of a very large scale, its ability to erode the bed or deposit sediment was small.

Photo 7.3 shows the situation at the middle part of the Stava River after two months when we did the field survey. There is an obvious difference in flood marks between the one on the left bank and the one on the right bank, the level difference is called super-elevation. Photo 7.4 shows the situation just upstream of the Romano bridge where the channel cross-sectional area is the narrowest. The debris flow climbed over the upstream old Romano Bridge but the downstream new one was only slightly damaged at the handrail on the left-hand bank side, meaning that the debris flow involving houses and trees passed through the narrow space of only about 15 m in span under the bridge. Because this location is at the channel bend, the heights of the flood marks on both sides of the channel are largely different. The difference in flood marks on both sides in the channel upstream of the bridges, however, would have been marked when the bore front passed the section before the bridges gave rise to the backwater effects to the section. Due to the channel constriction by the abutments of the bridge, the debris flow would have been built up as soon as its front came to the bridge section. Then, it would have proceeded upstream as a rebounding bore. This was speculated based on the fact that the flood marks around the tree stems on the left bank were higher and clearer at the downstream facing side than at the upstream facing side.

Photo 7.4 Situation upstream of Romano Bridge (Section 13).

7.3.2 Reproduction of the debris flow in the Stava River and its verification

Estimation of flow velocity

We did the simple measurements of cross-section of the flow at the locations indicated by the numbers in Figure 7.25. Some interpolations were also done by reading the topographical map of 1/5,000 scale at the locations depicted by the numbers to which prime symbols are attached.

Figure 7.27 shows the longitudinal profile of the Stava River obtained from the topographic map. The reach upstream of section 13 has an almost constant slope of about 5°.

As mentioned earlier, the peculiarity of the flood marks is the evident difference in the elevations on both sides. Figure 7.28, which is comprised of three figures, shows these characteristics in detail. The middle one shows the heights of the flood marks h on the right and left banks. The height zero in this figure is at the lowest level at each cross-section. The difference between the altitude of the flood marks on both sides at each cross-section H' is correlated to the curvature of the centerline of the stream channel $1/r_{co}$ (r_{co} is the curvature radius) in the upper figure. The symbols R and L on the axis, indicating the magnitude of $1/r_{co}$ (left side ordinate axis), mean that the center of radius exists on the right and the left banks, respectively, and the symbol R on the axis, indicating H' (right side ordinate axis), means that the flood mark on the right-hand side bank is higher than that on the left-hand side bank, and L means the opposite. Thus it can be seen on Figure 7.28, that the flood mark on the outer bank of the channel bend is in general higher than that on the inner bank. This is the effect of

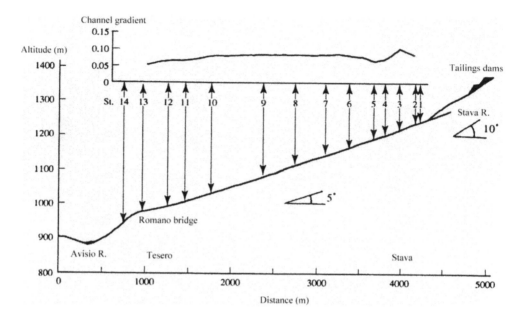

Figure 7.27 Longitudinal profile of the Stava River.

centrifugal force. The lower figure shows the longitudinal variation in the flow cross-sectional areas. Neglecting the natural variability, there is a decreasing tendency in the downstream direction between Section 4 and Section 10. The cross-section suddenly increases around Sections 10 to 10' and downstream of that it stays almost constant. The latter tendency would be due to the large resistance to flow by the houses on the flood plain and the backwater effects of the Romano Bridge.

Let us estimate the velocity of flow using the phenomena of super-elevation. According to the field measurements of the channel, the average cross-sectional shape from Section 4 to Section 10 is represented by a triangle with both side slopes being 1/3. As for the super-elevation at the channel bend for turbulent type debris flow, as stated in section 4.5, Lenau's formula is applicable, that is for a trapezoidal section as shown in Figure 7.29:

$$E_{\text{max}} = \frac{U^2}{2r_{co}g}(2mh_0 + b) \tag{7.4}$$

where E_{max} is the maximum of the super-elevation, U is the cross-sectional mean velocity, r_{co} is the curvature radius of the centerline of the channel, g is the acceleration due to gravity, and other variables appear in Figure 7.29. Herein, $b = 0$ and $m = 3$. Although the maximum difference in the measured flood marks on both banks around the measuring section is not necessarily equal to $2E_{max}$, it is assumed as equal to $2E_{max}$. By using Equation (7.4) and the upper figure in Figure 7.28, the cross-sectional mean velocity at each section along the channel is obtained as shown in Table 7.1, where in the reach 2–3 the cross-sectional shape is approximated as a rectangular.

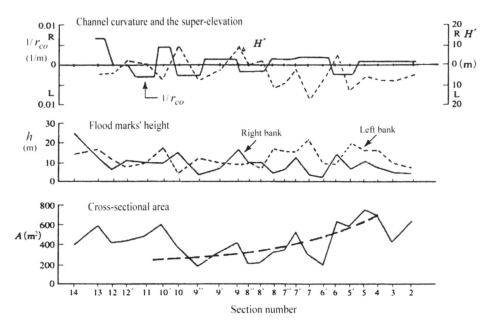

Figure 7.28 Changes in various quantities of the debris flow along the channel.

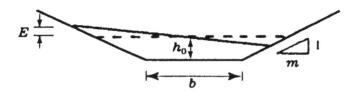

Figure 7.29 Super-elevation at a channel bend in a trapezoidal section.

Table 7.1 Mean velocity and the roughness coefficient.

Sectional reach	Mean velocity	Manning's roughness coefficient
2–3	18 m/s	
3–5	23	
6'–7	31	
7–8	25	0.04
8–9	22	
9–9"	22	
9"–10	18	
10–10'	11	0.08
10'–12	6.2	0.13
12'–13	6.8	0.12

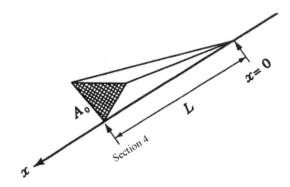

Figure 7.30 The mud mass assumed to start from Section 4.

Although the debris flow was as dense as 0.5 in solids concentration, the comprising material was very small and so the relative depth had a value of the order of 10^5. In such a case, as explained in section 2.6.3, the resistance law of the flow becomes almost identical with that of plain water flow, and hence Manning's resistance law is applicable. Manning's roughness coefficient in each reach between the sections was obtained by a reverse calculation from the data on velocity as tabulated in Table 7.1. The high roughness values downstream of Section 10 are attributable to the large cross-sectional areas in these reaches.

Numerical reproduction of flowing process

The mud mass produced by the collapse of the tailings dams collided with the mountain slope at the center of Stava area and changed direction. It became a bore-like flow around Section 4. Herein, the deformation of the hydrograph along the Stava River is discussed under the assumption that a triangular pyramid-shaped mud mass as shown in Figure 7.30 is suddenly given in the reach upstream of Section 4. The velocity is assumed to be given by Manning's formula and the effects of the acceleration terms are neglected. In such a case the flow can be routed by the kinematic wave method explained in Section 4.6.1. If the cross-section is triangular and the channel is assumed to be prismatic, then the cross-sectional area at an arbitrary position x is obtained by solving the following equation:

$$\frac{\partial A^*}{\partial t^*} + \frac{4}{3} A^{*1/3} \frac{\partial A^*}{\partial x^*} = 0 \tag{7.5}$$

under the initial condition:

$$\left. \begin{array}{ll} A^*(x^*, 0) = x^{*2}; & (0 \le x^* < 1) \\ A^*(x^*, 0) = 0; & (-\infty < x^* \le 0,\ 1 < x^* < \infty) \end{array} \right\} \tag{7.6}$$

where the following non-dimensional expressions are adopted:

$$(A^*, x^*, t^*) = (A/A_0, x/L, Ut/L) \tag{7.7}$$

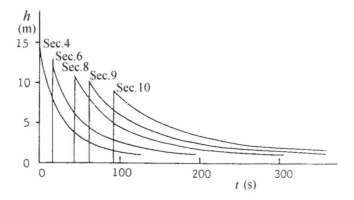

Figure 7.31 Depth versus time relationships at some sections.

where A_0 is the cross-sectional area of Section 4 at $t = 0$, U is the velocity of flow when the cross-sectional area is A_0.

The solution is:

$$x^* = A^{*1/2} + \frac{4}{3}A^{*1/3}t^* \tag{7.8}$$

and the location of the forefront is given by:

$$\left. \begin{array}{l} t_s^* = (1 - A_s^{*3/2})/A_s^{*4/3} \\ x_s^* = A_s^{*1/2} + (1 - A_s^{*3/2})/A_s^* \end{array} \right\} \tag{7.9}$$

where the suffix s means the forefront.

Given that the volume of the triangular pyramid in Figure 7.30 was estimated to be $185{,}220\,\mathrm{m}^3$, $A_0 = 700\,\mathrm{m}^2$ and $n_m = 0.02$, thus $L = 794\,\mathrm{m}$, and U at Section 4 is equal to 27.3 m/s. Then, the maximum cross-sectional area is calculated from the lower equation in Equation (7.9). The result is shown by the broken line in the lower figure in Figure 7.28. Although the actual measured cross-sections largely fluctuate, in general, the tendency to decrease downstream matches the field data, and it shows the attenuation of the flow area or discharge in such a sudden flow of short duration downstream is very large.

Figure 7.31 shows the temporal variation in the depth at the center of the cross-section and Figure 7.32 shows that of the discharge. If the average flow area in the reach between Section 10 and Section 13 is assumed, based on the data shown in the lower figure in Figure 7.28, to be $500\,\mathrm{m}^2$ and the maximum velocity in this reach is assumed to be 7 m/s from Table 7.1, the maximum discharge in this reach becomes $3{,}500\,\mathrm{m}^3/\mathrm{s}$. This value is a little smaller than the calculated result at Section 10 shown in Figure 7.32, but considering that the calculation in Figure 7.32 is done ignoring attenuation due to the storage effects of irregularly changing cross-sections along the channel, it may be concluded that the flow routing by this method is satisfactory.

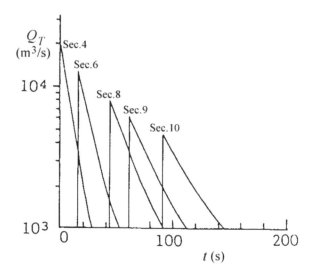

Figure 7.32 Hydrographs of the debris flow at some sections.

Verifications of reproduction

In Figure 7.33 the arrival times of the debris flow forefront calculated by the kinematic wave theory and that obtained from the velocity calculated by super-elevation as given in Table 7.1 are compared. Because the calculation by the former method starts from Section 4, the time of starting by the former method at Section 4 is set on the line indicating the latter method (the dotted line). As shown in Figure 7.33, both results agree well from Section 4 to Section 10, but downstream of Section 10 the discrepancy is conspicuous. This is due to the effects of the extreme bend around Section 10′ and the houses in the valley downstream of Section 10′ which are not taken into account in the kinematic wave approximation. The agreement in the cross-sectional areas calculated by the kinematic wave method with the field measurements and the agreement in the peak discharges at Section 10 routed by the kinematic wave method with the one obtained by the combination of kinematic wave method and the velocity estimation by super-elevation are already mentioned. Thus, the reproduction by the kinematic wave method is verified in view of variations in flow areas, discharges and translation of forefront.

There was a seismograph record that seemed to correspond to the ground vibrations due to the debris flow at Cavalese, located 3.7 km from collapsed tailings dam. We calculated the power spectra of that record and found that they distributed in very narrow band of 1–4 Hz and that the predominant frequency of oscillation varied with time. The predominant frequencies could be divided into three categories; from 12:23:35 p.m. to 12:23:55 p.m. frequencies were 2–3 Hz, from 12:25:35 to 12:25:55 frequencies were 1–2 Hz, and from 12:27:50 to 12:30:00 frequencies were 1–4 Hz. The first two groups continued for about 20 seconds and they had very high powers, whereas the last group had continued for about 100 seconds and the power was comparatively low. If the first group was induced by the collision of the mud mass

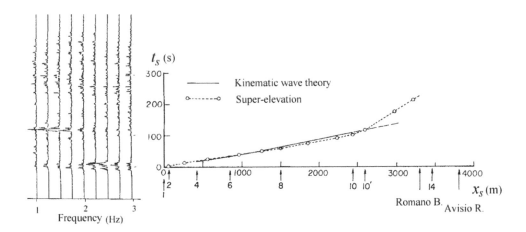

Figure 7.33 Advance of the bore front and the correspondence to the seismic record at Cavalese.

with the mountain slope around Section 2 of the Stava area, it is possible to match the occurrence time of that vibration to the origin of the graph on the right showing the arrival time versus distance graph in Figure 7.33. Then, the second large vibration of 1 to 2 Hz corresponds to the time that the debris flow crashed in the extreme bend near Section 10′. This provides a reasonable estimate of the velocity of the debris flow. The third, weaker vibration continues for a long period and may be explained as that generated when the debris flow passed through the narrow gorge downstream of the Romano Bridge (Section 13). The bore front arrived at the Romano Bridge about 225 seconds after it passed Section 1. The duration of the debris flow at Section 10, which overflowed the channel works is estimated to be about 200 seconds from Figure 7.31. The duration at the Romano Bridge must not be much different, and the duration of the vibration of the third group is of the same order. Consequently, the whole duration of the debris flow from flowing into the Stava River (Section 1) to flowing out to the Avisio River is estimated to be about 425 seconds. This value agrees with the duration of the vibration in the seismic record and also corresponds to the accounts of witness.

7.4   DISASTERS CAUSED BY THE ERUPTION OF THE NEVADO DEL RUIZ VOLCANO

7.4.1   Outline of the disasters

On 13 November 1985, after 140 years of dormancy, an eruption occurred from the crater at the summit of the Nevado del Ruiz volcano, Colombia, and it accompanied a small-scale pyroclastic flow. Because the altitude of the summit is as high as 5,400 m, the area higher than 4,800 m was covered by the ice cap (glacier). The rapid melting of the surface of the ice cap due to the coverage by pyroclastic flow triggered disastrous volcanic mudflows (lahars) in several rivers which originate from the volcano. Among

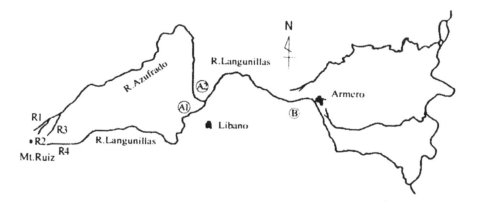

Figure 7.34 Lagunillas River system and the location of Armero.

them the one in the Lagunillas River was the largest, and furthermore, the river basin contained the city of Armero on its alluvial fan. The mudflow flooded on all over the city and 21,000 of the 29,000 inhabitants were killed.

We did a field survey from 19 December 1985 to 3 January 1986. The entire picture of our survey is written in the report (Katsui 1986), so that, here, only the phenomena associated with the mudflow that thoroughly destroyed Armero are explained. Because the survey was done in a short period and the disasters were widely ranged, the survey was necessarily rough.

Figure 7.34 shows the rough pattern of the Lagunillas River system and the location of Armero City. The mudflows originated in the Azufrado and Lagunillas Rivers joined near Libano and continued down the Lagunillas River. The river channel before the disaster suddenly changed direction to go south-east just upstream of Armero. Because the capacity of the channel was too small for the mudflow to smoothly follow the river course, the major part of the mudflow overflowed from the channel bend to directly hit Armero City. Figure 7.35 shows the area of mudflow flooding and deposition. The flooding flow took two directions; the main flow went straight ahead towards the center of the city, and then it met the hill off the urban area and gradually changed direction towards the south-east, the other branched flow took the north-east direction and then went northwards. The branching was due to a hilly topography at the central part of the fan, but even in this relatively high area, sediment was deposited thickly enough to completely bury the ground floor of houses. The sediment deposited area was as large as about 30 km^2 and the average thickness of the deposit seemed to be about 1.5 m.

The following photos may help to understand the situation of the mudflow from the origin to the deposition: Photo 7.5 shows the situation at the source area of the Azufrado River. The Arenas crater is nearby and the ice cap scarcely remains on the steep slope.

Photo 7.6 shows the fan top area of Armero and the Lagunillas River just upstream of the fan indicating very high flood marks along the channel. The bending section A-A

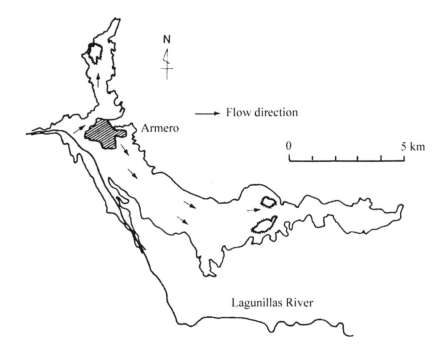

Figure 7.35 The area of sediment deposition and the flow direction.

Photo 7.5 The source area of the Azufurado River.

has a curvature radius of about 150 m. We went to this section to measure the super-elevation, the flow depth and the width of the channel, and by applying Equation (7.4), we calculated that the maximum discharge that passed through this section was 28,660 m³/s. At section B-B, which is the straight reach downstream of section A-A,

Photo 7.6 Armero fan top and the flood marks in the Lagunillas River.

Photo 7.7 Situation of the Armero fan.

we adopted 0.04 as the Manning's roughness coefficient, which is equal to the case of debris flow in the Stava River, and obtained the maximum discharge as 29,640 m^3/s.

Photo 7.7 is a view from the Armero site to upstream direction. The lower right part of the photograph is the hilly area that divided the flooding flow into two. The

Photo 7.8 This area was the central part of the city.

Photo 7.9 Deposit a little distant from the main flow.

main flow went to the lower left and around the arrow mark we can see the traces of houses that were swept away. Photo 7.8 shows the deposit in the main flow area. There were many public facilities such as the national bank and the school. We found a large boulder about 10 m in diameter around here. Photo 7.9 shows the deposit a little distant from previous photo. The surface layer of thickness 1 m or so is comprised of very fine material and beneath that is a mixture of coarse and fine materials.

7.4.2   Reproduction of the phenomena

Estimation of water supply to respective rivers

Because the cause of mudflow was the melting of ice cap due to the pyrocalstic flow, for the reproduction of the mudflow it is necessary to estimate the quantity of melt water. Herein, the quantity of heat supply per unit time from the pyrocalstic flow $\rho_p C_a T_a Q_a$ is simply considered to be thoroughly expended to melt ice. Then, if the heat of fusion is q_m cal/g, the water production rate in unit time Q_{w0} is given as:

$$Q_{w0} = \frac{\rho_p C_a T_a}{\rho q_m} Q_a \qquad (7.10)$$

where ρ_p is the density of the pyroclastic material, C_a is the specific heat of the pyrocalstic material, T_a is the temperature of the pyrocalstic flow, Q_a is the discharge of the pyrocalstic flow. In Equation (7.10) the temperature of the produced water should be 0°C, but in fact, the water should be considerably hotter, and therefore T_a should be the temperature difference between the pyrocalstic flow and the water. The pyrocalstic flow is, however, very hot and therefore in practice, T_a is equal to the temperature of the pyroclatic flow.

If the discharge of the pyroclastic flow given to the total Lagnillas River basin is written as Q_{aL}, and $\rho_a/\rho = 2.65$, $C_a = 0.2$ cal/g°C, $q_m = 80$ cal/g, and $T_a = 800$°C

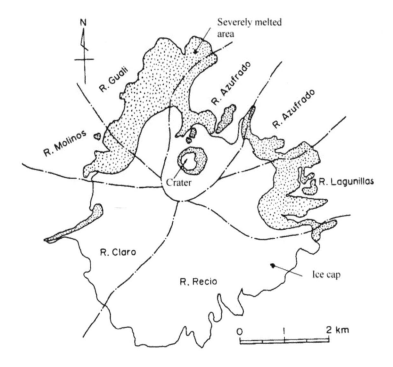

Figure 7.36 Ice cap areas severely melted by the pyroclastic flow.

are assumed, the water supply rates to the respective sub-basins will be given from Equation (7.10) as:

$$Q_{w0i} = 5.3 r_i Q_{aL} \qquad (7.11)$$

where the suffix i means the i-th sub-basin, and r_i is the affected area ratio of i-th sub-basin to the total affected area of the Lagnillas River basin.

To calculate Q_{w0i} from Equation (7.11), Q_{aL} and r_i should be given. But, because no data for the direct estimation of Q_{aL} is available, it can only be assumed and verified by the results of simulation. The area affected by the severe ice melting can be found from the data of surveys of ice cap before (1983) and after the eruption. Figure 7.36 shows that data (INGEOMINAS 1985), in which the dotted zones are assessed to be severely affected. From Figure 7.36, the affected areas of the Azufrado River and the Lagunillas River are 1.85 km^2 and 1.60 km^2, and hence, the values of r_i are 0.536 and 0.464, respectively. Therefore, as an example, if Q_{aL} is assumed to be 700 m^3/s, the water supply to the Azufrado and the Lagunillas Rivers becomes 1,989 m^3/s and 1,721 m^3/s, respectively.

Calculation of the mudflow hydrographs in the respective rivers

Water supplied to the Azufrado and Lagunillas rivers severely eroded their river courses. The longitudinal channel gradients at the source areas of these rivers were obtained from the topographical map, and were found to be about 12° and 19°, respectively. Therefore, the supplied water containing the material of the pyrocalstic flow should have generated an immature debris flow in the Azufrado River and a debris flow in the Lagunillas River, by eroding the riverbeds. Although no data for the riverbed materials in the rivers is available, according to our rough survey of the riverbed materials in the Molinos River channel that also originates from the Ruiz Volcano, the mean particle size of the coarse fraction was about 10 cm and the voids between the coarse fraction were filled up with plenty of fine materials. A similar composition would hold true for the materials in the Azufrado and Lagnillas rivers. Because the perennial water stream existed on the riverbed due to the normal melting of the ice cap, and it rained severely at the time of eruption, the riverbed sediment is assumed to be saturated by water.

In the Azufrado River, a stretch of 7 km from the summit of the volcano, and in the Lagunillas River a stretch of 5 km from the summit of the volcano, the channel gradients are very steep so the debris flow discharges and the solids concentrations will develop downstream. However, the channel gradients become shallower downstream of these steep reaches. Then, the coarse particle's concentration in the flow becomes over loaded and deposition of coarse particles will occur. The fine fraction is also entrapped in the deposit but the fine constituent in the upper layer can keep its high concentration while in the depositing process. The channel gradients of both rivers again become steep downstream and the ability of the flow to erode the bed is restored, hence the concentration of the coarse fraction again becomes large, and simultaneously, the fine materials in bed are entrained into the flow to increase the concentration of the fine fraction. Thus, the concentrations of the coarse particle fraction become large in one place and small in another during the passage of many steep and mild reaches, but the concentrations of fine particle fraction do not change much but rather monotonously increase downstream. Thus, the mudflow that struck Armero contained a large amount of fine fraction.

The above mentioned processes of mudflow generation and development can be reproduced by applying the governing equations explained in section 3.1.2, in which the Manning's resistance law is applicable because the ratio of flow depth to the representative diameter is very large. The parameter values may be assumed as $C_* = 0.64$, $C_{*L} = 0.256$, $C_{*F} = 0.384$, $C_{*DL} = 0.5$, $\tan \phi = 0.75$, $d_L = 10$ cm, $\sigma = 2.65$ g/cm^3, $n_m = 0.04$, $B = 50$ m, $D = 20$ m, $\delta_e = \delta_d = 10^{-4}$. We calculated the discretizing space and time in meshes of $\Delta x = 200$ m, $\Delta t = 1.0$ s.

After some trial calculations the supplied water discharges were determined: For the Azufrado River 831 m^3/s, 831 m^3/s and 391 m^3/s for the respective three sub-basins (see Figure 7.34), and for the Lagunillas River 1,500 m^3/s. The sediment supplied as the pyroclastic flow and then contained in the supplied water to the rivers is neglected. The duration of the melting of the ice was set to fifteen minutes, following the witness report for the Molinos River basin.

Figure 7.37 shows the supplied water discharge to the respective rivers (hatched parts) and the calculated mudflow hydrographs at the positions A1, A2 and B illustrated in Figure 7.34. Although the discharge in the Lagunillas River was smaller, the mud flood in the Lagunillas River arrived at the junction point A1 earlier than the flood in the Azufrado River arrived at the junction point A2 because the traveling distance in the Lagunillas River is shorter than that in the Azufurado River. The effect of time lag remained even at the fan top of Armero where the hydrograph had two peaks. The hydrograph at the position B comprises one part of the upstream boundary conditions, and Figure 7.38 shows this hydrograph together with other boundary conditions at

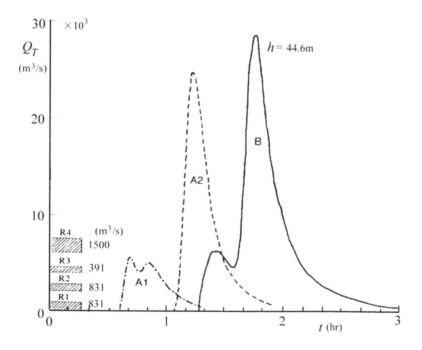

Figure 7.37 The calculated hydrographs of the mudflow. The hatched ones are supplied hydrographs.

the position B. The peak discharge at the fan top was calculated as 28,600 m³/s and this is in good agreement with the field measurement. The calculated C_F value is very large and this reproduces the flow as a very thick mudflow.

The total flowed out solids volume by these calculations becomes about 20.4×10^6 m³. However, the bulk volume of the deposit around Armero was estimated to be about 45×10^6 m³. If the calculated volume of sediment is deposited with $C_* = 0.64$, then the bulk volume becomes 32×10^6 m³. This value is a little smaller than the estimation in the field, but the estimation in the field assumes an average thickness of deposit as 1.5 m without a precise survey. Thus it may be concluded that the calculated sediment volume is a reasonable estimate. The actual time lag between the occurrence of the eruption and the appearance of the mudflow at Armero was said to be about two hours. In the calculation, the mudflow that flowed down the Lagunillas River arrived at Armero after about one hour and twenty minutes and that down the Azuflado River arrived after about one hour and thirty-five minutes. The calculated flood seems to arrive a little earlier than the actual one. This difference is related to the adoption of $n_m = 0.04$, so that more precise discussion on the process of traveling in narrow gorge is necessary.

Reproduction of the mudflow flooding and deposition on the Armero fan

According to the calculated solids concentrations, the mudflow that ran out to the Armero fan should have had the fine fraction dispersed uniformly all over the depth, with the coarse fraction concentrated in the lower part, in the manner illustrated in

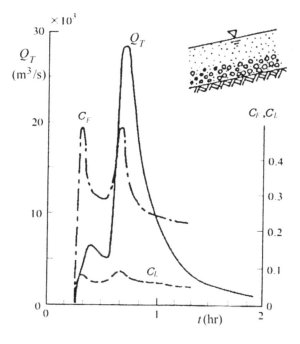

Figure 7.38 Hydrograph and the sediment concentrations at the fan top.

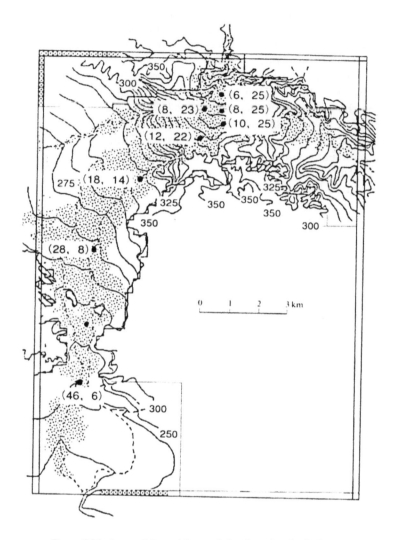

Figure 7.39 Area of deposition and the domain of calculation.

Figure 7.38, that is a kind of immature debris flow whose fluid phase is composed of very dense mud.

The area of deposition is shown in Figure 7.35, but Figure7.39 shows the area in more detail with the contour map of the topography, and it also shows the domain of the calculation to reproduce the depositing processes below. The in-flow boundary conditions are the ones illustrated in Figure 7.38. The time of the sudden increase in the discharge, $t = 16$ minutes in Figure 7.38, is set to be $t = 0$ minutes in the analysis of the flooding.

We analyze the two-dimensional flooding processes of the mudflow, taking the material to be a one-phase continuum and assuming that the momentum conservation equations are given by Equations (3.115) and (3.116), in which the resistance terms

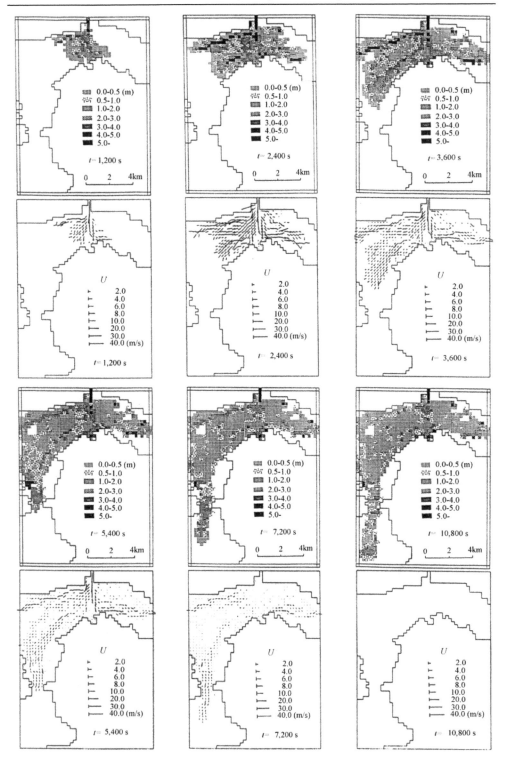

Figure 7.40 Flow depth plus deposit thickness and the velocity vector.

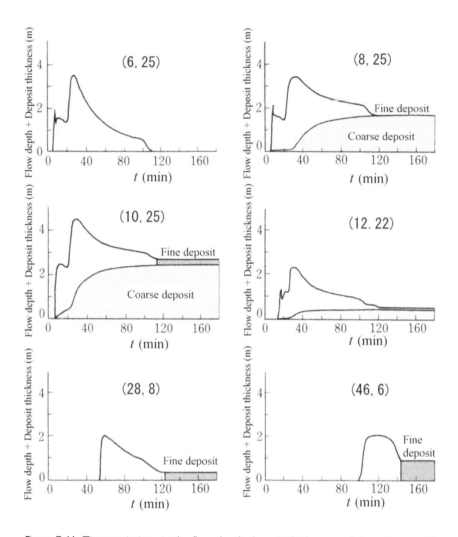

Figure 7.41 Time variations in the flow depth, deposit thickness and deposit material.

are given by Equations (3.124) and (3.125). The continuity equation is given by (5.71), in which the depositing rate of coarse particles is described by Equation (3.134).

As long as Equations (3.124) and (3.125) are used as the resistance law, the flow will not stop. This contradicts the actual phenomenon, so, a threshold velocity U_{TH} is introduced. If the velocity of flow $(u^2 + v^2)^{1/2}$ becomes slower than that threshold value the flow will stop. This treatment might be considered expedient, but as in the case of viscous debris flow, a very thick mud flow will not be able to continue its motion with very slow velocity.

The calculation was carried out using the same parameter values as the case of flow routing in the river channel except for the mesh sizes: $\Delta x = \Delta y = 250$ m, $\Delta t = 1$ s. The threshold velocity was set at $U_{TH} = 0.4$ m/s.

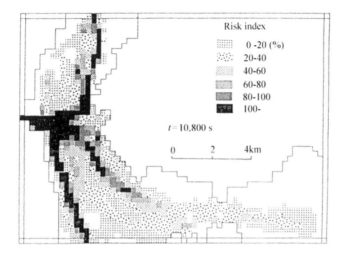

Figure 7.42 Distribution of the risk indices of house destruction.

Figure 7.40 shows pair graphs of the temporal variations of surface stage (flow depth plus deposit thickness) and the velocity vector after the mudflow appeared at the fan top. Armero city should have been destroyed within twenty minutes and by three hours all the dynamic phenomena should have finished. The flow was split into three branches, one along the original Lagunillas River, and other two along the small streams distributing on the fan. The major flow did not follow the Lagnillas River but followed a tiny stream named Viejo River that had been flowing east originating from Armero city.

The time sequential changes in the flow depth, deposit thicknesses and their materials at some representative places indicated in Figure 7.39 are shown in Figure 7.41. The maximum flow depth at the mesh (6,25), which is located in the river channel around the fan top, reaches 3.5 m, but virtually no deposition takes place due to the high ability to transport sediment there. On the other hand, at the meshes (8, 25) and (10, 25), located in the urban area of Armero, although the flow depths are almost the same as or a little larger than that in (6, 25), the thickness of the deposit of the coarse sediment is over 2 m. This result corresponds well with the actual situation.

Both the peak flow depth and the thickness of deposit decrease downstream and the volume ratio of coarse material to fine in the deposit becomes smaller downstream. Further downstream, at mesh (28, 8), the deposit is comprised almost entirely of fine material.

The index for the destruction or washing away of Japanese wooden houses has been given as Equation (7.1). Substituting the density of mudflow into ρ_T in Equation (7.1), we can obtain the time-varying risk indices for the houses on the Armero fan. Figure 7.42 shows the distribution of the maximum index values at the respective positions. Although the houses of Armero do not have the same structures and materials as Japanese houses, these values should represent the general degree of the power operated on the houses and we can understand why a great many houses were swept away.

7.5 SEDIMENT DISASTERS IN VENEZUELA

7.5.1 Outline of the disasters

From 8 to 19 December 1999, continuous severe rainfall that amounted to more than twice the annual rainfall, caused disasters in the eight states of northern Venezuela. The state of Vargas, located on the northern slopes of the Cordillera de la Costa mountain range, was the most severely damaged by landslides, debris flows, and flash floods induced by concentrated rainfall from 15 to 16 December. The official statistics of damage compiled by the Civil Defense Agency of the state of Vargas at the time of our reconnaissance in March 2000 were as follows: Affected houses, 40,160; completely collapsed houses, 20,000; affected persons, 214,000; injured persons, 2,700; dead persons, 248; missing persons, 2,850; refugees, 43,569. The media and local governments, however, estimate the lives lost as 25,000 to 50,000 and the International Red Cross makes it as high as 30,000. The reason for the wide range in the death toll is due to the lack of census data and difficulties in recovering bodies from under thick rocky debris or deep in the sea. This disaster is said to be the worst ever to occur in Latin America in the twentieth century (Takahashi et al. 2001; Takahashi et al. 2001a).

The state of Vargas is long and narrow and faces the Caribbean Sea to the north. It is separated from Caracas to the south by the high Cordillera de la Costa mountain range. The central part, the most severely devastated area, is shown in Figure 7.43. The distance from the coast to the ridge is about 10 kilometers, and the altitude of the ridge is between 2,000 and 2,700 m, therefore the northern slopes of the mountain range have an average steepness of 10 to 15 degrees. The major rivers shown in Figure 7.43 form alluvial fans adjacent to the coast, but the rest of this coastal zone area consists of steep mountain fronts rising directly from the shoreline. At first glance, these alluvial fans provide the only relatively flat areas that are suitable to live on, but they are by no means safe. Towns, however, have been constructed on these fans, and a road hugging the steep mountain fronts along the coast connects these towns.

Although December is normally the dry season on the Caribbean coast area, a cold front that lingered for about twenty days along the coast caused a total rainfall of 1,207 mm at the Maiquetia airport. For the three days from 14th to 16th, 914 mm of rainfall was recorded. The strongest rainfall was along the coast, it was weaker in the higher mountain areas. This situation is indirectly supported by the fact that, in the Caracus area on the opposite side of the mountain ridge, no severe flooding occurred.

The geological setting of this area has the general trend of a belt-like formation parallel to the Caribbean coastline. The zone, about 1.5 km wide on the coast, is a 'mélange' consisting mainly of calcareous phyllite. The phyllite has been heavily weathered to create clayey materials and forms steep slopes of 25 to 40 degrees that face the sea. Because this zone contains much clay, water permeability is low. Vegetation in this zone is poor, mostly sparse shrubs and cactuses, perhaps due to human activity. The roots of these types of vegetation are only 20 to 30 cm deep. This environment may help the occurrence of very dense shallow landslides. To the south of this zone, schist and granitic gneiss zones form the main parts of the mountains. Dark colored sands, derived from landslides in the phillite zone, cover the narrow beaches at the foot of the steep mountain front, whereas white sands, gravel, and boulders derived from the granitic gneiss areas cover the alluvial fans of the major rivers.

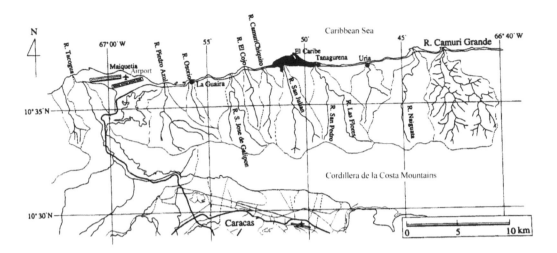

Figure 7.43 Central part of the state of Vargas and metropolitan Caracas.

The characteristic geological and rainfall distributions described above therefore must be the main factors that determined the distributions of the numbers and positions of the landslides and debris flows. The landslide density was very high in the mainly weathered phillite hilly area along the coast; 18% by the area ratio at the maximum. The very dense landslide area, however, is almost completely between Maiquetia to the east and the Naiguata River to the west. This situation correlates well with the decrease in the rainfall to the west and east of this zone.

Landslides that occurred on the schist and gneiss slopes were a kind of rockslides, and according to my inspection from a helicopter, many of their scars are connected with riverbeds. Little debris remains on these scars or the riverbeds. The landslide density of these slopes is much lower than in the former area. Presumably this is because these areas are covered with good vegetation and the rainfall was less than in the coastal area.

The total volume of deposited sediment on the alluvial fans of the seven rivers shown in Figure 7.43; El Cojo, Camuri Chiquito, San Julian, Cerro Grande, Uria, Naiguata and Camuri Grande is estimated to be about 6–7 million m^3. Almost all the debris originated from the schist and gneiss areas and it far exceeded the volume of newly produced sediment due to landslides. According to witness reports heard during the reconnaissance, there was a severe rainfall in 1951 that left a vast amount of sediment in the upstream reaches of these rivers. This time, the riverbeds in the upstream reaches were almost completely eroded to the bedrock, therefore, the majority of the runoff sediment from these rivers should have been this accumulated sediment.

Beside the aforementioned natural conditions, various social conditions played a very important role in worsening the disasters. The central part of the state of Vargas is separated from the metropolitan Caracas by a high mountain range, but is easily accessible from Caracas via a pass. Maiquetia, the principal international airport of Venezuela, and La Guaira, a major trading seaport, are located in this state and connected by an express way of about 20–30 km with Caracas. Due to these advantageous geographical conditions and the state's beautiful coastal and mountain settings, the alluvial fans of the major rivers located between the foot of mountain range and the coast have been highly developed as resort towns. Development of the area has caused an inflow of people for employment. These newcomers could not find flat land suitable for building their houses, so many shantytowns were constructed on very steep slopes highly vulnerable to landslides near the developed alluvial fans or on canyon bottoms that are clearly vulnerable to floods. Luxurious high-rise hotels and condominiums, yacht harbors, and slums constitute the strange, characteristic scenery of this area. These towns were the main sites of the disasters.

One typical pattern of damages is due to landslides on the very steeply sloping areas. Because the landslide density in these areas is very high, as is the density of the houses constructed, these houses had high probability of being struck by falling earth blocks or encountering the sliding of the ground itself on which they were built. Some houses were destroyed by the collapse of houses above them as if they were falling dominoes. The other pattern arose in the narrow canyon bottoms between the steep mountain slopes, which were filled with many low-income houses. These were obliterated by debris flows that exceeded the entire width of the canyon. This type of damage typically occurred along the Cerro Grande and Uria rivers. The Cerro Grande River is the one that enters the Tanaguarena township, shown in Figure 7.43, where a reach, about 800 m long and 200 m wide, immediately upstream of the fan-top forms a canyon plain. This was occupied by many houses, leaving only a narrow channel that could drain away only a normal flood. Almost all of these houses were swept away by the debris flow. At the mouth of the Uria River, a long stretch of the coastal cliffs had been hollowed out to form a canyon 700 m long and 220 m wide. This plain was completely covered by a shantytown, Carmen de Uria. Two of its five central blocks were obliterated by debris flows. The houses that remained were by no means safe as the ground floor was buried by debris or destroyed. More than five thousands people are said to have been killed in this town.

The alluvial fans on which many luxurious high-rise hotels were located were also not safe. These typical examples can be seen on the fans of the San Julian and Camuri Grande rivers. The fan of the San Julian after the disasters can be seen on the photograph at the front of this chapter. This fan was swept by several debris flows which left a covering deposit 1.2 km² wide and a few meters thick, that included boulders 2 to 3 meters in diameter. Many lower situated houses between hotels and condominiums were flushed away or buried to the roof, and some high-rise concrete buildings partly collapsed because of being hit by large boulders (Photo 7.10). Photo 7.11 shows the Camuri Grande alluvial fan, where the sediment deposited was almost completely composed of fine sand and almost all the buildings were buried at least to the ceiling of ground floor.

The residential areas on the fans of the San Julian, Cerro Grande and Uria rivers were obliterated by sediment including large boulders, whereas the residential areas of Naiguata and Camuri Grande were covered by sand, although the damages were not slight. To clarify the reasons for these differences, the longitudinal profiles of the main channels of these rivers are compared in Figure 7.44. The altitudes and distances from the river source are standardized to the height of the source of the San Julian River, 2,170 m. Figure 7.44 shows that every river channel is steeper than the slope

Photo 7.10 Building collapsed due to the direct strikes by boulders.

Photo 7.11 Alluvial fan of the Camuri Grande River.

of the debris flow generation limit (15°) in the upstream reaches. All these rivers actually produced debris flows. Arrows show the positions of the upstream limits of the residential areas in these river basins. In order, the residential areas in the Uria, Cerro Grande and San Julian basins are closer to the debris flow generation reach

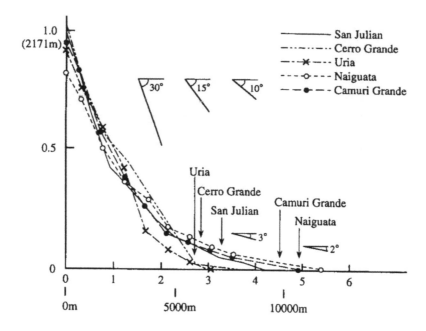

Figure 7.44 Longitudinal profiles of the rivers and positions of upstream limits of residential areas.

and are on land steeper than 4° through which large-scale debris flow can pass. In contrast, the residential areas on the Camuri Grande and Naiguata fans are far from the debris flow generating reach, and are flatter than 2° to which debris flows do not reach.

Resemblances in topography and the hazard areas recall the disaster of 1938 in Kobe district, Japan that is called the Great Hanshin Flood Disaster. The total sediment volume yielded from the Rokko mountain area is estimated as $5–7.7 \times 10^6 \, \text{m}^3$, similar to that from the seven major Vargas rivers, but the damage in Hanshin was much less: 616 people died, 1,410 houses were obliterated, 854 houses were buried, and 2,213 houses were partly destroyed. The mountain slopes of the basins have similar values with those of the rivers in Vargas, therefore, phenomena similar in quality as well as quantity should have happened in the mountain areas. Hence, the differences in damages were presumably due to the fact that at that time the developed areas in the Hanshin district were limited to the flat zones distant from the outlets of the gorges. The relative distance of the mountain ridge from the coast standardized to the heights of the respective mountains is shorter in the Vargas rivers even for Naiguata and Camuri Grande than in the Hanshin rivers. Damages in the Hanshin district were mainly due to the flooding of the flood flow that contained much sand and woody debris. These facts provide important data for considering the location of safer residential areas.

7.5.2 Reproduction of debris flow hydrograph and others in the Camuri Grande River

The SERMOW model that is introduced in Chapter 6 is applied to estimate the debris flow hydrograph and other characteristics in the Camuri Grande River basin. The whole Camuri Grande River basin, as illustrated in Figure 7.45, consists of the two sub-basins; the Camuri Grande River basin, which has an area of 23.8 km^2 originating from a ridge of about 2,250 m in altitude and the Miguelena River basin, which has an area of 19.0 km^2. These two rivers join near the fan top and the integrated river enters the fan to discharge into the Caribbean Sea. The main stem of the river is 12.3 km long and the average gradient is 10°. The canyon width, B_0, is determined from aerial photographs and our reconnaissance data, and varies as shown in Figure 7.45.

The basin was divided into 525 sub-basins for a flood runoff analysis, and the river channels were divided into 1157 sections by a constant distance of $\Delta x = 100$ m to calculate the various hydraulic quantities in each section; 728 sections in the Camuri Grande and 329 in the Miguelena Rivers. Our field survey also found that before the flood the riverbed had stored sediment about 5 m thick whose particle size distribution is shown in Figure 7.46. In the calculation, the particle size was divided into 15 groups. The fraction smaller than 100 μm (10% of the total particles) was considered as only one group ($k_1 = 1$) of the fine fraction which constitutes the muddy interstitial water. Prior to the implementation of calculation, a 5 m thick riverbed was laid on the bedrock except for the sections whose channel slopes are steeper than the angle of repose and no sediment was laid for the steeper sections.

The estimated rainfall intensities are given in Figure 7.47. The daily rainfall from 0:15 a.m. 15 December to 23:15 p.m. 16 December amounted to 281.5 mm and the maximum hourly rainfall was 40.8 mm from 5:15 to 6:15 a.m. on 16 December. According to the witness reports the debris flow occurred at about 6 a.m. corresponding to the occurrence of the maximum estimated rainfall intensity. Taking the amount of antecedent rainfall into account, the intensities shown in Figure 7.47 are considered entirely effective to cause runoff.

The spatial grid-size in the calculation was set at $\Delta x = 100$ m. Time intervals were changed depending on the flow velocity; $\Delta t = 0.1$ s for velocities of 60 cm/s or more; $\Delta t = 0.2$ s for those of 40–60 cm/s; $\Delta t = 0.4$ s for 20–40 cm/s; $\Delta t = 1$ s for 10–20 cm/s; $\Delta t = 2.5$ s for 5–10 cm/s; and $\Delta t = 5$ s for 0–5 cm/s.

Values of the other quantities were set as: $C_* = C_{*L} = C_{*F} = C_{*Lmax} = C_{*Fmax} = 0.65$, $\sigma = 2.65$ g/cm^3, $\rho = 1.0$ g/cm^3, $\tan \phi = 0.7$, and $d_m = 0.5$ m. The degree of saturation in the bed and Manning's roughness coefficient were set as $s_b = 1.0$ and $n_m = 0.03$ for the river reach $B_0 > 50$ m; $s_b = 0.8$ and $n_m = 0.04$ for $B_0 = 20$–40 m; and $s_b = 0.8$ and $n_m = 0.05$ for $B_0 = 10$ m. The equivalent roughness coefficient on the slope n_e is assumed to be 1.0, and the coefficient in the formulae for the erosion and deposition velocities, K and δ_d are assumed to be 0.5 and 0.0002, respectively. As the downstream boundary condition, the sea surface level at 500 m off the river mouth was set as constant.

Because the number of landslides was small, the sediment supply from the side banks was neglected; sediment runoff is attributed only to bed erosion.

Figure 7.48 shows temporal variations in the rates of flow and sediment discharge and in the mean diameters calculated at the confluence of the Camuri Grande and

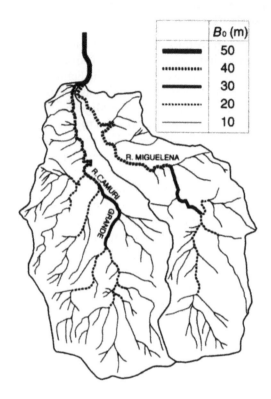

	B_0 (m)
▬▬	50
••••••••••	40
▬▬	30
•••••••••	20
———	10

Figure 7.45 River channel system and channel widths used in the simulation.

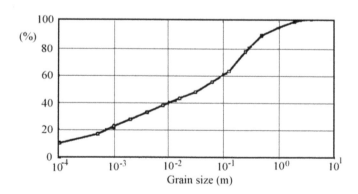

Figure 7.46 Particle size distribution in the riverbed material.

the Miguelena Rivers. A little before 6:00 a.m. on December 16, a debris flow ran out from the Camuri Grande River. It had a peak discharge of about $700 \, \text{m}^3/\text{s}$ and a sediment concentration almost equal to C_*. A little after that there was a debris flow from the Miguelena River which had a peak discharge of about $200 \, \text{m}^3/\text{s}$. The

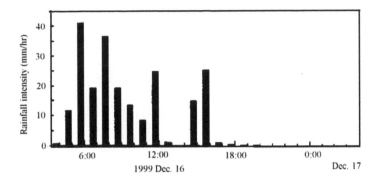

Figure 7.47 Estimated hourly rainfall in the Camuri Grande basin.

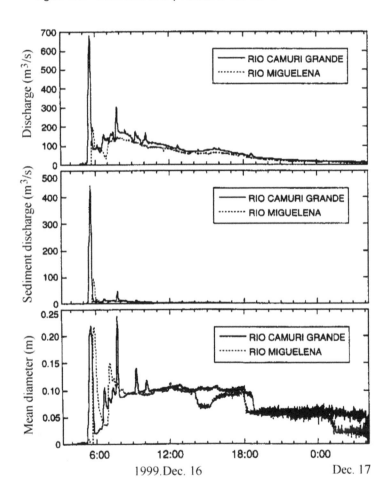

Figure 7.48 Calculated hydrographs, sediment graphs and mean particle diameters.

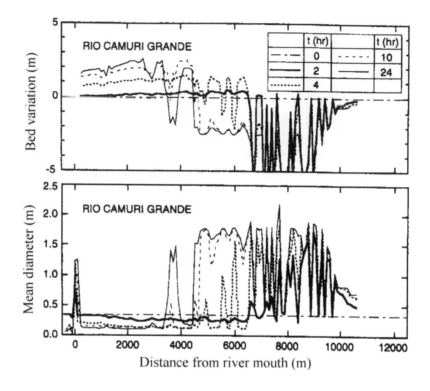

Figure 7.49 Time varying longitudinal distributions of riverbed variations and mean diameter.

durations of these debris flows were less than thirty minutes. After these debris flows, flash floods with comparatively thin sediment concentrations occurred for more than ten hours. The mean particle size in the first debris flow was calculated as larger than 20 cm. Although occasionally the mean particle size in the after floods was large, it was generally 5 to 10 cm, coinciding with the mean particle size in the river bed.

The calculated longitudinal distribution of riverbed variations and the mean diameter on the surface of the riverbed along the main channel of the Camuri Grande River are given in Figure 7.49. In Figure 7.49, $t = 0$ is 3:15 a.m. on 16 December. At $t = 2$ hours, severe erosion took place within the upstream river reach, but neither erosion nor deposition occurred in the downstream reach. The characteristic saw-tooth pattern reflects the scattered junctions with tributaries where the runoff sediment is temporarily stored. At $t = 4$ hours, a rise in the riverbed of about 2 m is seen from the river mouth to about 4 km upstream. This agrees with our findings in the field investigation. The tendency for the particle mean diameter to become small in the downstream reach also agrees with those findings. The calculated results, however, show that particles with a diameter larger than 1 m are deposited at the river mouth, and, although how these large particles were deposited at the river mouth is not known, such a large particle deposit actually occurred at the fan-top (about 1 km upstream of the river mouth). This discrepancy must be due to the ignoring of particle segregation within the flow;

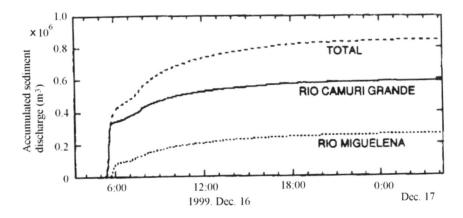

Figure 7.50 Substantial sediment yields from the two rivers and their total.

particle segregation within a debris flow works to concentrate the largest boulders to the forefront, and this part will be deposited at the fan-top.

The calculated substantial volumes of sediment runoff from both rivers are shown in Figure 7.50. The final substantial volume is 850 thousand cubic meters (bulk volume, $1,300 \times 10^3$ m^3). The estimated bulk sediment volume deposited on the fan is 1,618 thousand cubic meters, obtained by comparisons of the ground level of the fan before and after the flood. The difference is attributable to various reasons; the error included in the estimation by the comparison of ground levels, the premise of the calculation that the thickness of the deposit on the riverbed before the flood is 5 m, and the effects of landslides being neglected. Because the area of sediment deposition on the fan is about 0.79 km^2, the calculated runoff sediment volume corresponds to the deposit of 1.6 m mean thickness. This value has the feeling of reality as compared with impressions gained on the spot.

7.5.3   Reproduction of sediment flooding on the Camuri Grande fan

The river channel of the Camuri Grande River, as shown in Figure 7.51, changes its direction perpendicularly to the east immediately upstream of the fan, and after joining the Miguelena River it changes again, perpendicularly to the north. Thus, the river channel has a crank shape, making it difficult for a flood flow to pass through. This unusual shape might have been introduced to detour round the site of the Simon Bolivar University on the left-hand bank of the river. To the north of the university campus, as shown in Photo 7.11, there are resort hotels and housing complexes. On the right-hand side of the upstream part there are concrete apartment houses, and to the north of them is the slum district.

The debris flow that contained plenty of boulders larger than 1 m (possibly the front part of debris flow) stopped within the crank part upstream of the confluence with the Miguelena River. The succeeding debris flow and flood flow overflowed

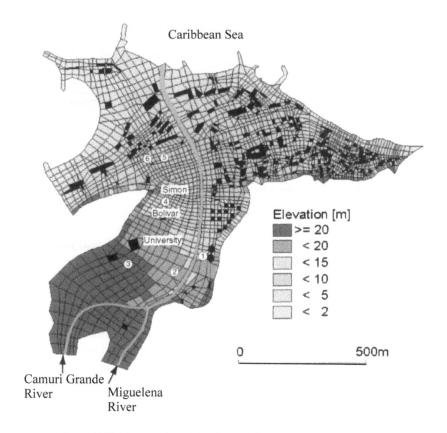

Caribbean Sea

Elevation [m]

>= 20
< 20
< 15
< 10
< 5
< 2

0   500m

Camuri Grande /
River   Miguelena
                 River

Figure 7.51 Topography of the Camuri Grande fan before flooding.

the channel and then became shallower by the deposition of boulders and continued straightforward into the campus of the university, where many buildings were
destroyed or buried up to the roof. A part of succeeding flood passed through the
cranked channel and joined with the flood from the Miguelena River. This flow soon
joined with the flow that passed through the university campus, and helped by a gradual channel bend it overflowed onto the right bank area to bury the ground floor of
the apartment buildings and destroy the slum district downstream.

Photo 7.12 shows the situations shot at the locations marked by the numbers in
Figure 7.51. Number 1 is a four-story apartment house on the right bank. Because
the ground floor is completely buried by sand, the deposit thickness is about 3 m.
Number 2 is a university building beyond an athletic field near the river side. Little
sediment is deposited on the ground and the river bank in front is broken. Number
3 is a three-story schoolhouse of the university that is located at the most upstream
part of the fan. In front of the house wall facing the flow plenty of boulders and
woody debris accumulated and the ground floor is buried. The steel-frame schoolhouse
photographed in Number 4 is destroyed. Plenty of woody debris and sand mixed with
some large boulders are deposited suggesting that it was destroyed by a flow with a

Photo 7.12 Situations of sediment deposition at some places on the fan.

rather dilute sediment concentration. Number 5 shows a car parked in the forest area of the campus. It is buried up to about half of its height, suggesting that the thickness of deposit in the forest area is about 80 cm. The building in Number 6 is close to the university's gate facing the main national road. The deposit thickness is about 1 m and it is comprised of sand finer than 1 mm.

Figure 7.52 shows the thickness distribution of the deposit on the fan obtained by comparison of the two photographic maps (1/1,000), before and after the flood. From Figure 7.52, the sediment volume deposited on this fan is estimated as 1,620 thousand cubic meters. The deposit thickness in the forest area in Figure 7.52 is more than 5 meters, in reality, as shown in Photo 7.12, it is far shallower, suggesting some errors in the reading of the aerial photos. A thick cover of woods could cause such an error. Moreover, Figure 7.52 shows a thick deposit in the northeast part of the fan that is extended to the east as if one is extending arms, but this deposition is due to sediment runoff from the basin that emerges from the mountain adjacent to this area, and therefore, the sediment volume in this area should be extracted from the estimated

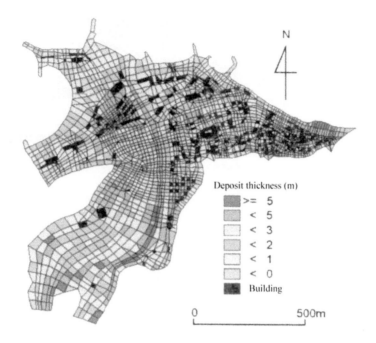

Figure 7.52 Distribution of deposit thickness obtained by reading aerial photographs.

runoff sediment volume. Thus, the estimation of 1,620 thousand cubic meters would be an overestimation.

Fundamental equations

The method of calculation using two-dimensional flooding equations has already been explained in section 5.3.2, in which the conservation equations of momentum are given by Equations (3.115) and (3.116) and the continuity equations for the entire mass, for the coarse fraction, and for the fine fraction are given by Equations (5.71), (5.72), and (5.73), respectively. The shear stresses on the bed are given corresponding to the types of sediment transportation by Equations (3.120)–(3.125). Herein, the particle distributions are not discussed but the spatial and temporal variations of the mean particle diameters are discussed using the continuity equation for the particle number, Equation (5.85).

Implementation method of the calculation and the boundary conditions

One of the calculation methods that can comparably well reflect the alignment of the river channel, other topographies of the flooding area and various significant matters is a finite volume method, in which the objective area is divided into arbitrary unstructured meshes. This method is used here, and the meshes depicted in Figures 7.51 and 7.52 are the results of division into meshes.

Figure 7.53 illustrates the arrangement of unknown variables in the numerical calculation. The entire area is covered by a Cartesian coordinate, in which the positive

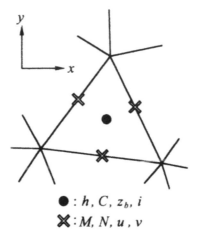

$\bullet : h, C, z_b, i$

$\times : M, N, u, v$

Figure 7.53 Arrangement of variables on a mesh area.

x is towards the east and the positive y is towards the north. The variables M, N, u and v are defined on the midpoints of the boundaries of the area, and variables h, C_L, and z_b are defined on the centroid of the area. The flow flux between the adjacent mesh areas of a particular area is calculated at the time steps n, $n+2$, $n+4$, and so on, and flow depths in the respective areas are calculated at the time steps $n+1$, $n+3$, $n+5$, and so on using discretized fundamental equations, i.e. a leap-frog method. The details of the finite volume expressions of the fundamental equations can be found elsewhere (Nakagawa *et al.* 2001).

The upstream boundary conditions are the hydrograph, sediment graph, and temporal variation in the mean particle diameter, obtained by the flood routing described earlier, given at the inflow points of the Camuri Grande River and the Miguelena River shown in Figure 7.52.

A free-dropping type condition is used as the outflow boundary condition at the river mouth and at the shoreline:

$$M \text{ or } N = (2/3)^{3/2} h \sqrt{gh} \tag{7.12}$$

The constant time interval, $\Delta t = 0.2\,\text{s}$ is used for the entire calculation period of twenty-five hours; from 3:15 a.m. on 16 December to 4:15 a.m. on 17 December. The numerical quantities in the calculation are the same as the flood routing except that $\delta_d = 0.001$ and a constant Manning roughness coefficient of $n_m = 0.03$ are used.

Results of calculation

The time-varying deposit areas and deposit thicknesses calculated are shown in Figure 7.54. The origin of time in Figure 7.54 is set to 3:15 a.m. on 16 December. A drastic change can be seen between $t = 2$ hours and 4 hours, corresponding to the peak discharge of the debris flow at about 6 o'clock. After that time, although the deposit thicknesses gradually increase, the area of deposition does not change much. Finally

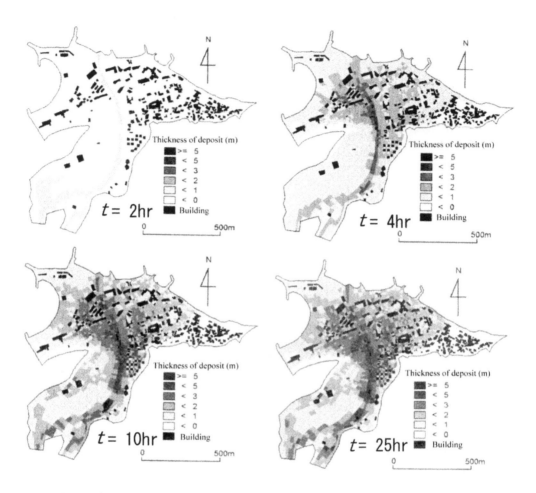

Figure 7.54 Sediment deposited area and the thicknesses of deposit by the calculation.

at $t = 25$ hours, the deposit thicknesses along the channel of the Camuri Grande River and at around the central part of the fan where deposition expanded in a fan shape are thicker than those at other places. If this final stage in the calculation is compared with Figure 7.52, the thicknesses in the forest area in the university campus and in the area of houses across the main national road connecting the cities on the Caribbean Coast in Figure 7.52 are thicker than the calculations. This discrepancy, however, as mentioned earlier, is probably due to poor accuracy in the reading of aerial photographs. The calculation does not predict the deposition around the area at the east end of the Camuri Grande community where many small houses are located, whereas in reality the area is very deeply buried by sediment. This is the area affected by the debris flows and landslides that had no relation to the Camuri Grande River as mentioned earlier. Consequently, the calculated results of the sediment deposited area and the thickness of deposit agree fairly well with the actual situation.

7.6   DEBRIS FLOW DISASTERS AT ATSUMARI, HOUGAWACHI OF MINAMATA CITY

7.6.1   Outline of the disasters

A stationary Baiu front was activated by the inflow of wet air and brought a continuous severe rainfall from 18 to 20 July 2003 around the wide area of Kyushu, Japan. From the night of the 18th to the dawn of the 19th, it rained severely in Fukuoka Prefecture in the northern part of Kyushu. This caused an inundation in Fukuoka City and several debris flows in Dazaifu City. Then, from the midnight of the 20th, the center of the rainfall moved to the south, and it rained in Minamata City, Kumamoto Prefecture. The maximum hourly rainfall was 91 mm and the total twenty-four hours rainfall was 397 mm. The hyetograph and the accumulated rainfall amount are shown in Figure 7.55.

At about 4:20 a.m. on 20 July, a debris flow induced by a large-scale landslide occurred in the Atsumari River, in the Atsumari district of Hougawatchi, Minamata City. Fifteen people were killed, five had minor or serious injuries, seventeen houses were completely destroyed, and one house was half destroyed. The entire view of this debris flow is shown on the back cover page of this book, and the situation of sediment deposition just upstream of confluence where the Atsumari River meets the Hougawati River is shown on photograph at the front of Chapter 4.

The Atsumari River is a small ravine with a basin area of 1.14 km² and a length of 3.2 km. This ravine has been designated by the government as susceptible to debris flow. The basic geology of this basin is tuff breccia and andesite covers it. The area higher than 400 m in altitude (the highest point is 586 m), where the thickness of andesite is large, is comparatively flat, but from 300 m to 400 m where tuff breccia appears on the riverbed the side slopes are steeper than 30°, and in the area whose altitude is lower than 300 m, the slopes become flat again. Reflecting these characteristic topographies, the longitudinal slope of the river at the reach downstream of the lower consolidating work is about 7°, there were houses here, and the upstream of this reach to an altitude of 400 m the river channel becomes a steep V shaped one, and further upstream the channel gradient becomes flat again and the channel cross-section becomes widely open. Figure 7.56 shows the longitudinal profile of the Atsumari River.

The debris flow was generated by a large-scale landslide that occurred on the slope of about 30° at about 1,200 m upstream of the river from the confluence with the Hougawachi River. The landslide had a depth 10 m, a width 70 m and a length of 120 m, and a total volume of 50 to 100 thousand cubic meters. Photo 7.13 shows that landslide, and Figure 7.57 shows the results of reconnaissance by Mizuno *et al.* (2003).

The transformation of the landslide to a debris flow is similar to the case of the Harihara River explained in section 3.2. Both of the landslides occurred in comparatively similar regions and the geology was also similar. However, in contrast to the Harihara River case, in which the debris flow contained much fine material and the landslide occurred four hours later than the cessation of the rainfall, the debris flow in the Atsumari River contained plenty of big boulders and less fine material, and the landslide occurred in the midst of the severest rainfall. These characteristic differences were brought about by the different weathering situations of the slid slopes. The slope

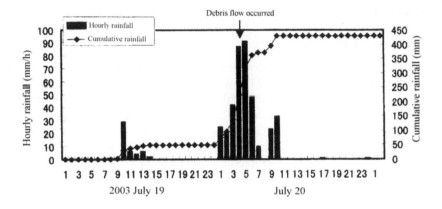

Figure 7.55   Rainfall from 19 to 20 July at Minamata (Fukagawa station).

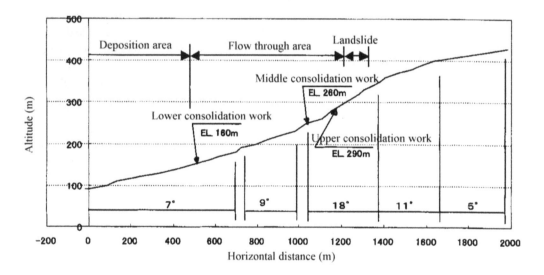

Figure 7.56   Longitudinal profile of the Atsumari River (Mizuno et al. 2003).

surface of andesite at the Atsumari River should have had many fissures to be able to produce many large blocks, and the seepage flow should have flowed very rapidly within these fissures forming lateral groundwater flow on the boundary between the weathered surface and fresh andesite responding to the severest rainfall intensity. The fallen trees seen on the slid surface in Photo 7.13 were produced after the main slide transformed into debris flow.

Although there remains a slight trace of running on the opposite slope, the landslide seems not to have formed a landslide dam (natural dam). The slid earth blocks destroyed the upper, middle and lower bed consolidation structures sequentially, while transforming into a debris flow, and eroded the bed material, that would not be much,

Photo 7.13 Landslide that induced debris flow.

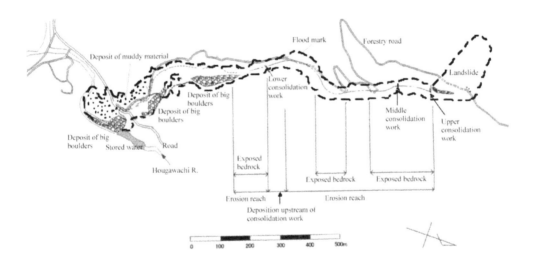

Figure 7.57 Debris flow traces in the Atsumari River (Mizuno *et al.* 2003).

to the bedrock. The high flow velocity is speculated by the very high traces at the outer banks of every channel bends. For example, Photo 7.14 is viewing upstream from the position of the lower consolidation work, the trace on the right bank of the channel is higher than the trace on the left bank by about 10 m, that difference is due to the nearly 90° channel bend shown in Figure 7.57.

As shown in Figure 7.57 deposition begins from about 100 m downstream of the lower consolidation work. The deposits on the left banks are the natural levee-like

Photo 7.14 Situation around the lower consolidation work.

deposits formed by the overflowing from the channel onto the terrace that existed before the debris flow. At about 400 m downstream of the lower consolidation work, the right bank forms a terrace about 10 m different in elevation from the channel bed, and over that terrace plenty of muddy materials were deposited and many houses were destroyed. The upstream part of this terrace constitutes the outer bank of the gradual channel bend, where the difference in elevation between the surface of terrace and the channel bed is only 1–2 m. Therefore, this muddy deposit would be the rear part of the debris flow that was pushed away to the outer bank by the preceding deposition of large boulders on the left bank.

7.6.2   Reproduction of the processes of debris flow

Because the debris flow at Minamata City is a typical one induced by a landslide, it gives a chance to validate the simulation model presented in section 3.2.4. Here, an example of application by Satofuka (2004) is introduced, in which, the same as the model in section 3.2.4, the existence of sediment on the riverbed and the effects of flood flow in the river are neglected. Actually, there should have been a thin sediment bed throughout the river and it was almost thoroughly eroded to the bedrock, but it would not much affect the scale of debris flow in comparison to the volume of the landslide. The effects of the flood flow that existed in the river prior to the debris flow is unknown, and a simulation that takes those effects into account is an interesting theme for further works. In the simulation, as explained earlier, the upper part of the landslide earth block is treated as a bundle of circular cylinders and the lower layer saturated by water is considered to be liquefied sequentially by the action of shearing force during motion, the stoppage and depositing processes of this liquefied part is, however, ignored. This is the clear defect in reproducing the depositing situations around the spot of the disaster, but, here, the simulation focuses on the processes of transformation into debris flow and the flow in the river channel. However, when the situation is attained in which there remains no more liquefied layer beneath a cylinder and the velocity of that cylinder becomes small, the cylinder is considered to be united with the riverbed; the cylinder is considered as stopped, and as an alternative to recognize the cylinder, the riverbed height is raised by the height of the cylinder.

The topographical data of the basin were obtained from the Asia Air Survey, Co. Ltd.

Conditions for calculation

The shape of the landslide is estimated, as illustrated in Figure 7.58(a), by comparing the contour lines before and after the slide, where the contour lines before the landslide were estimated from the circumferential topography. The distribution of the slide depths thus obtained is shown in Figure 7.58(b). The estimated slide volume is 67.5 thousand cubic meters and this is almost comparable with the field measurement. The initial heights of the circular cylinders were determined by these data as illustrated in Figure 7.59.

The mesh size in the calculation is 10 m in both x and y directions and the time step is 0.05 seconds. The radius of the cylinder is 4.85 m, the initial center to center distance of two adjacent cylinders; D_1, is 9.24 m, the distance, D_2, in which the attractive force becomes the maximum is 9.70 m, and the largest distance, D_3, within which only the attractive force operates is 10.16 m. The solids concentration in the cylinders is 0.5; the degree of saturation in the unsaturated part of the cylinder is 0.3; the representative particle diameter is 10 cm; Manning's roughness coefficient, $n_m = 0.04$; the particle density is 2.65 g/cm^3; the density of the interstitial fluid is 1.0 g/cm^3; the kinetic friction coefficient between the earth block and the riverbed, $\mu_k = 0.5$; the internal friction angle of the particles $\phi = 30°$; and the adhesive force between

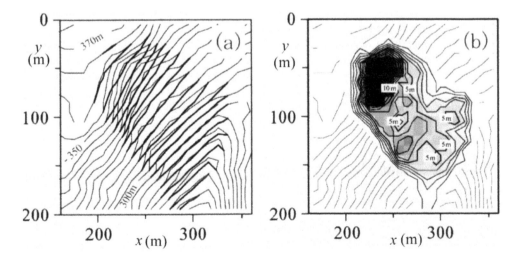

Figure 7.58 Shape of the landslide.

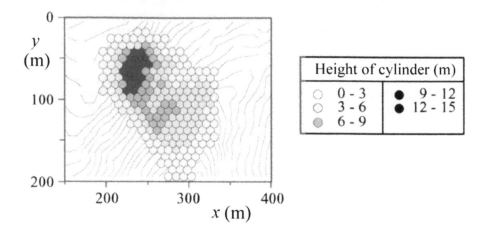

Figure 7.59 The initial arrangement of cylinders and their heights.

the cylinders, $c_b = 179$ Pa, which is found after some trial and error calculations so that the cylinders behave as a bundle at least initially. As many boulders whose diameters are more than 5 m are contained in the debris flow deposit, setting the representative particle diameter in debris flow as 10 cm has no convictive reason, but it is determined to assure that the debris flow is of the stony type.

The case 1 simulation assumes that, similar to the case of the Horadani debris flow in section 7.2, the slid earth block is already completely liquefied at the spot of the landslide, and instead of setting the height of the cylinders to zero, the initial depths of flow are assumed to have the distribution as depicted in Figure 7.58(b). The mass of liquid having such a spatial depth distribution is unconstrained at once.

In Case 2, the cylinder heights distribute as shown in Figure 7.59 and the initial depths of liquefaction beneath all the cylinders are set at a small value of 5 mm. The cylinders are completely saturated by water, and the coefficient β in the equation for a liquefying velocity, Equation (3.89) is set at 0.12.

In the series of simulations in Case 3, the heights of cylinders and the initial thicknesses of the liquefied layer are the same as Case 2 except that β is set at 0.012 and the unsaturated thicknesses in the cylinders are changed depending on the sub-cases. The thickness of unsaturated layer in Case 3-0 is 0 m and that in Case 3–1 is 1 m; the last digit attached to case number represents the thickness of the unsaturated part of the cylinder in meter. If the height of a cylinder is shorter than the prescribed thickness of the unsaturated layer, that cylinder is considered entirely unsaturated. The difference between Case 3-0 and Case 2 is only the difference in β value.

The total calculating time for Case 1, Case 2, Case 3-0 and Case 3–1 is 200 seconds and in other cases it is 300 seconds.

The results of calculation and some discussions

Figure 7.60 shows the discharge hydrographs at $x = 900$ m, in which, different to the previous descriptions, x is the distance measured downstream from the landslide site, so

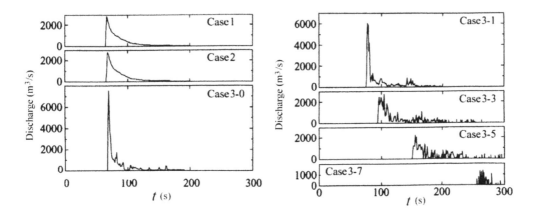

Figure 7.60 Hydrographs at $x = 900$ m.

$x = 900$ m approximately corresponds to the starting site of the deposition in the actual case. Time zero in Figure 7.60 is the instant of landslide. The cumulative discharge of the liquefied layer and the volume of the cylinders in every time step that passed through the position $x = 900$ m were recorded and the total volumes of the respective fractions that ran off by the end of calculations were obtained as shown in Figure 7.61. The results of Case 1 and Case 2 are almost the same. Because the progressing speed of liquefaction in Case 2 was fast, the entire slid earth block transformed into a debris flow in a short time and in a short distance, resulting in a hydrograph similar to that in Case 1. In contrast, in the gradual liquefaction in Case 3-0, at the time of passing the section $x = 900$ m, about half of the volume was left as earth blocks, and because they accumulated in the front of debris flow, the peak discharge became about the twice as large as that in Cases 1 and 2. By comparing the cases in Case 3, it can be seen that the larger the ratio of the unsaturated part, the smaller the peak of the debris flow becomes and the later the debris flow arrives at $x = 900$ m.

Figure 7.62 shows the time-varying flow depths and cylinder heights in the respective cases, and Figure 7.63 shows the traces of debris flow at the end of calculations. From these figures, it can be seen that in Cases 1, 2, 3–3 and 3–5 the debris flow flows along the original river channel of the Atsumari River and the flow cannot ride over the terrace on the right-hand side, whereas in Cases 3-0 and 3–1 the flow splits into two; one that overrides onto the terrace and one that follow the original river coarse, agreeing approximately to the actual situation. In Cases 3–7 and 3–10 flow stops upstream of $x = 900$ m. This is due to the non-liquefaction in the major part of the cylinder. Figure 7.64 shows the distributions of the thicknesses of deposition in Case 3–10. The majority of the slid earth blocks deposit upstream of $x = 500$ m and a natural dam-like topography is formed. Actually, no such topography was formed. Therefore, it is presumed that the earth block had been saturated by water before the landslide took place. However, due to the slow progress of liquefaction, transformation into a debris flow should have proceeded gradually in the river channel.

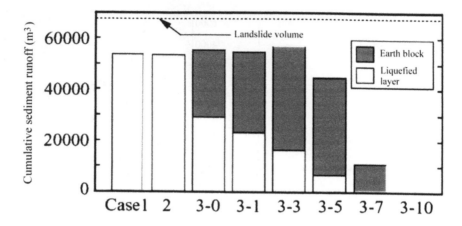

Figure 7.61 Cumulative sediment runoffs at $x = 900$ m.

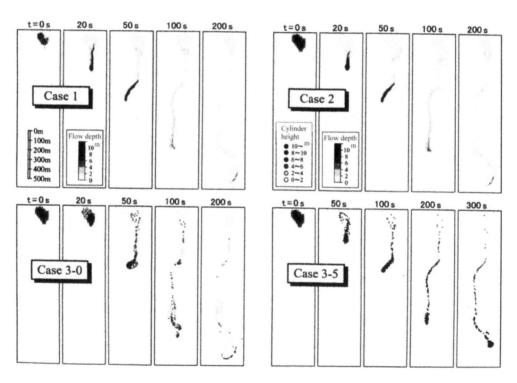

Figure 7.62 Distributions of depths and cylinder heights.

From these results of calculations, it was made clear that the processes of transformation of an earth block into a debris flow can be well reproduced by the method
presented here if only the thicknesses of the saturated and unsaturated parts in the
earth block and the velocity of liquefaction in the saturated part are appropriately

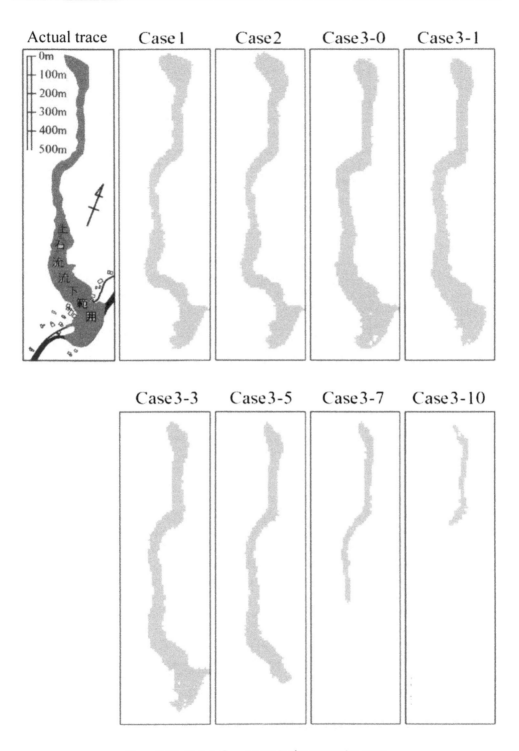

Figure 7.63 Debris flow traces in the respective cases.

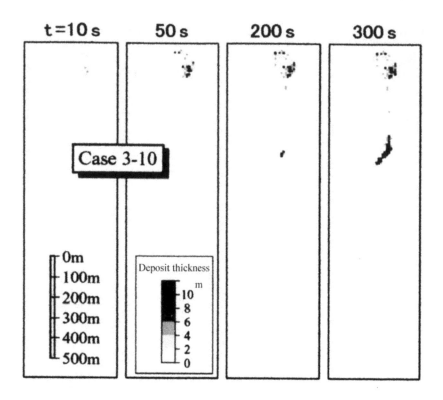

Figure 7.64 Time-varying riverbed rises in Case 3–10.

given. Especially in Case 3-0, the actual situation that the flow is separated into the one that overrides onto the terrace and the one that follows the normal river channel is well reproduced. Moreover, in Case 3-0, as shown in Figure 7.62, after 100 to 200 seconds the calculated earth block's gathering positions rather well coincide with the places of boulders' deposition in Figure 7.57. However, this simulation overlooked the deposition process of the liquefied part, the effects of flood flow already in the river, the distributions of particle diameter, and especially the effects of many large boulders. Therefore, there remain many points that need improvement.

Chapter 8

Countermeasures for debris flow disasters

The photograph shows a solid-frame-type grid dam installed in the Shiramizudani Ravine that originate from the Yakedake Volcano. Continual surveys that aim to make clear the structure's performance to control debris flow and to get more appropriate design criteria are on-going.

INTRODUCTION

In contrast to the disasters arising in the area of direct strike of the causative phenomena such as earthquakes and ground subsidence, the disasters due to the propagation of hazardous agents such as debris flows, flood flows and tsunamis affect the populated area via the processes of causative phenomena's generation and then their propagation. Therefore, as well as the reduction of vulnerability in the inhabited area, the countermeasures for the disasters belonging to the latter category can be selected from, the removal of causative phenomena if possible, the control of the hazardous agents underway, and the defense works surrounding the inhabited area. The civil engineering methods against the disasters are called the structural countermeasures or the hard countermeasures; even tree planting on bare slopes may be categorized as belonging to structural countermeasures because it needs some engineering works. The measures to enhance the proof strength of the area of human activity against the disasters may be hard countermeasures such as the reinforcing of buildings, the safer arrangement of houses and other exposed objects in the area, and the preparation of preventive structures or green belt. The so-called nonstructural countermeasures or soft countermeasures, such as the regulation of developing the hazard-prone area and the avoidance of risk by taking refuge, are equally important.

The areas susceptible to debris flow disaster are, in general, narrow and not suitable for installing large structures. Moreover, setting only one structure, be it a check dam or a diversion channel, is often by no means sufficient to make a debris flow harmless. In addition, it is said that more than 80 thousands debris flow prone ravines exist in Japan; vast amounts of cost and time would be necessary to make all these ravines safe even at a minimum level. Therefore, an integration of the effective hard and soft countermeasures is important.

In this chapter, various hard and soft countermeasures against debris flows are discussed, focusing on the objectives of the installation and the specific design methods based on the engineering principles that have been made clear in the previous chapters.

8.1 METHODS TO PREVENT DEBRIS FLOW GENERATION

8.1.1 Hillside works

Because debris flow arises from the occurrence of a landslide or the erosion of a steep riverbed, if it is possible to prevent the generation of a landslide and the accumulation of sediment on a riverbed, then debris flow will never occur. However, in Japan many recently-planted forests contain conifers that are too densely planted and meagerly grown due to insufficient tree thinning. It is easy for these to become a landslide because the root systems to bind the land do not develop well. Also, after the clear cutting of these forests (selective cutting is not adopted for economic reasons), the roots will decay within 5–10 years making the surface of the slope very weak against land-sliding. Sliding of weak slopes not only causes debris flow with plenty of woody debris but also the bare land left after the slide becomes an active source of sediment production that causes sediment accumulation on riverbed. Therefore, the appropriate afforestation of bare slopes and the appropriate conservation of these slopes will restrain the occurrence of debris flow.

The afforestation of once denuded steep mountain slopes, however, is difficult if not impossible, and generally, it requires the following tasks: preventing movement of the surface soil by bench-cutting the slope; adding fertile soil; retaining water; then planting trees that easily enroot; and finally restoring to the original forest physiognomy. This work is called hillside works, and in Japan, there are many examples from the nineteenth century, but, nowadays, few are done except for the restoration of landslide scar.

Photo 8.1 shows the time sequential situations of the restoration of a landslide scar; the landslide induced the Harihara River debris flow in July 1997. The left-hand slide photo shows the situation in March 1998 before the implementation of hillside works. The middle one is in March 2001 just as the engineering works were finishing. The right-hand side one is in August 2003, the grass and shrubs are growing thickly and, in due time, the landslide scar will become obscure.

8.1.2 Drainage works

Deep-seated large-scale landslides are caused by the rise of the ground-water level and sometimes it transforms into debris flow. The debris flow that occurred in the

Photo 8.1 Hillside works on the landslide scar in the Harihara River basin.

basin of the Harihara River is one such typical case. Although drainage works that consist of collecting wells and drainage pipes are the standard method to stabilize slowly moving or intermittently moving landslide masses, such engineering works are seldom used to withhold the occurrence of landslide-induced debris flow because the identification of a sudden juvenile landslide position in advance is difficult. Drainage is important to prevent reactivation of a landslide on the once slid slope and it is a standard engineering method, as can be seen in Photo 8.1. If the concentration of water toward the gully bed is avoided, the onset of a gully bed erosion type debris flow can, in principle, be prevented. However, because the gully is formed by the concentration of water, laying a new drainage channel other than the natural drainage system would be impractical. A peculiar example of a drainage system is the tunnel excavated into the wall of the crater-lake of the Kelut Volcano in Indonesia. This volcano erupts about once every fifteen years and each time the water stored in the lake blasts off to generate a volcanic debris flow, or lahar. Because some 150 thousand people live at the foot of the mountain, preventing the debris flow generation due to over-spilled water is very important. It is well known that the larger the stored water volume the larger the magnitude of the debris flow became, and the excavation of a drainage tunnel was proposed as early as 1901, but it was so difficult work that it was finally completed in 1928. Meantime, in the 1919 eruption, the stored water volume was 40 million cubic meters and the induced debris flow reached 37.5 km from the crater, killing 5,160 people. In the 1951 eruption, thanks to the function of the drainage tunnel, the stored water was only a million cubic meters, and therefore, the debris flow reached only 6.5 km and the number of casualties was only seven. However, that eruption lowered the base of crater by about 80 m and it destroyed the tunnel. Then, when the next eruption occurred, in 1966, the stored water volume was 20 million cubic meters. In that case, debris flow reached 31 km and 286 people were killed. A new tunnel was excavated before the eruption of 1990, when the stored water volume was 2 million cubic meters. Although a small-scale debris flow was generated, no one was killed by debris flow. I visited the crater just before the 1990 eruption and also after the eruption in November 1991. In 1991, the plentiful water in the crater before the eruption was dried up and the tunnel was destroyed, so that it was difficult even to find its entrance on the crater wall. Photo 8.2 views the over-spilling mouth of the crater-lake from the outer side of the crater, the distant ridge is an inside part of the outer rim of the crater.

The drainage tunnel idea is not limited to the one excavated into this crater-lake. One was also constructed in the natural dam created by the debris avalanche in 1980 at Mount St. Helens to drain water from Spirit Lake; and a drainage channel was urgently laid down on the body of the natural dam created by the Ontake landslide in 1984 that choked the Otaki River.

8.1.3 Groundsill and bed girdle

As a method for stabilizing the sediment bed against surface water flow, some bed consolidating structures have been built; the structures having small drops between the upstream and downstream beds are called groundsills, and those having no drop are called bed girdles. Takahashi *et al.* (1977) checked the effectiveness of setting a group of bed girdles on a steep slope channel bed. Bed girdles were set in the sediment

Photo 8.2 The cut edge of the crater lake of Mount Kelut.

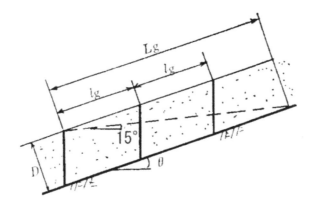

Figure 8.1 Arrangement of bed girdles.

bed with various intervals as shown in Figure 8.1. A sediment of $d_m = 3$ mm was laid $D = 4$ cm deep on a flume bed, 20 cm wide and 17° in slope, and surface water flow was applied suddenly from the upstream end. Whether debris flow occurred or not was examined and the sediment concentration in the flow at the downstream end of the flume was measured. The experiments revealed that even in the case of very small spacing of $l_g = 25$ cm the onset of debris flow could not be prevented, nevertheless when l_g was small the sediment concentration in flow was a little smaller than in the larger spacing cases. Figure 8.2 shows the relationship between the sediment concentration and the spacing of bed girdles, where L_g is the distance from the crest of a girdle to the intersection of a straight line (the broken line in Figure 8.1) having the critical slope for

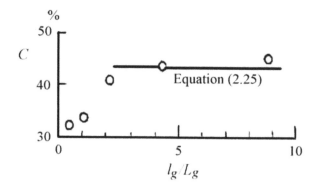

Figure 8.2 Sediment concentration versus spacing of girdles.

the onset of debris flow ($15°$) with the flume bed, i.e. $L_g = D/\tan(\theta - 15°)$. According to Figure 8.2, if $l_g/L_g > 2$, the sediment concentration in the flow is almost identical to the case with no bed girdles. Namely, if $l_g/L_g > 2$, although the thorough washout of accumulated sediment can be prevented by the sediment remaining under the broken line in Figure 8.1, the debris flow discharge and its sediment concentration at the front part cannot be reduced. Because L_g becomes smaller the steeper the channel slope, the effectiveness of bed girdles becomes minimal for steeper slope channels unless the spacing of the girdles is set very small. Watanabe *et al.* (1980) also did similar experiments and concluded that groundsills and bed girdles could reduce the total volume of runoff sediment but they had no effect on reducing the height of the debris flow front.

The example referred to in section 6.4.3 suggests that the stair-like bed profile can restrain the onset of a debris flow by a small-scale flood runoff, but it increases the scale of the debris flow once it is generated. Using groundsills to create an artificial stair-like profile may restrain the onset of a debris flow, and reduce the total volume of runoff sediment if spacing between structures is short. If these artificial structures are not destroyed, then the peak discharge will not be increased, unlike the natural step morphology, but if the spacing between the structures is long its effects would be limited because the local bed slope becomes almost equal to the average topographical slope. Whether such inefficient engineering works are adopted or not, along with other works such as check dams downstream, must be examined by comparing the cost to benefit.

8.2   DEBRIS FLOW CONTROL BY CLOSED-TYPE CHECK DAM

8.2.1   Sediment depositing process behind check (sabo) dam

When dam is sufficiently high

When a debris flow collides with a vertical wall, except for splashes that are churn up by the air entrapped in the flow, it can reach as high as the energy height it possessed

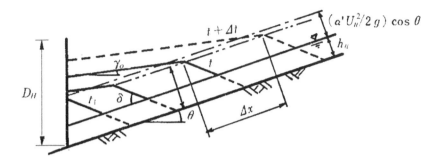

Figure 8.3 The process of debris flow deposition behind a high sabo dam before filling up.

just before the collision. Therefore, if the height of a dam is higher than this height, the debris flow will stop and deposit behind the dam, at least initially. The condition for this situation to arise is, referring to the notations in Figure 8.3, written as follows:

$$D_H \cos\theta \geq h_u + \frac{\alpha' U_u^2}{2g} \cos\theta \qquad (8.1)$$

where α' is the energy coefficient and U_u is the cross-sectional mean velocity of the approaching debris flow. We, here, different from the normal definition of 'high' and 'low' dams, use the term 'high sabo dam' to mean a sabo dam that is higher than one that satisfies Equation (8.1). Sabo is a Japanese word meaning sediment hazards prevention and the sediment checking dam is generally called a sabo dam throughout the world.

The surface slope of deposit γ_o is given by Equation (5.14), in which the unit width discharge of a squeezed surface water flow is obtained by subtracting the water discharge entrapped in the deposit from the water discharge in the approaching debris flow as:

$$q_o = q_t(1 - C_u) - (1 - C_u)C_u q_t / C_* \qquad (8.2)$$

where q_t is the unit width discharge of the approaching debris flow, and C_u is the sediment concentration in the approaching debris flow. This surface water flow is assumed to flow out from the drainage holes set in the dam body. Therefore, if a steady debris flow approaches, the deposit raises its height keeping a constant surface slope as shown in Figure 8.3.

After a certain time, the deposit height reaches the dam crest, some parts of the flow pass over the dam and the rest is deposited behind the dam. The depositional process in this stage can be modeled as shown in Figure 8.4. Namely, if debris flow continues, the surface slope of deposit becomes steeper and steeper setting the center of anti-clockwise rotation at the crest of the dam. After its slope attains the equilibrium

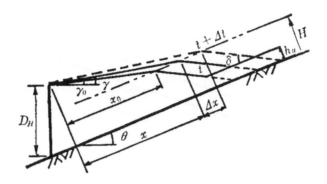

Figure 8.4 Depositing process after filling up.

slope that is able to transport sediment with the concentration in the approaching debris flow; i.e. the slope that satisfies Equation (2.25), the deposit keeps its form constant and the approaching debris flow passes through without deposition.

The analytical models for the depositing processes before and after the filling up of the dam can be found elsewhere (Takahashi 1983).

When dam is low

When the dam height cannot satisfy Equation (8.1), the debris flow passes over the dam immediately after the collision with the dam. The deposit behind the dam maintains a constant form having its surface slope equal to the equilibrium slope for the approaching debris flow. In this case, the sabo dam becomes an obstacle for the debris flow but it does not control the debris flow.

8.2.2 Erosion process of the deposit behind the sabo dam

The closed-type sabo dam, like the normal concrete dam, is often filled up before the onslaught of a debris flow. The traditional Japanese idea of constructing a concrete wall type sabo dam as a method to control debris flow considers that the equilibrium slope of the deposit after filling up by normal scale flood flows is about half of the original bed on which the dam is constructed; and when the debris flow comes down the dam can further store sediment making the depositing slope steeper in the manner shown in Figure 8.4. For this debris flow controlling function to be sustainable and maintenance free, the steeply deposited material must be naturally removed after each debris flow so as to recover the shallow slope that existed before the storing debris flow. In some cases this concept is fulfilled, actually however, many examples revealed that the flattening of the deposit surface is hardly fulfilled because the large boulders transported by the debris flow and deposited steeply behind the dam cannot be transported by a normal-scale flood flow.

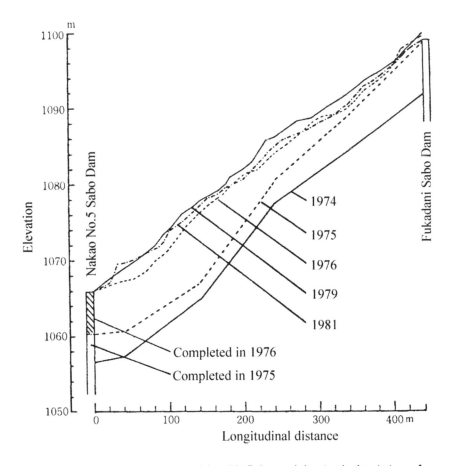

Figure 8.5 Debris flow deposition behind Nakao No.5 dam and the riverbed variations afterwards.

For example, Figure 8.5 shows the river bed variations upstream of the Nakao Number 5 Sabo Dam in the Ashiaraidani River (Sawada 1985). The construction of this dam commenced in 1975 and was completed in 1976. Meanwhile, the part under the height of 1,060 m was already complete when a debris flow occurred on 13 July 1975 and deposited to the height of crest at that time. The deposit thickness was, as shown in Figure 8.5, maximum in the reach within 200 m downstream of the Fukadani Sabo Dam and the Fukadani Sabo Dam was buried up to the crest. The average longitudinal valley slope in this reach is about 5°, corresponding to the debris flow depositing slope, we can see that the main part of the debris flow would have been deposited in this reach after passing over the Fukadani Sabo Dam even if the Nakao Number 5 Dam did not existed. Incidentally, the Fukadani Sabo Dam had been filled up long before and in the case of the 1975 debris flow the reach upstream of the Fukadani Sabo Dam was a little eroded.

Another debris flow occurred on 11 June 1976 when the Nakao Number 5 Dam was complete. The dam was immediately filled up. As far as this debris flow is concerned, this dam performed its function to check the debris flow very well. However, the riverbed variation thereafter is insignificant as shown in Figure 8.5 and the Fukadani Sabo Dam is still under the debris as of today (2003). The longitudinal surface slope of the deposit in between the Nakao Number 5 and Fukadani dams after 1976 is almost parallel to the previous stable slope as it was in 1974. According to the traditional design concept, the bed surface line of the Nakao Number 5 Dam, after checking the 1976 debris flow, similar to the other stair-like constructed series dams, should have been lowered within a short period to intersect with the foundation of the next upstream dam, i.e. the Fukadani Sabo Dam. As this designed bed erosion is not fulfilled, the Nakao Number 5 Dam can no longer control debris flow.

The classical concept of debris flow control does not pay much attention to the effects of the sediment released to the downstream areas, which was being deposited in the upstream reach of the dam to control the debris flow. If the dam is immediately upstream of a residential area, the sediment that is released after checking a debris flow might cause some hazards. Therefore, the investigations on the erosion of the debris flow deposit upstream, the re-deposition of the released sediment downstream and their effects are indispensable before the construction of a sabo dam. In this context, a one-dimensional mathematical riverbed erosion model and a two-dimensional model for the erosion upstream and deposition downstream of a sabo dam in respect to heterogeneous material are introduced before the discussions on the effects of the sabo dam to downstream.

Riverbed variation by the bed-load transport of heterogeneous material

One-dimensional riverbed variation was discussed in Chapter 6, but that method aimed to analyze a large and rapid variation due to a highly concentrated flow as debris flow, so that the model may not be suitable to discuss the detailed variation processes due to bed load transport such as the formation of an armor coat. Here, the discussion is specified to the phenomena related to bed load transport and a somewhat new model suitable to such phenomena is introduced (Takahashi et al. 1998a).

Similar to the previous method, we consider that if the sediment concentration in the flow is thinner than the equilibrium value, erosion takes place and conversely if it is denser than the equilibrium concentration deposition takes place. Further we consider that the larger the difference in the concentration from the equilibrium value the larger the erosion or deposition velocity becomes. Namely, the erosion or deposition velocity is written as:

$$i \propto (C_{B\infty} - C) \tag{8.3}$$

This formula represents the general erosion or deposition ability of the flow. In the case of a bed comprised of a heterogeneous material, however, the erosion or deposition rates of individual particles would depend on their diameters. Hence, to reflect such a mechanism, the erosion or deposition velocities of a group k particle i_k is written as:

$$i_k \propto (u_* - u_{*ck}) \tag{8.4}$$

Note that under the general tendency of erosion if u_* is smaller than u_{*ck} or under the general tendency of deposition if u_* is larger than u_{*ck}, then the group k particle is considered not to be eroded or deposited under the respective situations.

From these considerations, we describe the erosion and deposition velocities of group k particles as follows:

Under the general tendency of erosion, i.e. $(C_{B\infty} - C) > 0$:

$$i_k = \begin{cases} f_{bk}\delta_e \dfrac{C_{B\infty} - C}{C_*}|u_* - u_{*ck}| ; & (u_* > u_{*ck}) \\ 0 & ; \quad (u_* \leq u_{*ck}) \end{cases} \tag{8.5}$$

Under the general tendency of deposition, i.e. $(C_{B\infty} - C) \leq 0$:

$$i_k = \begin{cases} f_{0k}\delta_d \dfrac{C_{B\infty} - C}{C_*}|u_* - u_{*ck}| ; & (u_* < u_{*ck}) \\ 0 & ; \quad (u_* \geq u_{*ck}) \end{cases} \tag{8.6}$$

and,

$$i = \sum_k i_k \tag{8.7}$$

where f_{bk} and f_{0k} are the existence ratios of group k particles on the bed surface and in the flow, respectively.

If q_{buk} is the equilibrium bed load discharge for the case in which bed consists of only the group k particles, the equilibrium particle concentration in flow is given as:

$$C_{\infty k} = \frac{q_{buk}}{q_t} \tag{8.8}$$

Then, the total equilibrium sediment concentration is:

$$C_{B\infty} = \sum_k f_{bk} C_{\infty k} \tag{8.9}$$

and the transport concentration is:

$$C = \frac{1}{q_t} \sum_k q_{bk} \tag{8.10}$$

where q_{bk} is the bed load discharge of group k particles per unit width.

Substitution of above erosion and deposition velocities in the governing equations for a one-dimensional analysis enables the prediction of the riverbed variation. The governing equations are:

Continuity equation of water:

$$\frac{\partial h}{\partial t} + \frac{1}{B}\frac{\partial Q}{\partial x} = 0 \tag{8.11}$$

Equation of motion for water flow:

$$\frac{\partial u}{\partial t} + u\frac{\partial u}{\partial x} = -g\frac{\partial(h + z_b)}{\partial x} - \frac{u_*^2}{h} + \frac{\partial}{\partial x}\left(\varepsilon_z \frac{\partial u}{\partial x}\right) \tag{8.12}$$

where ε_z is the vortex viscosity and using the Kármán constant κ it is given as:

$$\varepsilon_z = \frac{\kappa}{6}u_* h \tag{8.13}$$

The friction velocity is given by using the logarithmic resistance law so as to consider the effects of the particles diameter distribution on the bed:

$$u_* = \frac{u}{6.0 + 5.75\log(h/k_r)} \tag{8.14}$$

where k_r is the equivalent roughness height and it is assumed to be equal to 90% of the particle diameter of the bed surface d_{90}.

Equation of bed variation:

$$\frac{\partial z_b}{\partial t} + i = 0 \tag{8.15}$$

Continuity equation of total bed load:

$$\frac{\partial(Ch)}{\partial t} + \frac{1}{B}\frac{\partial(CuhB)}{\partial x} + \frac{\partial z_b}{\partial t}C_* = 0 \tag{8.16}$$

Continuity equation for group k particles:

$$\frac{\partial(C_k h)}{\partial t} + \frac{1}{B}\frac{\partial(C_k uhB)}{\partial x} - C_* i_k = 0 \tag{8.17}$$

where C_k is the concentration of group k particles in flow, and:

$$f_{0k} = \frac{C_k}{C} \tag{8.18}$$

The variation of the grain size on the riverbed is given by Equations (6.49) and (6.51). Here we consider the case of $J = J_0 = C_*$, and noting that i_k is the erosion or deposition velocities of bulk volume, these respective equations become:

$$\frac{\partial f_{0k}}{\partial t} + \frac{i_k}{\delta_m} + \frac{\partial z_b}{\partial t}\frac{f_{bk0}}{\delta_m} = 0; \quad (\partial z_b/\partial t < 0) \tag{8.19}$$

$$\frac{\partial f_{0k}}{\partial t} + \frac{i_k}{\delta_m} + \frac{\partial z_b}{\partial t}\frac{f_{bk}}{\delta_m} = 0; \quad (\partial z_b/\partial t \geq 0) \tag{8.20}$$

where f_{bk0} is the existence ratio of group k particles in the layer beneath the bed surface layer.

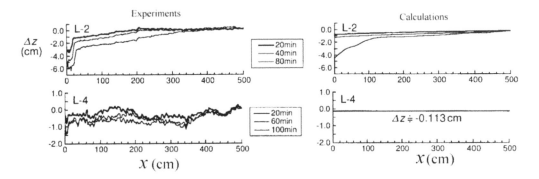

Figure 8.6 Temporal riverbed variations in the experiments and in the calculations.

The bed load discharge on the bed comprised of uniform material of size k; q_{buk}, is obtained herein by using Ashida and Michiue's (1972) formula:

$$q_{buk} = 17\sqrt{(\sigma/\rho - 1)gd_k^3}\,\tau_{*uk}^{3/2}\left(1 - \frac{\tau_{*cuk}}{\tau_{*uk}}\right)\left(1 - \frac{u_{*cuk}}{u_*}\right) \tag{8.21}$$

where τ_{*cuk} and u_{*cuk} are the non-dimensional critical tractive force and the critical shear velocity for the group k particles, respectively, and for a uniform particle bed these values are obtained from Iwagaki's (1955) critical tractive force equation.

The critical friction velocity of a group k particles on a heterogeneous material bed, is obtained by the non-modified Egiazaroff equation, Equation (6.28) for all of the diameter groups because, on a widely distributed particle bed, the small particles will be more easily trapped between coarser particles than on a less widely distributed bed, thus, Equation (6.29) would not be applicable.

These fundamental Equations (8.5) to (8.21) were applied to the experiments. The experiments used a flume of 6 m long and 20 cm wide, whose erodible bed part occupied 530 cm reach to which rigid bed reaches of 40 cm and 30 cm were attached upstream and downstream, respectively. The flume gradient and the initial slope gradient of the erodible bed were set at 0.025. Although various combinations of bed materials and the supplied sediment materials under various discharges were examined, here Case L-2, which used material having $d_m = 0.326$ cm and $(d_{84}/d_{16})^{1/2} = 1.575$ as the erodible bed material and material having $d_m = 0.156$ cm and $(d_{84}/d_{16})^{1/2} = 1.505$ as the supply material, and Case L-4, which used material having $d_m = 0.390$ cm and $(d_{84}/d_{16})^{1/2} = 2.194$ as the erodible bed material and material having $d_m = 0.156$ cm and $(d_{84}/d_{16})^{1/2} = 1.505$ as the supply material, are described. The supplied water discharges and supplied sediment discharges for both cases were 1,620 cm^3/s and 0.201 cm^3/s, respectively.

Figure 8.6 compares the temporal riverbed variations in the experiments with the calculations, $x = 0$ is at the upstream end of erodible bed. Although the severe local scour near the upstream end of erodible bed cannot be reproduced by the calculations, the general tendency of the riverbed degradation from upstream and the quantity of variations were rather well explained.

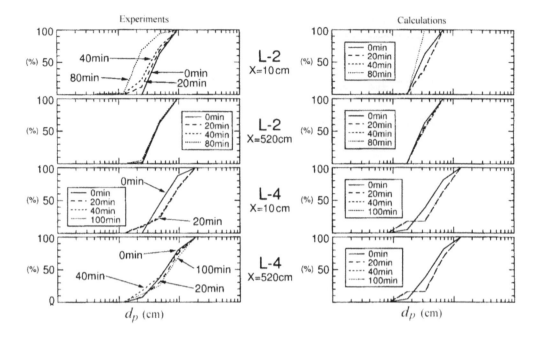

Figure 8.7 Temporal variations of particle size distributions on the bed surface.

Figure 8.7 shows the temporal variations of the particle size distributions on the bed surface at the positions $x = 10\,\text{cm}$ and $x = 520\,\text{cm}$. The calculations well reproduced the general characteristics. Especially in L-4, with the elapse of time, the ratio of medium size particles decreases and a bipolar distribution becomes conspicuous both in the experiment and calculation.

Two-dimensional riverbed variation model for a heterogeneous material

In the upstream reach of a sabo dam, the deposit is usually comprised of widely distributed material and the width of deposit surface is several times wider than the original valley bottom width due to riverbed rising. Hence, the erosion due to a normal scale flood flow forms a narrow stream channel and the general degradation of the deposit in the entire deposit width proceeds by the lateral fluctuation of the narrow stream channel within which the erosion takes place. This means that the one-dimensional analysis is insufficient. In this context, the above mentioned one-dimensional model is extended to two-dimension.

Setting the x axis to be the direction of the flow and the y axis to be the lateral direction, we describe the continuity and motion equations of flow as follows:

$$\frac{\partial h}{\partial t} + \frac{\partial(uh)}{\partial x} + \frac{\partial(vh)}{\partial y} = 0 \tag{8.22}$$

$$\frac{\partial u}{\partial t} + u\frac{\partial u}{\partial x} + v\frac{\partial u}{\partial y} = -\frac{1}{\rho}\frac{\partial p}{\partial x} - \frac{\tau_{bx}}{\rho h} + 2\frac{\partial}{\partial x}\left(\varepsilon_x\frac{\partial u}{\partial x}\right) + \frac{\partial}{\partial y}\left(\varepsilon_y\frac{\partial u}{\partial y}\right) \tag{8.23}$$

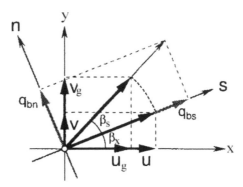

Figure 8.8 Flow direction and sediment transport direction.

$$\frac{\partial v}{\partial t} + u\frac{\partial v}{\partial x} + v\frac{\partial v}{\partial y} = -\frac{1}{\rho}\frac{\partial p}{\partial y} - \frac{\tau_{by}}{\rho h} + \frac{\partial}{\partial x}\left(\varepsilon_x\frac{\partial v}{\partial x}\right) + 2\frac{\partial}{\partial y}\left(\varepsilon_y\frac{\partial v}{\partial y}\right) \qquad (8.24)$$

where,

$$\frac{\tau_{bx}}{\rho h} = \frac{u}{\sqrt{u^2 + v^2}}\frac{u_*^2}{h} \qquad (8.25)$$

$$\frac{\tau_{by}}{\rho h} = \frac{v}{\sqrt{u^2 + v^2}}\frac{u_*^2}{h} \qquad (8.26)$$

$$u_* = \frac{\sqrt{u^2 + v^2}}{6.0 + 5.75\log(h/k_r)} \qquad (8.27)$$

Substitution of u_* (and therefore τ_*) obtained from Equation (8.27) into Equation (8.21) gives $C_{B\infty}$. The difference between $C_{B\infty}$ and the actual sediment concentration, C, gives the erosion or deposition velocities at the respective locations, that is given by the same formula to the one-dimensional analysis.

The continuity equation of group k particles is:

$$\frac{\partial(C_k h)}{\partial t} + \frac{\partial(C_k h u_g)}{\partial x} + \frac{\partial(C_k h v_g)}{\partial y} - C_* i_k = 0 \qquad (8.28)$$

where u_g and v_g are the x and y components of the sediment transport velocity, respectively. If the directions of flow and sediment transport are the same, u_g and v_g are equal to u and v, but generally, due to the effects of the lateral inclination of the bed and the secondary flow they are not equal. Hence, u_g and v_g should be given by some methods. We define the s-axis as the main flow direction and the n-axis perpendicular to it as shown in Figure 8.8. Then, the angle between the s-axis and the x-axis, β_x, is given by:

$$\beta_x = \arctan\frac{v}{u} \qquad (8.29)$$

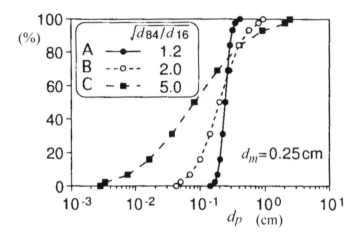

Figure 8.9 Particle size distributions used in the simulation.

Neglecting the effects of the secondary flow, we get the sediment discharge toward the n-axis by the following Hasegawa's formula:

$$q_{bn} = -q_{bs}\sqrt{\frac{\tau_{*c}}{\mu_s\mu_k\tau_*}}\frac{\partial z}{\partial n}$$

(8.30)

where q_{bn} and q_{bs} are the sediment discharges in the n and s directions, respectively, μ_s is the static friction coefficient of the particle, and μ_k is the dynamic friction coefficient of the particle. As the term in the root sign in Equation (8.30) does not significantly change spatially, we set the root term as a constant value K_B. Then, the direction of bed load transport makes an angle β_s to the direction of the s-axis, which is given as:

$$\beta_s = \arctan\left(-K_B\frac{\partial z_b}{\partial n}\right)$$

(8.31)

If the total sediment discharge does not change even though the lateral sediment motion exists, u_g and v_g are given as follows:

$$u_g = \sqrt{u^2 + v^2}\cos(\beta_x + \beta_s)$$

(8.32)

$$v_g = \sqrt{u^2 + v^2}\sin(\beta_x + \beta_s)$$

(8.33)

Numerical simulations were accomplished using these fundamental equations, in which at first a channel 14 cm wide and 2 cm deep was incised at the center of an erodible bed 5 m long and 60 cm wide and a water flow with a discharge 800 cm³/s was released to freely erode the channel. The flume slope was 1/20 and no sediment was supplied from upstream.

Four kinds of bed material were examined. The particle size distributions of the materials in Case A, Case B, and Case C are given in Figure 8.9 by the A, B and C curves.

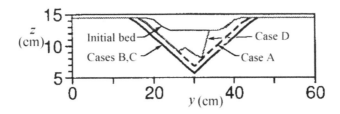

Figure 8.10 Cross-sections at $x = 250$ cm with $t = 60$ minutes by calculations.

The mean diameters of these three materials are all the same but only the patterns of distribution are different. The Case D simulation was done to examine the methods for treating the large particles in the calculation. So, even though the bed material was the same as Case C, 125 randomly selected cells in the calculation meshes were composed of only the largest group particles in material C and the rest of the cells were composed of the material minus the largest group particles from material C. Because the large boulders are not necessarily distributed uniformly on the bed, the uneven distribution may have a significant effect on the erosion process. All the calculations were accomplished by discretizing into $\Delta x = 5$ cm, $\Delta y = 2$ cm and $\Delta t = 0.02$ s.

Figure 8.10 shows the cross-sections at $x = 250$ cm with $t = 60$ minutes. The erosion rates in Cases B and C, in which the range of particle size distribution were large in comparison to Case A, were larger than in Case A. These results would correspond to the larger content of fine particles in the materials B and C than in the material A. At $t = 60$ minutes, the sediment transport was almost none and an armor coated bed was formed in every case. The uneven bed configuration in Case D was formed due to small erodibility of the cells containing only the big particles.

Figure 8.11 shows the contour lines and the mean diameters on the bed surface for Cases C and D. Even though the bed is composed of the same material, the predicted erosion rates and bed configurations are deferent depending on the methods used to consider the large particles. The total eroded sediment amount for Case C is about 4.5×10^4 cm^3 and that for Case D is about 2.0×10^4 cm^3. In these calculations the maximum particle size was about 2 cm and it was comparable to the lateral size of a mesh. To explain the change in the pattern of the stream channel, it would be necessary to have discretization of at least 1/10 of the channel width. Large boulders of such a scale are common on an actual bed. The calculation can be accomplished easily even when larger particles than a mesh size exist, but as shown in this example, we must remember that the results of calculations vary depending on the calculation method used. Much more data needs to be obtained to get the appropriate method.

8.2.3 Effects of sediment control by sabo dam on downstream

Discussions on the functions of a sabo dam to control sediment have been based only on the change in sediment discharge at the position of the dam, and little attention has been

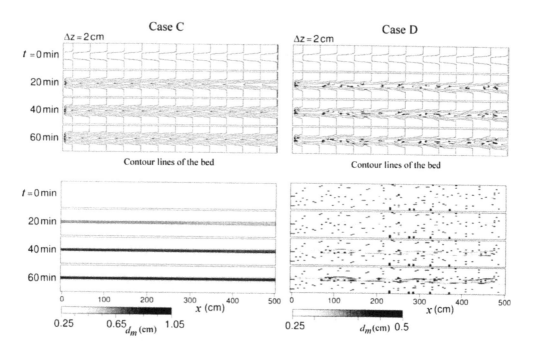

Figure 8.11 Contour lines of the bed and distributions of mean diameters on the bed obtained by different handling methods of the largest class particles.

paid to the effects of the deposited bed erosion behind the dam in the downstream direction. Depending on the circumstances, the sediment eroded from the deposit behind the dam may redeposit on the channel in the downstream reach and cause a serious problem there. Here, this problem is examined by experiments and by numerical simulations.

The experiments used a flume of effective length 5 m and width 40 cm. In the central part of the flume, a sabo dam, whose waterway section width is 17 cm, was installed and the erodible bed slopes upstream, i_1, and downstream, i_2, of the dam were set independently. The wing height of the dam was high enough for there to be no overflow. A straight channel with a constant width 10 cm and a depth 2 cm was incised in the initial bed. The bed slopes upstream and downstream of the dam were parallel to the initial bed slopes. The incised channel bottom height at the immediate upstream side of the dam was adjusted to the height of the crest of the waterway and the immediate downstream side of the dam was set lower by 2 cm, i.e. the dam had 2 cm head. The bed immediately downstream of the dam was paved with gravels about 4 cm in diameter to prevent the local scouring. Except for this reach, the bed was composed of a sediment mixture with a mean diameter of 2.83 mm and the standard deviation $(d_{84}/d_{16})^{1/2} = 2.70$. No sediment was supplied from upstream.

The numerical simulations for the two-dimensional model mentioned above, in which the sediment was considered uniform with $d_m = 2.8$ mm, were applied.

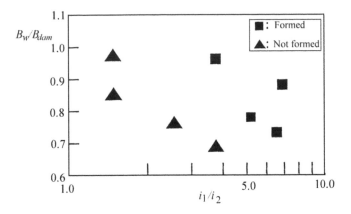

Figure 8.12 Bar formation criteria at the downstream reach of a sabo dam.

Experiments revealed that in some cases, where there was a large difference in the upstream and downstream channel slopes, large-scale bars were formed downstream of the dam. Once these bars were formed, they were hardly eroded throughout the period of prolonged water supply. The development of the bar caused the stream channel to change its course toward the side bank and soon the flow collided with the bank. This phenomenon may be harmful depending on the situation around the river.

Figure 8.12 shows the classification of the cases that formed and did not form a significant deposit at the downstream part of the dam. In Figure 8.12 B_{dam} is the dam's waterway width and B_w is the average width of the stream channel upstream of the dam. Because the cause of significant bar formation is the reduction in the capacity of sediment transport downstream, the bar formation cases are plotted in the range where i_2 is far smaller than i_1. Even in the case of quite a small difference in the upstream and downstream channel slope, if the channel contraction at the waterway is large, a bar is formed. This is due to a forced unevenness of the sediment transport immediately downstream of the contracted flow at the waterway. This phenomenon causes a sediment deposit immediately downstream of a slit dam, which has a deeper and narrower waterway than a normal closed-type dam, and the deposit may hinder the sediment flushing from the slit. It is desirable to investigate this beforehand when a slit dam installation is planned.

Figure 8.13 shows the case of a significant sediment deposit and compares the temporal bed's contour line changes between the experiment and the calculation. The thick lines in the experimental result are the water edges read from photographs taken in the experiment.

A significant sediment deposit arose in the area immediately after the dam and about 60 cm downstream. A bar that stood out of the water surface was formed. In the numerical simulation a bar was formed as well, but its developing speed was different from the experiment, meaning that further investigations on the sediment discharge formula and erosion or deposition velocity formulae are necessary.

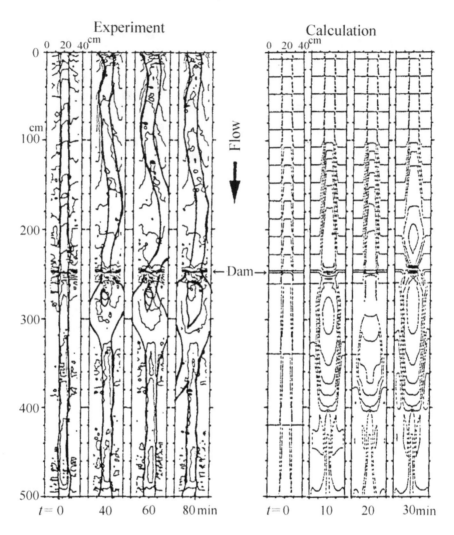

Figure 8.13 Temporal riverbed variations upstream and downstream of a sabo dam by flume experiment and numerical simulation. $i_1 = 1/20, i_2 = 1/80, Q = 800\,\mathrm{cm^3/s}$.

8.3 DEBRIS FLOW CONTROL BY OPEN-TYPE SABO DAMS

8.3.1 Kinds and sediment checking mechanisms of open-type sabo dams

The closed-type sabo dam mentioned in the previous section is easily filled up by sediment that is runoff by normal floods, and when a debris flow or a highly sediment concentrated flood flow come down, it often fails to check or control them; i.e. the closed-type sabo dam has a defect in the sustainability of its function. Furthermore, the stoppage of the sediment transport in the normal flood period by the closed-type sabo dam that is not yet filled up is said to be harmful to the ecosystem downstream due

Photo 8.3 Clogging of a grid-type sabo dam.

to the stabilization of the stream channels and the minimal sediment motion, because the ecosystem needs a moderate sediment transportation and exchange in particles.

Recent discussions of such defects, under the catchphrase 'sediment passing sabo', have attached importance to methods that can check the hazardous sediment but continually supply the safe and necessary sediment downstream. Sabo dams that have such functions are known as open-type sabo dams. These have large openings in the dam body through which the sediment runoff from the normal flood flow passes but when a debris flow or large-scale flood flow occurs it works like a not yet filled up closed-type sabo dam.

There are two kinds of open-type sabo dam. They vary in the mechanism checking the sediment. One is the slit dam, or large conduit dam, which has large openings in the dam body, but the ratio of the total width of openings to the dam length (width of the valley) is small so as to guarantee that backwater effects are generated when a debris flow or large flood flow occur. Sediment is trapped behind (upstream) the dam due to a large reduction in the flow velocity.

Another type is the grid-type sabo dam, where the individual open space between the pillars is normally less than the width of the opening of the slit dam but the ratio of the total open space to the dam length is large and virtually no backwater effect occurs even in the case of debris flow. For this type of dam, the sediment is checked by the clogging of the openings by the pinching stones. Photo 8.3 shows from downstream the pinching stones that clog the space between the steel pipes.

The sediment control function of the slit-type dam has been mainly discussed with reference to its use in the case of bed load transport (Mizuyama *et al.* 1990; Okubo *et al.* 1997; Masuda *et al.* 2002). Herein, the function of slit dam is not discussed. However, the following fact must be pointed out. In some cases, the slit dam is filled up during the ascending period of the flood, but in the recessing stage the stored sediment is eroded and flushed out. Although this phenomenon is referred to as a successful debris flow control because the sediment storing capacity before the flood is recovered when the flood ceased (Fukuda *et al.* 2002), whether this phenomenon really brought benefit to the downstream river must be examined carefully because the dam might merely delay the sediment transport without reduction of the total sediment runoff.

A considerable number of grid-type sabo dams, which are constructed by a steel-pipe framework as shown in the photograph of the front page of this chapter, have been constructed in Japan. There are some other types that use different framing structures, but because the principle of sediment checking is the same, the discussion here is focused on this jungle gym type. The investigations on the debris flow checking function of the grid-type sabo dam commenced with the experiments to find a suitable spacing between pipes, using small-scale models (Ashida and Takahashi 1980). From this investigation the following design criterion was obtained:

$$\left.\begin{array}{l} l_s/d_{\max} = 1.5 \sim 2.0 \\ l_s/d_b > 2.0 \end{array}\right\} \tag{8.34}$$

where l_s is the spacing between pipes, d_{max} is the diameter of the maximum size class particles accumulating in the front of the debris flow, and d_b is the diameter of the maximum size class particles transported as a bed load by a normal scale flood flow. For the determination of d_{max}, a field survey of the particle size distributions in the basin is necessary, but d_b can be determined by the application of the critical tractive force formula.

Mizuyama *et al.* (1995) considered that the trap efficiency of grid dam should not only depend on l_s/d_{max} but also on the sediment concentration and the velocity of the forefront of the debris flow. Then, by analyzing their experimental data they obtained following result:

$$P = 1 - 0.11\left(\frac{1}{d_{95}} - 1\right)^{0.36} C^{-0.93} \tag{8.35}$$

where P is the reduction rate of the peak discharge by passing through the dam and this is defined as:

$$P = 1 - (Q_{pd}/Q_p) \tag{8.36}$$

in which Q_{pd} is the peak discharge after passing through the dam and Q_p is the peak discharge without dam. Equation (8.35) does not contain velocity and this means that as far as their experiments are concerned effects of velocity is negligible.

Mizuno *et al.* (2000) tried to do the debris flow trap simulation by using the discrete particle model. Mizuno *et al.* (2000) and Mizuyama *et al.* (1995) intended to define the trap efficiency deterministically by giving the debris flow conditions and the properties of dam.

Takahashi *et al.* (2000a) considered that a small difference in the arrival time of individual boulders would markedly affect the blocking rate, and therefore, the phenomena should be intrinsically stochastic. The following describes their investigations.

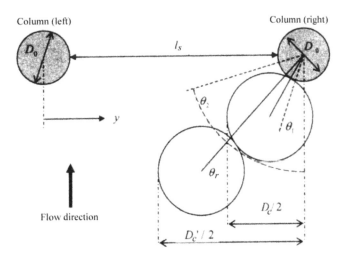

Figure 8.14 The bridging process by the particle contact.

8.3.2 Blocking model of grid-type dam

Aiming to clarify the process of grid blocking, a planer grid made of copper bars 3 mm in diameter was installed in a flume 10 cm wide and 18° in slope. Uniform particles 1 cm in diameter were flowed down the flume under various particle concentrations and water flow rates. The grid makes five lateral and five vertical square spaces in the channel cross-section, each square space has a 1.7 cm lateral and 1.7 cm vertical in net spacing. Therefore, l_s/d_{max} is 1.7, which satisfies the empirical clogging condition, given by Equation (8.34).

Observations in the experiments revealed that there are two types of clogging. One is due to the simultaneous arrival of two or more particles, and the other is due to the arrival of particles while the previously arrived one is rotationally moving around and touching with a column. The rotational motion of a particle around a column acts to narrow the free space between that particle and the other adjacent column and on some occasions, depending on the position and number of rotating particles, the free space becomes too narrow to pass another particle that arrives a little later than the rotating ones. Thus, these already rotating and newly arrived particles make an arch bridge between the two adjacent columns to block the space.

Figure 8.14 shows a snap shot of an instant of the second type. The shaded circles are the copper columns forming the grid plane perpendicular to the flow and the two open circles are the particles rotating around the central axis of a column. These particles are in contact with each other and they contact with the column directly and indirectly via the contact with another particle. It should be noted that once the angle between the normal line to the grid plane and the segment connecting the center of the column and the center of the particle in contact with and rotating around the column becomes larger than a critical value, θ_2, the rotating particle together with the other particles in contact with it can no longer act to narrow the space, i.e. all the particles pass through the spacing. Similarly, if a segment connecting the centers of two arbitrarily selected particles among the particles indirectly contacting with a column makes an angle larger than θ_2, the outer particle (the one further from the column indirectly attaching) together with other outside contacting particles pass through the spacing. Herein this angle is called the detaching angle and, from the experiments, it is about 60–65°.

If a particle collides with a column or with other particles rotating around the column at an angle of θ_1 and begins rotating around the column with an angular velocity of ω, the period t_c for that particle from collision to detachment is given by $(\theta_1 - \theta_2)/\omega$. Writing the angle between the flow direction and the segment connecting the center of the particle and that of the column at time $t_1(t_1 < t_c)$ as θ_r, the virtual diameter of the column (the contacting particles act to enhance the virtual diameter of the column) is given by:

$$D'_c = (D_c + d_p)\sin(\theta_1 + \omega t_1) + d_p \qquad (8.37)$$

where D'_c and D_c are the virtual diameters of the column before and after the collision of the particle, respectively. If this particle collides directly with the column, then $D_c = D_0$.

The regression analysis of the experimental data based on the equation of angular momentum conservation gives the following relationship:

$$\omega = 1.278\frac{u_p}{d_p}\sin\theta + 0.094 \qquad (8.38)$$

where u_p is the velocity of particle just before the collision with a column or with another particle.

Consider that i pieces of particles come down to the grid plane during the period of T_L by random time intervals t_i at random lateral positions y_i. The velocity of the particles and their concentration are constants. When a particle collides with the column or with other particle rotating around the column, the new virtual column diameter D'_c is calculated, and then by comparing the length l_s and $D'_c/2$, we judge whether the space between two adjacent columns is clogged or not. Namely, i trials are done within the period T_L, and for each trial whether the spaces between columns are clogged is checked. At last $(t = T_L)$ the number of clogged spaces is counted. If these trials are repeated n times under the same conditions, the clogging probability F_{cp} in the period T_L is given by:

$$F_{cp} = (\text{number of clogged spaces})/n \qquad (8.39)$$

If such trials are done under different values of T_L, the relationship between T_L and F_{cp} is obtained. Figure 8.15 is the comparisons of these clogging probabilities found by the calculations with the experiments. Although there is a tendency that the larger the clogging probability the larger the difference between the calculations and experiment becomes, the general characteristic change in clogging probability is explained by the model introduced.

However, to make this relationship usable in the computer simulation of debris flow blockage, the instantaneous blocking probability in a small time interval of Δt is necessary. If this probability, P_r, is assumed constant under the constant u_p and C,

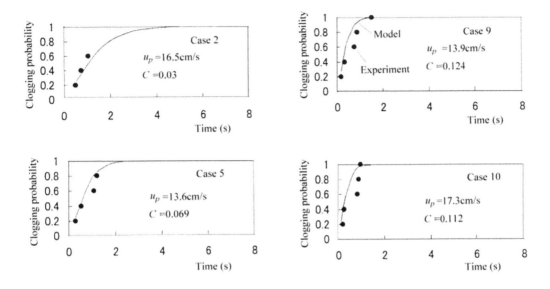

Figure 8.15 Comparison of clogging probabilities between the model and the experiments.

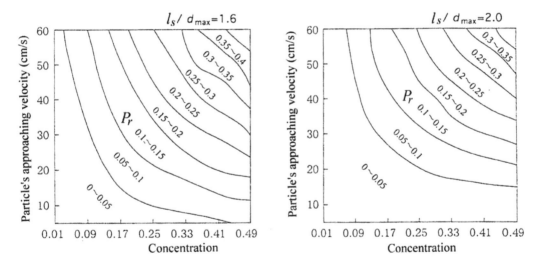

Figure 8.16 Calculated instantaneous blocking probabilities.

the relationship between P_r and F_{cp} is given by $F_{cp} = 1 - (1 - P_r)^n$, where $n = T_L/\Delta t$. Therefore,

$$P_r = 1 - (1 - F_{cp})^{1/n} \tag{8.40}$$

The calculated results of the instantaneous blocking probability P_r under the conditions of $l_s/d_{max} = 1.6$ and 2.0 are shown in Figure 8.16. This figure evidently shows that P_r depends on both the particle concentration and the approaching velocity of the particles, and under the same concentration and particle's approaching velocity, P_r depends on l_s/d_{max}, the larger the value of l_s/d_{max}, the smaller the blocking probability becomes.

The actual grid-type sabo dam has been constructed three-dimensionally like a jungle gym structure, having many grid faces along the river channel. The observations during the laboratory experiments using such a structure revealed that the blockade of the second row grid space is caused by a similar process to that for the first row except that the particle velocity becomes small due to the effect of collision with the column in the first row. Because the particle velocity just upstream of the second row is diverse, it is assumed to be the product of u_p and a random number between 0 and 1, where u_p is the particle's approaching velocity to the first row. The particle concentration is considered not to change. Then, the blocking probability of the second row P_{r2} is obtained from Equation (8.40). Therefore, the total blocking probability for a dam having two rows is described as:

$$P_{2\,row} = P_{r1} + (1 - P_{r1})P_{r2} \tag{8.41}$$

Similarly, the instantaneous blocking probability for a dam having n rows is given as:

$$P_{n\,row} = \sum_{n=1}^{n-1} P_{rn} + \left(1 - \sum_{n=1}^{n-1} P_{rn}\right) P_{rn} \tag{8.42}$$

8.3.3   Model for debris flow controlling by a grid-type sabo dam

Experiments for the functions of grid-type dam

Laboratory experiments were carried out using a tilting flume 5 m long, 10 cm wide and 20 cm deep with a slope of 18°, whose rigid bed was roughened by pasting a sediment mixture of 0.3 cm in mean diameter. The experimental material was laid in a stretch 1.5 m long and 10 cm deep within the upstream reach, and saturated with water prior to the experimental run. At the downstream end of the flume a prescribed kind of open-type sabo dam was set. Then, a debris flow was generated by supplying water at a rate of 300 cm³/s for fifteen seconds. A part of the generated debris flow was checked by the dam but the rest flowed out of the flume. By capturing this flowed out debris flow by a sampler, the discharge of water plus sediment, sediment discharge and sediment size distributions were measured.

Two kinds of experimental material were used. The size distributions are shown in Figure 8.17. The maximum and mean diameters of Material A are 1 cm and 0.24 cm, and those of Material B are 0.8 cm and 0.22 cm. These materials are almost identical except for the largest class diameters. There were four kinds of dam models, as shown in Figure 8.18. The effective widths of spacing were 1.6 cm for all the models. To give a standard for the comparison of dam's efficiency the experiment without a dam was also carried out. Materials A and B were used for every dam model and each case was repeated three times under a prescribed condition to cover against chance fluctuations. The experimental run numbers for the different experimental conditions are shown in Table 8.1.

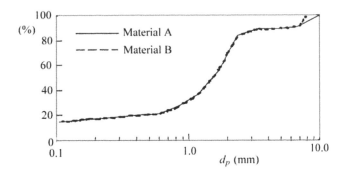

Figure 8.17 Size distributions in the experimental materials.

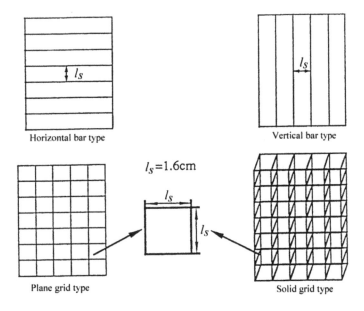

Figure 8.18 Schematic diagrams of dam models.

Table 8.1 Kinds of experiments.

Run no.	Dam type	Material
1~3	None	A
4~6	None	B
7~9	Plane grid	A
10~12	Plane grid	B
13~15	Solid grid	A
16~18	Solid grid	B
19~21	Vertical bar	A
22~24	Horizontal bar	A

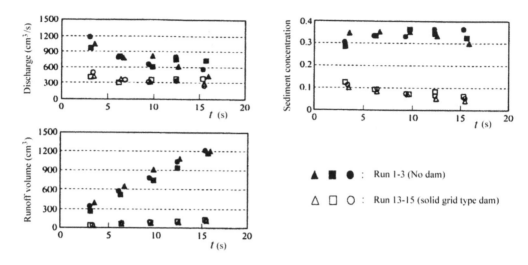

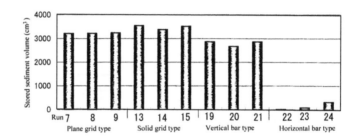

Figure 8.19 Characteristics of runoff debris flows with and without dams.

Figure 8.20 Stored sediment volumes behind the dams.

Debris flows in the experiments were the typical stony-type and they accumulated the largest particles in the forefront. Except for the case of the horizontal bar type dam, as soon as the forefront arrived at the dam all the opening spaces were choked. The succeeding flow bulged at the dam face and the majority of the particles were deposited forming a deposit area that rebounded upstream. Even in the case of the solid grid-type dam, the overwhelming majority of particles were checked by the first row grid face and the engaging between particles was tight. In the case of the horizontal bar type dam, the debris flow passed almost as if it flowed down without a dam.

Figure 8.19 compares the temporal changes in discharge, sediment concentration and total amount of sediment that passed through the solid grid-type dam with those values when no dam was installed. As can be seen in Figure 8.19, all the discharge, sediment concentration and total amount of sediment runoff were significantly reduced by the installation of the solid grid-type dam. Except for the horizontal bar type, these characteristics were similar for all the dam types.

Figure 8.20 compares the sediment volumes stored by the different type dams for Material A. The reason for smaller deposited volume in the case of the plane vertical

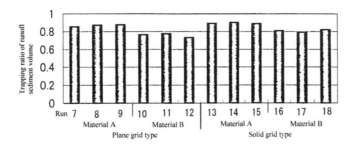

Figure 8.21 Difference in the efficiency of grid dams by the difference in maximum sizes of materials.

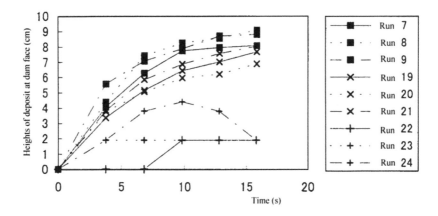

Figure 8.22 Temporal changes in deposit heights at the dam face.

bar type dam in comparison to the plane grid-type dams is due to a slightly weaker engaging between particles at the front row, the sporadic collapses of the deposit structure cause the passing of sediment through the dam.

As mentioned earlier, the ratio of the distance between the columns to the maximum particle diameter in the debris flow affects the clogging probability. Figure 8.21 demonstrates how the difference in the maximum particle size affects the trapping ratio of the runoff sediment volume for the cases of the plane grid-type and the solid grid-type dams. Here the trapping ratio of the runoff sediment volume is defined as the ratio of the total volume of the stored sediment to the sum of the total volume of stored sediment and the total volume of passed through sediment. From Figure 8.21 we can see that for both the plane grid-type and the solid grid-type dams the trapping ratio of the runoff sediment volume becomes small if the maximum size of the particles in the debris flow becomes small. Namely, the efficiency of the grid dam becomes larger, the larger the maximum size of the particles contained in the runoff debris flow.

Figure 8.22 shows the temporal changes in deposit heights at the dam face for the grid-type, vertical bar type and horizontal bar type dams. The tendency of space clogging is larger in the order of grid-type, vertical bar type and horizontal bar type, and therefore, the deposit heights are larger in this order. Whereas the final deposit height for the horizontal bar type dam reached at most to the height of the lowest bar, the final deposit height for the vertical bar type dam reached as high as a little lower than that of the grid-type dam. As mentioned earlier, the engaging between the particles for the case of the vertical bar type dam is a little weaker than the case of the grid-type dam, and this causes the sporadic collapse of the deposit resulting in a slightly lower deposit height than in the case of the grid-type dam. Namely, the vertical bars play the major role in checking the large particles in the forefront of the debris flow and the addition of horizontal bars enhances the engaging between particles that stabilize the deposited sediment.

Numerical simulation of debris flow control by a grid-type dam

Let us consider the case in which a debris flow is generated by the occurrence of surface water flow or by the supply of an already liquefied surface landslide mass, and it bulks up downstream by the erosion of a steeply deposited debris bed in a gully. When there are plans to control debris flow in such a gully by installing a grid-type sabo dam, the performance of the dam must be assessed beforehand. Thus, it is necessary to integrate the debris flow routing model with the grid clogging model and the rebounding deposit behind the dam model. In the investigation of the performance of the grid dam it is particularly important to estimate how large the particles are that are concentrated in the front part. Hence, the debris flow routing model must be the one that can predict the particle segregation processes.

The debris flow routing method that considers the focusing of large particles toward the forefront was introduced in section 4.6, so this is method used here. The debris flow depositing process behind a grid dam can be predicted by obtaining the clogging probability mentioned in section 8.3.2. However, the clogging model did not consider the effect of the horizontal bar that would become crucial if the clearance between a horizontal bar and the surface of deposit just upstream of the dam becomes less than one particle diameter of the maximum size. To take this effect into account, when the clearance becomes smaller than the maximum diameter of debris flow material, this clearance value is considered as the effective grid spacing alternative to the previous effective spacing obtained by considering only the vertical posts. Thus, if the spacing is judged to be clogged in a time step, the bed level just upstream of the dam is raised with the height equal to the diameter of the largest particle and the riverbed variation upstream of the dam is calculated.

The numerical simulations to reproduce the phenomena that appeared in the experiments were carried out using the method described above. Here, the case of the solid grid-type dam (Run 13-Run 15) is explained (Takahashi *et al.* 2002). The calculation was accomplished by setting $\Delta t = 0.02$ s and $\Delta x = 10$ cm. The particle sizes were classified into five groups, and in each time step the segregation of the particles was calculated together with the debris flow characteristics such as discharge, velocity, depth, and particle concentration, particularly the concentration of the maximum diameter class particles at the forefront. Then, the velocity and the concentration of the largest class particles were substituted into the clogging probability model and the instantaneous blocking probability was obtained under the condition $n = 1,000$ and $\Delta t = 0.02$ s. If this probability was larger than an independently generated random

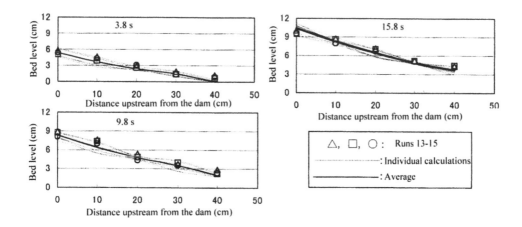

Figure 8.23 Experimental and calculated bed levels behind the dam at different times.

number between 0 and 1, the dam was considered to be blocked. If it was judged blocked, the bed level just upstream of the dam was raised with the height equal to the diameter of the largest particle, and the riverbed variation upstream of the dam was calculated. The parameter values used in the calculation were $\rho = 1.0\,\mathrm{g/cm^3}$, $\sigma = 2.65\,\mathrm{g/cm^3}$, $\tan\varphi = 0.75$, and $s_b = 1$.

Figure 8.23 compares the longitudinal riverbed level changes at 3.8, 9.8 and 15.8 seconds after the first clogging of the dam. Because the value of the random number that gives the standard to judge whether the dam is clogged changes depending on the initially selected random number, the calculation was repeated ten times for each experimental case. The dotted line shows the respective ten results and the thick line is the average of these respective calculations. As is clear in Figure 8.23, the calculated results coincide well with the experiments.

Figure 8.24 compares the calculated debris flow hydrographs, cumulative sediment runoff volumes, and particle concentrations with the measured ones at the downstream end of the flume.

From the results shown in Figures 8.23 and 8.24 it can be seen that the blocking phenomena of a grid-type dam and its effects downstream can be rather well reproduced by the numerical simulations introduced herein.

8.3.4   Determination of the optimum spacing and the optimum position to install

Method to determine the grid spacing depending on the particle size distributions

In the previous subsection the debris flow developing process and particle segregation process were discussed by the methods introduced in section 4.6. But, for the assessment of grid dam's performances under the various conditions of sediment transportation as debris flow, immature debris flow and bed load transport, the general method; SERMOW, introduced in section 6.2.1 is more appropriate to use. In section

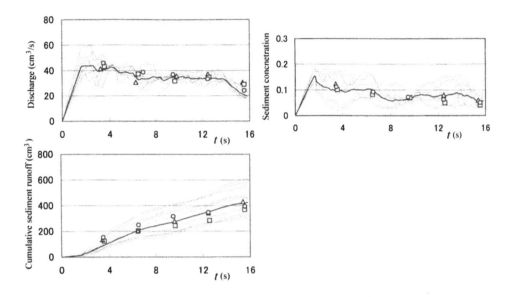

Figure 8.24 Calculated and experimental hydrographs, sediment concentrations and cumulative sediment runoff just downstream of the dam.

6.2.1, however, the particle segregation phenomenon is neglected, so that, after calculating the flow routing using the method described in section 6.2.1, the calculation of the process of particle segregation should be added during the same time step by the integration of Equations (4.28), (2.23) and (4.41).

When a grid-type sabo dam is installed, the optimum spacing between the frames and the optimum position to set along the river must be decided beforehand, by comprehensively considering the distance from the source area of debris flow, the river channel form, characteristics of bed material, and other factors. However, in some circumstances the possible installation position may be limited. In such cases, for the dam to perform its function as well as possible, only the optimum spacing must be chosen.

Consider that a grid-type sabo dam will be constructed at the position indicated in Figure 8.25. The dam will check debris flow if l_s/d_{max} is ess than 1.5–2.0, the smaller that value the more surely the dam will check the debris flow. But, if the value of l_s/d_{max} is too small, then the dam frame work will have more members than it needs, and this is a waste of money and materials. Moreover, the short spacing will check the large particles and drifting woods that runoff in a normal scale flood flow, which results in the consumption of storing capacity for the coming debris flow. If, on the other hand, the largest size particles that rarely exist in the basin are chosen to determine d_{max}, the concentration of these particles in the forefront of debris flow will be too small to clog the spacing and the debris flow containing large boulders, even though the size is less than predetermined d_{max}, may pass through the dam. Consequently, the difference in debris flow control efficiencies by setting d_{90} and d_{95} as d_{max} is experimented by the

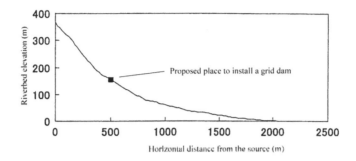

Figure 8.25 The longitudinal profile of the river and the position a grid dam will be installed.

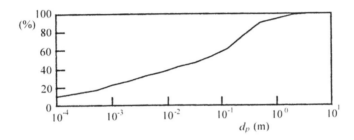

Figure 8.26 Particle size distributions in the riverbed.

numerical simulations and from that we obtain a criterion to determine d_{max} in the design of a grid dam.

We assume that the riverbed material has the size distributions as shown in Figure 8.26. The development of debris flow and the size segregation within the flow along the river channel are calculated by the method referred to earlier in this sub-section. Particle size is classified into fifteen groups, the maximum size is 4 m and the minimum size is 100 μm. Only one particle size group is considered to be the fine fraction, i.e. 100 μm, and the existence ratio of the maximum size particles and the minimum size particles are 2% and 10%, respectively. The value of d_{90} is 0.5 m and the value of d_{95} is 1 m. In the calculation, both l_s/d_{90} and l_s/d_{95} are set at 1.6; for Case 3 l_s is 0.8 m and for Case 2 l_s is 1.6 m.

We assume that the riverbed comprised of the material shown in Figure 8.26 is 5 m thick, however, in the reach having a steeper slope than the angle of repose of the material the bedrock is considered to crop out and no deposit exists. For the sake of simplicity, water is supplied only from the upstream end by the hydrograph shown in Figure 8.27. The river width upstream of the dam site is set 10 m and that downstream of the dam site is 20 m, and the dam length is 8 m.

Figure 8.28 shows the debris flow hydrographs and sediment graphs just down-stream of the dam site in the cases of no dam (Case 1) and in Cases 2 and 3. When a grid dam is installed, either of Case 2 or Case 3, the peak discharge is cut to less than half and the peak arising time delays. In the sediment graphs, there remains a peak in

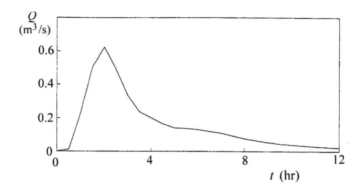

Figure 8.27 Hydrograph supplied from the upstream end.

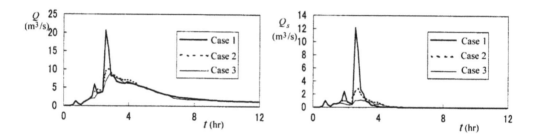

Figure 8.28 Debris flow discharges and sediment discharges just downstream of the dam site.

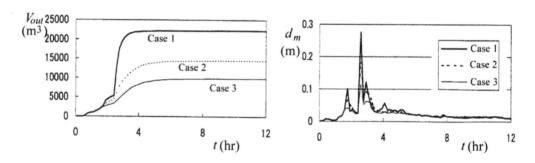

Figure 8.29 Cumulative sediment runoffs and mean diameters just downstream of the dam site.

Case 2 but no evident peak exists in Case 3. From these results, it can be seen that choosing the spacing corresponding to d_{90}, will stop the peak sediment discharge.

Figure 8.29 shows the cumulative sediment runoffs and the mean diameters of runoff sediment just downstream of the dam site. In Case 2, 35% of the runoff sediment to the dam site was captured by the dam, whereas in Case 3, 60% was captured. In Case 3, because many large boulders were captured, the mean particle size that passed through the dam was smaller than that in Case 2.

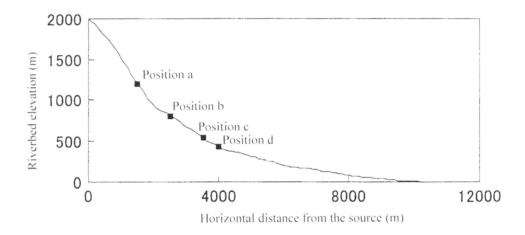

Figure 8.30 Longitudinal profile of the river channel and possible dam sites.

These results clearly show that if the position of a possible installation of a grid dam is decided beforehand, numerical simulations can give the optimum spacing between frames.

The selection of an appropriate position to install a grid-type dam

Consider that the river channel has the longitudinal profile as shown in Figure 8.30. The riverbed material is assumed to be the same as before and the spacing between the frames in this time is predetermined. Therefore, the spacing is set to 0.8 m that is obtained as the optimum corresponding to d_{90}. The supplied water from the upstream end will entrain the riverbed sediment and a debris flow develops along the channel. Because the degree of convergence of large stones toward the forefront and the peak sediment discharge depend on the distance of travel, the efficiency of the dam to control debris flow will depend on its location.

The candidate sites for the installation of a grid dam are set at a, b, c, and d whose distances from the source are 1,500 m, 2,500 m, 3,500 m and 4,000 m, respectively. A preparative calculation which predicts the situations of debris flow at the respective positions is carried out. Figure 8.31 shows the results of the calculation in which the peak sediment discharges become larger toward the downstream direction, but at the positions c and d the peak values are almost the same. Namely, the debris flow has fully developed upstream of position c.

Hence, numerical simulations for the case of no dam (Case 1), the cases of installation at positions c (Case 2), b (Case 3) and a (Case 4) are carried out and the dam functions in the respective cases are compared. Figure 8.32 is the hydrograph of the supplied water at the upstream end. The bed material is assumed to be laid 5 m thick. The river width is set to 10 m from the source to location d and that downstream is 20 m. The calculation is done with $\Delta x = 100$ m.

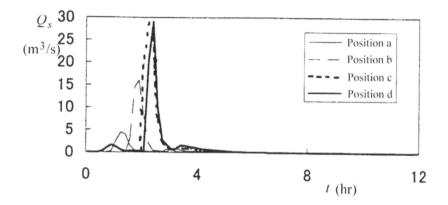

Figure 8.31 Sediment graphs at respective positions.

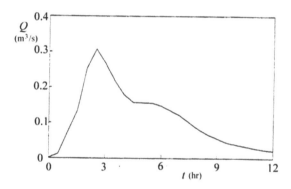

Figure 8.32 Hydrograph given at the upstream end.

Figure 8.33 shows the hydrographs and sediment graphs immediately downstream of position d in the respective cases. In comparison with the case without a dam (Case 1), the peak flow rate is reduced by a large amount. Case 2 has the biggest reduction, then Case 3 and Case 4. In Case 2 the peak flow discharge is reduced by about 66%, but in Case 4 it is only reduced by 12%. A similar tendency is found for the sediment discharge. The reason for such a tendency is as follows: At position a, the debris flow is still not sufficiently developed and the degree of accumulation of large particles in the forefront is low. Therefore, the dam at position a is not easily clogged and plenty of sediment associated with water is passed through the dam. The flow thus passed through the dam continues to develop to a debris flow containing large boulders in front down to the position d. Hence, the hydrograph and sediment graph at position d in Case 4 are not much different from those in Case 1. Whereas, if a grid dam is set at position c, the debris flow has developed before it reaches that position and it is easily checked by the dam. Once the forefront is checked the succeeding debris flow will be deposited by the rebounding deposition. Therefore, the debris flow as well as sediment discharges are greatly reduced at position d.

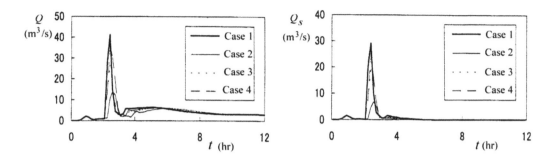

Figure 8.33 Hydrographs and sediment graphs at the position d in the respective cases.

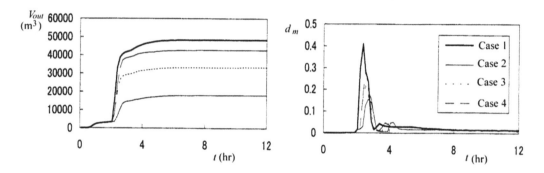

Figure 8.34 Cumulative sediment runoffs and mean diameter of particles in flow at immediately downstream of the position d.

Figure 8.34 shows the cumulative sediment runoffs and the mean particle diameters in the flow immediately downstream of position d. The functions described earlier are clearly shown in these figures.

The above numerical simulations reveal that the optimum position of grid dam installation is where debris flow arrives at its most developed stage. By the combination of these simulations for the selection of the optimum position and for the determination of the optimum spacing between frames, we can design most efficient dam structure and the best place to install it.

8.4 MAKING DEBRIS FLOW HARMLESS BY CHANNEL WORKS AND TRAINING WALLS

8.4.1 Design of countermeasures on the fan of the Camuri Grande River

In a mountain area, alluvial fans are precious flatlands where people can live. Because these fans have been formed by the flooding of debris flows and flood flows, they

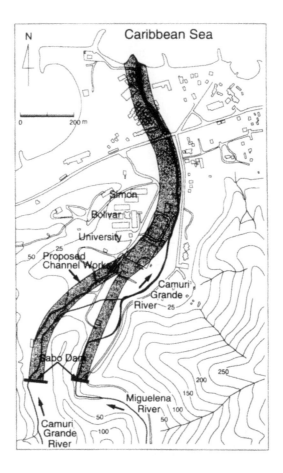

Figure 8.35 Location of sabo dams and plan view of the channel works.

have a high risk of suffering from flooding hazards, and hence the safety of the river penetrating the fan is very important. Normally, a sabo dam is installed immediately upstream of the fan top, at the debouche of mountain ravine, and the downstream ravine is reinforced by channel works that combine revetment works and groundsills. Here, taking the fan of the Camuri Grande River that was explained in section 7.5 as an example, one particular plan, the combination of sabo dams and channel works is considered and its effectiveness is assessed using the numerical simulations (Nakagawa *et al.* 2001).

The conceptual arrangement of these facilities is shown in Figure 8.35. Five different combinations were examined as shown in Table 8.2. In Case 1 there is no sabo dam but channel works 2 m deep exist. Case 4 has the combination of 20 m high and 10 m high closed-type sabo dams in the Camuri Grande River and the Miguelena River, respectively, and channel works 3 m deep. The conditions for other cases are clear in Table 8.2.

Table 8.2 Conditions for calculation.

Sabo works		Case No.				
		Case 1	Case 2	Case 3	Case 4	Case 5
Sabo dam	Camuri Grande	No	10 m	20 m	20 m	20 m
height in	Miguelena	No	10 m	10 m	10 m	10 m
Bank height of the Channel		2 m	2 m	2 m	3 m	5 m

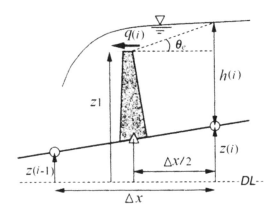

Figure 8.36 Definition of the gradient used in calculating the discharge over the dam.

First, we examine the effectiveness of the sabo dams. The method used in the calculation of the hydrograph, sediment graph, and particle composition at the fan-top is the one-dimensional routing, SERMOW, described in Chapter 6, in which the fundamental equations are discretized by $\Delta x = 100$ m, the unknown variables h, C and z are calculated at a time step on the boundaries of Δx, and variables q_t defined on the mid-point of Δx is calculated at the next time step: leap-frog method. The sabo dam is set at the calculation point of the flow discharge per unit width, $q_t(i)$, as shown in Figure 8.36. The flow surface gradient, θ_e, that is necessary for the calculation of the flow discharge is set equal to the gradient obtained by the stage difference between the flow surface stage immediately upstream of the dam and the dam crest:

$$\theta_e = \tan^{-1}\left\{\frac{z(i) + h(i) - z1}{(\Delta x/2)}\right\} \tag{8.43}$$

where $z(i)$ is the riverbed elevation at the point of stage definition immediately upstream of the dam, $h(i)$ is the flow depth at that point, and $z1$ is the elevation of the dam crest.

The flow depth on the crest of the dam h' that is used for the calculation of the discharge over the dam is given by the following equation:

$$h' = \begin{cases} h(i) + z(i) - z1 \; ; & h(i) + z(i) - z1 > 0 \\ 0 & ; & h(i) + z(i) - z1 \leq 0 \\ h(i) & ; & z(i) > z1 \end{cases} \tag{8.44}$$

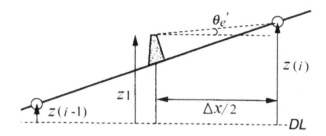

Figure 8.37 Gradient used in calculating equilibrium sediment concentration at the dam.

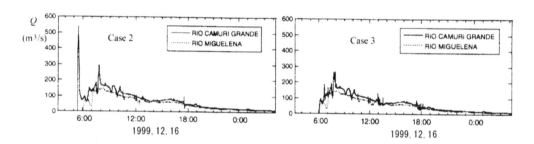

Figure 8.38 Hydrographs at just upstream of the junction of the rivers.

The gradient θ'_e, needed to calculate the equilibrium sediment concentration at the dam crest is evaluated by referring to Figure 8.37 as:

$$\theta'_e = \tan^{-1}\left\{\frac{z(i) - z1}{(\Delta x/2)}\right\} \tag{8.45}$$

When θ'_e is less than zero, the equilibrium concentration is set as zero.

As for the depositing velocity in the backwater reach of the dam, Equation (6.34) that takes the settling velocity of particles into account is used, only when the shear velocity is less than the settling velocities of the respective group particles and the flow depth is deeper than 1 m and the velocity of the flow is less than 50 cm/s.

The rise of riverbed behind the dam causes the riverbed width to broaden. To take this effect into consideration, in the case of the 10 m dam height, the riverbed width is assumed to be 100 m in the reaches 300 m (3 mesh distance) immediately upstream of the dam, whereas in the case of the 20 m high dam, the 400 m reach (4 mesh distance) from just upstream of the dam, the riverbed width is assumed to be 100 m. The flow width (channel width) in these reaches is assumed to be equal to the riverbed width.

The calculating conditions are the same as those in section 7.5.2.

We compare the calculated results for Cases 2 and 3. Figure 8.38 shows the hydrographs in the respective rivers just upstream of the junction of the rivers. In Case 2 a rather conspicuous discharge peak of 550 m³/s remains in the Camuri Grande River. If there is no dam the peak discharge there is about 700 m³/s (see Figure 7.48), therefore,

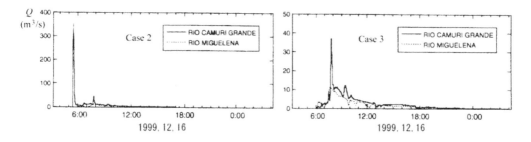

Figure 8.39 Sediment graphs at just upstream of the river junction.

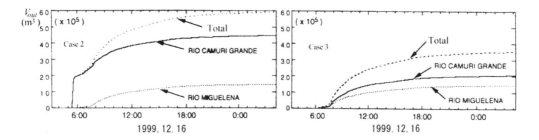

Figure 8.40 Sediment volume that overflowed the dams in the respective rivers.

the dam reduces the peak by about 150 m³/s. In Case 3, the debris flow peak that occurred in the Camuri Grande River at about 6:00 a.m. in Case 2 completely disappears. Therefore, if the dam is 20 m high, it can completely capture the debris flow. The sediment runoff from the Miguelena River is far less than that from the Camuri Grande River and the 10 m high dam can sufficiently check the sediment.

Figure 8.39 shows the sediment graphs in the respective rivers just upstream of the junction. Because the evident peak sediment discharge of about 350 m³/s remains in the Camuri Grande River in Case 2, the debris flow should have overflowed the 10 m high sabo dam. Whereas, in Case 3, the peak sediment discharge in the Camuri Grande River radically drops to about 1/10 of that in Case 2. Moreover, the peak does not correspond to the debris flow at 6:00 a.m. but it appears around 8:00 a.m.

Figure 8.40 shows the cumulative runoff sediment in net volumes from the respective rivers. The sediment volume runs off in Case 3 is about half of that in Case 2. Table 8.3 summarizes the effects of sabo dams in terms of the sediment trapping ratio, where V_{out} is the sediment volume that overflowed the dam, and $V_{out-cut}$ is the sediment trapping ratio defined by the ratio of the trapped sediment volume to the total runoff sediment volume in Case 1. This table shows that the 10 m high sabo dam in the Camuri Grande River can trap 24.8% of the total runoff sediment and the trapping ratio can be enhanced to 65.2% if dam height is raised to 20 m.

Figure 8.41 shows the temporal changes in the mean diameters of the particles in the flow that overflowed the respective dams. If Case 2 is compared with the case of no sabo dam (see Figure 7.48), the coarse particles that runoff around 6:00 a.m. from

Table 8.3 Effectiveness of sediment control by sabo dams.

	Case No.				
	Case 1	Case 2		Case 3	
River	V_{out} (m^3)	V_{out} (m^3)	$V_{out\text{-}cut}$ (%)	V_{out} (m^3)	$V_{out\text{-}cut}$ (%)
Camuri Grande	912,000	686,000	24.8	317,000	65.2
Miguelena	397,000	223,000	43.8	223,000	43.8
Total	1,309,000	909,000	30.6	540,000	58.7

V_{out}: Sediment volume outflowed from the dam.
$V_{out\text{-}cut} = (V_{out\ (case\ 1)} - V_{out})/V_{out\ (case\ 1)} \times 100$.
$V_{out\ (case\ 1)}$: V_{out} for Case 1.

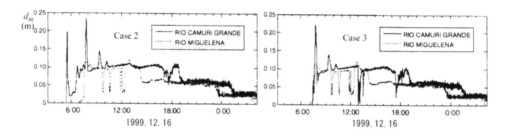

Figure 8.41 Mean diameter of the particles in flow that overflowed the respective dams.

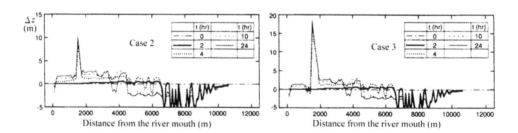

Figure 8.42 Riverbed variations in the Camuri Grande River in respective cases.

the Miguelena River disappear, whereas in the Camuri Grande River particles almost the same size as the case of no sabo dam, runoff. In Case 3, sediment runoffs around 6:00–8:00 a.m. are scarce and the particle size is about that of the smallest group. But, after 8:00 a.m., the effect of the sabo dam to the mean diameter is not evident in both cases.

Figure 8.42 shows the riverbed variations in the Camuri Grande River. In both cases, at 5:15 a.m., two hours after the time origin, severe bed erosion takes place in the upstream reach but still no variation occurs around the dam site; 1500 m from the river mouth. After four hours, deposition at the dam proceeds at once, then deposition

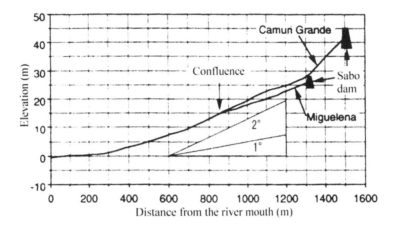

Figure 8.43 Longitudinal profile of the channel works.

continues gradually. In Case 2 the dam is finally filled up, but in Case 3 there still seems to be some capacity to store sediment.

Although the effectiveness of the sabo dam is good, for checking all the runoff sediment, it needs higher dams and therefore, coping with a potential debris flow only using sabo dams will be economically and environmentally unfavorable. In this context, the method in which a sabo dam checks some of the sediment and channel works cope with the rest would be feasible. Normally, channel works are expected to transfer sediment to the downstream area safely, but if no ample gradient to transport sediment is available, deposition in the channel arises and in some cases the flow carrying the high concentration of sediment spills over the channel causing more hazards.

The Camuri Grande River pours into the Caribbean Sea immediately after penetrating the alluvial fan, and the channel gradient cannot be steep enough. Furthermore, at the downstream end due to the backwater effects, sediment is easily deposited and the rebounding deposition causes the channel to be buried.

Herein, we consider the channels as shown in Figure 8.43, whose depths are already mentioned and the widths are 50 m and 40 m, respectively in the Camuri Grande River and the Miguelena River before joining, and 80 m downstream of the confluence. These widths are obtained by assuming that the channel can convey the flow of a normal scale flood with a depth of about 2 m. Because the channel width on a rather narrow fan area is subject to natural and social constraints, its width cannot be changed freely. Therefore, the performances of channel works under fixed channel widths are examined by changing the depth of channel as shown in Table 8.2.

The longitudinal profile of the channel beds are shown in Figure 8.43, they basically obey the natural topographic gradient, but in some parts artificial excavation is inevitable due to designed alignment of the channel. The channel gradient in the major parts is about 2°, and in the reach of 300 m from the river mouth it is about 1°. In about a 200 m reach immediately downstream of the sabo dam in the Camuri Grande River, the channel gradient is a little steeper as shown in Figure 8.43.

An imaginary sea basin is provided as the outflow boundary condition near the river mouth. Namely, to take the effect of backwater into account, a water basin of 200 m long is attached to the river mouth. The downstream end of the pool is a wall having the same height as the mean sea level. If the water stage in the pool becomes higher than that wall, the water in the pool outflows as a freely dropping flow with the overflow depth equal to the difference between the pool stage and the crest of the wall. The bottom of the pool has a 1/10 slope toward the offshore direction, and the maximum depth at the wall is about 20 m. In this basin, similar to the case of deposition behind the sabo dam, sedimentation due to the settling of particles is considered when the settling velocity of each particle diameter becomes larger than the local shear velocity in the basin.

Figure 8.44 shows the deposition area pattern and the deposit thickness distributions in all five cases at the final stage ($t = 25$ hours).

The sediment deposit area and thickness of the deposit are much greater in Case 1, in which no sabo dam exists and the depth of channel is shallow (2 m). In the channel, a deposit of less than 5 m (in some parts thicker than 5 m) is produced and the thickness of deposit is thicker at the area of river mouth than in other places. In the reproduction of the actual situation shown in Figure 7.54, a thick deposition occurred in the central part of the fan. But in Case 1, where the wide channel works are installed, although the flooding occurred over almost all the fan area, the particularly thick deposit in the central part of the fan disappeared.

In Case 2, due to installation of 10 m high sabo dams in the respective two rivers, some parts of the yielded sediment were checked but their quantity was not sufficient and flooding occurred over almost all the fan. The deposit thickness in the channel works was, however, considerably thinner than in Case 1.

In Case 3, the 20 m high sabo dam in the Camuri Grande River worked to drastically reduce the sediment runoff, so that the flooding area and thickness of deposit were remarkably decreased. However, flooding from the right-hand bank of the channel and at the downstream reach left a deposit of thinner than 1 m.

The situation in Case 4, where the channel depth was increased to 3 m, was not much different from the situation in Case 3.

In Case 5, where the channel depth was increased to 5 m, deposition was almost limited to within the channel works, except for the small scale flooding from a few points along the channel.

Therefore, the numerical simulations revealed that the alluvial fan of the Camuri Grand River becomes safe if sabo dams of 20 m high and 10 m high are installed in the Camuri Grande River and the Miguelena River, respectively, and in addition, if channel works: 5 m deep and 50 m wide in the Camuri Grande River upstream of the junction; 5 m deep and 40 m wide in the Miguelena River; and 5 m deep and 80 m wide in the Camuri Grande River downstream of the junction, are installed.

The examinations mentioned above were done under the assumption that the same scale debris flow and flood flow as the one in 1999 would occur again. Whether these scale phenomena should be the objective to design countermeasures is the another problem to be discussed. As referred to in section 7.5, in 1951 a rainfall event comparable with that in 1999 occurred. This generated a great many landslides producing a vast amount of riverbed accumulation. In the 1951 case, the sediment runoff to the alluvial fan is said to be not much, and almost all of the sediment stored in 1951

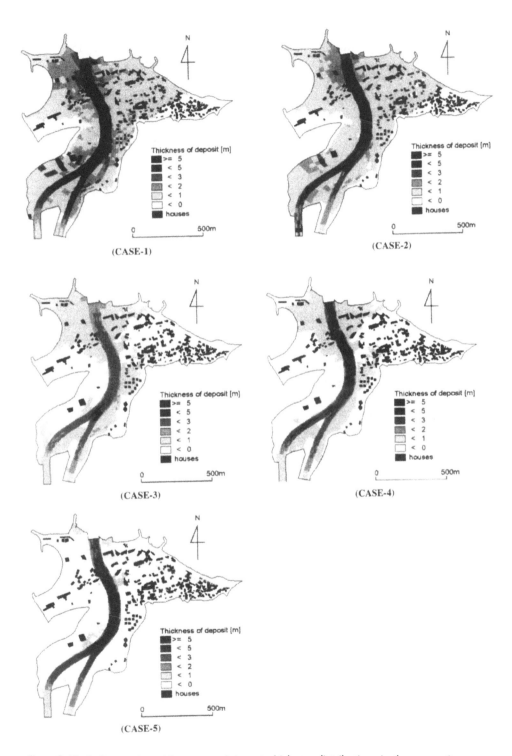

Figure 8.44 Sediment depositing area and deposit thickness distributions in the respective cases.

seems to have yielded by the 1999 flood. If this kind of phenomena; rainstorm $\rightarrow$ sediment storage in the upstream $\rightarrow$ rainstorm $\rightarrow$ sediment runoff by the erosion of stored sediment, is a repeating cyclic, then the phenomena similar to the 1999 event may repeat once in a hundred years or so. From the economical point of view, coping with this scale phenomenon only by the structural countermeasures as examined above may not be expedient. The comprehensive countermeasures that integrate the nonstructural countermeasures such as the regulation of land use and the warning and evacuation systems with the structural countermeasures would be advisable.

8.4.2 Management of debris flow by a training dike

One method to prevent destruction of the community on an alluvial fan from debris flow tries to confine it within the channel works as the example referred to above. The other method, depending on the topography and land use, tries to guide the flow towards a safer area or to store the sediment in the mountain side by constructing training dikes outside the inhabited area.

Herein, the results of our experiments to make clear the forms of sediment deposit produced by straight training dikes deflecting from the upstream flow direction are outlined.

The experimental facility has a straight debris flow generating flume of 10 cm wide and 18° in longitudinal gradient, and the flume is connected to a flooding board whose longitudinal gradient is adjusted to set the bed slope along the training wall at 7° irrespective to the deflection angle of the training wall. Therefore, the larger the deflection angle the steeper the longitudinal gradient of the flooding board becomes, so that the action of debris flow to push the wall becomes stronger in the case of a larger deflection angle. The debris flow material is a sediment mixture, of which mean diameter is 2.03 mm, $C_* = 0.7$, $\sigma = 2.6 \, \text{g/cm}^3$, $\tan \varphi = 0.8$. The material was laid on the flume bed in an area 4 m long and 10 cm deep, and after soaking with water the predetermined discharge of water flow was given from upstream to generate debris flow. The supplied water discharges were 1 l/s and 2 l/s. For the case of supplied water discharge is 1 l/s, if the theory written in section 2.2.2 is applied, the completely developed debris flow at the outlet of the flume must have the characteristics; $C_u = 0.427$, $u_u = 116 \, \text{cm/s}$, $h_u = 2.2 \, \text{cm}$. Because the flooding board gradient toward the flow is 7°, referring to the theory in section 5.1.1 and setting $\tan \alpha_i = 0.6$ and $\kappa_a = 0.5$, the arrival distance of the debris flow without the training wall should be $x_L = 124 \, \text{cm}$, and the time to stop counted from the instant that the debris flow front appeared at the outlet of the flume should be $t = 2.1 \, \text{s}$.

Figure 8.45 shows the cases for the water supply of 1 l/s. The broken lines in the figure are the outer edges (in plan view) and the surface stages of debris flow (in side view) in the respective elapsed times from the instant that the front appeared at the outlet of the flume, and the solid lines are the contour lines of the final deposit and the final longitudinal profile of deposit along the training wall in the plan and side views. Although there are some discrepancies case by case, the existence of the training wall seems to have little effect on the arrival distance of the debris flow. In the experiments, the front of the deposit gets a little longer by the effect of squeezed water, and the deposit height is a little elevated by the accumulation of successive flow. Hence, the theoretically obtained arrival distances seem to be a little shorter than the

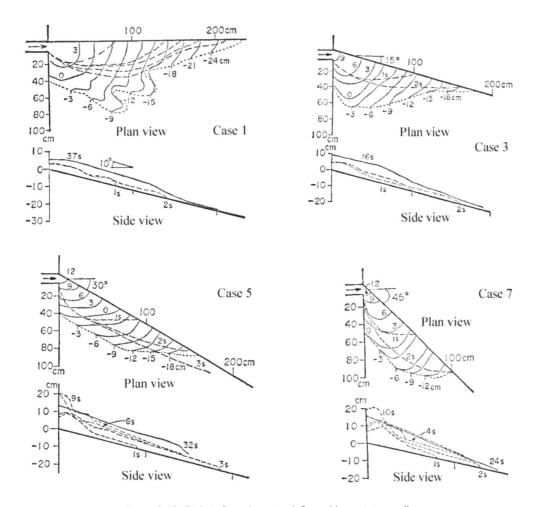

Figure 8.45 Debris flow deposits deflected by training walls.

experimental results, but in general the theory described in section 5.1.1 is applicable for the prediction of necessary length of training wall.

The necessary height of the training wall is indispensable information for the design. As is shown in the side view diagrams in Figure 8.45, the broken lines, in some cases at certain times, surpass the height of the solid lines. Thus, the training wall must be higher than the surface of the final deposit. This is due to the amplification of intermittent waves in the flow by the collision with the wall. Figure 8.46 shows that the maximum flow stages of the debris flow front at the outlet of the flume and the maximum flow stages during the period of the experimental runs vary with the deflection angle. The numbers attached to the symbols in Figure 8.46 represent the experimental cases. As the deflection angle becomes large, both stages become large.

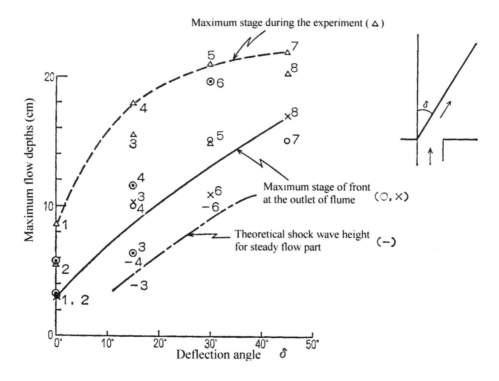

Figure 8.46 Maximum flow depths versus deflection angles.

It is considered that the shock by the collision of debris flow with wall becomes larger with the increasing deflection angle. Hence, we try to calculate the shock wave height by the application of Ippen's formula:

$$
\left.
\begin{aligned}
\frac{h_{sh}}{h_n} &= \frac{1}{2}\left(\sqrt{1 + 8F_r^2 \sin^2 \beta_1} - 1\right) \\[2ex]
\tan \delta &= \frac{\tan \beta_1 \left(\sqrt{1 + 8F_r^2 \sin^2 \beta_1} - 3\right)}{2\tan \beta_1 + \sqrt{1 + 8F_r^2 \sin^2 \beta_1} - 1}
\end{aligned}
\right\}
\tag{8.46}
$$

where h_n is the normal flow depth, h_{sh} is the depth after shock, β_1 is the angle of the shock wave, and F_r is the Froude number. Because the normal flow depth for the supplied water discharge of 1 l/s is 2.2 cm and that for the supplied water discharge of 2 l/s is 2.9 cm, h_{sh} by Equation (8.46) for the respective experimental cases can be plotted by the symbol (−) in Figure 8.46, and the average tendency of these points is shown by the chained line in the figure. In the cases other than those plotted in Figure 8.46, the flow downstream of shock wave becomes sub-critical and Equation (8.46) becomes inapplicable. The theoretically obtained shock wave heights are considerably smaller than the experimental results.

The shock wave height calculation mentioned above does not consider the roll waves. Now, we consider the amplification of roll waves. Because the Froude numbers corresponding to the normal flow depths 2.2 cm and 2.9 cm are 2.5 and 3.3, respectively, if Figure 4.16 is referred to, the depths including roll waves are obtained as 3.3 cm and 5.8 cm, respectively. Assuming these depths as the normal depths the shock wave calculation gives the amplified depths as shown in Figure 8.46 by the circles dotted in the center. The maximum stages of debris flow fronts at the outlet of the flume for the respective experimental cases are shown in Figure 8.46 by circles and × marks and the average tendency of these plotted values is indicated by the solid line. For the time being we consider the maximum stage of the debris flow before deposition is given by this calculation. In Figure 8.46, the maximum stages after the development of deposit are shown by triangles and the envelope of these marks is given by the broken line. Because these stages are approximately the sums of deposit heights and shock wave heights, the necessary height of the training wall is given by this broken line. The longitudinal profile of the deposit may be obtained by applying Equation (5.14), but the results change considerably depending on the choice of flow width for which no reliable basis is available at the moment. The alternative method would be the two-dimensional calculation of deposition but the shock wave height should be added to the calculated maximum height of the deposit to give the necessary height of the training wall.

8.5   DESIGN DEBRIS FLOWS FOR THE COUNTERMEASURE PLANNING

The prediction of the maximum discharge of debris flow that passes through a section is necessary for the design of structural countermeasures such as sabo dam and channel works and for the determination of necessary clearance under bridges. It is also indispensable information to know the arrival distance of debris flow front on a fan area.

If the maximum discharge of a stony debris flow Q_T is predicted, the velocity and flow depth are given from Equations (2.29) and (2.30) by the following equations:

$$U = \left(\frac{2}{5d_p}\right)^{2/5}\left[\frac{g\sin\theta}{0.02}\left\{C+(1-C)\frac{\rho}{\sigma}\right\}\right]^{1/5}\left\{\left(\frac{C_*}{C}\right)^{1/3}-1\right\}^{2/5}\left(\frac{Q_T}{B}\right)^{3/5} \quad (8.47)$$

$$h = \left(\frac{2}{5d_p}\right)^{-2/5}\left[\frac{g\sin\theta}{0.02}\left\{C+(1-C)\frac{\rho}{\sigma}\right\}\right]^{-1/5}\left\{\left(\frac{C_*}{C}\right)^{1/3}-1\right\}^{-2/5}\left(\frac{Q_T}{B}\right)^{2/5} \quad (8.48)$$

The prediction of total runoff sediment volume is also important for the determination of the storage capacity of the structures and the predictions of the hazardous area and the scaling of hazards and risks within the hazardous area. Therefore, prediction of the potential debris flow hydrograph and other characteristics are essential for coping with the hazards.

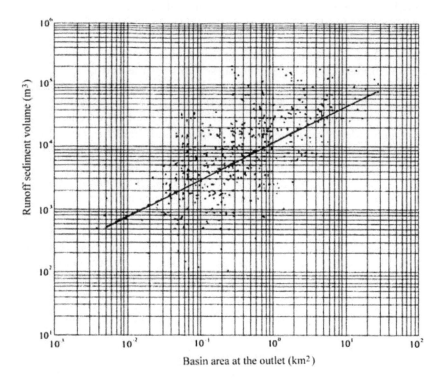

Figure 8.47 Runoff sediment volume versus basin area.

8.5.1 Method based on the previous data

Mizuhara (1990) investigated the relationships between the runoff sediment volume of debris flows and some factors considered to be related using the Japanese data from 1972 to 1985.

Figure 8.47 shows the relationship between the runoff sediment volume and the basin area upstream of the outlet of mountain torrent, 663 data points are used. Although there is a scattering of data of about two orders of magnitude, the average tendency is given by:

$$V_{out} = 1.14 \times 10^4 A_d^{0.583} \tag{8.49}$$

where V_{out} is the runoff sediment volume in m^3, and A_d is the basin area in km^2.

As the area of debris flow generation is the sub-basin whose channel gradient is steeper than 15°, the relationship may be more reasonable if such areas are considered as the relating factor. The results of such consideration become:

$$V_{out} = 1.46 \times 10^4 A_{15}^{0.583} \tag{8.50}$$

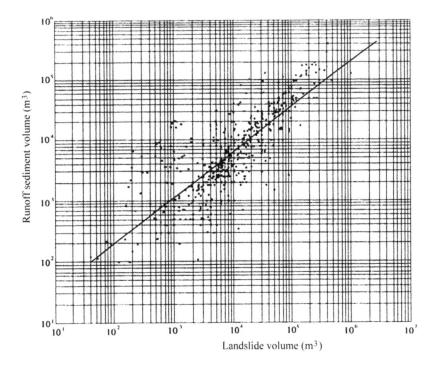

Figure 8.48 Runoff sediment volume versus landslide volume.

where A_{15} is the area of sub-basin whose channel gradient is steeper than 15°. The scattering of data in Equation (8.50) is, however, almost the same with that in Equation (8.49).

Figure 8.48 shows the relationship between the runoff sediment volume and the volume of landslides that occurred in the same basin. The relationship is given as:

$$V_{out} = 6.59 V_{la}^{0.750} \tag{8.51}$$

where V_{la} is the volume of landslides in m³. This relationship suggests that if the landslide volume exceeds 2,000 m³ the landslide volume is larger than the runoff sediment volume, and if the landslide volume is less than 2,000 m³ the runoff volume exceeds the sediment volume produced by the landslides. The reason for this is considered to be that large-scale landslides seldom transform completely into debris flow but small-scale landslides easily transform into debris flow and moreover they erode the riverbed downstream to develop into large-scale debris flows. In these data, however, the sediment volume that ran off as debris flow would have been the sum of the individual debris flows that occurred in the respective branches in the basin, and the landslide volume would have also been the sum of the respective volumes of many landslides in the basin. It is necessary to investigate the relationships between the runoff sediment volume of one debris flow and the landslide volume that induced that debris flow.

Table 8.4 Sediment volume per unit length

| Valley shape | Geology | | |
	T_s (m^3/m)	T_v (m^3/m)	R (m^3/m)
V	0.8	4.8	1.6
U	15.8	16.7	15.3
Plane	16.4	41.1	22.7

Sample number = 62.
T_s: Sedimentary rock, T_v: Volcanic rock, R: Granite and metaorphic rock.

Table 8.5 Sediment accumulation width.

| Valley shape | Geology | | |
	T_s (m)	T_v (m)	R (m)
V	3.00	5.18	5.96
U	11.04	9.44	0.62
Plane	12.63	14.56	12.78

8.5.2 Prediction of total sediment runoff by field investigation

One often finds that the bedrock is revealed for a long distance after the occurrence of a debris flow. Therefore, the sediment runoff as debris flow could be estimated if the sediment volume accumulating on the bedrock is known. For example, in the guide book for the fundamental survey related to the implementation of the Sediment Disaster Prevention Act, they recommend adopting the assumption that the objective sediment volume is the product of the possible distance of the erosive reach L_{me}, erosion depth D_e and width of erosion B'. Even if the deep-seated, large-scale landslide case is omitted, the estimation of these contributing factors together with the sediment volume of shallow, small-scale landslides is a formidable task.

Takahashi (1982a) found that, in the basin of the Kizu River, the accumulated sediment volume per unit length of the torrent bed, V_T, was dependent on geology and the cross-sectional shape of the valley. On average, these relationships were arranged as shown in Table 8.4. The flow width is not necessarily the same as the width of sediment accumulation, and therefore, the erodible sediment volume V_{Te} is obtained by $V_{Te} = (B'/B_0)V_T$, where B' is the surface flow width and B_0 is the sediment accumulating width. According to the field survey, B_0 was also dependent on geology and the cross-sectional shape of the valley as shown in Table 8.5. B' is assumed to be given by the regime theory ($B' = 5Q^{1/2}$, Q is the water flow discharge obtained from the rational formula), where B' does not exceed B_0.

In estimating the sediment volume from newly produced landslides, the relationship between the total rainfall amount and the landslide number per unit area of slopes steeper than 30° obtained for the Ise Bay Typhoon that brought the severest disasters to the Kizu River Basin was used. Then, the estimated numbers of landslides

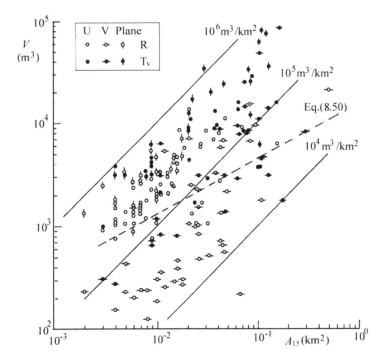

Figure 8.49 Estimated sediment runoff by debris flow versus basin area whose channel is steeper than 15° in the Kizu River Basin.

induced by the rainfall of the probability of once in 100 years (645 mm) were, for the normally vegetated slopes steeper than 30°, 0.9/ha and, for the sapling and denuded forests, 1.1/ha. Furthermore there was a rather good correlation between the individual landslide area and the depth of the slide, and the individual landslide area was correlated with the geology of the basin. Consequently, because the prevailing geology and the situation of vegetation were different for each sub-basin, the expected landslide volume per one slide was obtained for the respective sub-basins. The sediment volume that would be newly produced by thus occurring landslides V_l and the remaining sediment volume after the previous landslides V_r were added to the erodible sediment volume V_{Te} in the respective sub-basins. Figure 8.49 shows the total sediment volumes, $V_{su} = V_{Te} + V_l + V_r$, for the respective sub-basins correlated with the basin areas whose channel slopes are steeper than 15°. The specific sediment yield scatters in the range of $10^4 - 10^6$ m³/km². The broken line in Figure 8.49 is Equation (8.50). Even in one particular Kizu River Basin, the estimated runoff sediment volumes for respective sub-basins scatter as widely as for the nationwide different river basins.

8.5.3 Theoretical prediction of debris flow scale

The methods to predict debris flow hydrograph and other characteristic values have already been discussed in detail for the respective processes of generation; due to

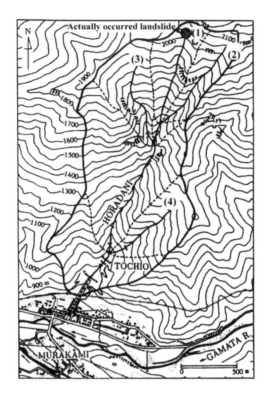

Figure 8.50 Hypothetical landslide positions.

erosion of steep riverbed; induced by the liquefaction of a landslide earth block; and due to the collapse of a natural dam. The sediment runoff models based on a similar idea as the kinematic wave flood runoff model and named SERMOW and SERMOW ver.2 were also introduced and proved to be feasible. As examples of these methods some actual debris flow disasters were reproduced numerically. These methods are, of course, applicable to the case of the determination of design scale debris flow. Therefore, the methods are not discussed again, but herein, as an expansion of the discussion of the Horadani debris flow in section 7.2, the effects of the position of the landslide and the timing of many landslide occurrences to the magnitude of the debris flow are discussed, and some guidance for the determination of design debris flow are obtained.

Effects of landslide position and the order of many landslides occurrences

In the Horadani basin there were 21 landslide scars in 1996 including the one that occurred in 1979 that generated the Horadani debris flow. In 1979 there was only one landslide at the position (1) in Figure 8.50. If a landslide had occurred at the source areas of other sub-basins, the debris flow hydrograph would have different scales and forms. Hence, we calculated the debris flow hydrographs at the fan top caused by the respective landslides at (1) to (4), in which the manner and magnitude

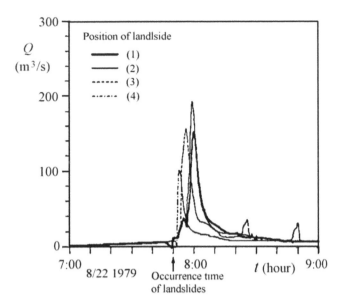

Figure 8.51 Effect of landslide position on the debris flow hydrograph.

of the sediment and water supply to the riverbed were the same as the case of landslide at position (1) that was explained in section 7.2. The riverbed conditions before the debris flow were also the same. Figure 8.51 shows the results of calculation. There are evident differences between the cases. Because the debris flow hydrograph generated by the erosion of bed material depends on the sediment volume on the bed and the water discharge in the channel, the debris flow hydrograph produced by the landslide at position (4) has the smallest peak discharge, in which the channel length is the shortest. On the other hand, the debris flow generated by the landslide at position (2) has the highest peak discharge because it has the longest channel length. The arrival time of the debris flow induced by the landslide at position (4) is the earliest because the length to travel is the shortest.

The number of landslide is not necessarily only one. But, when many landslides occur there may be some time lags between the landslides. Figure 8.52 shows the results of calculations, where in one case landslides occur at positions (1) and (2) at the same time and in another case the landslide at position (2) occurs ten minutes later than the landslide at position (1). In the former case the peak discharge is very large and in the latter case the peak discharge becomes smaller and the hydrograph has two peaks; the first peak is due to the landslide at position (1) and second peak is due to the landslide at position (2). The duration of debris flow for the latter case becomes longer.

Prediction of location, time and sediment runoff volume of landslide

The discussion above makes clear that the determination of designs for debris flow must contain the possible landslide position, scale and occurrence time. However,

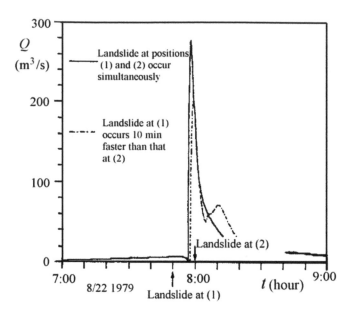

Figure 8.52 Effects of time lag between the two landslides.

the state-of-the-art answer as to 'when' and 'where' a landslide may occur is still far from satisfactory stage. Herein, an investigation aiming to approach the answer for the proposition is introduced (Takahashi and Nakagawa 1989). The problem of predicting how large landslide will occur is also very difficult problem, but as discussed in section 7.2, the scale of shallow landslide does not much affect the hydrograph far downstream, so that this problem will not be discussed herein.

The structure of the soil layers is generally divided into A, B and C layers from the surface, each including many sub-layers: the A layer mainly consists of humic soil; the B layer consists of weathered soil; and C layer is still not weathered much. The division into layers herein is not based on a strict definition in pedology but merely represents the fact that layers of soil differ in permeability and in other aspects as well. Landslides which occur in severe storms are mainly those at layers A or B. The main cause of such a landslide would be the effects of parallel seepage flows which appear from the bottom of each layer resulting from the discontinuous permeability at each layer boundary. The depths of surface landslides are so shallow that the slip surfaces can be considered to parallel to the slopes. Therefore, the slide mechanism may be analyzed by applying the theory on the stability of infinite uniform slopes.

Consider soil layers parallel to a slope with parallel seepage flows as shown in Figure 8.53. The operating shearing stress at the boundary between the A and B layers τ_A and the resisting stress τ_{AL} against the operating stress are described as follows:

$$\tau_A = g \sin\theta \cos\theta [D_A(1 - \lambda_A)\sigma_A + \{(D_A - WF_A - H_A)s_{0A}$$
$$+ WF_A s_A + H_A\}\lambda_A\rho + H_s\rho] \tag{8.52}$$

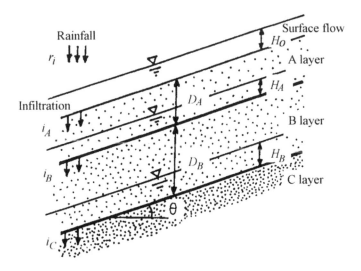

Figure 8.53 The structure of surface soil layer and seepage flows through the slope.

$$\tau_{AL} = g \cos^2 \theta[(D_A - H_A)(1 - \lambda_A)\sigma_A + \{(D_A - WF_A - H_A)s_{0A}$$
$$+ WF_A s_A\}\lambda_A \rho + H_A(1 - \lambda_A)(\sigma_A - \rho) + H_o \rho] \tan \varphi_A + c_A \qquad (8.53)$$

Similarly, at the boundary between the B and C layers:

$$\tau_B = \tau_A + g \sin \theta \cos \theta[D_B(1 - \lambda_B)\sigma_B + \{(D_B - WF_B - H_B)s_{0B}$$
$$+ WF_B s_B + H_B\}\lambda_B \rho] \qquad (8.54)$$

$$\tau_{BL} = \tau_A \tan \varphi_B / \tan \theta + g \cos^2 \theta[(D_B - H_B)(1 - \lambda_B)\sigma_B + \{(D_B - WF_B - H_B)s_{0B}$$
$$+ WF_B s_B\}\lambda_B \rho + H_B(1 - \lambda_B)(\sigma_B - \rho)] \tan \varphi_B + c_B \qquad (8.55)$$

where λ is the void ratio, WF is the distance of the wetting front measured from the upper boundary of the layer, s is the degree of saturation, s_0 is the initial degree of saturation, c is the cohesive strength, suffixes A and B mean the values in A and B layers, respectively. In the above equations if $WF \geq D - H$, $WF = D - H$, if $H_A \geq D_A$, $H_A = D_A$, and if $H_B > D_B$, instead of Equation (8.55),

$$\tau_{BL} = (\tau_{AL} - c_A) \tan \varphi_B / \tan \varphi_A + g \cos^2 \theta D_B(1 - \lambda_B)(\sigma_B - \rho) \tan \varphi_B + c_B \qquad (8.56)$$

is used.

When a landslide occurs in either the A or the B layer, it slides on the slip surface that appears at the boundary of either layer (Takahashi 1981a). The factor of safety against landslide SF is given by:

$$SF_A = \tau_{AL} / \tau_A \qquad (8.57)$$

$$SF_B = \tau_{BL} / \tau_B \qquad (8.58)$$

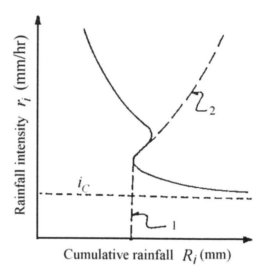

Figure 8.54 The critical line to induce landslide.

Because H, H_o and WF are functions of time, the safety factors are also functions of time.

Assuming the infiltration rates in the respective layers are constant, the degrees of saturation above the wetting fronts in the respective layers are constant, and the lateral saturated seepage flow obeys Darcy's law, the cumulative rainfall amount until the safety factors become 1 under constant rainfall intensity is obtained. These calculations were repeated for different rainfall intensities to obtain the critical lines on which SF_A or SF_B become equal to 1 on the plane of Cartesian coordinate whose two axes are the rainfall intensity and the cumulative rainfall amount, respectively. The schematic diagram of that critical line is shown in Figure 8.54.

In the region to the right of this critical line, landslide can occur. The critical line is asymptotic to the $r_i = i_C$ line in the extent of large R_i, where i_C is the infiltration rate to the C layer. Therefore, if the rainfall intensity is weaker than i_C, a landslide does not occur no matter how large the cumulative rainfall amount. Hence, the regional difference in the threshold rainfall for a landslide depends on i_C as well as the properties of the surface soil. The broken line 1 in Figure 8.54 corresponds to $i_C = 0$, in which case, as long as the slope gradient is steeper than the critical value, landslide inevitably occurs when R_i exceeds a critical value. The broken line 2 corresponds to the case of very strong A layer due to bondage by tree roots system and so on. If the surface of slope is covered by a very strong layer, while R_i is small, landslide does not occur even under very strong rainfall intensity. According to the sensitivity analyses of Equations (8.53) to (8.58), the difference in the values of θ, ϕ, c, D, and i largely affects the shape and position of critical line in Figure 8.54. Therefore, the accurate estimation of these parameter values is very important.

Because the actual slopes in the basin have complex features due to the distribution of troughs and ridges, and landslides tend to occur in the concave slopes, the above

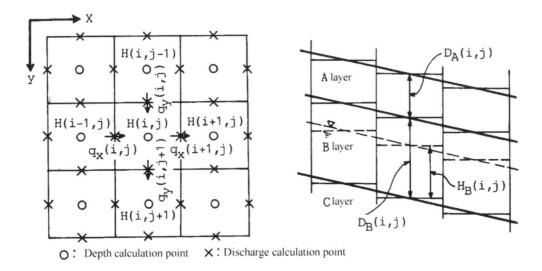

O: Depth calculation point   X: Discharge calculation point

Figure 8.55 Discretization of the top soil layer.

theory for uniform slope must be modified to be applicable to general topographical configurations. Here, the top soil layers of the whole area are discretized into a set of two-storied rectangular solids as shown in Figure 8.55, in which the horizontal scale of each solid corresponds to the average area of a unit landslide and the first and second stories from the top correspond to A and B layers, respectively. Concerning water flow, the planar concentration and diversion of surface as well as seepage flows are considered to occur only between the corresponding layers in adjacent meshes; no water exchange between A and B layers and so on, and in the calculation of the safety factor for landslide in the respective mesh the representative values of slope, thicknesses of layers, etc are substituted in Equations (8.57) and (8.58). The hydraulic gradient I ($=\tan\theta$) is considered to be equal to the topographic gradient obtained by the difference in the center altitudes of respective meshes. Namely, the x-wise hydraulic gradient I_x ($=\tan\theta_x$) of mesh (i, j) is obtained from the differential coefficient of the spline curve connecting the central points of three adjacent meshes. Then, the steepest gradient that is used in the calculation of safety factor is given by $I = (I_x^2 + I_y^2)^{1/2}$.

The surface flow in a mesh is calculated by the continuity equation:

$$\frac{\partial h_o}{\partial t} + \frac{\partial q_{ox}}{\partial x} + \frac{\partial q_{oy}}{\partial y} = r_i - i_A \tag{8.59}$$

and the Manning's equation:

$$q_{ox} = \frac{1}{n}(h_o \cos\theta_x)^{5/3}\left(\frac{I_x}{\sqrt{1+I_x^2}}\right)^{1/2} \tag{8.60}$$

$$q_{oy} = \frac{1}{n}(h_o \cos \theta_y)^{5/3} \left(\frac{I_y}{\sqrt{1+I_y^2}} \right)^{1/2} \tag{8.61}$$

where q_o is the surface flow discharge per unit width. The infiltration rate i_A becomes equal to i_B if the A layer is entirely saturated to the surface of slope, and both A and B layers are saturated i_A becomes equal to i_C.

The vertical infiltration under an unsaturated condition takes place forming a wetting front that moves downward. Although there are some detailed discussions on its mechanism that take the variations and distributions in the degree of saturation within the soil layer into account, herein, considering the balance of accuracy for other parameter values and so on, the degree of saturation in the layer above the wetting front is assumed constant. Then, the distances traveled by the wetting front WF in the respective layers between the time t_0 and t satisfies:

$$\int_{t_0}^{t} (i_A \text{ or } i_B)dt = (s - s_0)WF\lambda \tag{8.62}$$

where i_A becomes the smaller one of r and i_A when no surface water flow exists and it becomes i_A when surface water exists, i_B is equal to i_B when saturated lateral seepage flow exists on the bottom of the A layer and when no saturated seepage flow exists it becomes i_A or r depending on the existence of surface water flow.

If the wetting front arrives at the water surface of lateral seepage flow on the boundary between the layers, the depth of lateral seepage flow changes. The flows in the respective layers are obtained by the continuity equation:

$$\lambda_{ef}\frac{\partial H_{sub}}{\partial t} + \frac{\partial q_x}{\partial x} + \frac{\partial q_y}{\partial y} = (i_A - i_B) \text{ or } (i_B - i_C) \tag{8.63}$$

and the Darcy's equation:

$$q_x = kH_{sub}\frac{I_x}{\sqrt{1+I_x^2}} \tag{8.64}$$

$$q_y = kH_{sub}\frac{I_y}{\sqrt{1+I_y^2}} \tag{8.65}$$

where k is the lateral transmission coefficient of saturated seepage flow, H_{sub} is the depth of seepage flow, and λ_{ef} is the effective void ratio for seepage flow and it is assumed to be:

$$\lambda_{ef} = \lambda(1 - s) \tag{8.66}$$

Predictions of water flow on the surface as well as subsurface and the safety factor based on the analysis of water flow by use of Equations (8.52) to (8.66) enable the determination of the mesh that can slide under a given rainfall. The upstream boundary condition for the analysis is given as the lateral flow discharge is zero at the boundary mesh on the perimeter of the basin, but the downstream boundary condition needs some ingenuity to take account of the topographical connecting condition between the slope and the stream channel. The initial condition must be given by the initial wetting status of the slope and the discharge in the stream channel.

If a possible sliding mesh is identified, the slid earth mass may move down slope following the route connecting the direction of the steepest gradient in each mesh. Depending on the gradient, the state of vegetation along the route, and the properties of the moving earth mass, the slid earth mass will come down to the stream channel or stop on the slope. At the moment there is no method to determine whether the earth mass stops or arrives at the stream channel and continues to be transported downstream. Therefore, it may be reasonable to consider that the earth mass will reach the stream channel if the average gradient of the motion route is steeper than 20°. At that time, whether the earth mass is liquefied on the route is another difficult problem to precisely answer. From the view point of hazard mapping due to debris flow, as assumed in the analysis of the Horadani debris flow, it would predict the safe side to assume that the mass is liquefied on the route of motion.

As the surface water flow depths on the mesh along the stream channel are obtained by this model, whether debris flow due to erosion of the channel bed occurs is judged by referring to the condition for debris flow generation introduced in section 3.1.1:

$$h_o/d_p \geq k^{-1} \text{ or } q_* \geq 3 \text{ and } \theta \geq 15° \tag{8.67}$$

Because this model, different from other flood runoff analysis models previously used in this book, faithfully reflects the behaviors of water on and within the slope and also in the stream channel, application of this model should obtain the correct hydrograph at any location in the basin. To examine this, a flood runoff analysis in the Shimotani basin was done by this model. Based on the topographic map of 1/2500 scale, the basin was discretized by a mesh 12.5 m square. The thickness of the B layer was determined, by referring to the survey data, as a function of mesh's location from the ridge: if it is near the ridge, $D_B = 66/\cos\theta_{i,j}$ cm; if it is in the middle between the ridge and the stream channel, $D_B = 51/\cos\theta_{i,j}$ cm; and if it is near the stream channel, $D_B = 42/\cos\theta_{i,j}$ cm. The thickness of the A layer was set equal to $0.4D_B$. The other parameters were; $i_A = k_A = 2 \times 10^{-4}$ m/s, $i_B = k_B = 1.2 \times 10^{-5}$ m/s, $i_C = 0.55 \times 10^{-5}$ m/s ($= 20$ mm/hr), $\lambda_A = 0.4$, $\lambda_B = 0.3$, $s_{0A} = s_{0B} = 0.3$, $s_A = s_B = 0.8$. For the mesh corresponding to the perennial stream channel, $i_C = 0$ was assumed. To create the initial channel flow condition, prior to giving the actual rainfall, the constant rate rainfall of 12 mm/hr was given for two hours.

The output hydrograph obtained at the outlet of the basin by inputting actual rainfall data, however, had a smaller peak discharge than the actual runoff data and the shape of the hydrograph was acute so the total runoff volume became too small in comparison to the actual data. This means that the water loss into the C layer in the model was too large, but even if i_C value was reduced significantly, the calculated results were not well improved. Then, we considered that not all the water that infiltrated into the C layer was lost but some part of i_C; $p_h i_C$, ran off rather early via the pipe

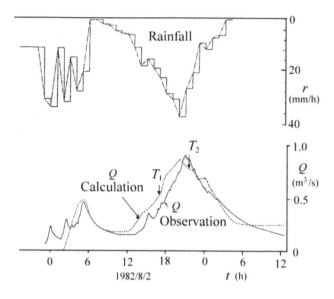

Figure 8.56 Comparison of hydrographs between the data by observation and calculation.

system developed in the C layer. If the pipe system is parallel to the slope surface, the pipe flow would be analyzed by the following continuity and motion equations:

$$\frac{\partial h_h}{\partial t} + \frac{\partial q_{hx}}{\partial x} + \frac{\partial q_{hy}}{\partial y} = p_h i_C \tag{8.68}$$

$$q_{hx} = h_h v_h \tan \theta_x / \sqrt{\tan^2 \theta_x + \tan^2 \theta_y} \tag{8.69}$$

$$q_{hy} = h_h v_h \tan \theta_y / \sqrt{\tan^2 \theta_x + \tan^2 \theta_y} \tag{8.70}$$

where h_h is the depth of pipe flow if it is considered as a planar flow, v_h is the nominal velocity of flow whose depth is h_h, q_{hx} and q_{hy} are the unit width discharge of pipe flow in the x and y directions, respectively. Setting $p_h = 0.6$, $i_C = 0.67 \times 10^{-5}$ m/s, $v_h = 10^{-2}$ m/s, and the same values as the previous calculation for other parameters, we obtained a hydrograph that rather well fitted to the actual data as shown in Figure 8.56.

To examine the applicability of the model to the prediction of landslide, we analyzed the data from the Takoradani basin and Shimotani basin in the Kizu River Basin. The geology of these basins is weathered granite and these are covered by artificial forest of Japanese cedar. There are many small-scale surface landslides in these basins, the majority of them originated from the severe rainfall in 1959 accompanied by the Ise Bay Typhoon and the number of landslides has increased since then. The field investigation just after the typhoon revealed that the average slide area was about 100 m² and the depth of slip surface was less than about 1 m. The adopted

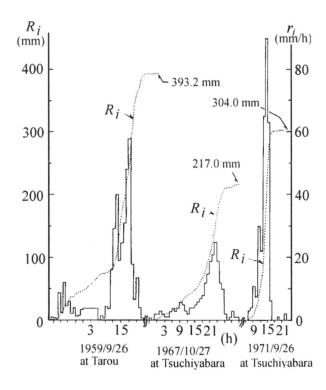

Figure 8.57 Previous conspicuous rainfalls in the Kizu River Basin.

parameter values other than those used in the calculation of the hydrograph were: $i_C = 0.6 \times 10^{-5}$ m/s, $c_A = 1{,}000$ kgf/m^2 ($= 9800$ Pa), $c_B = 400$ kgf/m^2 ($= 3920$ Pa), and $\tan \varphi_A = \tan \varphi_B = 0.7$. Very large c_A value was adopted to assure the landslide occurrence in the B layer.

Three kinds of rainfall data were examined as shown in Figure 8.57. Calculations were conducted by discretizing into the horizontal mesh of 12.5 m × 12.5 m and $\Delta t = 60$ s. However, if surface water flow appeared, the calculation became unstable unless $\Delta t = 1$ s or less was adopted. Because the surface flow and the pipe flow as assumed in this model did not affect the stability of the slope, for the sake of computing time saving, calculations of these flows were abbreviated.

Figure 8.58 shows the distributions of mesh whose safety factor for sliding reached less than 1 under the rainfall condition of the Ise Bay Typhoon (rainfall in 1959). In the same figure the distributions of landslide scars recognized in the aerial photograph taken in 1966 are also shown. Many landslides occur coinciding with the mesh whose SF is smaller than 1, but there are some landslides in the mesh satisfying $SF > 1$, and also there are many meshes satisfying $SF < 1$ without a landslide. There may be many causes for these ambiguities, anyway, landslides occur in some part of meshes whose smallest SF values attained during one rainfall event smaller or larger than 1.

The SF values of respective meshes at the time of maximum rainfall intensity in the event of the Ise Bay Typhoon were obtained and the relationship between those values

Figure 8.58 Existing landslides and the meshes experienced *SF* < 1 in 1959 rainfall in the Takoradani basin.

and the rates of landslide occurrence were obtained as shown in Table 8.6. The rate of landslide occurrence decreases with increasing *SF* (stability increases) as expected.

By comparing the aerial photographs taken in 1966 and 1970, it became clear that no conspicuous landslides occurred during that period. The severest rainfall in this period was the one in the October 1967, whose hyetograph is shown in the center graph of Figure 8.57. The calculation of *SF* values under this rainfall condition revealed there was no mesh satisfying *SF* < 1.

Figure 8.59 shows the distributions of landslides recognized in the aerial photograph taken in 1973 and meshes that attained *SF* < 1 in the occasion of rainfall in 1971 that is the most conspicuous rainfall between 1970 and 1973, whose hyetograph is given in Figure 8.57. The number of landslides is more than that in Figure 8.58, and

Table 8.6 Probability for landslide versus SF rank.

	Takoradani basin		Shimotani basin
SF	Ise Bay Typhoon	September 1971	Ise Bay Typhoon
<0.8	}0.030	}0.035	0.086
0.8–1.0			0.048
1.0–1.2	0.034	0.026	0.073
1.2–1.4	0.031	0.015	0.043
1.4–1.6	0.014	0.012	0.014
1.6–1.8	0.018	0.006	0.012
>1.8	0.000	0.002	0.000

Figure 8.59 Existing landslides and the meshes experienced SF < 1 in 1971 rainfall in the Takoradani basin.

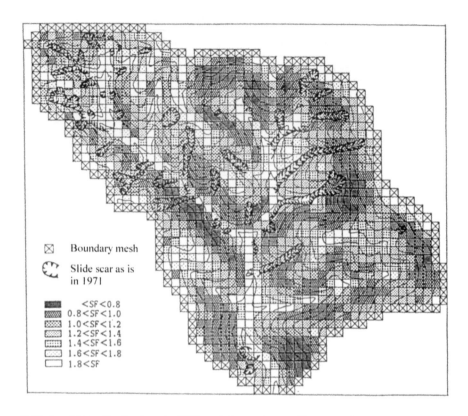

Figure 8.60 Landslide distributions in 1971 and the minimum SF's attained in the Ise Bay Typhoon in the Shimotani basin.

presumably it was increased by the rainfall in 1971. Because the probability of slide occurrence even in the mesh satisfying $SF < 1$ is small, if one considers that the location of a landslide is determined probabilistically, the chance that the slide locations in the two rainfall events overlap, would be small. Hence, we obtained the new landslide occurrence rates by the rainfall in 1971, omitting the meshes already slid by the rainfall of the Ise Bay Typhoon, as shown in Table 8.6. The resemblance of occurrence rates between the two rainfall events suggests that this consideration is reasonable.

A similar calculation was done in the Shimotani Basin under the rainfall condition of the Ise Bay Typhoon. Figure 8.60 shows the results. Here the distributions of meshes are demonstrated by scaling into seven levels the smallest SF values attained during the entire rainfall period in the respective meshes (These distributions are similar to those at the time of maximum rainfall intensity). In the same figure the distributions of landslide scars indicated in the recent topographic map are shown, which should include the landslides that occurred in the occasion of the Ise Bay Typhoon and the ones that occurred in 1971. The landslide occurrence rates using the indicated slides were calculated as shown in Table 8.6, whose values are about two times of those in Takoradani in each rainfall event. Because the existing landslide scars should be the results of two conspicuous rainfall events as mentioned, the landslide occurrence rates in an event may have been about the same values to those in Takoradani. This conclusion suggests that the relationship between the SF values and the landslide occurrence rates in Takoradani would have some universality at least in the Kizu River Basin.

The total expected landslide volume in a basin by a future rainfall event is obtained by the following formula:

$$\text{(Landslide volume)} = \sum \{(\text{landslide occurrence rate in a SF rank})$$
$$\times (\text{mesh number of that SF rank}) \times (\text{area of each mesh})$$
$$\times (\text{landslide depth})\}$$

where Σ means the summing up for all SF ranks.

As shown in Figure 8.59, the majority of landslides occurred halfway up the slope. If future landslides will occur in similar locations on the slope, based on the field survey, the expected depth of slides is about 70 cm. Application of this formula to the Takoradani Basin for the case of the Ise Bay Typhoon gives 7,656 m^3. The total number of meshes is 2,535, that gives the expected specific sediment production as 19,330 m^3/km^2. This value is within the range of estimated specific sediment yield in the Kizu River Basin shown in Figure 8.49. Similarly, if the same scale rainfall event as that in September 1971 occurs and the landslide occurrence rate in each SF rank is the same as that for the Ise Bay Typhoon, the landslide volume is calculated as 5,380 m^3, more or less similar to the actual landslide volume of 6,719 m^3.

A similar discussion was applied to the landslide and debris flow that occurred in the Choryuji district in Nagasaki for the case of the Nagasaki disasters in July 1982 (Takahashi and Nakagawa 1986).

The properties of the surface soil in the basin were assumed uniform and determined based on the field survey as follows: $D_A = 20/\cos\theta_{i,j}$ cm, $D_B = 50/\cos\theta_{i,j}$ cm, $i_A = k_A = 2 \times 10^{-4}$ m/s, $i_B = k_B = 3 \times 10^{-5}$ m/s, $i_C = 0.6 \times 10^{-5}$ m/s, $\lambda_A = 0.4$, $\lambda_B = 0.3$, $s_{0A} = s_{0B} = 0.3$, $s_A = s_B = 0.8$, $c_A = 1{,}000$ kgf/m^2 (=9800 Pa), $c_B = 600$ kgf/m^2 (5880 Pa), $\tan\varphi_A = \tan\varphi_B = 0.7$. The basin was discretized on the topographic map of 1/2500 into a square mesh of 12.5 m. The rainfall data is given in Figure 8.61. The calculation was accomplished in sixty second intervals before the surface water flow appeared and the time interval was changed to one second after the surface water flow appeared. The Manning's roughness coefficient of the ground surface was set to 0.05. No pipe flow was considered.

The calculated discharges of surface water flow Q, seepage flow in the A layer Q_A, and seepage flow in the B layer at the outlet of the basin are shown in Figure 8.61.

The filled in black meshes in Figure 8.62 are the one whose SF value became less than 1. The variations of SF in the B layer for the meshes marked (A), (B) and (C) in Figure 8.62 are shown in Figure 8.61. The time at which SF_B becomes less than 1 is a little later than the time of peak surface discharge and it coincides with the time zone of actual debris flow occurrence.

The dotted meshes in Figure 8.62 are the meshes determined by Equation (8.67) as to generate debris flow, and many those meshes coincide with the actual traces of debris flows. Herein, the representative diameter of bed material was set to 6 cm. The predicted debris flow occurrence time is a little earlier than the appearance of peak surface flow discharge. It may be a little too early but it gives approximately

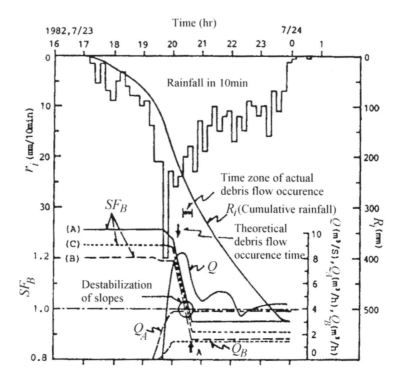

Figure 8.61 Rainfall data and the calculated time variations in discharges and stability of slopes in Choryuji district, Nagasaki Prefecture.

reasonable result. One possible reason for the earlier occurrence of debris flow in the prediction than in the actual occurrence would be attributable to the neglect of the stair-like bed topography that contributes to increase the threshold water discharge for the generation of debris flow, as was discussed in section 6.4.3.

Referring to the discussions on landslide in the Kizu River Basin and Choryuji basin, the start time of the landslide should approximately coincide with the time at which the expected number of landslide just surpasses 1. Then, on the whole meshes in the Shimotani basin, the rainfall shown in Figure 7.56 was given to count the SF variations of meshes. We picked out only the meshes whose SF became less than 1 scaling into the ranks $SF < 0.8$ and $0.8 < SF < 1$, and multiplied the landslide occurrence probabilities for the respective ranks to the numbers of picked out meshes belonging to respective ranks. We defined the instant at which the sum of the products surpasses 1 as the time of the start of the landslide in the basin, T_1. The landslide number was assumed to increase monotonously until it became the maximum that occur at the time when the total of the SF values of all the meshes becomes the minimum, T_2. T_1 and T_2 in Figure 8.56 were determined in this manner. The thus determined T_1 and T_2 seem to be reasonable because T_1 appears in the second wave of rainfall and T_2 appears just after the maximum rainfall intensity. The location of each landslide may be decided by randomly picking up a mesh whose SF is less than 1. Therefore, the

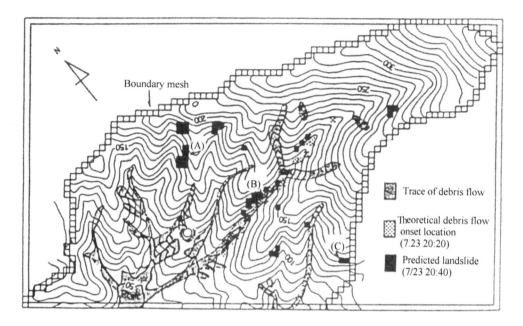

Figure 8.62 Predicted and actual landslides and debris flows in Choryuji.

location changes in every trial and, depending on the location of the landslide, the debris flow discharge and timing to reach the outlet of the basin will change. For the determination of designs for debris flow several trials are necessary.

The landslide probability in this method is small and it causes considerable variations in the location and time of landslide occurrence and in the resulting sediment yield to the outlet of the basin. The reason for the small probability is due to the great number of meshes where $SF < 1$. Considering this problem, Michiue and Fujita (1990) pointed out that the assumption of equal permeability in the vertical and lateral directions results in too much water storage in the soil layer and causes many meshes having an $SF < 1$. They adopted a lateral permeability in the saturated seepage flow 10 times larger than the vertical unsaturated permeability. The number of $SF < 1$ meshes decreased considerably and the landslide probability as well as the hitting ratio of prediction increased. The adoption of a larger permeability in the lateral direction corresponds to considering pipe flows in the A and B layers, and it would be reasonable. But, at the moment, there is no method to determine in situ permeability exactly.

8.6 DEBRIS FLOW PRONE RAVINES AND HAZARDOUS AREA

8.6.1 Debris flow prone ravine

The method to quantitatively predict the sediment transport phenomena under arbitrary rainfall conditions in arbitrary basins has already been explained. By the application of this method, we can judge whether the objective ravine is susceptible to debris flow occurrence. However, because there are too many ravines to be checked, to apply this simulation method to all the ravines from the start of assessment is not practical. Alternatively, at first, under some rough criteria, the possibility of debris flow occurrence should be judged and only the ravines judged to have a high potential for debris flow should be examined by detailed quantitative method to consider the concrete countermeasures against the potential debris flow.

For the debris flow occurrence due to bed erosion by surface water flow, as explained in section 3.1.1, the potential to give rise to a debris flow can be judged by whether Equation (8.67) is satisfied or not. Namely, if a ravine has a high chance of satisfying Equation (8.67), then it has a high potential of a debris flow caused by bed erosion. Then, in an particular ravine, knowing the representative diameter of deposit on bed from field survey, we obtain the minimum surface water flow discharge q_c which satisfies Equation (8.67) at a position upstream of which the channel gradient is steeper than 15°. On the other hand, by an appropriate flood runoff analysis, we can obtain the surface water runoff, q_o, at the position where the aforementioned q_c is obtained. Defining $X = q_o/q_c$, if $X \geq 1$, then there is some potential that a debris flow will arise.

A little more detailed discussion on X is given below. The surface water flow depth h_o is given by

$$h_o = \left(\frac{f}{8g \sin \theta}\right)^{1/3} \left(\frac{Q}{B}\right)^{2/3} \tag{8.71}$$

where f is the resistance coefficient of flow and it is approximately given on a very steep channel by $f = 1.12 \sin \theta$, Q is the surface water flow discharge, and B is the channel width. Therefore, the critical water flow discharge to give rise to a debris flow, Q_c, at the assessing position is, from Equation (8.67):

$$Q_c = \left(\frac{8 \sin \theta}{fk^3} gd_p^3 B^2\right)^{1/2} \tag{8.72}$$

The surface water flow discharge is obtained from the rational formula:

$$Q = \frac{1}{3.6} f_R r_e A_{15} \tag{8.73}$$

where f_R is the runoff coefficient, r_e is the effective rainfall within the arrival time of the flood in mm/hr, and A_{15} is the basin area at the assessing position in km². For the very strong rainfall as that which generates debris flow, f_R would be considered equal to 1. The arrival time of the flood is given by Kadoya and Fukushima (1976) as follows:

$$t_p = 290 A_{15}^{0.22} r_e^{-0.35} \tag{8.74}$$

Table 8.7 Risk indices and debris flow occurrence ratios (Shodo Island, July 1974).

Risk index X	Number of basin	Debris flow arising basins	Occurrence ratio
$X \geqq 12.1$	13	7	0.538
$12.1 > X \geqq 6.1$	23	8	0.348
$6.1 > X \geqq 3.0$	57	21	0.368
$3.0 > X \geqq 1.5$	85	37	0.435
$1.5 > X \geqq 0.76$	42	1	0.024
$0.76 > X$	16	0	0
Total	236	74	0.314

Table 8.8 Risk indices and debris flow occurrence ratios (Shodo Island, September 1976).

Risk index X	Number of basin	Debris flow arising basins	Occurrence ratio
$X \geqq 12.1$	10	8	0.800
$12.1 > X \geqq 6.1$	22	15	0.682
$6.1 > X \geqq 3.0$	60	24	0.400
$3.0 > X \geqq 1.5$	89	36	0.404
$1.5 > X \geqq 0.76$	42	12	0.286
$0.76 > X$	17	1	0.059
Total	240	96	0.400

The relationship between the rainfall intensity within the flood concentration time and that in 60 minutes, r_{e60}, is given by the following formula:

$$r_e = \frac{100}{t_p + 40} r_{e60} \tag{8.75}$$

If the stream channel width for the critical discharge, Q_c, and that for the flood runoff discharge, Q, at the assessing point are equal and given by the regime relationship, $B = 5Q^{1/2}$, X is given by Q/Q_c. The relationship between the thus obtained X and debris flow occurrence was examined for the two disasters that occurred on Shodo Island in July 1974 and September 1976. As for the representative diameter, a result of size distribution analysis in a sample obtained from a ravine in the 1974 case was adopted for all the ravines; i.e. 4 cm. Then, assuming $k = 0.7$ we obtain $Q_c = 0.33 \, \text{m}^3/\text{s}$.

Tables 8.7 and 8.8 are the results of applications. There are tendencies that the larger the X values the larger the debris flow occurrence ratios become, and furthermore, the occurrence ratio abruptly changes with approximately $X = 1$ as the boundary. However, because there should be large variations in the parameter values between basins as well as in a particular basin, the boundary value, $X = 1$, itself has not such a crucial meaning. For example, if the representative diameter of the bed

Table 8.9 Risk indices and debris flow occurrence ratios (Nagasaki, July 1982).

Risk index Y	Number of basin	Debris flow arising basins	Occurrence ratio
$Y > 3.0$	11	7	0.64
$3.0 > Y > 1.5$	14	7	0.5
$1.5 > Y > 1.0$	16	3	0.1
$1.0 > Y$	102	0	0
Total	143	17	0.12

material becomes two times larger, X becomes 1/8. This means that the difference in the characteristic parameter values of the basins plays a very important role.

The risk index can also be defined using the basin area at the assessing point. If this area is described as A_{15c}, it is given under certain effective rainfall intensity r_e as follows:

$$A_{15c} = \frac{3.6}{f_R r_e} \left(\frac{8 \sin \theta}{f \kappa^3} g d_p^3 B^2 \right)^{1/2} \tag{8.76}$$

Therefore, the other presentation of risk index Y is defined by $Y = A_{15}/A_{15c}$.

We obtained the relationship between Y and debris flow occurrence ratio taking the Nagasaki Rainfall Disaster in July 1982 as an example. Table 8.9 shows the results, in which $d_p = 10$ cm, $B = 5$ m, and $r_e = 150$ mm/hr are assumed based on field survey, then, $A_{15c} = 4.6$ ha. This table confirms that Y can be also a good index to judge the risk of a debris flow arising.

According to the guidelines of Japanese Ministry of Construction for the investigations on the debris flow prone ravines and hazardous area due to the flooding of debris flow (1999), a ravine that has more than 5 ha at the assessing point upstream of which the channel gradient is steeper than 15° and the bed seems to have abundant sediment, is judged to be a ravine prone to debris flow. This criterion is almost the same as, similar to the Nagasaki case previously mentioned, the debris flow prone ravine is the one whose representative bed particle diameter is 10 cm and debris flow is generated by rainfall stronger than 150 mm/hr. However, many debris flows actually occur even from weaker rainfall. If in the same criterion debris flow occurs by the rainfall of 75 mm/hr, it is equivalent to consider that the representative diameter is 6.3 cm or less. This representative particle size may be within the reasonable estimation. Hence, the Japanese criterion that considers the debris flow prone ravine is the one whose A_{15} is larger than 5 ha irrespective of the actual representative diameter would be reasonable.

The aforementioned criterion is valid for the bed erosion type debris flow. But, debris flow may be induced by landslide as well. The same guideline referred to above adds the possibility of debris flow occurrence due to landslide by taking the situations of slopes into account. Namely, the situations I (thickness of surface layer, existence of colluvial deposit, existence of weathered rock zone, existence of volcanic debris zone or pyroclastic deposit zone, existence of volcanic ash zone, existence of shattered zone, and existence of tertiary or quaternary zones), case history of considerably large (more than $1,000\,m^2$) landslide, existence of perpetual spring, and the situations II (existence of bare slope, deforested area, new crack, and sliding scarp) are considered as the factors that increase the risk of a debris flow induced by a landslide. Although those factors are difficult to assess quantitatively, the existence of colluvial deposit such as talus and landslide deposit, weathered rock zones such as decomposed granite and solfataric clay, volcanic debris or pyroclastic material zones such as weathered agglomerate, tuff breccia, volcanic ash and loam, fractured rock zones, bare slopes, deforested area, and perpetual spring should be noted as factors to give rise to a landslide. Especially, the appearance of a new crack should be considered as the signal of imminent danger for a slide.

8.6.2 Hazardous zone by debris flow

Hazard map based on case history

If there are case histories in the objective area, they should be the first thing taken into account. The delineation of the hazardous zone and the description of hazards such as the distributions of deposit thickness can be rather simply done if the disaster occurred recently and the traces and memories of residents are fresh. Even the disaster occurred in a long time ago and no memory of residents or records are available, if the characteristic deposit topography and geology of debris flow, such as debris flow lobes, natural levee-like ground swell, and inverse grading in the deposition layer is identified, a hazard map can be made based on these observations. It must be noted here that these observations were produced under the circumstances of the depositing area at that time. If the topographical condition, land utilization and situation of countermeasures are different from the case history, or the hazard map under the different scale phenomena is necessary, information from these old observations is not sufficient and can even mislead to make a false hazard map.

Figure 8.63 is the hazard map prepared about two years before the eruption of Mount Saint Helens in 1980. This was made based on the geological investigations of the case histories tracing back 4,500 years (Crandel and Mullinaux 1978).

The actually occurred phenomena and their affected area are shown in Figure 3.25. There are some discrepancies between the expected and actual phenomena: no lava flow occurred; the scale of pyroclastic flow was less than expected; hazards due to an unexpected debris avalanche and blast were severe and widely affected; no mud flow occurred in the Kalama and Lewis Rivers (mud and debris flows occurred in Pine Creek and the Muddy River, which are the tributaries of the Lewis River); mud flow in the Toutle River was larger than expected and it clogged the Columbia River; etc. However, it must be emphasized that, based on the hazard map, some preparations for the mitigation of potential hazards were implemented; off-limits into the red zone, evacuation from the red zone, draw down of water stage in the reservoirs along the Lewis River, and so on. It was lucky that the unexpected debris avalanche did not run down into the Lewis River. In the case of the eruption of Mount Fugen, Unzen Volcano in 1990, the pyroclastic flow was unexpected. But, some evidences of pyroclastic flow were discovered afterwards. The difference in the two cases should be taken as a lesson.

Figure 8.64 is the hazard map of the Nevado del Ruiz Volcano, Colombia that was prepared about one month before the eruption in 1985 by INGEOMINAS (This figure

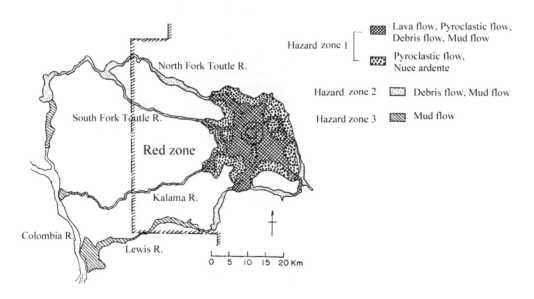

Figure 8.63 Hazard map prepared before eruption.

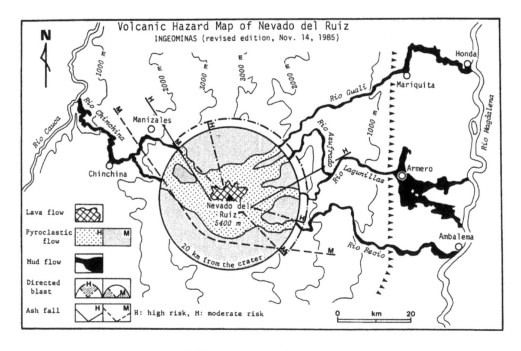

Figure 8.64 Hazard map of Nevado del Ruiz.

is a revised form published dated the day after the eruption, but the original one is not so much different from this). There are of course some discrepancies from the actual phenomena: no lava flow occurred; blast and pyroclastic flow occurred but their scales were smaller than expected; no mud flow occurred in the Recio River. However, in the Chinchina and the Guali Rivers mud flows occurred approximately the same as expected, and also in the Lagunillas River the hazardous zone was almost correctly predicted. Especially, it is noted that Armero City that was annihilated by mud flow was entirely inside the predicted hazardous zone.

Armero City was located on the alluvial fan of the Lagnillas River. The alluvial fan is in general a location that tends to suffer from even small-scale flood flow. Moreover, because the Lagnillas River brings together the two major rivers originating from the area closest to the crater of volcano, the Azufrado and the Lagnillas Rivers, the mud flow becomes very large to cause severe repeated disasters. Actually, in the case of the eruption in 1845, Armero lost more than 1,000 inhabitant, the traces of that disaster gave important data to make the hazard map.

Although the prepared hazard map was rather accurate as shown in Figure 8.65 and it was handed over to the local government one month before the tragedy, it was

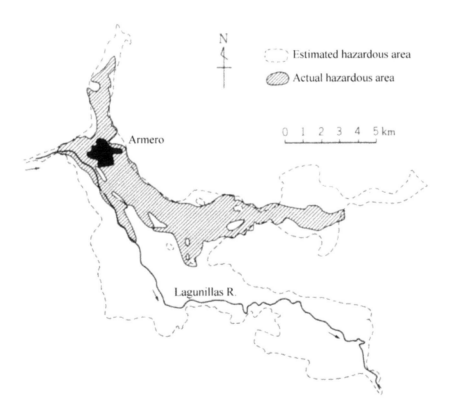

Figure 8.65 Actual deposition of mud flow in Armero area and the estimated hazardous area before the disaster.

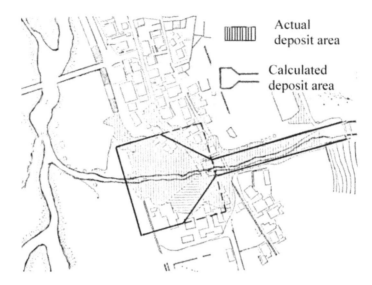

Actual
deposit area

Calculated
deposit area

Figure 8.66   Comparison of the calculated hazard zone and the actual deposit at Horadani.

not effectively used and the mud flow killed more than 20 thousands people. The arrangement of systems to make effective use of a hazard map is important in the mitigation of disasters.

Execution of hazard map by numerical simulation

Figure 8.66 compares the modeled deposit area with the actual deposit in the case of the Horadani debris flow discussed in section 7.2, in which the estimation of the deposit area was conducted by the simplified method depicted in Figure 5.14. Although there are some protrusions from the delineation especially along roads having a downward slope, the result seems to be acceptable.

The mathematical simulation model for the flooding and deposition of debris flow is discussed in detail in Chapters 5 and 7. This simulation model can be applied to prepare a hazard map associated with the appropriate selection of objective debris flow. Herein, only the calculated results of two different cases are explained.

Figure 8.67 shows the processes of flooding and deposition by the numerical simulation, in which the debris flow hydrograph at the mouth of the Horadani basin is the one calculated by assuming that two landslides at positions (1) and (2) simultaneously occurred at 7:50 a.m. (see Figure 8.52). The peak discharge in this case is about twice that in the case of only one landslide occurred at position (1), the actually occurred case. The total runoff sediment volume in this simulation is about 70,000 m^3, whereas in the actual case it was about 50,000 m^3.

At 8:05 a.m., the time immediately after the passage of peak discharge, plenty of sediment is deposited in the channel works and the debris flow overflows from both banks; in particular the overflow from the left bank is extensive. This situation

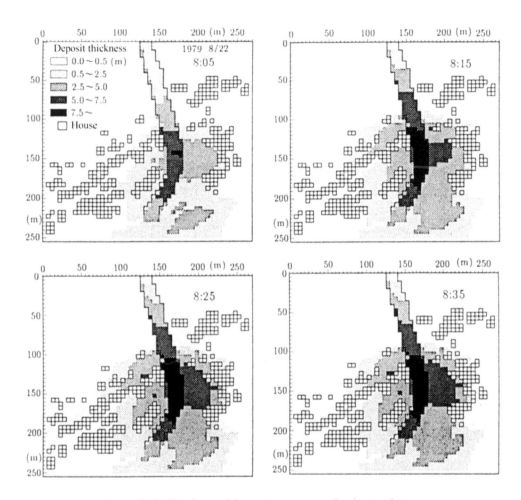

Figure 8.67 Flooding and depositing processes by the simulation.

is not much different from the actual case shown in Figure 7.24. At 8:15 the deposit thickness in the channel and the overflow from both banks has increased, and at 8:25 the deposit on the left bank side developed. Obvious changes cannot be seen between 8:25 and 8:35, which means that the flooding and deposition had almost ceased by 8:25. The comparison of the final deposit thicknesses and areas of deposition between Figure 8.67 and Figure 7.24 reveals that the deposit thickness and flooding area on the left bank in Figure 8.67 are larger than in Figure 7.24. However, the difference in the flooding area seems not much irrespective of the difference in the peak discharge and total runoff sediment volume. This is presumably due to the topographic characteristics on this fan.

Figure 8.68 shows the situations of deposit due to the mud flow associated with the eruption of Mount Tokachidake on 24 May 1926 and the distributions of population in 1985. We reproduced this mud flow by numerical simulation (Takahashi *et al.* 1990).

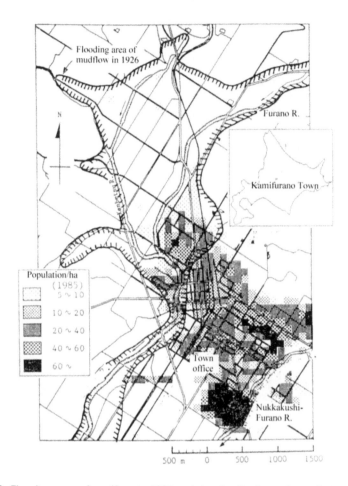

Figure 8.68 Flooding area of mudflow in 1926 and the distributions of population at present.

The mudflow hydrograph given to the Furano River is as shown in Figure 8.69 (Miyamoto *et al.* 1989), and the reproduction method is the same one used in reproducing the Armero mudflow explained in section 7.4. The parameter values were: $n_m = 0.04$, $d_L = 10$ cm, $C_{*DL} = 0.7$, $\sigma = 2.65$ g/cm^3, $\rho = 1.0$ g/cm^3, tan $\alpha_i = 0.75$, $\delta_d = 10^{-4}$ and $U_{TH} = 0.4$ m/s. The concentrations of fine and coarse particle fractions in the inflowing mudflow were 31.5% and 15.5%, respectively.

Figure 8.70 shows the distributions of the flooding area and thicknesses of deposit after 60 minutes from the beginning of flooding. The calculated results coincide with the actual situation in Figure 8.68. In this case, the central part of Furano Town is outside the thick deposit.

By the numerical simulation method, it is possible to make a hazard map for an imaginary phenomenon that has never occurred but that might be very hazardous if it did occur under a different scenario. For example, Figure 8.71 shows the results

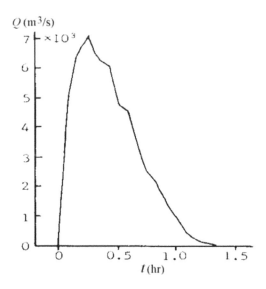

Figure 8.69 Mudflow hydrograph used in the calculation.

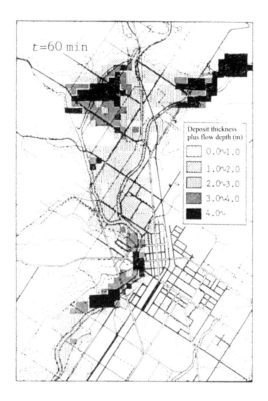

Figure 8.70 Flooding of mudflow from the Furano River.

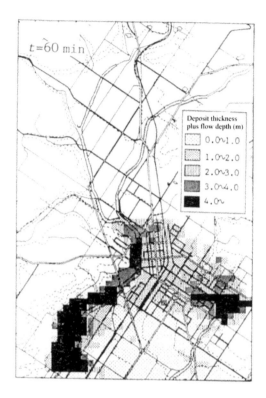

Figure 8.71   Flooding of mudflow from the Nukkakushi-Furano River.

of calculation for the case that the same scale mudflow occurred in the Nukkakushi-Furano River. In this case, almost the whole Kamifurano Town will suffer from severe flooding.

8.7   PREDICTION OF DEBRIS FLOW OCCURRENCE BY RAINFALL

If one uses the theoretical method for the prediction of the magnitude of debris flow mentioned in section 8.5.3, although there remains the problem of accuracy, one can predict the potential of a debris flow arising under an arbitrary rainfall condition. However, as there are so many debris flow prone ravines, it is not practical to apply this method in real time for respective basins for the object to issue warnings or advice evacuation. For this purpose, some method more direct and easier for the residents to feel the possible occurrence of debris flow is preferable.

The landslide's position and time prediction theory explained in section 8.5.3 makes clear the threshold rainfall condition for the occurrence of a landslide, in which it is defined by a kind of hyperbolic curve on a coordinate system whose two axes are the rainfall intensity and the cumulative rainfall amount, respectively, as shown

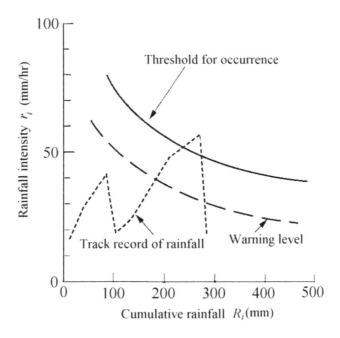

Figure 8.72 Conceptual diagram of the threshold for debris flow occurrence.

in Figure 8.54. Because the surface landslide easily transforms into debris flow, the threshold for a landslide is also the threshold for a debris flow.

Consider that the solid line in Figure 8.72 is the threshold line. The distance of this line from the origin of the coordinate system changes depending on the characteristics of the individual basin: high infiltration rate, cohesive soil, mild gradient on slope and in river channel, scarce sediment accumulation on the bed steeper than 15°, and small basin area for the channel steeper than 15° are the factors that set the threshold curve distant from the origin, and the reverse characteristics are the factors that set the threshold curve close to the origin.

In a particular district, collect the obvious rainfall records that induced or are supposed to have nearly induced debris flow and draw the track records of those rainfalls as the dotted line in Figure 8.72. This zigzag line having kinks in every one hour is sometimes called a 'snake line'. Mark the point on the snake line that induced debris flow at the position corresponding to the time of debris flow occurrence (the rainfall intensity at the time of debris flow occurrence and the cumulative rainfall until that time specify the position of the mark). If it is possible to plot some such marks and further draw the snake lines that nearly induced debris flow, as many as possible, one will be able to draw a threshold line observing the tendency of marks and other snake lines that did not give rise to debris flow. According to the guidelines of the Japanese Ministry of Construction (at present, Ministry of Land, Infrastructure and Transport), the threshold line is a straight line declining towards the right. But, as explained in section 8.5.3, it would be physically correct to represent it by a hyperbolic curve that does not intersect with the abscissa (R_i axis).

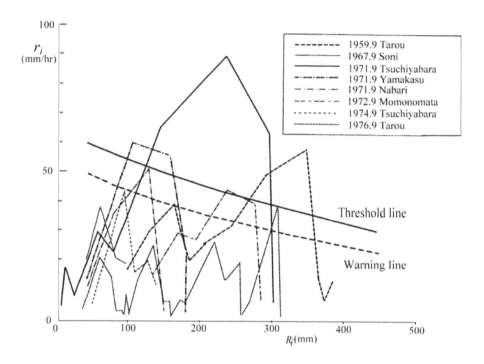

Figure 8.73 The threshold line for the Kizu River Basin.

Once a district or basin-specific threshold curve is obtained, one can issue a warning or advice evacuation when the real time plotting of the track record of rainfall seems to intersect with the threshold curve in the following period. If a reliable rain forecast is available, then a warning can be issued or evacuation advised with some lead time. But, under the present status of no reliable rain forecast, issuing the warning after the confirmation of snake line intersection with the threshold line might be too late for safety evacuation. One possibility for avoiding this problem is to draw a warning level line below the threshold line as depicted by the broken line in Figure 8.72. If the snake line passed this line, then the advice for immediate evacuation should be issued.

Figure 8.73 shows the threshold and warning lines which I made using the rainfall and debris flow records in the Kizu River basin. The records: 1959 Tarou; 1971 Tsuchiyabara, Yamakasu and Nabari; and 1972 Momonomata, are the rainfall that generated debris flow and the others did not generate debris flow.

Similarly, I obtained the threshold lines at several Japanese districts as shown in Figure 8.74 (Takahashi 1981a).

Figure 8.75 plots the relationship between the maximum hourly rainfall and the cumulative rainfall at the time when maximum intensity occurred in many heavy rainfalls in Japan. The closed circles indicate cases of obvious sediment hazards and the open circles indicate cases of no sediment hazards. Because the majority of landslides occur at the time of maximum rainfall intensity, the line approximately separating the

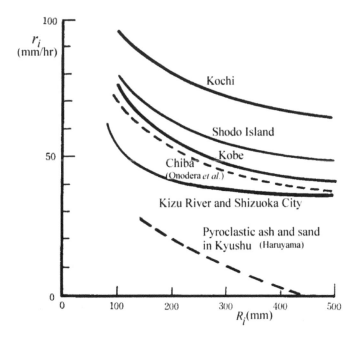

Figure 8.74   Several threshold lines in Japan.

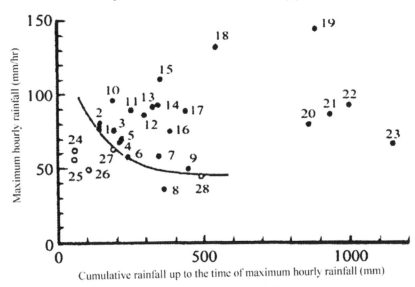

Figure 8.75   Relationship between the maximum hourly rainfall and the cumulative rainfall at the time of the maximum intensity 1. Kobe, 2. Minamiyamashiro, 3. Kure, 4. Saiko, 5. Arita R., 6. Uetsu, 7. Kizu R., 8. Tenryu R., 9. Hanshin, 10. Okuetsu, 12. Tanzawa, 13. Shodo Island, 14. Kochi, 15. Amakusa, 16. Moji, 17. Aso, 18. Kochi, 19. Isahaya, 20. Kurobe R., 21. Shodo Island, 22. Owase, 23. Omaru R., 24. Kobe, 25. Uetsu, 26. Kobe, 27. Shodo Island, 28. Shodo Island.

two cases would be the general threshold line for landslide occurrence in the weakest areas of Japan. The line resembles those threshold lines for the Kizu River basin and Shizuoka city in Figure 8.74. This presumably means that, except for peculiar soils such as Shirasu (pyroclastic ash and sand in Kyushu, see Figure 8.74), the threshold line for the Kizu River basin and Shizuoka city is applicable to most weak slope areas.

One difficult problem in determining the threshold line is the treatment of a discontinuance in rainfall; from when the snake line must be drawn as the beginning of a continuous rainfall and how long a discontinuation of rainfall separates the before and after continuous rainfalls as two independent rainfall events. Particularly, if the threshold line is the one that intersects with the abscissa, otherwise this problem is strictly answered, the snake line for a long intermittent rainfall will necessarily intersect with the threshold line even if the rainfall intensity is weak. One practical and actually adopted method to avoid this difficulty considers the half-life period of the effect of antecedent rainfall; the effect of antecedent rainfall to cause landslide diminishes with time regardless of whether the rainfall is continuing or disrupted. Then, a new problem arises, how the half-life period is determined, the physical basis of the half-life time is vague.

The discussion in section 8.5.3 clearly shows that the water content in the soil layer controls the stability of the slope. The 'tank model' which was developed in Japan to analyze flood runoff describes the relationship between the rainfall, the water content in the soil layer and the runoff discharge. This resembles the soil layer model introduced in section 8.5.3. Suzuki et al. (1979) applied this tank model to the discussion on the threshold of landslide occurrence. They used a three-storied tank to analyze flood runoff from a small watershed (17.6 ha) in Rokko Mountain and determined the appropriate parameter values of respective tanks. Then, entering the hyetograph of a rainfall that induced sediment disasters, they found that landslides occurred when the water storage in the first and second tank exceeded 35 mm and 50 mm, respectively. Moreover, they gave the rainfalls that induced sediment disasters respectively at Shodo Island and Kochi district to the same tank model, and found that the threshold to induce landslides at Shodo Island was the same as Rokko Mountain, whereas at Kochi the threshold water storages were 70 mm in the first tank and 90 mm in the second tank.

Michiue (1982) used the three-storied tank shown in Figure 7.16 (not the same as the one used by Suzuki et al.) that was used to represent the average flood runoff characteristics in granite areas in Japan (Ishihara and Kobatake 1978) in the case of sediment disasters at Kure. He claims that if the total water storage in the three tanks exceeds 70 mm cliff failures occur and if the storage in the first tank exceeds 80 mm or the runoff discharge exceeds 20 mm/hr, debris flow occurs. Further, Michiue and Hinokidani (1990) obtained the threshold to landslides in the granite area of the San-in district from the same tank was 125 mm in the total water storage in the three tanks. Figure 8.76 is one example of their analyses.

The prediction of landslide by the application of the tank model has a similar meaning as the prediction by the application of the threshold rainfall concept. Figure 8.77 is the three-storied tank obtained as the one suitable to analyze flood runoff from the Shimotani basin in the Kizu River Basin, in which the rainfall in every fifteen minutes is used as the input. There is no observed record of runoff in this basin for the severe rainstorm that generated debris flow or landslides, so that the coefficients of outflow

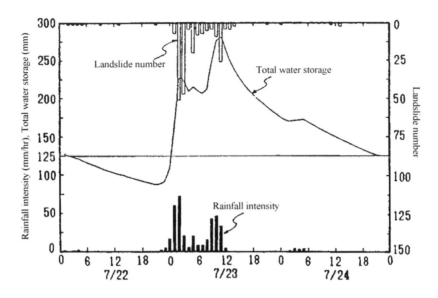

Figure 8.76 Application of tank model for the San-in rainfall disaster in 1983 (Hamada city).

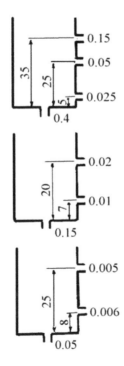

Figure 8.77 Tank model for Shimotani.

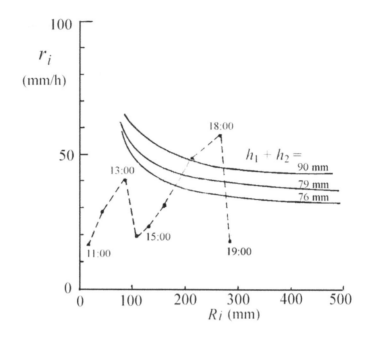

Figure 8.78 Water storage in the first and second tanks under constant intensity of rainfall.

from the holes in the base of the first tank and other holes in the second and third tanks are unreliable but herein they are assumed correct.

A rainfall of uniform intensity was continually given to this tank until the water storage in the first and second tanks became a prescribed value $(h_1 + h_2)$. Changing the rainfall intensities and $(h_1 + h_2)$ values, the relationships between the rainfall intensity and the cumulative rainfall amount were obtained for respective $(h_1 + h_2)$ values as shown in Figure 8.78. For this model tank, if the rainfall is interrupted for more than three hours, even the antecedent very strong rainfall having an intensity of 40 mm/hr and amounting 200 mm does not affect the value of $(h_1 + h_2)$. Whereas, a relatively weak rainfall of 10 mm/hr that continues without interruption shortly attains the pre-scribed $(h_1 + h_2)$ value. Taking these characteristics of the model tank into account, a series of rainfall stronger than 10 mm/hr without an interruption of more than three hours was considered as one unit rainfall for the case of the Ise Bay Typhoon and the snake line of that unit rainfall was drawn by the broken line in Figure 8.78. The times of breaking of snake line are attached near the respective points. According to the data, many landslides occurred from 17:00 to 19:00. Therefore, among the three lines of $(h_1 + h_2)$ in Figure 8.78, the one of $(h_1 + h_2) = 79$ mm seems appropriate to adopt as the threshold.

As mentioned above, landslides occurred at Rokko Mountain and Shodo Island when the total water storage of the first and second tanks reaches more than 85 mm, whereas at Shimotani landslides seem to occur when the storage reaches 79 mm. As shown in Figure 8.74 the threshold line for the Kizu River Basin is a little lower than that for Kobe (Rokko Mountain), so that one will agree with this difference in the water storage heights between the two thresholds. Of course, the model tanks are different and the time steps of input to respective model tanks are also different, and so some ambiguity is left in the discussion. There is no physical base to consider that the storage in the first and second tanks corresponds to the water storage in the A and B layers but it is intuitively understood that there would be a close relationship. Moreover, the resemblance in the shape and position of the line indicating $(h_1 + h_2) = 79$ mm in Figure 8.78 and the threshold line for the Kizu River basin in Figure 8.74 on the same $(r - R)$-plane is noteworthy. This suggests that the two methods have similar significance. The tank model method has an advantage in that it is not necessary to note the interruption of rainfall, but it also has disadvantages that the risk of landslide can only be known after calculation and it is rather difficult to appeal to the intuition of inhabitants.

The Meteorological Agency of Japan uses a three-storied tank model similar to the one shown in Figure 7.16 for the districts allover Japan and the total volume of water stored in these three tanks is defined by them as the 'Soil Water Index'. The agency announces the current state of sediment disaster generation risks based on these soil water indices. The standard values for announcing risks are different depending on the localities.

References

Allen, J.R.L., 1984, Bulk self-fluidization in sedimentary structures; their character and physical basis, Elsevier: 315–317.

Arai, M., 1991, *Fundamental study on the processes of flowage and deposition of muddy type debris flow*: Doctoral thesis, Kyoto Univ.: 114p (*in Japanese*).

Arai, M. and Takahashi, T., 1983, A method for measuring velocity profiles in mud flows: Proc. 20th Cong. IAHR, 3: 279–286.

Arai, M. and Takahashi, T., 1986, The mechanics of mud flow: Proc. JSCE, 375/II-6: 69–77 (*in Japanese*).

Arai, M. and Takahashi, T., 1988, Depositing process of mud flow on gentle slope bed plunging from steep slope: Proc. 6th Congress of APD, IAHR: 83–90.

Arai, M., Sawada, T. and Takahashi, T., 1998, An application of image analysis technique on the velocity of debris flow observation: Proc. 3rd Intern. Conf. on Hydroscience and Eng., IAHR: 1–9.

Arai, M., Liu, S. and Takahashi, T., 2000, A discussion on mechanism of viscous debris flow via surface velocity analysis: Annual J. Hydr. Eng., 44: 693–698 (*in Japanese*).

Armanini, A. and Gregoretti, C., 2000, Triggering of debris-flow by overland flow; A comparison between theoretical and experimental results: Proc. 2nd Intern. Conf. on Debris-Flow Hazards Mitigation: 117–124.

Armanini, A., Dalri, C., Fraccallo, L., Larcher, M. and Zorzin, E., 2003, Experimental analysis of the general features of uniform mud-flow: Proc. 3rd Intern. Conf. on Debris-Flow Hazards Mitigation: 423–434.

Ashida, K., 1987, Case study on the choking of river channel due to landslide-Totsugawa daisasters in 1889-: *in Prediction and countermeasures of secondary disasters*, 2, Zennkoku Bousai Kyoukai: 37–45 (*in Japanese*).

Ashida, K. and Michiue, M., 1972, Study on hydraulic resistance and bed-load transport rate in alluvial streams: Proc. JSCE, 206: 59–69 (*in Japanese*).

Ashida, K. and Takahashi, T., 1980, Study on debris flow control-Hydraulic function of grid-type open dam-: Annuals, DPRI, 23B-2: 433–441 (*in Japanese*).

Ashida, K., Takahashi, T. and Sawada, T., 1976, Sediment yield and transport on a mountainous small watershed: Bull. DPRI, Kyoto Univ., 26(3): 119–144.

Ashida, K., Takahashi, T., Sawada, T., Egashira, S. and Sawai, K., 1977, On sediment disasters in Shodo island: Research Report of Damages done by Typhoon No.17 in 1976: 109–115 (*in Japanese*).

Ashida, K., Takahashi, T. and Mizuyama, T., 1977a, Study on the initiation of motion of sand mixtures in a steep slope channel: J. JSECE, 29(4): 6–13 (*in Japanese*).

Ashida, K., Takahashi, T. and Mizuyama, T., 1978, Study on bed load equations for mountain streams: J. JSECE, 30(4): 9–17 (*in Japanese*).

Ashida, K., Takahashi, T. and Arai, M., 1981, Study on debris flow control; debris flow in bends of rectangular section: Annuals, DPRI, 24B-2: 251–263 (*in Japanese*).

Ashida, K., Egashira, S., Ando, N., 1984a, Generation and geometric features of step-pool bed forms: Annuals, DPRI, 27B-2: 341–353 (*in Japanese*).

Ashida, K., Egashira, S., Kamiya, H. and Sasaki, H., 1985, The friction law and moving velocity of a soil block on slopes: Annuals, DPRI, 28B-2: 297–307 (*in Japanese*).

Ashida, K., Takahashi, T. and Sawada, T., 1986, Runoff process, sediment yield and transport in a mountain watershed (15): Annuals, DPRI, 29B-2: 291–307 (in Japanese).

Awal, R., Nakagawa, H., Kawaike, K., Baba, Y. and Zhang, H., 2008, Prediction of flood /debris flow hydrograph due to landslide dam failure by overtopping and sliding, Annuals, D.P.R.I, 51: 603–611.

Awal, R., Nakagawa, H., Kawaike, K., Baba, Y. and Zhang, H.,2009, Three dimensional transient seepage and slope stability analysis of landslide dam, Annuals, D.P.R.I., 52: 689–696.

Bagnold, R.A., 1954, Experiments on a gravity-free dispersion of large solid spheres in a Newtonian fluid under shear: Proc. R. Soc. London, A 225: 49–63.

Bagnold, R.A., 1966, The shearing and dilatation of dry sand and 'singing' mechanism: Proc. Roy. Soc. London, A295: 219–232.

Bagnold, R.A., 1968, Deposition in the process of hydraulic transport: Sedimentology, 10: 45–56.

Bakhtiary, A.Y., 1999, *Computational mechanics of bed-load transport at high bottom shear*: Doctoral dissertation submitted to Kyoto Univ.: 1–94.

Bathurst, J.C., 2002, Physically-based erosion and sediment yield modeling: the SHETRAN concept, Summer, W and Walling D.E. (ed.) Modelling erosion, sediment transport and sediment yield, IHP-VI Technical Document in Hydrology, No.60, UNESCO, Paris: 47–67.

Bathurst, J.C., Graf, W.H. and Cao, H.H., 1985: Bed load discharge equations for steep mountain rivers: Intern. Workshop on Problems of Sediment Transport in Gravel Bed Rivers, Colorado State Univ.: 26p.

Benda, L.E. and Cundy, T.W., 1990, Predicting deposition of debris flows in mountain channels: Canadian Geotech. J., 27: 409–417.

Blackwelder, E., 1928, Mudflow as a geologic agent in semiarid mountains: Bull. Geol. Soc. Am., 39: 465–484.

Campbell, C.S., 1990, Rapid granular flows: Annual Rev. of Fluid Mech,, 22: 57–92.

Campbell, C.S., 2001, Granular flows in the elastic limit: Spec.Publ. Int. Ass. Sediment, 31: 83–89.

Campbell, C.S. and Brennenn, C.E., 1985, Computer simulation of granular shear flows: J. Fluid Mech., 203: 449–473.

Campbell, C.S., Cleary, P.W. and Hopkins, M., 1995, Large scale landslide simulations; global deformation, velocities and basal friction: J. Geophys. Res., 100: 8267–8273.

Chen, C., 1987, Comprehensive review of debris flow modeling concept in Japan: Rev. Eng. Geol., 7: 13–29.

Chen, C., 1988, General solutions for viscoplastic debris flow: J. Hydraul. Eng., 114: 237–258.

Chigira, M., 2001, Large scale landslide: *in Handbook of disaster reduction science* (DPRI ed.), Asakura Shoten, Tokyo: 190–191 (*in Japanese*).

Chow, V., 1959, *Open-channel hydraulics*: McGraw-Hill.

Costa, J.E. and Schuster, R.L., 1988, The formation and failure of natural dams: Geol. Soc. Am. Bull., 100: 1054–1068.

Coussot, P. and Meunier, M, 1996, Recognition, classification and mechanical description of debris flows, Earth-Science Reviews, 40: 209–227.

Coussot, P., Laigle, D., Arattano, M., Deganutti, A. and Marchi, L., 1998, Direct determination of rheological characteristics of debris flow: J. Hydraul. Eng., 124(8): 865–868.

Crandell, D.R. and Mullinaux, D.R., 1978, Potential hazards from future eruptions of Mount St. Helens Volcano, Washington: Geological Survey Bull. 1383-C: C1–C26.

Crosta, G.B., Cucchiaro, S. and Frattini, P., 2003, Validation of semi-empirical relatiobships for the definition of debris-flow brhavior in granular materials: in Debris-Flow Hazards Mitigation, Rickenmann & Chen (eds), Millpress, Rotterdam: 821–831.

Daido, A., 1971, On the occurrence of mud-debris flow: Bull., DPRI, 21(2): 109–135.

Davies, T.R.H., 1982, Spreading of rock avalanche debris by mechanical fluidization: Rock Mechanics, 15: 9–24.

Denlinger, R.P. and Iverson, R.M., 2001, Flow of variably fluidized granular masses across three-dimensional terrain 2. Numerical predictions and experimental tests: J. Geophys. Res., 106 B1: 553–566.

Egashira, S., Miyamoto, K. and Itoh, T., 1997, Constitutive equations of debris flow and their applicability: 1st Intern. Conf. on Debris-Flow Hazards Mitigation, ASCE: 340–349.

Erismann, T.H., 1979, Mechanisms of large landslides, Rock Mech., 12: 5–46.

Fleming, R.W., Ellen, S.D. and Algus, M.A., 1989, Transformation of dilative and contractive landslide debris into debris flows. An example from Marin County, California: Engineering Geology, 27: 201–223.

Foda, M.A., 1993, Landslides riding on basal pressure waves, Continuum Mech. and Thermodynamics.

Fukuda, K., Matsumoto, K., Taniguchi, K., Tsubakishita, Y., Shimizu, M. and Mizuyama, T., 2002, Research of sediment control by slit sabo dam during floods-Case study on the Babadani branch of the Kurobe River-: J. JSECE, 54(6): 25–34 (in Japanese).

Furuya, T., 1980, Landslides and landforms: in Landslides, slope failures and debris flows (Takei, A. ed.), Kajima Shuppan, Tokyo: 192–230 (in Japanese).

Garde, R.J. and Ranga Raju, K.G., 1963, Regime criteria for alluvial streams: J. Hydraulics Div., ASCE, 89(6): 153–164.

Gidaspow, D., 1994, Multiphase flow and fluidization: Academic Press, London: 1–467.

Gilbert, G.K., 1914, The transportation of debris by running water: Prof. Paper, 86, USGS.

Goguel, J., 1978, Scale-dependent rockslides mechanisms, with emphasis on the role of pore fluid vaporization, in 'Rockslides and Avalanches, 1, Natural Phenomena', Elsevier, Amsterdam: 693–705.

Gotoh, H. and Sakai, T., 1997, Numerical simulation of sheet flow as granular materials: J. Waterway, Port, Coastal and Ocean Eng., ASCE, 123(6): 329–336.

Graf, W.H. and Acaroglu, E.R., 1968, Sediment transport in conveyance systems (Part 1) A physical model for sediment transport in conveyance systems: Bull. IAHS, XIII, No.2, 20–39.

Habib, P., 1975, Production of gaseous pore pressure during rock slides, Rock Mechanics, 7: 193–197.

Hampton, M.A., 1975, Competence of fine-grained debris flows: J. Sedimentary Petrology, 45(4): 834–844.

Hanes, D.M. and Inmann, D.L., 1985, Observations of rapidly flowing granular-fluid materials: J. Fluid Mech., 150: 357–380.

Harada, E., 2001, Computational mechanics of the size segregation in bed load and drift sand on heterogeneous bed materials: Doctoral dissertation submitted to Kyoto Univ.: 1–126 (in Japanese).

Hashimoto, H. and Tsubaki, T., 1983, Reverse grading in debris flow: Proc. JSCE, 336: 75–84 (in Japanese).

Hashimoto, H., Tsubaki, T. and Nakayama, H., 1978, Experimental consideration on debris flow: Proc. 33rd National Meeting of JSCE: 574–575 (in Japanese).

Heim, A., 1882, Der Bergsturz von Elm, Deutsch: Geol. Gesell. Zeitschr., 34: 74–115.

Hirano, M. and Hashimoto, H., 1993, The field measurement of debris flow in the Mizunashi River and the characteristics of flow: in Investigations of debris flows and pyroclastic flows *at Unzen* (Hirano.M. ed.), Research Report for Grant-in-Aid for Scientific Research: 9–22 (*in Japanese*).

Hirano, M., Hashimoto, H. and Kohno, T., 1994, Field measurements of debris flows in the MIzunashi River in 1993: *in Investigations of debris flows and pyroclastic flows at Unzen* (Hirano, M. ed.), Research Report for Grant-in-Aid for Scientific Research: 13–24 (*in Japanese*).

Hirano, M., Hashimoto, H., Fukutomi, K., Taguma, K. and Pallu, M.S., 1992, Non-dimensional parameters governing hyper-concentrated flow in an open channel: Proc. Hydraul Eng., 36: 221–226 (*in Japanese*).

Hoshino, M., Onozuka, R., Asai, K., Inazawa, Y. and Hisamatsu, F., 1998, Landslide disaster at Sumikawa, Hachimantai in May, 1997 (2nd report) – Behaviors of landslide, debris avalanche, and debris flow and characteristics of topography-: Report of Geographical Survey Institute, Japan, 90: 50–71 (in Japanese).

Howard, K.E., 1973, Avalanche mode of motion; implication from lunar examples, Science, 180: 1052–1055.

Hsü, K.J., 1975, Catastrophic debris stream (Sturzströms) generated by rockfalls: Geol. Soc. Am. Bull., 86: 129–140.

Hungr, O. and Morgenstern, N.R., 1984, High velocity ring shear tests on sand, Geotechnique, 34-3: 415–421.

Hungr, O., Morgan, G.C. and Kellerhals, R., 1984, Quantitative analysis of debris flow torrent hazards for design of remedial measures: J. Can. Geotech., 21: 663–677.

Hutchinson, J.H., 1988, Morphological and geotechnical parameters of landslides in relation to geology and hydrology: Proc. 5th Intern. Symp. On Landslide: 3–35.

Imaizumi, F., Tsuchiya, S., Ohsaka, O., 2003, Flow behavior of debris flows in the upper stream on mountainous debris torrent, J. JSECE, 56(2): 14–22 (*in Japanese*).

INGEOMINAS, 1985, Mapa del casquet de hielo del Volcan Nevado del Ruiz: Informe preminar, Memoria explicative: 1–7.

Ishihara, Y. and Kobatake, S., 1978, Study on synthetic flood hydrograph, Annuals, DPRI, 21B2: 153–172 (*in Japanese*).

Ishikawa, Y., 1999, Morphological and geological features of debris flows caused by earthquakes: J. JSECE, 51 (5): 35–42 (*in Japanese*).

Iverson, R.M., 1997, The physics of debris flows: Reviews Geophysics, 35(3): 245–296.

Iverson, R.M. and Denlinger, R.P., 1987, The physics of debris flows-a conceptual assessment: Proc. Symp. Erosion, Sediment. Pacific. Rim, IAHS Publ., 165: 155–165.

Iverson, R.M. and Denlinger, R.P., 2001, Flow of variably fluidized granular masses across three-dimensional terrain 1. Coulomb mixture theory: J. Geophys. Res., 106 B1: 537–552.

Iverson, R.M., Reid, M.E. and LaHusen, R.G., 1997, Debris flow mobilization from landslide: Ann. Rev. Earth Planet. Sci., 25: 85–138.

Iverson, R.M., Reid, M.E., Iverson, N.R., LaHusen, R.G., Logan, M., Mann, J.E. and Brien, D.L., 2000, Acute sensitivity of landslide rates to initial soil porosity: Science, 290: 513–516.

Iwagaki, Y., 1955, *Study on the mechanics of soil erosion*: Doctoral Thesis, Kyoto Univ.: 215p (*in Japanese*).

Japan Society of Erosion Control Engineering, 2000, On the method to delineate the hazardous area due to sediment runoff; a proposal, 14p (*in Japanese*).

Jenkins, J.T. and Savage, S.B., 1983, A theory for the rapid flow of identical, smooth, nearly elastic particles: J. Fluid Mech., 130: 187–202.

Johnson, A.M., 1970, *Physical processes in geology*: Freeman, Cooper, San Francisco: 1–577.

Johnson, P.C. and Jackson, R., 1987, Frictional-collisional constitutive relations for granular materials with application to plane shearing: J. Fluid Mech., 176: 67–93.

Juien, P.Y. and Lan, Y., 1991, Rheology of hyperconcentrations: J. Hydraul. Eng., 117: 346–353.

Kadoya, M. and Fukushima, A., 1976, Concentration time for floods in small or medium river basins: Annuals, DPRI, 19B2: 143-152 (*in Japanese*).

Kan, Z. (ed.), 1996: *Debris flow hazards and their control in China*: Science Press: 1–118 (*Chinese with English translation*).

Kan, Z., 1994, Observation and study of debris flow patterns in Jiangjia ravine: *in Japan-China joint research on the prevention from debris flow hazards* (Takahashi, T. ed.), Reserch report of the grant-in-aid for scientific research, no.03044085: 30–41.

Kaneko, T. and kamata, K., 1992, Disscussion of the arrival distances of pyroclastic flows at Unzen in 1991 by Energy line/cone concept: Kazan, 37(1): 35–46 (*in Japanese*).

Katsui, Y. (Organizer), 1986, The 1985 eruption of Nevado del Ruiz Volcano, Colombia, and associated mudflow disaster: Report of Natural Disaster Scientific Research, B-60-7: 1–102 (*in Japanese*).

Kent, P.E., 1966, The transport mechanism in catastrophic rock falls, J. Geology, 74: 79–83.

Kitou, K., Hirano, M. and Hashimoto, H., 1993, Characteristics of dry granular flow in steep slope channel: Ann. J. Hydraul. Eng., JSCE, 37: 617–622 (*in Japanese*).

Knapp, R.T., 1951, Design of channel curves fro supercritical flow: Trans. ASCE, 116, Paper 2435: 296–325.

Kobayashi, S., 1995, *The occurrence characteristics and disasters of slush-flow*: Res. Report of Grant-in-Aid for Scientific Res.: 1–92 (*in Japanese*).

Krieger, J.M., 1972, Rheology of monodisperse lattices: Adv. Coll. Int. Sci., 3: 111–136.

Legros, F., 2002, The mobility of long-runout landslides, Engineering Geology, 63: 310–331.

Lenau, C.W., 1979, Super critical flow in bends of trapezoidal section: J. Engineering Mech. Div., ASCE, 105(EM1): 43–54.

Kobayashi, Y., 1994, Effect of basal guided waves on landslides, PAGEOPH, 142No.2: 329–346.

Lun, C.K., Savage, S.B., Jeffrey, D.J. and Chepurniy, N., 1984, Kinetic theories for granular flow; inelastic particles in Couette flow and slightly inelastic particles in a general flow field: J. Fluid Mech., 140: 223–256.

Major, J.J. and Pierson, T.C., 1992, Debris flow rheology; experimental analysis of fine-gained slurries: Water Resources Res., 28(3): 841–857.

Major, J.J. and Iverson, R.M., 1999, Debris-flow deposition: effect of pore-fluid pressure and friction concentrated at flow margins: Geol. Soc. Am. Bulletin, 111(10): 1424–1434.

Masuda, S., Mizuyama, T., Fujita, M., Abe, H., Oda, A. and Otsuki, H., 2002, Fundamental study about sediment runoff control by a series of slit sabo dams: J.JSECE, 54(6): 39–41 (*in Japanese*).

Matsumura, K., Takahama, J., Shima, D., Michiue, M., Miyamoto, K., Kitae, M., 1997, Experimental investigation on dry granular flow: Lecture Meeting of JSECE: 180–181 (*in Japanese*).

Melosh, H.J., 1980, Acoustic fluidization-a new geologic process?, J. Geophys. Res., 84: 7513–7520.

Michiue, M., 1982, Tank model method: *in A study on forecasting and mitigation of secere rain storm damage* (Takasao, T. representative researcher): Report of Natural Disaster Scientific Research: 63–67 (*in Japanese*).

Michiue, M. and Fujita, M., 1990, Prediction method by the stability analysis of infinitely long slope: *in Investigation on the prediction of landslide due to severe rainfall* (Michiue, M. representative researcher): Report of Natural Disaster Scientific Research: 75–85 (*in Japanese*).

Michiue, M. and Hinokidani, O., 1990, Prediction of the time of landslide occurrence by tank model: *in Investigation on the prediction of landslide due to severe rainfall* (Michiue, M. representative researcher): Report of Natural Disaster Scientific Research: 94–97 (*in Japanese*).

Middleton, G.V., 1970, Experimental studies related to problems of flysch sedimentation: *in Flysch sedimentology in North America* (Lajoie, J. ed.): Geol.Assoc. Can. Spec. Pap. 7: 253–272.

Miyamoto, K., 1985, *Study on the grain flows in Newtonian fluid*: Doctoral thesis, Ritumeikan Univ. (*in Japanese*).

Miyamoto, K., Suzuki, H., Yamashita, S. and Mizuyama, T., 1989, Reproduction of 1926 volcanic mudflow at Tokachidake: Annual J. Hydraulic Engineering, JSCE, 33: 361–366 (*in Japanese*).

Mizuhara, K., 1990, Relationships between the runoff sediment volume of debris flows and the possible relating factors: *in Predictions of debris flow occurrence and magnitude* (Kataoka, J. representative investigator): Report of Natural Disaster Scientific Research: 48–53 (*in Japanese*).

Mizuno, H., Mizuyama, T., Minami, T. and Kuraoka, S., 2000, Analysis of simulating debris flow captured by permeable type dam using distinct element method: J. JSECE, 52(6): 4–11 (*in Japanese*).

Mizuno, H., Sugiura, N., Terada, H., Uchida, T., Haramaki, T., Sokabe, M., Sakurai, W., Nishimoto, H., Osanai, N., Takezawa, N. and Doi, Y., 2003, The debris flow disasters caused by localized rainfall of seasonal rain front in Kyushu region in July, 2003 (prompt report): J. JSECE, 56(3): 36–43 (*in Japanese*).

Mizuyama, T., 1977, Study on the bed load transport in mountain rivers: Doctoral dissertation, Kyoto Univ. (*in Japanese*).

Mizuyama, T., 1980, Sediment transport rate in the transition region between debris flow and bed load transport: J. JSECE, 33(1): 1–6 (*in Japanese*).

Mizuyama, T., Abe, S., Yajima, S. and Ido, K., 1990, Application of a two-dimensional river bed routing method to slit sabo dams: J. JSECE, 42(5): 21–28 (*in Japanese*).

Mizuyama, T., Kobashi, S. and Mizuno, H., 1995, Control of passing sediment with grid-type dams, J. JSECE, 47(5): 8–13 (*in Japanese*).

Mizuyama, T., Mori, T., Sakaguti, T., Inoue, K., 2011, Japanese natural dams and the countermeasures, Kokonnshoin: p186 (*in Japanese*).

Morgan, R.P.C., Quinton, J.N., Smith, R.E., Govers, G., Poesen, J.W.A., Auerswald, K., Chisei, G., Torri, D. and Styczen, M.E., 1998, The European Soil Erosion Model (EUROSEM); a dynamic approach for predicting sediment transport from fields and small catchments, Earth Surface Processes and Landforms, 23: 527–544.

Mugiono, R., 1980, *A short description of the telemetric system installed at the Kelut Volcano*: Gadja Mada University Press: 1–39.

Muramoto, Y., Uno, T. and Takahashi, T., 1986, Investigation of the collapse of the tailings dam at Stava in the northern Italy: Annuals, DPRI, 29A: 19–52 (*in Japanese*).

Murano, Y., 1965, On the debris flow of Mt. Tokachi: J. JSECE, 59: 14–23.

Nakagawa, H., Takahashi, T., Sawada, T. and Satofuka, Y., 1996, Design hydrograph and evacuation planning for debris flow: Annuals DPRI, 39B-2: 347–371 (*in Japanese*).

Nakagawa, H., Takahashi, T., Satofuka, Y. and Kawaike, K., 2001, Sediment disasters caused by the heavy rainfall in the Camuri Grande River basin, Venezuela, 1999-Reproduction of sediment runoff, flooding and deposition and evaluation of effectiveness of the sabo works by means of numerical simulation-: Annuals DPRI, 44B-2: 207–228 (*in Japanese*).

Nishii, A., 2009, The role of landslides in controlling the distribution of linear depressions on granitic mountains in the Hida Range, J. of geography, 118(2): 233–244 (in Japanese)

O'Brien, J.S. and Julien, P.Y., 1987, Laboratory analysis of mudflow properties: J. Hydraul. Eng., 114(8): 877–887.

Ohsumi Work Office, 1988. *Debris flow at Sakurajima*: Ohsumi Work Office, Ministry of Construction: 1–64.

Ohyagi, N., 1985, Definition and classification of sediment hazards: *in Prediction and countermeasures of sediment hazards* (Japanese Soc.Soil Mech. Foundation Eng. ed.), Japanese Soc.Soil Mech. Foundation Eng., Tokyo: 5–15 (*in Japanese*).

Okubo, S., Mizuyama, T., Kaba, M. and Ido, K., 1997, Sediment control by a series of slit sabo dams: J. JSECE, 50(2): 14–19 (*in Japanese*).

Okuda, S., Suwa, H., Okunishi, K., Nakano,K. and Yokoyama, K., 1977, Synthetic observation on debris flow, Part 3. Observation at valley Kamikamihorizawa of Mt. Yakedake in 1976: Annuals DPRI, 20B-1: 237–263 (*in Japanese*).

Okuda, S., Suwa, H., Okunishi, K., Yokoyama, K., Nakano,K. and Ogawa, K., 1978, Synthetic observation on debris flow, Part 4. Observation at valley Kamikamihorizawa of Mt. Yakedake in 1977: Annuals DPRI, 21B-1: 277–296 (*in Japanese*).

Okuda, S., Suwa, H., Okunishi, K., Yokoyama, K., Ogawa, K. and Hamana, H., 1979, Synthetic observation on debris flow, Part 5. Observation at valley Kamikamihorizawa of Mt. Yakedake in 1978: Annuals DPRI, 22B-1: 157–204 (*in Japanese*).

Okuda, S., Suwa, H., Okunishi, K., Yokoyama, K., Ogawa, K., Hamana, H. and Tanaka, S., 1980, Synthetic observation on debris flow, Part 6. Observation at valley Kamikamihorizawa of Mt. Yakedake in 1979: Annuals DPRI, 23B-1: 357–394 (*in Japanese*).

Okuda, S., Suwa, H., Okunishi, K., Yokoyama, K. and Ogawa, K., 1981, Synthetic observation on debris flow, Part 7. Observation at valley Kamikamihorizawa of Mt. Yakedake in 1980: Annuals DPRI, 23B-1: 357–394 (*in Japanese*).

Okuda, S., Okunishi, K., Suwa, H., Yokoyama, K. and Yoshioka, R., 1985. Restoration of motion of debris avalanche at Mt. Ontake in 1984 and some discussions on its moving state, Annuals DPRI, 28B-1: 491–504 (*in Japanese*).

Okura, Y., Ochiai, H. and Sammori, T., 2002, Flow failure generation caused by monotonic liquefaction: Proc. Intern. Symp. Landslide Risk Mitigation and Protection of Cultural and Natural Heritage, Kyoto: 155–172.

Peng, M. and Zhang, L.M., 2012, Breaching parameters of landslide dams, Landslide, 9 No.1:13–31.

Phillips, R.J., Armstrong, R.C., Brown, R.A., Graham, A.L. and Abbott, J.R., 1992, A constitutive equation for concentrated suspensions that accounts for shear-induced particle migration: Phys. Fluids A4(1): 30–40.

Pierson, T.C. and Scott, K.M., 1983, Downstream dilution of a lahar: Transition from debris flow to hyperconcentrated streamflow: Water Res. Res., 21(10): 1511–1524.

Pierson, T.C. and Costa, J.E., 1987, A rheologic classification of subaerial sediment-water flows: *in Debris flows/avalanches: process, recognition, and mitigation* (Costa, J.E. and Wieczorek, G.F. eds.), Rev. Eng. Geol., 7, Geolo. Soc. Am: 1–12.

Pilotti, M. and Bacchi, B., 1997, Distributed evaluation of the contribution of soil erosion to the sediment yield from a watershed, Earth Surface Processes and Landforms, 22: 1239–1251.

Rodine, J.D. and Johnson, A.M., 1976, The ability of debris, heavily freighted with coarse clastic materials, to flow on gentle slopes: Sedimentology, 23: 213–234.

Sabo Work Office, Jinzu River System, Ministry of Construction and Tiiki Kaihatsu Consultant, Co., Ltd., 1979, *Survey report of the damaged area and the landslides in the basin due to debris flow on 22 August*: 128p (*in Japanese*).

Sakurai, W., 2012, Implementation of countermeasure works against the large-scale sediment disasters in Kii Peninsula, Sabo, 111: 8–11 (*in Japanese*).

Sallenger, Jr. A.H., 1979, Inverse grading and hydraulic equivalence in grain-flow deposits: J. Sedimentary Petrology, 49(2): 553–562.

Sassa, K., 1997, Landslide-induced debris flow – mechanism of undrained loading on deposit-, Chikyu Monthly, 220: 652–660 (*in Japanese*).

Sassa, K., 2003, Sientific development and internationalization of sabo and landslide investigations: J. Japan Soc. Erosion Control Eng., 55(5): 1–3 (*in Japanese*).

Sassa, K., Fukuoka, H. and Wang, F.W., 1998, Possible long run-out mechanism of the landslide mass: Landslide News, 11: 11–15.

Satofuka, Y., 2004, Numerical simulation of the debris flow at the Atsumari River, Minamata City, 2003: Annual J. Hydraulic Engineering, JSCE, 48: (*in Japanese*).

Savage, S.B. and Jeffrey, D.J., 1981, The stress tensor in a granular flow: J. Fluid Mech., 110: 255–272.

Savage, S.B. and McKeown, S., 1983, Shear stress developed during rapid shear of concentrated suspensions of larger spherical particles between concentric cylinders: J. Fluid Mech., 127: 453–472.

Savage, S.B. and Sayed, M., 1984, Stress developed by dry cohesionless granular materials sheared in an annular shear cell: J. Fluid Mech., 127: 453–472.

Sawada, T., 1985, *Study on sediment runoff in mountainous basin*: Doctoral thesis, Kyoto University: 1–149 (*in Japanese*).

Sawada, T. and Takahashi, T., 1994, Sediment yield on bare slopes, Proc. International Symposium on Forest Hydrology, Tokyo: 471–478.

Schlumberger , 1882, Über den Muhrgang am 13 August 1876 im Wildbache von Faucon bei Barcelonnte (Niederalpen): *in Studien über die Arbeiten der Gebirge* (Demonzy, P. ed.): 289–299.

Sharp, R.P. and Nobles, L.H., 1953, Mudflow of 1941 at Wrightwood, Southern California: Bull. Geol. Soc. Am., 64:547–560.

Shen, N.H. and Ackermann, N.L., 1982, Constitutive relationships for fluid-solid mixtures: J. Eng. Mech. Div. ASCE, 108 (ME5): 748–763.

Shen, N.H. and Ackermann, N.L., 1984, Constitutive equations for a simple shear flow of a disc shaped granular material: Int. J. Eng. Sci., 22: 829–843.

Shieh, C. and Tsai, Y., 1997, Experimental study on the configuration of debris-flow fan: Proc. 1st Intern. Conf. on Debris-Flow Hazardos Mitigation: 133–142.

Shih, B., Shieh, C. and Chen, L., 1997, The grading of risk for hazardous debris-flow zones: Proc. 1st Intern. Conf. on Debris-Flow Hazardos Mitigation: 219–228.

Shreve, R.L, 1966, Sherman landslide, Alaska, Science, 154: 1639–1643.

Shreve, R.L., 1968a, The Blackhawk landslide, Spec. Paper Geol. Soc. Am., 108: 47p.

Shreve, R.L., 1968b, Leakage and fluidization in air-layer lubricated avalanches, Geol. Soc. Am. Bull., 79: 653–658.

Smart, G.M., 1984, Sediment transport formula for steep channels, ASCE J. Hydr. Eng. 110(3): 267–276.

Straub, S., 2001, Bagnold revisited, implications for the rapid motion of high-concentration sediment flows: Spec. Publ. Int. Ass. Sediment., 31: 91–109.

Suwa, H., 1988, Occurrence of debris flows and variations of gully forms: *in Forefront of disaster geomorphology*: Commemorative project committee for the retirement of Prof. Okuda: 83–104 (*in Japanese*).

Suwa, H., 1988a, Focussing mechanism of large boulders to a debris-flow front: Trans. Japanese Geomorphological Union, 9(3): 151–178.

Suwa, H., Okunishi, K., Okuda, S., Takahashi, H., Hasegawa, H., Takada, M. and Takaya, S., 1985, Depositional characteristics of the debris avalanche at Mt. Ontake in 1984, Annuals DPRI, 28B-1: 505–518 (*in Japanese*).

Suwa, H. and Okuda, S., 1982, Sedimentary structure of debris-flow deposits, at Kamikamihori fan on Mt. Yakedake: Annuals, DPRI, 25B-1: 307–321 (*in Japanese*).

Suwa, H. and Sawada, T., 1994, Comparison between the viscous debris flow and the stony one: *in Japan-China joint research on the prevention from debris flow hazards* (Takahashi, T. ed.), Research Report of the Grant-in-Aid for Scientific Res., No.03044085: 56–67.

Suwa, H., Okunishi, K., Okuda, S., Takahashi, H., Hasegawa, H., Takada, M. and Takaya, S., 1985, Depositional characteristics of the debris avalanche at Mt. Ontake in 1984: Annuals, DPRI., 28B-1: 505–518 *(in Japanese)*.

Suwa, H., Sawada, T., Mizuyama, T., Arai, M. and Takahashi, T., 1997, Observational study on viscous debris flows and countermeasures against them: Proc. Intern. Symp. On Natural Disaster Prediction and Mitigation, Kyoto: 401–406.

Suzuki, M., Fukushima, Y., Takei, A. and Kobashi, S., 1979, The critical rainfall for the disasters caused by debris movement: J. JSCEC, 31(3): 1–7 *(in Japanese)*.

Takahashi, T., 1977, A mechanism of occurrence of mud-debris flows and their characteristics in motion: Annuals, DPRI, 20B-2: 405–435 *(in Japanese)*.

Takahashi, T., 1978, Mechanical characteristics of debris flow: J. Hydraul. Eng., ASCE, 104: 1153–1169.

Takahashi, T., 1980, Debris flow on prismatic open channel: J. of Hydraulic Engineering, ASCE, 106, HY3: 381–396.

Takahashi, T., 1980a, Study on the deposition of debris flows (2)-Process of formation of debris fan-: Annuals, DPRI, 23B-2: 443–456 *(in Japanese)*.

Takahashi, T., 1981, Flood and sediment disasters associated with the eruption of Mt. St. Helens: J. JSECE, 33(3): 24–34 *(in Japanese)*.

Takahashi, T., 1981a, Estimation of potential debris flows and their hazardous zones; Soft countermeasures for a disaster: J. Natural Disaster Science, 3(1): 57–89.

Takahashi, T., 1982, Study on the deposition of debris flows (3); Erosion of debris fan: Annuals, DPRI, 25B-2: 327–348 *(in Japanese)*.

Takahashi, T., 1982a, Debris flow prediction: *in Sediment and flood hazards mitigation on an alluvial fan* (Ashida, K. Representative investigator), Report for Natural Disaster Scientific Research: 78–79 *(in Japanese)*.

Takahashi, T., 1983, Debris flow: *in Sediment disasters in river and their countermeasures* (Ashida, K., Takahashi, T. and Michiue, M. eds.), Morikita Shuppan, Tokyo: 55–149 *(in Japanese)*.

Takahashi, T., 1983a, Debris flow and debris flow deposition: *in Advances in the mechanics and the flow of granular materials* (Shahinpoor, M. ed.), II, Trans Tech Publications, Rockport: 699–718.

Takahashi, T., 1987, High velocity flow in steep erodible channels: Proc. IAHR Congress, Lausanne: 42–53.

Takahashi, T., 1989, Okuetsu rainstorm disaster- choking of the Mana River-: *in Prediction and the countermeasures against the secondary disasters*, Zenkoku-bousai-kyoukai, 3: 7–23 *(in Japanese)*.

Takahashi, T., 1991, *Debris flow*: Monograph of IAHR, Balkema, Rotterdam: 1–165.

Takahashi, T., 1993, Fluid mechanical modeling of the viscous debris flow: Proc. Pierre Beghin Intern. Workshop on Rapid Gravitational Mass Movements: Open file.

Takahashi, T., 1993a, Debris flow initiation and termination in a gully: Hydraulic Eng. '93, ASCE: 1756–1761.

Takahashi, T., 2001, Mechanics and simulation of snow avalanches, pyroclastic flows and debris flows, Spec. Publs., Int. Ass. Sediment, 31: 11–43.

Takahashi, T., 2001a, Processes of occurrence, flow and deposition of viscous debris flow: *in River, Coastal and Estuarine Morpho-dynamics* (Seminara, G. and Blondeaux, P. eds), Springer-Verlag, Berlin: 93–118.

Takahashi, T., 2002, A process-based sediment runoff model for steep and high mountain basin: Proc. 5th Taiwan-Japan Joint Seminar on Natural Hazards Mitigation: 1–16.

Takahashi, T., 2006, *Mechanisms of sediment runoff and countermeasures for sediment hazards*, Kinmirai Sha: 420p *(in Japanese)*.

Takahashi, T. and Yoshida, H., 1979, Study on the deposition of debris flows (1)-Deposition due to abrupt change of bed slope-: Annuals, DPRI, 22B-2: 315–328 *(in Japanese)*.

Takahashi, T. and Egashira, S., 1986, Investigation of mud flow: *in The 1985 eruption of Nevado del Ruiz Volcano, Colombia, and associated mudflow disaster* (Katsui, Y. ed.), Report of Natural Disaster Scientific Research No.B-60-7: 61–95 (*in Japanese*).

Takahashi, T. and Nakagawa, H., 1986, Prediction of occurrence and volume of surface landslides: Annual J. of Hydraulic Engineering, JSCE, 30: 199–204 (*in Japanese*).

Takahashi, T. and Kuang, S., 1988, Hydrograph prediction of debris flow due to failure of landslide dam: Annuals, DPRI, 31B-2: 601–615 (*in Japanese*).

Takahashi, T. and Nakagawa, H., 1989, Prediction of the sediment yield from a small basin in case of heavy rainfall: Annuals, DPRI, 32B-2, : 689–707 (*in Japanese*).

Takahashi, T. and Nakagawa, H., 1991, Prediction of stony debris flow induced by severe rainfall: J. JSECE, 44(3): 12–19 (*in Japanese*).

Takahashi, T. and Nakagawa, H., 1992, Prediction of flood hydrograph due to collapse of a natural dam by overtopping: Annuals, DPRI, 35B-2, 231–248 (*in Japanese*).

Takahashi, T. and Nakagawa, H., 1993, Estimation of flood/debris flow caused by overtopping of a landslide dam: Proc. XXV Cong. IAHR, Tokyo, III: 117–124.

Takahashi, T. and Nakagawa, H., 1994, Natural dam formation and the disaster – A possible explanation of one extreme event-: Proc. Intern. Workshop on Floods and Inundations related to Large Earth Movements, Trent: A8.1–A8.12.

Takahashi, T. and Tsujimoto, H., 1997, Mechanics of granular flow in inclined chute: J. Hydraul. Coast. Environment. Eng. JSCE, 565/II-39: 57–71 (*in Japanese*).

Takahashi, T. and Tsujimoto, H., 1999, Granular flow model of avalanche and its application: J. Hydroscience and Hydraulic Eng., JSCE, 17(1): 47–58.

Takahashi, T. and Tsujimoto, H., 2000, A mechanical model for Merapi type pyroclastic flow: J. Volcanol. Geothermal. Res., 98: 91–115.

Takahashi, T. and Satofuka, Y., 2002, Generalized theory of stony and turbulent muddy debris-flow and its practical model: J. JSECE, 55(3): 33–42 (*in Japanese*).

Takahashi, T., Hamada, S. and Yoshida, H., 1977, Experimental study on the initiation mechanism of debris flow: Proc. 14th Symp. on Natural Disaster Science: 115–118 (*in Japanese*).

Takahashi, T., Nakagawa, H. and Kuang, S., 1978, Estimation of debris flow hydrograph on varied slope bed: Erosion and Sedimentation in the Pacific Rim, IAHS Publ. 165: 167–177.

Takahashi, T., Nakagawa, H. and Kanou, S., 1985, Risk estimation against wash away of wooden houses by a flooding flow: Annuals, DPRI, 28B-2: 455–470 (*in Japanese*).

Takahashi, T., Nakagawa, H. and Nishizaki, T., 1986, Two-dimensional numerical simulation method to estimate the risk of a flood hazard caused by a river bank breach: Annuals, DPRI, 29B-2: 431–450 (*in Japanese*).

Takahashi, T., Nakagawa, H. and Sato, H., 1988, Formation of debris flow fan by fully developed and immature debris flows: 32nd Conf. on Hydraulics, JSCE: 497–502 (*in Japanese*).

Takahashi, T., Nakagawa, H., Higashiyama, M. and Sawa, H., 1990, Assessment of evacuation systems for water or mud floods; a combined simulation of flooding and the action of residents: J. Natural Disaster Science, 12(2): 37–62.

Takahashi, T., Nakagawa, H. and Yamashiki, Y., 1991, Formation and erosion of debris fan that is composed of sediment mixtures: Annuals, DPRI, 34B-2: 355–372 (*in Japanese*).

Takahashi, T., Nakagawa, H., Harada, T. and Yamashiki, Y., 1992, Routing debris flows with particle segregation: J. Hydraulic Engineering, ASCE, 118(11): 1490–1507.

Takahashi, T., Nakagawa, H., Satofuka, Y. and Tomita, T., 1997, Mechanics of the viscous type debris flow (2) – Flume experiments using well-graded sediment mixture: Annula, DPRI, 40IDNDR S. I.: 173–181 (*in Japanese*).

Takahashi, T., Nakagawa, H., Satofuka, Y. and Ogata, M., 1998, Mechanics of the viscous type debris flow (3)-Formation and propagation of the debris flow surge-: Annuals, DPRI, 41B-2: 265–276 (*in Japanese*).

Takahashi, T., Nakagawa, H., Satofuka, Y., Okumura, H. and Yasumoto, D., 1998a, Study on the erosion process in mountain river: Annuals, DPRI, 41B-2: 237–252 (*in Japanese*).

Takahashi, T., Nakagawa, H. and Satofuka, Y., 2000, Newtonian fluid model for viscous debris-flow: Proc. 2nd Intern. Conf. on Debris-Flow Hazards Mitigation: 255–267.

Takahashi, T., Nakagawa, H., Satofuka, Y. and Wang, H., 2000a, Stochastic model of blocking for a grid-type dam by large boulders in a debris flow, Annuals, DPRI, 43B-2: 287–294 (*in Japanese*).

Takahashi, T., Inoue, M., Nakagawa, H. and Satofuka, Y., 2000b, Prediction of sediment runoff from a mountain watershed: Annual J. Hydraulic Engineering, JSCE, 44: 717–722 (*in Japanese*).

Takahashi, T., Chigira, M., Nakagawa, H., Onda, Y., Maki, N., Aguirre-Pe, J. and Jáuregui, E., 2001, Flood and sediment disasters caused by the 1999 heavy rainfall in Venezuela: Research Report of Natural Disasters: 1–141.

Takahashi, T., Nakagawa, H., Satofuka, Y. and Kawaike, K., 2001a, Flood and sediment disasters triggered by 1999 rainfall in Venezuela; A river restoration plan for an alluvial fan: J. Natural Disaster Science, 23(2): 65–82.

Takahashi, T., Nakagawa, H., Satofuka, Y. and Wang, H., 2002, Simulation of debris flow control by a grid-type sabo dam: Annual J. Hydraulic Engineering, JSCE, 46: 689–694 (*in Japanese*).

Takei, A., 1987, Aritagawa disasters in 1953: *in Prediction and countermeasures of secondary disasters*, 2, Zenkoku Bousai Kyoukai: 47–71 (*in Japanese*).

Takei, K. and Mizuhara, K., 1982, Sedimentation of debris flow on alluvial fan: *in Study report on prevention and mitigation of the sediment disaster on alluvial fans* (Ashida, K. ed): 15–25 (in Japanese).

Tognacca, C., Bezzola, G.R. and Minor, H.E., 2000, Threshold criterion for debris-flow initiation due to channel-bed failure: Proc. 2nd Intern. Conf. on Debris-Flow Hazards Mitigation: 89–97.

Tsubaki, T., Hashimoto, H. and Suetsugi, T., 1982, Grain stresses and flow properties of debris flow: Proc. JSCE, 317: 70–91 (*in Japanese*).

Varnes, D.J., 1978, Slope movement types and processes: *in Landslides analysis and control* (Scguster, R.L and Krizek, R.J. eds.), NAS Sp. Rep. 176: 11–33.

Van Rompaey, A.J.J., Vestraeten, G., Van Oast, K., Govers, G. and Poesen, J., 2001, Modelling mean annual sediment yield using a distributed approach, Earth Surface Processes and Landforms, 26: 1221–1236.

Voight, B., Janda, R.J., Glicken, H. and Douglass, P.M., 1983, Nature and mechanics of the Mount St. Helens rockslide-avalanche of 18 May 1980, Geotechnique, 33: 243–273.

Wang, Y., 1993, personal communication.

Wang, G. and Sassa, K., 2002, Pore pressure generation and motion of rainfall-induced landslides in laboratory flume tests: Proc. Intern. Symp. Landslide Risk Mitigation and Protection of Cultural and Natural Heritage, Kyoto: 45–60.

Watanabe, M., Mizuyama, T., Uehara, S. and Suzuki, H., 1980, Experiments on the facilities against debris flow: Doboku Gijutu Shiryou, 22-2: 8–14 (*in Japanese*).

Willson, K.C., 1966, Bed-load transport at high shear stress, ASCE J. Hydr. Eng. 92(6).

Wilson, C.J.N., 1980, The role of fluidization in the emplacement of pyroclastic flows; an experimental approach, J. Volcanology and Geothermal res., 8: 231–249.

Wischmeier, W.H. and Smith, D.D., 1978, Predicting rainfall erosion losses; A guide to conservation planning, USDA Agric. Hdbk, No.537.

Wu, J., Kan, Z., Tian, L. and Zhang, S., 1990: *Observational investigation of debris flow in the Jiangjia Gully*, Yunnan: Science Publishing Co.: 1–251 (*in Chinese*).

Yagi, N. and Yatabe, R.: Strength of undisturbed weathered granite soil, Report for the research project for the sediment disasters in weathered granite area, 1986 (*in Japanese*).

Yano, K. and Daido, A., 1965, Fundamental study on mud-flow: Bull. DPRI, 14: 69–83.

Yamada, T., Minami, N., Osanai, N. and Mizuno, H., 1998, Actual state of debris flows induced by deep-seated slope failure on July 10, 1997 at the Harihara River basin, Kagoshima Prefecture, Japan: J. JSECE, 51(1): 46–54 (*in Japanese*).

Yamamoto, S., Ishikawa, Y., Miyoshi, I. and Mizuhara, K., 1999, Soil characteristics and fluidity of debris flows at the Gamahara River, at the Harihara River and at the Hachimantai Area: J. JSECE, 51(5): 28–34 (*in Japanese*).

Notations

A	acceleration $(= dc/dt)$
A	area of debris flow deposition
A_d	drainage area
A_g	area of the bare-slope that is connected to the talus
A_{15}	drainage area whose channel gradient is steeper than $15°$
A_{15c}	critical A_{15} which generate debris flow under the rainfall intensity of r_e
B	channel width
B'	width that seems to be eroded to form debris flow
B_d	submerged length of a big boulder at the half depth of the draft
B_{dam}	width of the waterway of sabo dam
B_{dead}	width of dead water zone in a cross-section
B_{do}	channel width downstream of channel slope change
B_f	width of the central planer portion on debris flow fan
B_m	width of main flow channel
B_p	width of house perpendicular to the direction of flow
B_{up}	channel width upstream of channel slope change
B_0	canyon width
Ba	Bagnold number
C_t	fluctuating velocity of particle
C	particle concentration in volume
C_a	specific heat of pyroclastic material
$C_{B\infty}$	equilibrium solids concentration in bed load
C_{ct}	grain concentration in laminar granular flow that has nearly a constant value
C_D	drag coefficient
C_d	solids concentration downstream of change in slope
C_e	equilibrium coarse particle concentration in viscous debris flow
C_F	volume concentration of the fine fraction in the interstitial fluid
C_k	volumetric concentration of k-th grade particle in flow
C_L	volume concentration of the coarse fraction in the entire debris flow material
$C_{L\infty}$	equilibrium concentration of coarse particle fraction
C_l	particle concentration in the particle mixture layer of immature debris flow
C_{lim}	minimum of concentration in the particle mixture layer of immature debris flow
C_m	average solids concentration in debris flow upstream and downstream of a change in slope

C_r	celerity of roll wave
$C_{s\infty}$	equilibrium transport sediment concentration in immature debris flow
C_{tr}	transport concentration
C_u	solids concentration in approaching debris flow
$\overline{C}$	average grain concentration in the specific range of flow depth
C_1	particle concentration in volume in the sediment accumulation after shearing motion
C_2	particle concentration in volume larger than that the dislocation of particles cannot take place
C_3	particle concentration in volume larger than that particles move with enduring contact with each other
C_4	particle concentrations in volume larger than that immature debris flow changes to debris flow
C_5	maximum particle concentration in flow in individual particle motion
C_*	concentration of solids when packed
C_{*DL}	volume concentration of coarse fraction in the static bed after deposition
C_{*F}	volume concentration of the fine fraction in the static bed
C_{*Fmax}	volume concentration of fine particles in the maximum compacted state
C_{*k}	volume ratio of fine particles to the volume of void space ($k = 1 \sim k_1$)
C_{*L}	volume concentration of coarse particle fraction on the bed
C_{*Lmax}	volume concentration of the coarse particles in the maximum compacted state
C_∞	equilibrium particle concentration in (mature) debris flow
$C_{\infty k}$	equilibrium concentration of class k particles
D	thickness of sediment accumulation on a gully bed
D_A, D_B	thicknesses of A and B layers, respectively
D_b	distance between the centers of two cylinders composing earth block
D_d	diameter of big boulder transported as if floating on the surface of debris flow
D_e	possible depth of riverbed erodable to develop debris flow
D_f	diffusion coefficient for the thickness of deposit
D_H	dam height
D_{max}	maximum height of natural dam
D_0	diameter of a column that composes the frame of grid dam
D_1	diameter of rigid cylinder of which assemblage consists the moving earth block
D_2	distance between cylinders in which the attractive force becomes the maximum
D_3	largest distance between cylinders within which attractive force operates
E	erosion rate of a slope
E_{max}	maximum of the super-elevation
$\mathbf{F}$	vector representing force
F	external force operating to the mass detaching from flow
F_b	friction at the bottom of flow
F_c	index to describe the risk for washing out of houses ($= u_p h_p$)
F_{cp}	clogging probability of grid type dam
$\mathbf{F}_i$	interaction force between solid phase and liquid phase per unit volume

F_p	index value whether fine particles can be stored in the skeletal structure of coarse particles
Fr	Froude number
F_{ra}	Froude number in the after flow
F_s	safety factor for slippage
F_{side}	friction force operating from the side banks
F_x	interaction force to x direction
F_z	interaction force to z direction
F_1	Froude number of flow at the entrance of the bend
H	thickness of earth block
H'	difference between the stages of the flood marks on right-bank and left-bank
H_A, H_B	lateral seepage flow depths on the bases of A and B layers, respectively
H_a	average depth of flow in the cross-section
H_{gw}	seepage flow depth in the natural dam body
H_o	surface flow depth on slope
H_p	height of a slide measured from the foot of the slope
H_s	flow surface stage $(= z_b + h)$
H_{sub}	depth of seepage flow
H_t	total depth of flow as the sum of steady flow depth and the roll wave height
H_w	water stage behind the natural dam
H_{ws}	initial overspill depth on the natural dam immediately after the slippage
I	gradient of channel $(=\sin\theta)$
I_x, I_y	hydraulic gradient to x and y directions, respectively
K	numerical coefficient which appears in the erosion formula for unsaturated bed
K_s	numerical coefficient which appears in the side bank erosion formula
L	length of earth block
L'	length of liquefied part within earth block
L_a	arrival distance of an earth block
L_B	base length of natural dam
L_{Bsc}	base length of triangular natural dam having the volume of V_{sc}
L_{me}	possible distance of erosion of the bed along the channel
L_T	crest length of natural dam
$L*$	total length of a deposit of debris avalanche
M	x-wise component of unit width discharge (flux) $(=uh)$
M_b	mass of an earth block composing cylinder
M_V	critical bearing moment of the yypical wooden houses in Japan
N	y-wise component of unit width discharge (flux) $(=vh)$
N_b	number of particle per unit area in the layer that is face to face with the bottom
N_{Bag}	Bagnold number $(= T_c/T_{fq})$
N_{mud}	$= T_t/T_c$
N_p	number of particles in flow above the height z
N_{Rey}	Reynolds number $(= T_t/T_{fq})$
N_{Sav}	Savage number
P	reduction rate of peak discharge by passing through the open-type sabo dam
$\mathbf{P}_c$	stress tensor due to collision of particle in granular flow

P_{ds}	pressure in solid phase
P_f	pressure in liquid phase in excess over hydro-static one
$P_{n\,row}$	instantaneous blocking probability of grid dam having n rows
P_r	instantaneous blocking probability of grid dam having only one row $(= P_{1row})$
P_1	submerged weight of d_1 particle
P_2	dispersive force due to the collision of d_1 particle with other particles in the layer above
P_3	dispersive force due to the collision of d_1 particle with other particles in the layer below
P_4	drag force operating to the d_1 particle
p_v	pore water pressure
Q, Q_w	water discharge in a channel
Q_a	discharge of pyroclastic flow
Q_c	minimum surface water discharge to generate debris flow
Q_{Lx}, Q_{Ly}	discharges of coarse particle fraction in unit area to the x and y directions
Q_p	peak discharge
Q_{pd}	peak discharge of debris flow after passing through the sabo dam
Q_s	sediment discharge
Q_T	water plus sediment discharge
Q_{w0}	water production rate by the effect of pyroclastic flow in unit time
R	an expression of resistance coefficient
R_i	cumulative rainfall
Re_*	particle Reynolds number $(= u_* d_p / \nu)$
Re	Reynolds number
S	area of the unit slope or the cross-sectional area of the talus
SF_A, SF_B	safety factor for slippage in A and B layers, respectively
S_0	bottom area of an earth block composing cylinder
T	total of shearing stresses in flow (only in Chapter 1)
T	granular temperature
T_a	temperature of pyroclastic flow
T_c	shear stress due to particle collision
T_{ds}	shear stress in solid phase
T_f	shear stress in fluid phase
T_l	life span of a natural dam
T_{fq}	shear stress due to deformation of fluid
T_p	cycle of roll waves
$\mathbf{T}_f$	stress tensor of liquid phase
$\mathbf{T}_s$	stress tensor of solid phase
T_{sq}	quasi-static Coulomb friction stress
T_t	macro-turbulent mixing stress
U	sectional mean velocity
U_c	critical velocity for the nominal commencement of deposition
U_e	velocity of earth block
U_o	velocity of the surface water flow over the deposit or velocity of earth block at the foot of a slope
U_s	velocity of the mass detaching from flow

U_{TH}	threshold velocity of mudflow to stop
V	volume of landslide
V_d	total volume of deposit
V_F	volume of fine sediment in pillar-shaped space in the flow having a height h and a unit bottom area
V_f	traveling velocity of debris flow front
V_k	total volume of kth grade particles on bed
V_L	volume of coarse sediment in pillar-shaped space in the flow having a height h and a unit bottom area
V_l	producible sediment volume by landslide
V_{la}	total volume of landslides in a basin
V_{out}	cumulative sediment runoff
V_r	remaining landslide sediment on slope
V_s	substantial volume of solids contained in pillar-shaped space in the flow having a height h and a unit bottom area
V_{sc}	critical soil volume of landslide larger than that a triangular natural dam is formed
V_{sp}	solids volume in the surface layer of riverbed
V_{su}	total producible sediment volume $(= V_{Te} + V_l + V_r)$
V_T	sediment volume per unit length of torrent bed
V_{Te}	erodable sediment volume per unit length of torrent bed
V_t	volume of surface layer of the bed
V_1	volume of a natural dam body
V_2	capacity of reservoir produced by a natural dam
W	width of landslide
WF_A, WF_B	distances of wetting front measured from the upper boundaries of respective A and B layers
X	risk index for the occurrence of debris flow in terms of flood runoff discharge
X_Q	non-dimensional presentation of river discharge by using natural dam height
X_f	coordinate moving with the velocity of V_f
Y	risk index for the occurrence of debris flow in terms of drainage area
Z	non-dimensional height $(= z/h)$
$\mathbf{a}$	acceleration of particle motion
a	distance measured downward from the surface of the sediment layer
a_c	depth in the sediment layer where $\tau = \tau_r$ is satisfied
a_i	numerical coefficient used to describe particle repulsive pressure by Bagnold
b	center to center distance between adjacent particles
$\mathbf{c}$	instantaneous velocity vector of particle
c	cohesive strength
c_A, c_B	cohesive strengths of A and B layers, respectively
c_b	adhesive force between the earth block composing cylinders
d_b	diameter of the maximum size class particles transported as bed load
d_k	diameter of particle in k-th grade
d_L	mean diameter of the coarse particles
d_m	mean diameter
d_{max}	diameter of the maximum size class particles in the front of debris flow

d_{mL}	mean diameter of coarse particles
d'_{mL}	mean diameter of movable particles on the bed
d_n	particle diameter n% is smaller than that in the mixture
d_p	particle diameter
d_1	diameter of a particular particle focused attention to its motion
d_*	non-dimensional particle diameter $(= sgd_p^3/\nu)$
e	restitution coefficient of particles
f	Darcy-Weisbach's resistance coefficient
f_{bk}	existence ratio of k-th grade particles on bed to the total particles
f_{bk0}	existence ratio of k-th grade particles in the layer beneath the bed surface layer
f_{bLk}	existence ratio of k-th grade particles within the total coarse particles
f_{in}	absolute value of attractive or repulsive forces between the two earth block composing cylinders
f_{inm}	the maximum attractive force between the two cylinders
f_p	particle velocity distribution function
f_R	runoff coefficient
f_s	shear stress between the two earth block composing cylinders
f_{sx}, f_{sy}	x and y-wise shear stress fractions between the cylinders composing earth block
f_x, f_y	x and y-wise attractive or repulsive forces working between the cylinders composing earth block
f_{0k}	existence ratio of class k particles in flow
g	acceleration due to gravity
g_0	radial distribution function $(= \lambda + 1)$
h	local mean depth of flow
h'	flow depth on the crest of dam
h_a	depth of the debris flow following the earth block
h_b	maximum wave height of roll waves
h_c	height of the force operating point from the ground
h_{fr}	depth at the debris flow front
h_g	depth variation along the outer bank of bending channel
h_h	depth of pipe flow under the ground
h_l	thickness of particle mixture layer in immature debris flow
h_{li}	thickness of liquefied layer inside the earth block
h_n	normal flow depth
h_o	water flow depth on gully bed or on the deposit surface
h_p	inundated flow depth
h_s	thickness of water saturated part in the earth block
h_{sb}	thickness of unsaturated part in the earth block
h_{sh}	shock wave height
$h_{s\,min}$	the smaller h_s value of the two contacting cylinders
h_u	depth of the approaching flow
h_w	roll wave height
h_0	surface stage of the mass detaching from flow
h_1	thickness of particle suspension layer in hybrid debris flow (upper layer)
h_2	thick ness of particle collision layer in hybrid debris flow (lower layer)

h_∞	asymptotic flow depth in the uniformly traveling debris flow
i	erosion (>0) or deposition (<0) velocity
i_A, i_B, i_C	infiltration rate of A, B, and C layers, respectively
i_c	changing rate of the earth block to the liquefied layer
i_{gk}	erosion rate of ith group particles on the talus
i_k	erosion or deposition velocity of class k particles
i_s	side bank erosion velocity
i_{sb}	deposition velocity in bulk
i_{sbk}	bed erosion or deposition rate of kth groupparticles
i_{sb0}	erosion velocity of bed when the bed is composed of uniform material; d_{mL}
i_{sml}, i_{smr}	mean wall recession velocities of the left- and right-hand side banks, respectively
i_1, i_2	bed slope of upstream and downstream of a sabo dam, respectively
k	lateral transmission coefficient of saturated seepage flow
k'	coefficient to represent resistance law
k_e	number of grades in the classification of particle size
k_g	coefficient to describe the sediment supply from the bare-slope to the talus
k_r	equivalent roughness height
k_s	coefficient of permeability
k_v	coefficient to represent the volume of a particle
l	turbulent mixing length
l'	mean free travel distance of a particle before colliding with other particle
l_d	apparent draft of a big boulder transported protruding its body from the flow surface
l_g	spacing of riverbed girdle or the length of a talus along the stream channel
l_l, l_r	height of left-hand and right-hand side walls above the surface of flow, respectively
l_s	spacing between pipes in grid type sabo dam
m	mass of a particle
m_s	ass to be deposited by detaching from flow
n_e	equivalent roughness coefficient of slope
n_m	Manning's roughness coefficient of channel
n_p	number of particles having velocity $\mathbf{c}$
p	static pressure in fluid phase
p'	effective pressure transmitted by inter-particle contact
p_{bed}	fluid pressure on the bed
p_c	particle repulsive pressure
p_h	numerical coefficient to describe pipe flow rate under the ground
p_i	coefficient less than 1 to describe the initiation of depositing process
p_s	quasi-static pressure transmitted via s skeletal structure of moving particles
p_v	pore water pressure
q	total flow rate of water plus sediment in uniti width
q_B, q_b	bed load discharge per unit width
q_{bk}	bed load discharge of class k particles per unit width
q_{bn}	unit width bed load discharge toward n axis
q_{bs}	unit width bed load discharge toward s axis
q_{buk}	equilibrium bed load discharge on the bed consists of only the class k particles

q_{bx}, q_{by}	x and y components of bed load discharge per unit width
q_c	minimum surface water discharge per unit width to generate debris flow
q_{hx}, q_{hy}	x and y-wise pipe flow discharges under the ground, respectively
q_{in}	input discharge per unit length of the river channel
q_m	heat of fusion of ice
q_o	unit width water discharge appears on the surface of sediment bed
q_{ox}, q_{oy}	surface flow discharge in unit width to x and y directions, respectively
q_s	sediment discharge per unit width
$q_{sg\,in}$	supply rate of sediment to the talus from the upper bare-slope in unit width
$q_{sg\,out}$	erosion rate of the talus by the flow in the river channel in unit width
q_{sk}	flow rate of k-th grade particles from the side banks per unit length of channel
q_{so}	unit width discharge of surface flow on the slope
q_t	total discharge of the sum of sediment and water per unit width
q_w	water inflow rate per unit width
q_{wd}	water discharge per unit width in debris flow downstream of slope changing point
q_{wm}	mean water discharge in debris flow upstream and downstream of change in slope
q_{wt}	water discharge per unit width in debris flow
q_{wu}	water discharge per unit width in debris flow upstream of slope changing point
q_x, q_y	lateral seepage flow discharges per unit width to x and y directions, respectively
q_0	water discharge per unit width supplied from the upstream end of the channel
q_*	non-dimensional unit width discharge $(= q_o/g^{1/2}d_p^{3/2})$
$\mathbf{r}$	instantaneous position of particle
r	water inflow rate per unit length of channel or the actual rainfall intensity
r_c	radius of earth block composing cylinder
r_{cc}	radius of channel curvature
r_{co}	radius of curvature at the center of channel
r_d	$(= d_1/d_m)^3$
r_e	effective rainfall intensity
r_i	rainfall intensity
r_0	excess rainfall intensity
s	degree of saturation in the void space between particles
s_A, s_B	degree of saturation in A and B layers, respectively
s_b	degree of saturation in the sediment bed
s_p	space between the two particles
s_0	initial degree of saturation in the dam body
s_{0A}, s_{0B}	initial degree of saturation in the A and B layers, respectively
t	time
t_p	arrival (concentrating) time of flood in the basin
u	local mean velocity of flow toward x direction
u'	non-dimensional expression of u $(= u/\sqrt{gh})$
u_a	velocity of the debris flow following the earth block
u_b	x-wise velocity of the earth block
u_g	x component of sediment transport velocity

u_l	mean velocity in the sediment mixture layer of immature debris flow
u_{li}	mean velocity of liquefied layer inside the earth block
u_o	velocity of surface water flow over the deposit
u_p	velocity of inundated flow
u_s	flow surface velocity
u_{sl}	slip velocity of particles on bed
u_u	mean velocity of approaching flow
u_*	friction velocity or shear velocity $(= gh\sin\theta)^{1/2}$
u_{*c}	shear velocity corresponding to U_c
u_{*ck}	critical friction velocity of the k-th grade particle
u_{*cuk}	critical shear velocity for the initiation of particle motion on the bed consists of uniform d_k particles
u_{*f}	shear velocity shared by fluid phase in debris flow
$\hat{u}_x$	x component of the velocity of fluid phses
$\hat{u}_z$	z component of the velocity of fluid phase
$\tilde{u}_x$	x component of the velocity of the center of mass of a mixture volume element
$\tilde{u}_z$	z component of the velocity of the center of mass of a mixture volume element
$\mathbf{v}$	velocity of particles in mean flow field
v	local mean velocity of flow toward y direction
v_b	y-wise velocity of earth block
v_h	nominal velocity of pipe flow under the ground
v_s	velocity of seepage flow
$\mathbf{v}_f$	velocity vector of fluid phase
$\mathbf{v}_s$	velocity vector of solid phase
v_f	propagating velocity of surge front on a soft deposit
v_g	y component of sediment transport velocity
v_{gz}	z-wise velocity of d_1 particle
v_2	mean velocity behind the surge front
$\hat{v}_x$	x component of the velocity of solid phase
$\hat{v}_z$	z component of the velocity of solid phase
w	velocity to z direction
w_s	settling velocity of a particle
w_s'	non-dimensional settling velocity of a particle $(= w_s/\sqrt{gh})$
x	distance measured along main flow direction
x_L	debris flow arriving distance from the slope change position
x_s	position of debris flow front on the longitudinal axis
z	height measured perpendicular to the bottom
z'	thickness of plug in Bingham flow $(= \tau_y/\rho g\sin\theta)$
z_b	erosion depth or deposit thickness
z_d	deposit thickness at the abrupt change in slope
z_1, z_2	depths downstream and upstream of a surge front, respectively
$z1$	elevation of dam crest from the datum line
α'	energy collection coefficient
α_e	ratio of the quasi-static skeleton pressure to the total pressure or the ratio of the length of liquefied layer to the length of slid earth block

α_i	collision angle between two grains
α_p	numerical coefficient
β	coefficient to describe the changing rate of the earth block into the liquefied layer or the gradient of the perspective line from the distal end to the scar of a landslide
β'	momentum correction factor
β_x	angle between s-axis (main flow direction) and x axis
β_s	angle between s-axis (main flow direction) and the direction of bed load transport
β_1	angle of the crest of shockwave originating from the entrance of the bend
γ	magnitude of local shear rate ($= du/dz$)
γ_d	surface slope of deposit
δ	flow deflection angle by training wall
δ_d	numerical coefficient which appears in the deposition rate formula
δ_e	numerical coefficient which appears in the erosion rate formula for saturated bed
δ_m	thickness of particle exchange layer
ε	specific weight of particle in fluid ($= \sigma/\rho - 1$)
$\varepsilon_x, \varepsilon_y, \varepsilon_z$	eddy diffusion coefficients to x, y, and z directions, respectively
η	rigid modulus or Bingham viscosity
η_a	specific viscosity of sediment mixture with water
θ	channel bed or flow surface slope
$\theta_{bx0}, \theta_{by0}$	x and y-wise inclination of the original bed surface, respectively
θ_c	flattest slope on which debris flow that comes down through the change in slope does not stop
θ_{ck}	critical slope gradient to stop the motion of an earth block
θ'_c	bend angle along the outer bank
θ_d	channel slope downstream of slope change
θ_e	water surface gradient at immediate upstream of a sabo dam
θ'_e	bed slope gradient at immediate upstream of a sabo dam
θ_s	slope gradient
θ_u	channel slope upstream of slope change
θ_x, θ_y	x and y-wise gradients of liquefied layer's surface
θ_w	surface gradient of an earth block or the flow
θ_{wx}, θ_{wy}	x and y-wise gradient of the earth block's surface
θ_{zx}, θ_{zy}	x and y-wise bed gradients
θ_0	initial bed slope before bed variation
θ_1	bed slope flatter than that dry granular flow becomes laminar (in Chapter 2)
θ_2	critical slope to generate a slide by the effect of parallel seepage flow to the surface
$\tilde{\theta}_1$	bed slope steeper than that a stony debris flow can be produced
κ	Kármán constant
κ_a	active earth pressure coefficient
κ_o	Kármán constant for plain water flow
λ	linear concentration ($= d_p/s$) defined by Bagnold
λ_A, λ_B	void ratios of A and B layers, respectively

λ_e	ratio of effective pressure in interstitial fluid to the total grain pressure
λ_0	porosity of natural dam body
μ	viscosity
μ_a	apparent viscosity of sediment mixture with water
μ_k	kinematic friction coefficient of earth block or of particles
μ_s	static friction coefficient of particles on sediment bed
ν	dynamic viscosity $(=\mu/\rho)$
ρ	fluid density
ρ_T	apparent density of debris flow material $(=((\sigma-\rho)C+\rho)$
ρ_a	density of debris flow following the earth block
ρ_{li}	density of liquefied layer
ρ_m	apparent density of interstitial fluid
ρ_p	density of pyroclastic material
ρ_t	density of earth block
ρ_{*b}	apparent density of the static bed $(=(C_*\sigma+(1-C_*)\rho s_b))$
σ	particle density
σ_A, σ_B	particle densities in A and B layers, respectively
σ_n	internal normal stress used by Johnson
τ	shear stress
τ_A	operating shear stress at the boundary between A and B layers
τ_{AL}	resisting stress at the boundary between A and B layers
τ_{bx}, τ_{by}	x and y-wise shear stresses working on the boundary between liquefied layer and the bottom of flow, respectively
τ_c	shear stress due to inter-particle collision
τ_f	shear stress allotted by the inter-granular fluid of the flow
τ_k	kinetic stress due to migration of particles in one layer to other layers
τ_r	resisting stress against motion within the sediment bed
τ_s	static skeletal shear stress transmitted between particles
τ_{sx}, τ_{sy}	x and y-wise shear stresses working on the boundary between earth block and the liquefied layer, respectively
τ_t	shear stress due to turbulent mixing
τ_μ	viscous shear stress
τ_y	yield stress
τ_*	non-dimensional shear stress
τ_{*c}	non-dimensional critical shear stress
τ_{*f}	non-dimensional expression of τ_f
τ_{*fc}	non-dimensional critical shear stress allotted by the interstitial fluid
τ_{*cuk}	non-dimensional critical tractive force on the bed consists of uniform d_k particles
τ_{0f}	the bed shear stress shared by turbulence
ψ	velocity coefficient defined by U/u_*
ψ	function representing the physical quantities in granular flow (in the section 2.4)
Ψ_d	downstream slope angle of dam body
Ψ_u	upstream slope angle of dam body
ϕ	internal friction angle

Φ non-dimensional sediment discharge

ϕ' coefficient representing the momentum change rate by the collision between the particle and the bed

ϕ_A, ϕ_B internal friction angles in A and B layers, respectively

ϕ_{bed} particle friction angle on the bed surface

ϕ_{sp} angle of repose for spherical particles

ω angular velocity of a particle around a column of grid type sabo dam

Index

Milton Keynes UK
Ingram Content Group UK Ltd.
UKHW051925141024
449569UK00027B/1359